The theory of
partial differential equations

The theory of partial differential equations

SIGERU MIZOHATA
Professor of Mathematics, Kyōto University

CAMBRIDGE
at the University Press
1973

CAMBRIDGE UNIVERSITY PRESS
Cambridge, New York, Melbourne, Madrid, Cape Town, Singapore, São Paulo, Delhi

Cambridge University Press
The Edinburgh Building, Cambridge CB2 8RU, UK

Published in the United States of America by Cambridge University Press, New York

www.cambridge.org
Information on this title: www.cambridge.org/9780521087278

Originally published in Japanese by Iwanami Shoten, Publishers, Tokyo 1965
English edition published by Cambridge University Press 1973
Re-issued in this digitally printed version 2008

A catalogue record for this publication is available from the British Library

Library of Congress Catalogue Card Number: 72-83593

ISBN 978-0-521-08727-8 hardback
ISBN 978-0-521-29746-2 paperback

Contents

Chapter [illegible] Green's functions and spectra

Preface

This book is based upon my lectures given at Kyōto and other universities. The original lectures covered some modern developments of the theory of partial differential equations, and the preliminaries which constitute chapters 1 and 2 in this book were merely outlined in the actual lectures. The reason why these rather lengthy chapters have been included is that the book was written not only for future analysts, but also for other unprepared young scientists or engineers who have a keen interest in the subject but little time for wading through existing copious literature merely to acquire basic knowledge sufficient to start reading the main body of this book.

As early as Autumn 1961, Professor S. Iyanaga, together with Professor K. Yosida, suggested that I should write a book of this kind. Since that time, I have tried several times to set down a workable arrangement without success. One early plan was to divide the contents of the book into two parts, and deal with the classical account in the first part, then proceed to the modern account of the same subjects in the second. This seemingly simple arrangement was in fact difficult because of the choice of subject-matter and the style of presentation; if the classical part, which might well be a foundation of the modern theory, were to be written without sufficient underlying notion against a certain historical background, it would, perhaps, be tedious for the reader.

For this reason, I introduced the modern treatment at the outset and then explained some classical problems in the light of recent developments. It is true to say that since the publication of L. Schwartz's monumental *Théorie des Distributions* (1950–1) a great deal of work has been carried out by many mathematicians using the new method. But, in actual fact, most of them were generalizations of problems which had already been (their rigour apart) treated before the publication of Schwartz's book by some other (or the same) mathematicians, but under restricted conditions. Aware of this situation, I have attempted to clarify wherever possible the basic ideas of comparatively old problems.

As for the organization of the contents, each chapter was written so as to be as self-contained as possible, assuring the individual reader's choice of access according to his interest and mathematical maturity. Various function spaces which appear in chapter 2 may not be familiar to a beginner, so that it is absolutely necessary for such readers to refer back to the definitions of these spaces from time to time until the ideas are truly grasped. For the reader's convenience, a list of symbols with simplified definitions, including these function spaces, is given at the end of the book.

In chapter 3, I deal with boundary value problems for elliptic partial differential equations. Relatively simple equations were listed to clarify the basic problems in this genre, where $\mathscr{D}_{L^2}{}^1(\Omega)$ and $\mathscr{E}_{L^2}{}^1(\Omega)$ play important rôles. Note that these spaces are both complete so that they are, of course, Hilbert spaces.

For example, in the last chapter of Courant–Hilbert: *Methoden der mathematischen Physik II*, $\mathscr{E}_{L^2}{}^1(\Omega)$ is defined as the completion by the metric of $\|u(x)\|+\sum\|\partial u(x)/\partial x_i\|$ of the function space of continuously differentiable functions in $\bar{\Omega}$. On the other hand, in this book, the same space is defined as a subset of the elements of $L^2(\Omega)$ of which all first order partial derivatives in the sense of distribution again belong to $L^2(\Omega)$, or in short are 'once differentiable within L^2'.

Some properties of the functions of this kind are given in chapter 2; more detailed properties are studied in chapter 3. The knowledge of Lebesgue integral is indispensable in chapter 3, but the elementary theory is sufficient. After reading chapter 3, the reader should glance at chapter 8 to study Green's functions and Green's kernels, and acquaint himself with the historical background of the theory established in chapter 3.

In chapter 4, starting from classical results, the idea 'hyperbolic type' is explained. This was a creation of Hadamard who used it to classify equations in a more intrinsic sense than his predecessors. One of the significant results of using his notion is that, given an initial value of C^∞-function for an elliptic partial differential equation, the initial value problem is not soluble in general without imposing appropriate conditions on the value.

Chapter 5 is devoted to the existence theorem of the solution of a general evolution equation. The alert reader may notice that this is entirely different from the Cauchy–Kowalewski existence theorem in its nature, illustrating an intrinsic difference between ordinary differential equations and partial differential equations.

Chapter 6 and chapter 7 introduce boundary value problems for hyperbolic equations with variable coefficients.

Finally, in 'Supplementary remarks', a brief sketch of the general boundary value problem for elliptic equations is given. If we wish to insist on the principle that completed theory must be stated in the light of a generalized form, this should be suitably incorporated in chapter 3, but I thought this was a little abrupt for the beginner, and also for the student of theoretical physics.

The bibliography at the very end gives references and suggestions for the interested reader who wishes to explore further in various genres.

On the whole, the approach is biased in favour of the method of functional analysis, starting from a rather abstract setting, then step by step, revealing interesting aspects of the theory. It is the author's hope that in this way the book communicates to the reader the beauties which are characteristic of the theory of partial differential equations. However, the author fears that it may be inordinate to expect the reader to appreciate the theory merely by reading a single volume with a limited scope such as this. As to style, it should be mentioned that, merely due to my taste, the proofs of the facts which are employed in this book are made as orthodox as possible, despite the widely known existence of more refined methods. As for the names of the established facts, the facts which are fundamental to future theories are stated under the title of 'lemmas', and finalized versions are collected under the title of 'theorems'.

During the preparation of this book, my colleagues, Mr Masaya Yamaguti and others, made various comments and gave invaluable advice related to the contents. I thank them all for their valuable criticisms. I should also like to express my thanks to Professor S. Iyanaga and Professor K. Yosida who really gave me the chance to write this book as a new title in the celebrated Iwanami series. My gratitude also goes to Mr Eiji Negishi, Mr Masahisa Makino, Mr Hideo Arai, who are editorial staff of Iwanami Shoten, for their efficient cooperation, and especially, to Mr Makino who, since the autumn of 1961, generously assisted me with a number of matters relating to the manuscript.

Sigeru Mizohata

May 1965
Kyōto

Preface to the English edition

I am very pleased to learn that *Theory of Partial Differential Equations* will appear in English.

As I have already said in the preface of the original Japanese edition, the intention of the book is to provide the prospective reader with recently developed methods centering around 'distribution' which have been extensively used in the fundamental theory of partial differential equations.

Although, remarkable progress has been made in various directions since the book was published in 1965 – especially in boundary value problems and the existence theory of solutions of general equations – I believe that the book still presents useful up-to-date information on some aspects of this fascinating mathematical discipline.

Finally, I am indebted to Mr Katsumi Miyahara for his painstaking translation of the book.

Kyōto University, Japan
September 1972 Sigeru Mizohata

CHAPTER 1

Fourier series and Fourier transforms

1 *Fourier series*

Let $f(x)$ be a continuous function, with period 2π, which we try to represent as follows:

$$f(x)=\tfrac{1}{2}a_0+\sum_{n=1}^{\infty}(a_n\cos nx+b_n\sin nx). \tag{1.1}$$

Let us assume that this right hand series is uniformly convergent. As is well known, we can integrate this term by term to obtain coefficients a_0, a_n and b_n. First we note that

$$\int_0^{2\pi}\cos mx\cos nx\,\mathrm{d}x=\tfrac{1}{2}\int_0^{2\pi}(\cos(m-n)\,x+\cos(m+n)\,x)\,\mathrm{d}x$$

$$=\begin{cases}0 & (m\neq n),\\ \pi & (m=n).\end{cases}$$

$$\int_0^{2\pi}\cos mx\sin nx\,\mathrm{d}x=\tfrac{1}{2}\int_0^{2\pi}(\sin(m+n)\,x-\sin(m-n)\,x)\,\mathrm{d}x=0.$$

$$\int_0^{2\pi}\sin mx\sin nx\,\mathrm{d}x=\tfrac{1}{2}\int_0^{2\pi}(\cos(m-n)\,x-\cos(m+n)\,x)\,\mathrm{d}x$$

$$=\begin{cases}0 & (m\neq n),\\ \pi & (m=n).\end{cases}$$

$$\int_0^{2\pi}\cos mx\,\mathrm{d}x=\int_0^{2\pi}\sin mx\,\mathrm{d}x=0\quad(m\neq 0).$$

In other words, the series of functions 1, $\cos x$ $\sin x$, $\cos 2x$, $\sin 2x,\ldots$ is an orthogonal system. Therefore, multiplying both sides of (1.1) by 1, $\cos nx$ and $\sin nx$ in turn, and integrating, we obtain

$$\left.\begin{aligned}\pi a_0 &= \int_0^{2\pi} f(x)\,dx\,,\\ \pi a_n &= \int_0^{2\pi} f(x)\cos nx\,dx\,,\\ \pi b_n &= \int_0^{2\pi} f(x)\sin nx\,dx\,,\end{aligned}\right\} \tag{1.2}$$

and thus we can calculate a_0, a_n, and b_n.

This poses the following problem. Suppose that $f(x)$ is a continuous function, and that we have determined a_0, a_n, b_n (the Fourier coefficients) by the use of (1.2). Then let

$$\varphi(x) = \tfrac{1}{2}a_0 + \sum_{n=1}^{\infty} (a_n \cos nx + b_n \sin nx)\,. \tag{1.3}$$

Is $\varphi(x)$ always meaningful and identical to the original $f(x)$, i.e. can we write $f(x) = \varphi(x)$? Generally speaking, the answer is in the negative. However, given $f(x)$, we can at least calculate the Fourier coefficients, and write

$$f(x) \sim \tfrac{1}{2}a_0 + \sum_{n=1}^{\infty} (a_n \cos nx + b_n \sin nx)\,.$$

We call this the *Fourier series* of $f(x)$.

Historically, the first mathematician to study the series rigorously was Dirichlet. Later, Riemann and Lebesgue laid the foundations of the theory. According to Lebesgue, given $f(x)$, the Fourier series can be defined when $f(x)$ is Lebesgue-integrable. In the following arguments, unless stated to the contrary, we consider $f(x)$ to be Lebesgue-integrable. We say simply that $f(x)$ is *summable*.

We now prove the following theorem.

THEOREM 1.1 *(Lebesgue)*. If $f_1(x)$ and $f_2(x)$ have the same Fourier coefficients, then $f_1(x) = f_2(x)$ except possibly on a set having measure 0.

Proof (By contradiction). When $f(x) = f_1(x) - f_2(x)$, the Fourier coefficients of $f(x)$ are all zero. Also, we assume $f(x)$ to be continuous and $f(x) \not\equiv 0$. As $f(x)$ is orthogonal to 1, $\cos nx$, and $\sin nx$ $(n = 1, 2, \ldots)$, so it is orthogonal to their linear combinations, which are trigonometric polynomials. So that

$$\int_0^{2\pi} f(x)\,\varphi(x)\,dx=0,$$

where $\varphi(x)$ is an arbitrary trigonometric polynomial.

Hence we may put $f(x)\geqslant m\,(m>0)$ on $[a, b]$ which lies within $[0, 2\pi]$. (We can, if necessary, replace $f(x)$ with $-f(x)$). $\varphi(x)$ is calculated from the following relations:

$$\varphi=\psi^n, \quad \psi(x)=1+\cos(x-\tfrac{1}{2}(a+b))-\cos\tfrac{1}{2}(a-b).$$

We note that φ is a trigonometric polynomial and therefore

$$0=\int_0^{2\pi} f\cdot\varphi\,dx=\int_a^b f\cdot\varphi\,dx+\int_{[0,\,2\pi]\setminus[a,\,b]} f\cdot\varphi\,dx,\dagger$$

where the second part of the integration does not exceed $\max|f(x)|(2\pi-(b-a))$ in its absolute value. In fact, in $[0, 2\pi]\setminus[a, b]$ we observe that $|\psi|\leqslant 1$, and it follows that $|\varphi|=|\psi|^n\leqslant 1$. On the other hand, the first part of the integration converges to $+\infty$ as n increases infinitely, the reason being that $f(x)\geqslant m$ on $[a, b]$, and $\psi\geqslant 1+\delta$ on $[a', b']$ where $[a', b']\subset[a, b]$ and $\delta>0$. Hence

$$\int_a^b f\cdot\varphi\,dx\geqslant\int_{a'}^{b'} f\cdot\varphi\,dx\geqslant m(1+\delta)^n\int_{a'}^{b'} dx=m(1+\delta)^n\,(b'-a').$$

This is a contradiction and hence we conclude $f(x)\equiv 0$.

Next, let $f(x)$ be summable. Write

$$F(x)=\int_0^x f(t)\,dt.$$

By our hypothesis ($f(x)$ is orthogonal to 1) we have $F(2\pi)=0$. We note the following relations:

$$0=\int_0^{2\pi} f(x)\cos mx\,dx=[F(x)\cos mx]_0^{2\pi}+m\int_0^{2\pi} F(x)\sin mx\,dx$$

$$=m\int_0^{2\pi} F(x)\sin mx\,dx,$$

† Translator's note: $[0, 2\pi]-[a, b]$ is the original expression, but the current Russian notation $\setminus$ is more widely used and less confusing.

$$0=\int_0^{2\pi} f(x)\sin mx\,dx=[F(x)\sin mx]_0^{2\pi}-m\int_0^{2\pi} F(x)\cos mx\,dx$$
$$=-m\int_0^{2\pi} F(x)\cos mx\,dx.$$

Therefore, the Fourier coefficients of $F(x)-C$ are all zero for some constant C. Using facts already proved, we have $F(x)-C\equiv 0$. Since $F(0)=0$, $F(x)\equiv 0$.

Now, recall the well-known theorem of Lebesgue $F'(x)=f(x)$ is true 'almost everywhere.' We finally† conclude that $f(x)=0$ 'almost everywhere' on $[0, 2\pi]$.

Q.E.D.

Note. The fact shows that the system of trigonometric functions $\{1, \cos x, \sin x, \ldots, \cos nx, \sin nx, \ldots\}$ is complete on $L^1[0, 2\pi]$.

2. *Dirichlet's Integrals*‡

First we state some important facts.

Abelian transform

Let $\{u_0, u_1, \ldots, u_n\}$, $\{v_0, v_1, \ldots, v_n\}$ be two number sequences where $u_0 \geqslant u_1 \geqslant u_2 \geqslant \cdots \geqslant u_n \geqslant 0$. If $\sigma_p=v_0+v_1+\cdots+v_p$ $(p=0, 1, 2, \ldots n)$, then
$$(\min_{p=0,\ldots,n} \sigma_p)\, u_0 \leqslant S \equiv u_0v_0+u_1v_1+\cdots+u_nv_n \leqslant (\max_{p=0,\ldots,n} \sigma_p)\, u_0.$$

Proof.

$$S=u_0v_0+\cdots+u_nv_n=u_0\sigma_0+u_1(\sigma_1-\sigma_0)+u_2(\sigma_2-\sigma_1)+\cdots+u_n(\sigma_n-\sigma_{n-1})$$
$$=\sigma_0(u_0-u_1)+\sigma_1(u_1-u_2)+\cdots+\sigma_{n-1}(u_{n-1}-u_n)+\sigma_n u_n,$$

† Translator's note: The theorem quoted here is: 'If $f(x)$ is any integrable function, its indefinite integral $F(x)$ has, almost everywhere, a finite differential coefficient equal to $f(x)$.' [See Titchmarsh: *Theory of Functions*. Oxford Univ. Press.]

‡ Translator's note: Dirichlet integral is *not* explicitly defined in the original context in this section. On pages 9 and 20 we probably have to note that

$$S_n=\frac{1}{\pi}\int_\alpha^{\alpha+2\pi} f(\theta)\frac{\sin(2n+1)\frac{1}{2}(x-\theta)}{2\sin\frac{1}{2}(x-\theta)}d\theta$$

is the *Dirichlet integral.*

and

$$u_0-u_1\geqslant 0,\quad u_1-u_2\geqslant 0,\cdots,u_n\geqslant 0.$$

It follows that

$$\min(\sigma_0,\sigma_1,\cdots,\sigma_n)\,u_0\leqslant S\leqslant\max(\sigma_0,\sigma_1,\cdots,\sigma_n)\,u_0.$$

Q.E.D.

Changing this to integral form, we obtain the following theorem.

THEOREM 1.2 *(Bonnet)*. Suppose $f(x)$ is Riemann-integrable on a finite interval $[a,b]$ and $\varphi(x)$ is a positive non-increasing function. Under these conditions, there exists a $\xi\in[a,b]$ such that

$$\int_a^b \varphi(x)\,f(x)\,\mathrm{d}x=\varphi(a+0)\int_a^\xi f(x)\,\mathrm{d}x.$$

Proof. If $\varphi(x)$ is discontinuous at $x=a$, we may replace the value at $x=a$, setting $\varphi(a)=\varphi(a+0)$ without changing the value of the integration which appears on the left hand side of the above equation.

We can, therefore, assume that $\varphi(a)=\varphi(a+0)$ at the outset. We partition $[a,b]$ into n equal intervals

$$a=x_0<x_1<x_2<\cdots<x_n=b,\quad h=x_i-x_{i-1}=(b-a)/n.$$

By the Abelian transform we have

$$\varphi(a+0)\,h\sum_{i=0}^{k'}f(x_i)\leqslant h\sum_{i=0}^{n-1}\varphi(x_i)\,f(x_i)\leqslant\varphi(a+0)\,h\sum_{i=1}^{k}f(x_i).$$

For an arbitrary $\varepsilon\,(>0)$, if h is sufficiently small, we establish

$$h\sum_{i=0}^{k'}f(x_i)\leqslant\max_\xi\int_a^\xi f(x)\,\mathrm{d}x+\varepsilon.$$

In fact, from the definition of integration we have

$$\left|h\sum_{i=0}^{k'}f(x_i)-\int_a^{a+(k'+1)h}f(x)\,\mathrm{d}x\right|\leqslant\varepsilon,$$

which is always true irrespective of the value of k' provided that h is sufficiently small. Therefore, for $h\to 0$ we have

$$\varphi(a+0)\,[\min_\xi\int_a^\xi f(x)\,\mathrm{d}x-\varepsilon]\leqslant\int_a^b\varphi(x)\,f(x)\,\mathrm{d}x\leqslant\varphi(a+0)\,[\max_\xi\int_a^\xi f(x)\,\mathrm{d}x+\varepsilon].$$

Now we know that $\varepsilon(>0)$ is arbitrary and that $\int_a^\xi f(x)\,dx$ is in fact a continuous function of ξ.

Q.E.D.

The following theorem is important for various applications.

THEOREM 1.3 *(Riemann–Lebesgue)*. Let $\psi(t)$ be summable on $[a, b]$. Then we have

$$\int_a^b \psi(t)\sin nt\,dt \to 0 \quad \text{as} \quad n\to\infty .$$

Note. The theorem can be generalized as follows:

$$\int_{-\infty}^{+\infty} \psi(t)\sin \lambda t\,dt \to 0 \quad \text{as} \quad \lambda\to+\infty ,$$

where λ is a real parameter, and $\psi\in L^1(-\infty, +\infty)$. This is also true when we replace $\sin\lambda t$ by $\cos\lambda t$.

Proof. We partition $[a, b]$ equally by taking endpoints $k\pi/n\,(k=0, \pm 1, \pm 2, \ldots)$. Then, we separate the integral into several parts

$$\sum_p \left(\int_{2p\pi/n}^{(2p+1)\pi/n} + \int_{(2p+1)\pi/n}^{(2p+2)\pi/n}\right) + \int_a^{a+\delta} + \int_{b-\delta'}^{b} .$$

Here $a+\delta$ is the first endpoint in the form of $2p/n$ lying to the right relative to a, and $b-\delta'$ is the last endpoint in the same form lying to the left relative to b. As $\delta, \delta' < 2\pi/n$, so $\delta, \delta'\to 0$ as $n\to 0$. Also, as $\psi(t)$ is summable, the two final integrals converge to 0 as $n\to 0$. On the other hand, if we put $t-(\pi/n)=t'$, we have

$$\sin nt = \sin n(t'+(\pi/n)) = -\sin nt' .$$

This means that the sum of the first and second integrals can be written as

$$\sum_p \int_{2p\pi/n}^{(2p+1)\pi/n} [\psi(t)-\psi(t+(\pi/n))]\sin nt\,dt .$$

Moreover, the absolute value of this integral does not exceed

$$\int_a^b |\psi(t)-\psi(t+(\pi/n))|\,dt .$$

By Lebesgue's lemma, if $\psi(t)$ is summable

$$\int_a^b |\psi(t+h)-\psi(t)|\, dt \to 0 \quad \text{as} \quad h\to 0.$$

It follows that the value of the integral in question converges to 0 as $n\to\infty$.

Q.E.D.

We have used Lebesgue's lemma, a result we shall use later.

We now give a proof.

LEMMA 1.1 *(Lebesgue).* Let $\psi \in L^p(a, b)$ where $p \geqslant 1$, and (a, b) is a finite or infinite interval. We have

$$\int_a^b |\psi(t+h)-\psi(t)|^p\, dt \to 0 \quad \text{as} \quad h\to 0.$$

Proof. It is sufficient to assume that (a, b) is finite, and $\psi(t)$ is bounded. We proceed under these assumptions. First, as $\psi(t)$ is measurable it can be represented as a measurable limit of a sequence of step functions.† Therefore, for an arbitrary $\varepsilon(>0)$ there exists a step function $\varphi(t)$ and

$$\int_a^b |\psi(t)-\varphi(t)|^p\, dt < \varepsilon.$$

Note that for $\varphi(t)$ the properties can be easily established. In fact, for an appropriate partition $a<x_1<x_2<\cdots<x_n=b$, $\varphi(t)$ is constant on each (x_i, x_{i+1}). Therefore, from

$$3^{1-p}\int_a^b |\psi(t+h)-\psi(t)|^p\, dt \leqslant \int_a^b |\varphi(t+h)-\varphi(t)|^p\, dt$$
$$+\int_a^b |\varphi(t+h)-\psi(t+h)|^p\, dt+\int_a^b |\varphi(t)-\psi(t)|^p\, dt$$

as $h\to 0$, the first term of the right hand side of the inequality tends to 0, and the second and the third are smaller than ε. ε is arbitrary (>0). Hence, the left hand term tends to 0 as $h\to 0$.

Q.E.D.

† Moreover, we can assume that the sequence is uniformly bounded.

Returning to the initial argument, we now consider the n-term sum of the Fourier series of $f(x)$

$$S_n = \tfrac{1}{2}a_0 + \sum_{p=1}^{n} (a_p \cos px + b_p \sin px). \tag{1.4}$$

We give a proof of the fact $S_n \to f(x)$. To this end we note that

$$\int_0^{+\infty} \frac{\sin x}{x}\, \mathrm{d}x = \tfrac{1}{2}\pi. \tag{1.5}$$

Care must be taken in dealing with the equality. We can readily see that as

$$\int_0^{+\infty} \frac{|\sin x|}{x}\, \mathrm{d}x = +\infty,$$

$|\sin x|/x$ is *not* summable in $(0, \infty)$, the integral is a so-called 'improper integral'.

In this sense we should write (1.5) as

$$\int_0^{\to +\infty} \frac{\sin x}{x}\, \mathrm{d}x = \lim_{A \to +\infty} \int_0^{A} \frac{\sin x}{x}\, \mathrm{d}x = \tfrac{1}{2}\pi. \tag{1.6}$$

We must verify the existence of the limit of (1.5); for this we use Bonnet's theorem (theorem 1.2). In fact,

$$\left| \int_A^{A'} \frac{\sin x}{x}\, \mathrm{d}x \right| = \left| \frac{1}{A} \right| \left| \int_A^{\xi} \sin x\, \mathrm{d}x \right| \leqslant \left| \frac{2}{A} \right| \to 0 \quad (A \to +\infty). \tag{1.7}$$

In order to conclude that the limit is in fact $\frac{1}{2}\pi$, we multiply by a converging factor $\mathrm{e}^{-\alpha x}$ $(\alpha > 0)$ to get

$$F(\alpha) = \int_0^{\infty} \mathrm{e}^{-\alpha x} \frac{\sin x}{x}\, \mathrm{d}x \quad (\alpha > 0).$$

It is obvious that

$$\int_0^{\to +\infty} \frac{\sin x}{x}\, \mathrm{d}x = \lim_{\alpha \to +0} F(\alpha),$$

where $F(\alpha)$ is continuously differentiable in $\alpha > 0$, and

$$F'(\alpha) = -\int_0^{\infty} \mathrm{e}^{-\alpha x} \sin x\, \mathrm{d}x.$$

The right hand term is $-1/(\alpha^2+1)$, therefore, $F(\alpha)=C-\arctan\alpha$ and $F(\alpha)\to 0\,(\alpha\to+\infty)$. Hence, $C=\frac{1}{2}\pi$ and $\lim_{\alpha\to+0} F(\alpha)=\frac{1}{2}\pi$.

Q.E.D.

We are now ready to prove the following well-known theorem.

THEOREM 1.4 *(Jordan–Lebesgue)*. Let $f(x)$ be a function which has period 2π, and is summable in $[0, 2\pi]$. We further assume that $f(x)$ is of bounded variation in the neighbourhood of x. Then we have

$$S_n(x)\to\tfrac{1}{2}[f(x+0)+f(x-0)] \quad (n\to\infty).$$

Moreover, if $f(x)$ is continuous in $[\alpha, \beta]$ and of bounded variation, then $S_n(x)$ converges uniformly to $f(x)$ in an arbitrary interval $[\alpha', \beta']$ which is entirely contained within $[\alpha, \beta]$.

Proof.

$$S_n=\frac{1}{2\pi}\int_{\alpha}^{\alpha+2\pi} f(\theta)\,\mathrm{d}\theta+\frac{1}{\pi}\sum_{p=1}^{n}\left[\cos px\int_{a}^{\alpha+2\pi} f(\theta)\cos p\theta\,\mathrm{d}\theta\right.$$
$$\left.+\sin px\int_{\alpha}^{\alpha+2\pi} f(\theta)\sin p\theta\,\mathrm{d}\theta\right]$$
$$=\frac{1}{\pi}\int_{\alpha}^{\alpha+2\pi} f(\theta)\left[\tfrac{1}{2}+\sum_{p=1}^{n}\cos p(x-\theta)\right]\mathrm{d}\theta$$
$$=\frac{1}{\pi}\int_{\alpha}^{\alpha+2\pi} f(\theta)\frac{\sin(2n+1)\frac{1}{2}(x-\theta)}{2\sin\frac{1}{2}(x-\theta)}\,\mathrm{d}\theta.\dagger$$

Here we have used the well-known formula

$$\tfrac{1}{2}+\sum_{p=1}^{n}\cos 2pt=\frac{\sin(2n+1)t}{2\sin t}. \tag{1.8}$$

In the foregoing expression of S_n, we put $\theta=x+2t$ and change the integral variable from θ to t, then we have

$$S_n=\frac{1}{\pi}\int_{\beta}^{\beta+\pi} f(x+2t)\,\frac{\sin(2n+1)t}{\sin t}\,\mathrm{d}t.$$

† Translator's note: The last integral is the ***Dirichlet integral.*** We note this here because otherwise the section title has no significance.

α is arbitrary, therefore β is also arbitrary. We put $\beta=-\frac{1}{2}\pi$, separate the integral to $\int_{-\frac{1}{2}\pi}^{0}+\int_{0}^{\frac{1}{2}\pi}$, and change t into $-t$ in the first integral. We now have

$$S_n=\frac{1}{\pi}\int_0^{\frac{1}{2}\pi}\frac{\sin(2n+1)t}{\sin t}[f(x+2t)+f(x-2t)]\,dt. \tag{1.9}$$

From (1.8) we note that

$$f(x)=\frac{1}{\pi}\int_0^{\frac{1}{2}\pi}\frac{\sin(2n+1)t}{\sin t}\cdot 2f(x)\,dt,$$

therefore (1.9) can be written as

$$\begin{aligned}\pi[S_n-f(x)]&=\int_0^{\frac{1}{2}\pi}\frac{\sin(2n+1)t}{\sin t}[f(x+2t)+f(x-2t)-2f(x)]\,dt\\&=\int_0^{\frac{1}{2}\pi}\frac{\sin(2n+1)t}{t}\varphi(t)\,dt,\end{aligned} \tag{1.10}$$

where

$$\varphi(t)=\frac{t}{\sin t}[f(x+2t)+f(x-2t)-2f(x)]. \tag{1.11}$$

Then we make $\delta>0$ small and write the right hand integral of (1.10) as

$$\pi[S_n-f(x)]=\int_0^{\delta}\frac{\sin(2n+1)t}{t}\varphi(t)\,dt+\int_{\delta}^{\frac{1}{2}\pi}\frac{\sin(2n+1)t}{t}\varphi(t)\,dt. \tag{1.12}$$

We observe that the second term of (1.12) converges to 0 as $n\to\infty$ by the Riemann–Lebesgue theorem 1.3. In fact, $\varphi(t)/t$ is summable in $[\delta,\frac{1}{2}\pi]$.

Now, as $f(x)$ is of bounded variation near the point x which we are now considering, $f(x+0)$ and $f(x-0)$ exist. The value of $f(x)$ can therefore be changed to give

$$f(x)=\tfrac{1}{2}[f(x+0)+f(x-0)].$$

In this way, as the value of $f(x)$ is affected only at a point, there is no change in its Fourier series.

Our problem now is merely to prove that the first term of the right hand integrals of (1.12) can be arbitrarily small when we make δ sufficiently small irrespective of the value of n. By our hypothesis, we conclude:

(1) $\varphi(t)$ is of bounded variation in $[0,\delta]$;

(2) $\lim\limits_{t\to+0}\varphi(t)=0.$

We put $\varphi(0)=0$, and (2) above indicates that $\varphi(t)$ is continuous at $t=0$. On the other hand, a function of bounded variation can be written as

$$\varphi(t)=p(t)-n(t),$$

where $p(t)$ and $n(t)$ are the positive and the negative variations respectively in $[0, t]$. Of course $p(t)$ and $n(t)$ are monotone-increasing in a wider sense. At the same time,

$$\lim_{t\to+0} p(t)=\lim_{t\to+0} n(t)=0.$$

Then we write

$$J_\delta=\int_0^\delta \frac{\sin(2n+1)t}{t}\, p(t)\,\mathrm{d}t=p(\delta)\int_\xi^\delta \frac{\sin(2n+1)t}{t}\,\mathrm{d}t \quad (0\leqslant\xi\leqslant\delta)$$

following Bonnet's theorem. Also,

$$\int_\xi^\delta \frac{\sin(2n+1)t}{t}\,\mathrm{d}t=\int_{(2n+1)\xi}^{(2n+1)\delta} \frac{\sin t}{t}\,\mathrm{d}t.$$

According to (1.6), for a certain A and an arbitrary h, we have

$$\left|\int_0^h \frac{\sin t}{t}\,\mathrm{d}t\right|\leqslant A$$

so that

$$|J_\delta|\leqslant 2A\cdot p(\delta).$$

Therefore,

$$\left|\int_0^\delta \frac{\sin(2n+1)t}{t}\,\varphi(t)\,\mathrm{d}t\right|\leqslant 2A[p(\delta)+n(\delta)]$$
$$\equiv 2A\times \underset{0\leqslant t\leqslant\delta}{\text{total variation}}\,[\varphi]. \tag{1.13}$$

From what has been said earlier about $\varphi(t)$ being continuous at $t=0$, the total variation $[\varphi]\to 0$ as $\delta\to 0$. We conclude that

$$S_n\to f(x)=\tfrac{1}{2}[f(x+0)+f(x-0)].$$

We now prove the remainder of the theorem. To do so we make $x\in[\alpha', \beta']$ a parameter; replacing $\varphi(t)$ by

$$\varphi(t;\, x)=\frac{t}{\sin t}\,[f(x+2t)+f(x-2t)-2f(x)].$$

Now, our problem is to show that the statements already proved are independent of the value of x. First, we consider (1.13). It is sufficient to note that

$$\underset{0\leqslant t\leqslant \delta}{\text{total variation}}[f(x+2t)+f(x-2t)-2f(x)]\leqslant \underset{t\in[x-2\delta,\, x+2\delta]}{\text{total variation}}[f(t)].$$

By our hypothesis, $f(x)$ is of bounded variation and continuous in $[\alpha, \beta]$, so for a sufficiently small δ the right hand term of the inequality can be taken as uniformly small in relation to $x\in[\alpha', \beta']$.

Next, we consider the second term of (1.12):

$$\int_{\delta}^{\frac{1}{2}\pi} \sin(2n+1)t\,\frac{\varphi(t; x)}{t}\,dt.$$

To apply the Riemann–Lebesgue theorem, we must return to the proof already given. The theorem itself is concerned with a *fixed* $\varphi(t)$, but in this case we wish to claim that $\varphi(t)$ tends uniformly to 0 as $n\to\infty$ for the family of functions which are subject to the parameter x. But, in this case, considering the form of $\varphi(t; x)$ and regarding the proof, we may readily conclude the uniformity (with respect to x). Here, we need to use the absolute continuity of the following indefinite integral: if $f(t)$ is summable, for an arbitrary ε, there exists a δ such that

$$\int_{e} |f(t)|\,dt<\varepsilon$$

when $me<\delta$ (e: an arbitrary measurable set).

Q.E.D.

3 *Application to the heat transfer equation*

We consider the following simple but historically important problem as an application of a Fourier series. The problem is that given the boundary condition

$$u(0, t)=u(\pi, t)=0 \quad (0\leqslant t<+\infty),$$

and the initial condition

$$u(x, 0)=f(x) \quad (0\leqslant x\leqslant \pi),$$

for the heat equation

$$\frac{\partial}{\partial t}u=\frac{\partial^2}{\partial x^2}u \tag{1.14}$$

within

$$\Omega=\{(x, t) \mid 0 \leqslant x \leqslant \pi, 0 \leqslant t<+\infty\},$$

find the solution $u(x, t)$ satisfying (1.14).

Let us find a solution in a formal way using known results. First we consider a continuous function $f(x)$ of bounded variation satisfying $f(0)=f(\pi)=0$. We put $f(x)=-f(-x)$ for $x \in[-\pi, 0]$ to extend the domain of the function $f(x)$ to $[-\pi, \pi]$. Notice that $f(x)$ is an odd function. Its Fourier series therefore reduces only to the terms of $\sin nx$.

Now, following the Jordan–Lebesgue theorem (1.4), we have a representation of $f(x)$ by its Fourier series

$$f(x)=\sum_{n=1}^{\infty} c_n \sin nx$$

and the right hand series is uniformly convergent. We define formally

$$u(x, t)=\sum_{n=1}^{\infty} \exp(-n^2 t)\, c_n \sin nx. \tag{1.15}$$

This can be shown to be a solution.†

We prove this as follows. It is clear that the right hand series in (1.15) converges. In fact from

$$c_n=\frac{2}{\pi} \int_0^\pi f(x) \sin nx\, \mathrm{d}x \rightarrow 0 \quad (n \rightarrow \infty) \quad \text{(Riemann–Lebesgue)}$$

we see $|c_n|<M$ $(n=1, 2, \ldots)$. Also $\sum \exp(-n^2 t)<+\infty$. Hence‡ this series converges for all $t>0$.

Actually, $u(x, t)$ has continuous partial derivatives for all orders with respect to (x, t). For example, the term by term differentiation of (1.15) can be estimated in its absolute values by

$$\sum \exp(-n^2 t)\, n|c_n| \leqslant M \sum_{n=1}^{\infty} n \exp(-n^2 t)<+\infty,$$

† Translator's note: The uniqueness of the solution is proved later.

‡ Translator's note: We see $|c_n|<M(n=1, 2, \ldots)$ where

$$M=\frac{2}{\pi} \int_0^\pi |f(x)|\, \mathrm{d}x,$$

i.e. c_n is uniformly bounded. Also $\Sigma \exp(-n^2 t)<+\infty$. Therefore, $\Sigma M \exp(-n^2 t)$ dominates the series (1.15). (There are no explanations about M in the original text.)

so that the differentiation can be justified, i.e.,

$$\frac{\partial u}{\partial x}=\sum_{n=1}^{\infty} n \exp(-n^2 t) c_n \cos nx .$$

For the same reason we have

$$\frac{\partial^2 u}{\partial x^2}=-\sum_{n=1}^{\infty} n^2 \exp(-n^2 t) c_n \sin nx .$$

Differentiation with respect to t is similarly performed i.e. taking an arbitrary $\delta(>0)$, term by term differentiation for the t in $u(x, t)$ in $\delta \leqslant t < \infty$ gives

$$\sum n^2 \exp(-n^2 t) |c_n \sin nx| \leqslant M \sum_n n^2 \exp(-\delta n^2) < +\infty .$$

since each term is uniformly estimated in t, the term by term differentiation is correct. The term by term differentiation at $t>0$ is therefore correct. Thus

$$\frac{\partial u}{\partial t}=-\sum n^2 \exp(-n^2 t) c_n \sin nx .$$

Moreover, we see that $u(x, t)$ is a solution for $t>0$; more significant still, we see that it is a function continuously differentiable for all orders.

Next, we wish to demonstrate that when $t \to +0$ we have uniformly for x

$$u(x, t) \to f(x).$$

Here, we note that the Fourier series $\sum c_n \sin nx$ is uniformly convergent, and also that $\{\exp(-n^2 t)\}$ $n=1, 2, \ldots$ is a positive decreasing sequence. By a lemma involving the Abelian transform (see the previous section) there exists N with

$$\left|\sum_{n=N+1}^{\infty} \exp(-n^2 t) c_n \sin nx\right| < \varepsilon \quad (t \geqslant 0)$$

for an arbitrary ε. We then separate into

$$|u(x, t)-f(x)| \leqslant \left|\sum_{n=1}^{N} (\exp(-n^2 t)-1) c_n \sin nx\right| + \left|\sum_{n=N+1}^{\infty} \exp(-n^2 t) c_n \sin nx\right| + \left|\sum_{n=N+1}^{\infty} c_n \sin nx\right| .$$

The last two terms $<2\varepsilon$ irrespective of t and the first term tends to 0 as $t \to +0$.

Q.E.D.

Note 1. The uniqueness of the solution can be proved as follows. If there are two such solutions, $u_1(x, t)$ and $u_2(x, t)$, consider $u(x, t)=u_1-u_2$. This must be the solution satisfying the initial and the boundary conditions ($u=0$). We then observe

$$E(t)=\int_0^\pi u(x, t)^2 \, dx .$$

This means that by differentiation

$$E'(t)=2\int_0^\pi u\cdot u_t \, dx=2\int_0^\pi u\cdot u_{xx} \, dx=-2\int_0^\pi u_x{}^2 \, dx\leqslant 0,$$

where $t>0$. To obtain this we use the fact that u is the solution of (1.14) and at the boundary ($x=0$, $x=\pi$) $u=0$.

Since $E(t)\geqslant 0$, and as $E(t)$ is non-increasing and $\lim_{t\to+0} E(t)=0$, we have $E(t)\equiv 0$.

Hence, $u(x, t)\equiv 0$.

Note 2. For the solution to the problem within $-\infty<x<+\infty$, instead of the given boundary condition, we must use a Fourier transform which we shall describe later.

4 *The system of orthogonal functions in L^2 (Ω).*

We are concerned with the fundamental properties of a Fourier series from more general aspects. Let Ω be an open set in n-dimensional Euclidean space R^n (It can be the entire R^n), and $L^2(\Omega)$ be the whole of the square-integrable functions, i.e. $f\in L^2(\Omega)$ when f is a complex-valued measurable function of $x=(x_1,\ldots, x_n)\in\Omega$ satisfying

$$\int_\Omega |f(x)|^2 \, dx<+\infty .$$

For the norm, we take

$$\|f(x)\|_{L^2}=\left(\int |f(x)|^2 \, dx\right)^{\frac{1}{2}} .$$

Of course, we identify $f(x)$ with $g(x)$ when they are equal almost everywhere in Ω.

As is well known, $L^2(\Omega)$ is *complete* (Riesz–Fischer theorem), i.e.,

if $\{f_n\}$, where $f_n\in L^2(\Omega)$, form a Cauchy sequence i.e. $\|f_n-f_m\|_{L^2}\to 0$ ($n, m\to\infty$), there exists an f ($\in L^2(\Omega)$) with $\|f_n-f\|_{L^2}\to 0$ ($n\to\infty$).

We do not give the proof of the theorem, being concerned with items which are not self-evident. Seeing this, we can simply observe

$$\int_{\Omega} |f_n(x)-f_m(x)|^2 \,\mathrm{d}x \to 0 \quad (n, m \to \infty),$$

but this does not lead us to conclude that $(f_n(x))$ is convergent almost everywhere.

The definition of an orthonormal system

We say that

$$\{\varphi_n(x)\} \quad (\text{where } \varphi_n \in L^2(\Omega))$$

form an *orthonormal* system if, and only if, they form an orthogonal system, i.e.,

$$\int_{\Omega} \varphi_n(x)\,\overline{\varphi_m(x)}\,\mathrm{d}x = 0 \quad (n \neq m)$$

and satisfy

$$\int_{\Omega} |\varphi_n(x)|^2 \,\mathrm{d}x = 1 \quad (n=1, 2, \ldots).$$

Suppose then $\{\varphi_n\}$ is an orthonormal system and consider a Fourier expansion of $f(x) \in L^2(\Omega)$:

$$f(x) \sim c_1\varphi_1(x) + c_2\varphi_2(x) + \cdots + c_n\varphi_n(x) + \cdots \tag{1.16}$$

where

$$c_n = \int_{\Omega} f(x)\,\overline{\varphi_n(x)}\,\mathrm{d}x. \tag{1.17}$$

c_n may be called the *Fourier coefficient*.

The definition of a complete orthonormal system

We say that an orthonormal system $\{\varphi_n\}$ is *complete* if and only if, for f $(\in L^2(\Omega))$,

$$\int f(x)\,\overline{\varphi_n(x)}\,\mathrm{d}x = 0 \quad (n=1, 2, \ldots)$$

is true only when $f=0$, i.e. $f(x)=0$ almost everywhere.

Now we return to (1.16). We claim that this right hand series is a convergent series in $L^2(\Omega)$. To see this we write

$$0 \leqslant \int \left| f(x) - \sum_{i=1}^{m} c_i\varphi_i(x) \right|^2 dx$$

$$= \int \left[f(x) - \sum_{i=1}^{m} c_i\varphi_i(x) \right] \left[\overline{f(x)} - \sum_{i=1}^{m} \bar{c}_i \overline{\varphi_i(x)} \right] dx$$

$$= \int |f(x)|^2 dx + \sum_{i=1}^{m} |c_i|^2 \int |\varphi_i(x)|^2 dx - \sum_{i=1}^{m} \bar{c}_i \int f(x)\, \overline{\varphi_i(x)}\, dx - \sum_{i=1}^{m} c_i \int \overline{f(x)}\, \varphi_i(x)\, dx$$

$$= \int |f(x)|^2 dx - \sum_{i=1}^{m} |c_i|^2 .$$

In short we have the following fundamental relation

$$\int \left| f(x) - \sum_{i=1}^{m} c_i\varphi_i(x) \right|^2 dx = \int |f(x)|^2 dx - \sum_{i=1}^{m} |c_i|^2 . \tag{1.18}$$

From this we have $\sum_{i=1}^{\infty} |c_i|^2 < +\infty$ and

$$\sum_{i=1}^{\infty} |c_i|^2 \leqslant \int |f(x)|^2 dx \quad \text{(Bessel's inequality)}. \tag{1.19}$$

Using the Riesz–Fischer theorem, we show the existence of

$$\varphi(x) = \lim_{m\to\infty} \sum_{i=1}^{m} c_i\varphi_i(x) \quad \text{in} \quad L^2(\Omega) \quad (\varphi \in L^2(\Omega)). \tag{1.20}$$

Again we warn the reader by writing 'in $L^2(\Omega)$' because this is not a type of convergence 'almost everywhere'.

The reason why we can apply the Riesz–Fischer theorem is seen by putting $\sum_{i=1}^{m} c_i\varphi_i(x) = f_m(x)$ and observing

$$\int |f_m(x) - f_{m'}(x)|^2 dx = \sum_{i=m+1}^{m'} |c_i|^2 \to 0 \quad (m\to\infty),$$

when $m' > m$.

Let us see if the Fourier series $\varphi(x)$ coincides with $f(x)$. First we note that the Fourier development of $\varphi \in L^2(\Omega)$ is nothing but $\varphi(x)$ itself. In fact,

$$\int \varphi(x)\, \overline{\varphi_n(x)}\, dx = \int \left(\lim_{m\to\infty} \sum_{i=1}^{m} c_i\varphi_i(x) \right) \overline{\varphi_n(x)}\, dx = \lim_{m\to\infty} \sum_{i=1}^{m} c_i \int \varphi_i(x)\, \overline{\varphi_n(x)}\, dx$$

$$= c_n .$$

Here we can see that from the Cauchy–Schwartz inequality, it is possible to change the limit of the sum sign to an integral sign.

Therefore,

$$\int [f(x)-\varphi(x)]\,\overline{\varphi_n(x)}\,dx=0 \quad (n=1,2,\ldots).$$

If $\{\varphi_n(x)\}$ is complete, we can claim $f(x)=\varphi(x)$, and from the fundamental relation (1.18), we see that

$$\sum_{i=1}^{\infty} |c_i|^2=\int |f(x)|^2\,dx \quad \text{(Parseval's equality)}. \tag{1.21}$$

Conversely, it is obvious that from (1.21) we can conclude $f(x)=\varphi(x)$ via (1.18).

Hitherto, we have assumed that $\{\varphi_i(x)\}$ is complete; now we take it as not complete. In this case, there exists an element $f\in L^2(\Omega)$ $(f\neq 0)$, orthogonal to any $\varphi_1(x)$.

The Fourier coefficients of this $f(x)$ are all 0, therefore, being $\varphi(x)=0$, $f\neq\varphi$.

To sum up we obtain:

THEOREM 1.5. The following three conditions are equivalent.

(*a*) An orthonormal system $\{\varphi_n\}$ is complete in $L^2(\Omega)$.

(*b*) The Parseval equality (1.21) is true for an arbitrary $f\in L^2(\Omega)$.

(*c*) The Fourier development of an arbitrary $f\in L^2(\Omega)$ represents $f(x)$.

Up to now we have been considering only the Fourier expansion of $f(x)$. Let us take a view of the situation in a reverse way. Let $\{\varphi_i\}$ be a complete orthonormal system. For an arbitrary $(c)=\{c_i\}$ satisfying the condition

$$\sum_{i=1}^{\infty} |c_i|^2<+\infty, \tag{1.22}$$

we observe

$$f(x)=c_1\varphi_1(x)+c_2\varphi_2(x)+\cdots+c_n\varphi_n(x)+\cdots,$$

where the right hand series has been defined as a limit in $L^2(\Omega)$ as we see in (1.22). In this case we can say that the Fourier series of $f(x)$ is in fact the right hand series.

From this, using Parseval's equality (1.21), we know that the Fourier series of $f(x)$ defines an isometric operator from the space $L^2(\Omega)$ onto the whole space l^2.

$$\|(c)\|_{l^2}=\left(\sum_{n=1}^{\infty}|c_n|^2\right)^{\frac{1}{2}}$$

Finally, some notes to qualify the rather overemphasized convergence of the series in $L^2(\Omega)$. In practice, according to the smoothness of f, we often estimate the decreasing order of c_n. If it is so, there is a property

(1) If $\sum c_n\varphi_n(x)$ is uniformly convergent in an open set ω of Ω, and f and φ_n are continuous in ω, then

$$f(x)=\sum_{n=1}^{\infty}c_n\varphi_n(x)\quad (x\in\omega).$$

The proof is simple, and therefore omitted.

(2) If it is not uniformly convergent, but, almost everywhere in ω

$$\sum_{n=1}^{\infty}|c_n|\,|\varphi_n(x)|<+\infty,$$

and, if

$$\sum_{n=1}^{\infty}|c_n|\,|\varphi_n(x)|\leqslant\Phi(x),\qquad \int_{\omega}\Phi(x)^2\,dx<+\infty,$$

then, from Lebesgue's lemma, we have

$$f(x)=\sum c_n\varphi_n(x),$$

almost everywhere for $x\in\omega$.

Example. From theorem 1.1 we know already that the system of trigonometric functions (see note there) is complete in $(0, 2\pi)$. In fact, this is because

$$L^2(0,2\pi)\subset L^1(0,2\pi).$$

Now, we consider $f(x)$ which is continuously differentiable with 2π as its period. The Fourier coefficients of $f(x)$ are

$$\pi a_n=\int_0^{2\pi}f(x)\cos nx\,dx=-\frac{1}{n}\int_0^{2\pi}f'(x)\sin nx\,dx=-\frac{\pi b_n'}{n},$$

$$\pi b_n=\int_0^{2\pi}f(x)\sin nx\,dx=\frac{1}{n}\int_0^{2\pi}f'(x)\cos nx\,dx=\frac{\pi a_n'}{n},$$

where a_n', b_n' are the Fourier coefficients of $f'(x)$. From this we see that the Fourier series of $f(x)$ is uniformly convergent. In fact, we see that it is bounded by

$$\sum_n |a_n|+|b_n| \leqslant \sum_n \frac{1}{n}(|a_n'|+|b_n'|) \leqslant \left(\sum_n \frac{1}{n^2}\right)^{\frac{1}{2}} (2\sum |a_n'|^2+|b_n'|^2)^{\frac{1}{2}}$$

using Bessel's inequality. In this case, therefore, the Fourier series of $f(x)$ is uniformly convergent and represents $f(x)$.

5 *Fourier's integral formula*

As shown in the Fourier series in section 2, if we denote the n-sum in the n term series as S_n, we have

$$\pi S_n = \int_\alpha^{\alpha+2\pi} \frac{\sin(2n+1)\frac{1}{2}(x-\theta)}{2\sin\frac{1}{2}(x-\theta)} f(\theta)\,d\theta \qquad \text{(Dirichlet's integral)},$$

where $f(\theta)$ is a function with period 2π. Let us re-write this, first noting that

$$\sin(2n+1)\tfrac{1}{2}(x-\theta) = \sin n(x-\theta)\cos\tfrac{1}{2}(x-\theta) + \cos n(x-\theta)\sin\tfrac{1}{2}(x-\theta).$$

From this

$$\pi S_n = \int_\alpha^{\alpha+2\pi} f(\theta)\sin n(x-\theta)\cot\tfrac{1}{2}(x-\theta)\cdot\tfrac{1}{2}\,d\theta + \tfrac{1}{2}\int_\alpha^{\alpha+2\pi} f(\theta)\cos n(x-\theta)\,d\theta,$$

where the second term of the right hand side tends to 0 as $n\to\infty$ by Riemann–Lebesgue theorem.

Therefore, if we assume that $f(\theta)$ is of bounded variation in the neighbourhood of $\theta = x$, we have

$$\lim \int_\alpha^{\alpha+2\pi} f(\theta)\sin n(x-\theta)\cot\tfrac{1}{2}(x-\theta)\cdot\tfrac{1}{2}\,d\theta = \tfrac{1}{2}\pi[f(x+0)+f(x-0)]$$

by theorem 1.4. In the above equation we replace $f(\theta)$ by

$$\varphi(\theta) = \begin{cases} \dfrac{2f(\theta)}{x-\theta}\tan\tfrac{1}{2}(x-\theta) & (|\theta-x|<\delta<\tfrac{1}{2}\pi), \\ 0 \quad \text{(elsewhere)}. \end{cases}$$

Then, we see that $\varphi(\theta)$ is of bounded variation in the neighbourhood of $\theta = x$, and $\varphi(x\pm 0) = f(x\pm 0)$.

Thus, referring to the Riemann–Lebesgue theorem, we have

$$\lim_{n\to\infty} \int_{\alpha}^{\alpha+2\pi} \frac{f(\theta)}{x-\theta} \sin n(x-\theta)\, d\theta = \tfrac{1}{2}\pi[f(x+0)+f(x-0)]. \tag{1.23}$$

Hitherto we have considered $f(\theta)$ as having period 2π, but now we assume

$$\int_{-\infty}^{+\infty} |f(\theta)|\, d\theta < +\infty .$$

Furthermore, we also assume $f(\theta)$ to be of bounded variation in an arbitrary bounded interval.

Consider any α satisfying the condition

$$\alpha < x < \alpha + 2\pi .$$

Then (1.23) is valid. Consider

$$\int_{-\infty}^{\alpha} \frac{f(\theta)}{x-\theta} \sin n(x-\theta)\, d\theta , \quad \int_{\alpha+2\pi}^{\infty} \frac{f(\theta)}{x-\theta} \sin n(x-\theta)\, d\theta .$$

According to the Riemann–Lebesgue theorem, the value of these integrals is convergent to 0 as $n \to \infty$.

Thus we obtain the following theorem:

THEOREM 1.6 *(Fourier's formula).*

$$\lim_{n\to\infty} \int_{-\infty}^{+\infty} \frac{f(\theta)}{x-\theta} \sin n(x-\theta)\, d\theta = \tfrac{1}{2}\pi[f(x+0)+f(x-0)],$$

where $f(\theta)$ is summable in $(-\infty, +\infty)$, and is of bounded variation in an arbitrary bounded interval.

Q.E.D.

We re-write the above formula as

$$F_n = \int_{-\infty}^{+\infty} \frac{f(\theta)}{x-\theta} \sin n(x-\theta)\, d\theta$$

and using

$$\frac{\sin n(x-\theta)}{x-\theta} = \int_0^n \cos v(x-\theta)\, dv$$

we then have

$$F_n = \int_{-\infty}^{+\infty} d\theta \int_0^n f(\theta) \cos v(x-\theta)\, dv.$$

We apply Fubini's theorem to obtain

$$F_n = \int_0^n dv \int_{-\infty}^{+\infty} f(\theta) \cos v(x-\theta)\, d\theta.$$

Theorem 1.6 indicates that F_n is convergent as $n \to +\infty$, but we can eliminate the condition: 'integral-valued and convergent'. To this end, first we choose z in the open interval (0, 1), such that

$$F_{n+z} - F_n = \int_n^{n+z} dv \int_{-\infty}^{+\infty} f(\theta) \cos v(x-\theta)\, d\theta,$$

replacing v with $n+v$,

$$= \int_0^z dv \int_{-\infty}^{+\infty} f(\theta) \cos(n+v)(x-\theta)\, d\theta,$$

Now, we separate the integral $\int_{-\infty}^{+\infty}$ into three parts

$$\int_{-\infty}^{+\infty} = \int_{-\infty}^{-\alpha} + \int_{-\alpha}^{\alpha} + \int_{\alpha}^{+\infty}.$$

If we take a sufficiently large value of α we can make the first and third terms of the integral $<\varepsilon(>0)$ irrespective of n and v. The second term converges to 0 as $n \to +\infty$ with respect to v. Therefore $F_{n+z} - F_n \to 0$, as $n \to \infty$.

So that we have now:

THEOREM 1.7 *(Fourier's double integral formula).*

$$\tfrac{1}{2}\pi[f(x+0)+f(x-0)] = \lim_{A \to +\infty} \int_0^A dv \int_{-\infty}^{+\infty} f(\theta) \cos v(x-\theta)\, d\theta, \qquad (1.24)$$

where we assume $f(\theta)$ satisfies the same conditions as the previous theorem 1.6.

Q.E.D.

6 *Fourier transforms*

In the modern analysis we often describe the 'inverse formula' of the Fourier transform in a different way, although it is substantially the same as (1.24). We now study this version.

First, we must clarify our present situation. Up to now we have assumed $f(\theta)$ is real-valued, but generally even when taken as complex-valued, what has been stated thus far still holds. In fact we may consider $f(\theta)=f_1(\theta)+\mathrm{i}f_2(\theta)$ as $f(\theta)$ divided into real and imaginary parts.

Then, the condition '$|f(\theta)|$ is summable' is equivalent to the condition '$f_1(\theta)$ and $f_2(\theta)$ are both summable'. This is also true when we replace 'summable' by 'of bounded variation'. After observing these facts, we consider $f(\theta)$ as a complex-valued function. Also, for convenience, we replace v by $2\pi v$ in (1.24), then as a result we have

$$\tfrac{1}{2}[f(x+0)+f(x-0)]=\lim_{A\to+\infty} 2\int_0^A [C(v)\cos 2\pi vx+S(v)\sin 2\pi vx]\,\mathrm{d}v, \tag{1.25}$$

where

$$C(v)=\int_{-\infty}^{+\infty} f(\theta)\cos 2\pi v\theta\,\mathrm{d}\theta,\quad S(v)=\int_{-\infty}^{+\infty} f(\theta)\sin 2\pi v\theta\,\mathrm{d}\theta.$$

We note that (1.25) has been obtained after the replacement of n in $\cos 2\pi nx$ and $\sin 2\pi nx$ in the Fourier series of $f(x)$ by a real parameter. Here we assume $v\geqslant 0$ and $v\leqslant 0$ are also meaningful. In this case we note that

$$C(-v)=C(v),\quad S(-v)=-S(v),$$

and use

$$\lim_{A\to+\infty}\int_{-A}^{A} [C(v)\cos 2\pi vx+S(v)\sin 2\pi vx]\,\mathrm{d}v \tag{1.26}$$

instead of (1.25). For $f(\theta)$ we define

$$\hat{f}(v)=\int_{-\infty}^{+\infty} f(\theta)\,e^{-2\pi \mathrm{i}v}\,\mathrm{d}\theta.$$

Obviously

$$\hat{f}(v)=C(v)-\mathrm{i}S(v).$$

Therefore, (1.26) can be written as

$$\lim_{A\to+\infty}\int_{-A}^{A} [C(v)-\mathrm{i}S(v)]\,\mathrm{e}^{2\pi \mathrm{i}vx}\,\mathrm{d}v=\lim_{A\to+\infty}\int_{-A}^{A} \hat{f}(v)\,\mathrm{e}^{2\pi \mathrm{i}vx}\,\mathrm{d}v$$

because $C(v)$ is an even and $S(v)$ is an odd function.

After changing the variable of integration, we finally obtain the following:

THEOREM 1.8. Let

$$f \in L^1(-\infty, +\infty),$$

where $f(x)$ is of bounded variation in an arbitrary bounded finite interval. If we define

$$\hat{f}(\xi) = \int_{-\infty}^{+\infty} e^{-2\pi i x\xi} f(x)\, dx, \tag{1.27}$$

we have

$$\tfrac{1}{2}[f(x+0)+f(x-0)] = \lim_{A \to +\infty} \int_{-A}^{A} e^{2\pi i x\xi} \hat{f}(\xi)\, d\xi. \tag{1.28}$$

Furthermore, if we assume that $\hat{f}(\xi)$ is summable and $f(x)$ is continuous, we have

$$f(x) = \int_{-\infty}^{+\infty} e^{2\pi i x\xi} \hat{f}(\xi)\, d\xi. \tag{1.29}$$

Q.E.D.

The $\hat{f}(\xi)$ defined in the above (1.27) is called the *Fourier transform* of $f(x)$ or the *Fourier image*. We can also write this as $(\mathscr{F} f)(\xi)$. (1.28) or (1.29) may be called the *inversion formulae* of a Fourier transform and this is of fundamental importance. (1.28) is rather complicated because generally $\hat{f}(\xi)$ is not summable. We see this from the following

EXAMPLE.

$$f(x) = \begin{cases} 1 & (|x| \leqslant A) \\ 0 & (|x| > A). \end{cases}$$

$$\hat{f}(\xi) = \int_{-A}^{A} e^{-2\pi i x\xi}\, dx = \left[\frac{e^{-2\pi i x\xi}}{-2\pi i \xi}\right]_{x=-A}^{x=+A} = \frac{\sin 2\pi A\xi}{\pi\xi}. \tag{1.30}$$

Therefore $\hat{f}(\xi)$ is *not* summable.

Some properties of the Fourier transform

In the foregoing section we defined $\hat{f}(\xi)$, the Fourier transform of $f(x)$. We now give some fundamental properties of $\hat{f}(\xi)$, and show that these properties remain true for a Fourier transform of broader scope.

(*a*) If $f(x)$ and $f'(x)$ are both continuous and summable, then performing an integration by parts in (1.27), and since $f(x) \to 0$ as $x \to \pm\infty$, we have

$$2\pi i\xi \hat{f}(\xi) = \int e^{-2\pi i x\xi} f'(x)\, dx, \tag{1.31}$$

$$2\pi |\zeta|\, |\hat{f}(\xi)| \leqslant \|f'(x)\|_{L^1}.$$

More generally, if we assume that $f(x)$ is continuously differentiable of order m, and $f(x), f'(x), \ldots, f^{(m)}(x)$ are all summable, then we have

$$\mathscr{F}[f^{(m)}(x)] = (2\pi i\xi)^m \mathscr{F}[f(x)], \tag{1.32}$$

$$|2\pi\xi|^m\, |\hat{f}(\xi)| \leqslant \|f^{(m)}(x)\|_{L^1}. \tag{1.33}$$

Roughly speaking, this characteristic signifies that the smoothness of $f(x)$ reflects the order of decrease of $\hat{f}(\xi)$ at infinity.

(*b*) $\hat{f}(\xi) = \int_{-\infty}^{+\infty} e^{-2\pi i x\xi} f(x)\, dx.$

If we differentiate this under the integral sign, we have

$$\hat{f}'(\xi) = \int_{-\infty}^{+\infty} -2\pi i x f(x)\, e^{-2\pi i x\xi}\, dx.$$

This can be correctly performed if we assume that $|x|\, |f(x)|$ is summable.

More generally, if not only $f(x)$ but also $x^m f(x)$ is summable, $\hat{f}(\xi)$ is continuously differentiable of order m. In this case we have

$$\frac{d^m}{d\xi^m} \hat{f}(\xi) = \mathscr{F}[(-2\pi i x)^m f(x)], \tag{1.34}$$

$$|\hat{f}^{(m)}(\xi)| \leqslant \|(2\pi x)^m f(x)\|_{L^1}. \tag{1.35}$$

This means that the larger the order of decrease of $f(x)$ as $x \to \pm\infty$, the higher the order of differentiability of $\hat{f}(\xi)$, the Fourier image of $f(x)$.

Note. From (*a*), if not only $f(x)$ but also $f'(x)$ and $f''(x)$ are continuous and summable, then $\hat{f}(\xi)$ is summable, and (1.29) is true in theorem 1.8.

(*c*) $\mathscr{F}[f(x-h)] = e^{-2\pi i h\xi} \hat{f}(\xi)$

In fact,

$$\int_{-\infty}^{+\infty} e^{-2\pi i x\xi} f(x-h)\,dx = \int_{-\infty}^{+\infty} e^{-2\pi i(x+h)\xi} f(x)\,dx = e^{-2\pi i h\xi} \hat{f}(\xi).$$

Now, we wish to state an important equality, Parseval's equality, for the Fourier series.

LEMMA 1.2 *(A special case of Plancherel's theorem).* Let $f(x)$ be continuously differentiable of order 2, and let $f(x)$, $f'(x)$ and $f''(x)$ be summable, then

$$\int_{-\infty}^{+\infty} |f(x)|^2\,dx = \int_{-\infty}^{+\infty} |\hat{f}(\xi)|^2\,d\xi. \tag{1.36}$$

Proof. As stated in the above note, from our hypothesis, $\hat{f}(\xi)$ is summable and (1.29) is true. Therefore

$$\int_{-\infty}^{+\infty} |f(x)|^2\,dx = \int_{-\infty}^{+\infty} f(x)\overline{f(x)}\,dx = \int_{-\infty}^{+\infty} f(x)\,dx \int_{-\infty}^{+\infty} \overline{\hat{f}(\xi)\,e^{2\pi i x\xi}}\,d\xi.$$

Also, from $f, \hat{f} \in L^1$, and Fubini's theorem, we have

$$\int_{-\infty}^{+\infty} |f(x)|^2\,dx = \int_{-\infty}^{+\infty} \overline{\hat{f}(\xi)}\,d\xi \int_{-\infty}^{+\infty} f(x)\,e^{-2\pi i x\xi}\,dx$$
$$= \int_{-\infty}^{+\infty} \overline{\hat{f}(\xi)}\hat{f}(\xi)\,d\xi = \int_{-\infty}^{+\infty} |\hat{f}(\xi)|^2\,d\xi.$$

Q.E.D.

This lemma is true if we assume $f \in L^2$. This can also be inferred from the results of the Fourier series. However, in this instance $f(x)$ is not necessarily summable and $\hat{f}(\xi)$, the Fourier transform, is not definable by an integral when we assume only $f \in L^2$. This is a disadvantage when we assume $f \in L^2$.

The advantage of doing so is that from our result we see that the Fourier transform is an isometric operator defined on a dense set in L^2, so that, by a well-known theorem we can extend this to an isometric operator on the whole L^2. We shall actually do this later when we treat a general n-dimensional case.

We can state now the partially proved theorem.

THEOREM 1.9. Let $f(x)$ and $g(\xi)$ be summable. In this case we write

$$\mathscr{F}f=\int_{-\infty}^{+\infty} e^{-2\pi i x\xi} f(x)\,dx,$$

$$\overline{\mathscr{F}}g=\int_{-\infty}^{+\infty} e^{2\pi i x\xi} g(\xi)\,d\xi,$$

and call them respectively, the *Fourier transform* and the *Fourier inverse transform*.

Using the above notations, assuming that f and g and their second-order derivatives are twice differentiable and summable, we have

$$\overline{\mathscr{F}}\mathscr{F}f=f \tag{1.37}$$

$$\mathscr{F}\overline{\mathscr{F}}g=g \tag{1.38}$$

and

$$\|\mathscr{F}f\|_{L^2}=\|f\|_{L^2}, \qquad \|\overline{\mathscr{F}}g\|_{L^2}=\|g\|_{L^2}. \tag{1.39}$$

Proof. There remains only the proof of (1.38). Let $g(\xi)$ be summable, then

$$\overline{\mathscr{F}}g=\overline{\mathscr{F}}\bar{g} \tag{1.40}$$

so that

$$\mathscr{F}\overline{\mathscr{F}}g=\mathscr{F}\cdot\mathscr{F}\bar{g}=\overline{\mathscr{F}}\mathscr{F}\bar{g},$$

We then apply (1.37) to obtain

$$\overline{\mathscr{F}}\mathscr{F}\bar{g}=\bar{\bar{g}}=g$$

Q.E.D.

7 *Several variables and functional spaces*

The results obtained in the previous sections are in fact still true for multiple variables; this can be proved by direct methods. Here we progress to several variables with Parseval's equality forming the basic idea. For convenience's sake we introduce some functional spaces which will be dealt with in later sections. For the time being the reader may regard them as notational devices; we shall study these spaces thoroughly in the second chapter by considering their topological nature. Generally speaking, when we study topological vector spaces, it is essential to choose spaces which are suitably and efficiently topologized with regard to the nature of the problems and our method of attacking them.

The studies of partial differential equations and functional spaces are indeed very close to each other; the study of partial differential equations necessitates the introduction of new functional spaces, and conversely, the study of functional spaces plays a significant rôle in the study of partial differential equations.

In the sequel we can see that all topological spaces are complete. As for notations employed, we follow L. Schwartz: *Théorie des distributions I, II.* (1950–1).

(1) $\mathscr{B}^m$: the space of $f(x)$ having continuous and bound partial derivatives (hereafter called simply derivatives) up to order m. The norm for the space is provided by

$$|f(x)|_m = \sum_{|\alpha| \leqslant m} \sup_{x \in R^n} |D^\alpha f(x)|, \quad \text{for} \quad f \in \mathscr{B}^m,$$

where

$$\alpha = (\alpha_1, \alpha_2, \ldots, \alpha_n), \quad |\alpha| = \alpha_1 + \alpha_2 + \cdots + \alpha_n;$$

$$D^\alpha f(x) = \left(\frac{\partial}{\partial x_1}\right)^{\alpha_1} \cdots \left(\frac{\partial}{\partial x_n}\right)^{\alpha_n} f(x).$$

We say $f \in \mathscr{B}^\infty$, or $f \in \mathscr{B}$ if and only if all partial derivatives of $f(x)$ of any order are bounded and continuous. In $\mathscr{B}^\infty$ there are countable semi-norms

$$p_m(f) = |f(x)|_m \quad (m = 0, 1, 2, \ldots).$$

In other words, there is a countable fundamental system of neighbourhoods

$$U_{mn} = \{f;\ p_m(f) < 1/n\}.$$

A complete topological vector space equipped with a countable fundamental system of neighbourhoods as its topology is called a *Fréchet space*.†

(2) $\mathscr{E}^m$: the space of functions continuously differentiable to order m. The topology for the space is given by

$$f_j(x) \to 0 \quad \text{in} \quad \mathscr{E}^m.$$

This means the left hand sequence and its derivatives to order m are all uniformly convergent to 0 on an arbitrary compact (bounded closed) set K. We write $|f_j(x)|_{\mathscr{E}^m(K)} \to 0 \quad (j \to \infty)$.

$\mathscr{E}^m$ is also a Fréchet space.

† See the definition 2.5 and the note attached to the definition 2.1 for the definition of the Fréchet space.

The support of a function

[First we explain a notational device and a convention.]†

We write $f \in L^1_{\text{loc}}$ if and only if a measurable function f defined on R^n is summable in a neighbourhood of every point in R^n. In this case we say that f is *locally summable*,‡ or simply f is *a function* following Schwartz.

Suppose f is a function. Then, any $x \in R^n$ can be classified into one of the following classes.

(1) $f(x)=0$ almost everywhere in a sufficiently small neighbourhood of x.

(2) For any neighbourhood V, however small, we have

$$\int_V |f(x)| \, dx > 0.$$

We call the set of all x having this property (2) 'the *support* of f'. Obviously, the support of a function is a closed set.

(3) $\mathscr{D}^m$: the whole set of functions in $\mathscr{B}^m$ having compact supports in $\mathscr{E}^m$. We shall not describe the topology of $\mathscr{D}^m$ here.

$\mathscr{D}$: The whole set of infinitely differentiable functions* having compact supports.

(4) $\mathscr{D}_{L^2}{}^m$: the whole set of $f \in L^2$ which also have all derivatives of order up to m belonging to L^2, where 'derivatives' mean derivative in the sense of the distribution. (We shall introduce this later.)

The following norm is taken in $\mathscr{D}_{L^2}{}^m$:

$$\|f(x)\|^2_{m, L^2} = \sum_{|\alpha| \leqslant m} \|D^\alpha f(x)\|_{L^2}. \quad **$$

Next, we consider the approximation of a given function by a smooth function. Let ρ be a function satisfying

(i) $\rho(x) \geqslant 0$, $\rho \in \mathscr{D}$, the support of $\rho \subset$ the unit sphere: $|x| \leqslant 1$

(ii) $\int \rho(x) \, dx = 1$.

For example, ρ defined as

$$\rho(x) = \begin{cases} C \cdot \exp\left(-\dfrac{1}{1-|x|^2}\right) & (|x|<1), \\ 0 & (|x| \geqslant 1), \end{cases}$$

† This is inserted for the continuity of this paragraph. (Translator)

‡ Cf. L. Schwartz, *Mathematics for the Physical Science* (Hermann-Addison-Wesley, 1966), p. 74, Example. Here the definition of locally summable is given as 'a function summable over any bounded set'. (Translator.)

* I.e. having differentials of any order. These are also continuous.

**For details see chapter 2, section 3.

satisfies (i) and (ii) where the constant C is taken as satisfying

$$\int \rho(x)\,\mathrm{d}x=1.$$

Then, we take $\varepsilon(>0)$ as a parameter and write

$$\rho_\varepsilon(x)=\left(\frac{1}{\varepsilon}\right)^n \rho\left(\frac{x}{\varepsilon}\right).$$

We note that $\rho_\varepsilon(x)$ satisfies (i) and (ii), but in this case, the support of $\rho_\varepsilon(x)$ is bounded as $|x|\leqslant\varepsilon$. Now, for $u\in L^1{}_{\mathrm{loc}}$, we define [the convolution of ρ_ε and u as]†

$$\rho_\varepsilon * u=\int \rho_\varepsilon(x-y)\,u(y)\,\mathrm{d}y.$$

We observe the following properties:

LEMMA 1.3.
(*a*) $\rho_\varepsilon * u\in\mathscr{E}^\infty$, i.e. infinitely differentiable.
(*b*) The support of $\rho_\varepsilon * u$ lies in the ε-neighbourhood of the support of u.
(*c*) For $u\in\mathscr{E}^m$ and $\varepsilon\to 0$, we have $\rho_\varepsilon * u\to u$ in $\mathscr{E}^m$.
(*d*) For $u\in L^p$, $p\geqslant 1$, we have $\rho_\varepsilon * u\to u$ in L^p.

Proof. (*a*) and (*b*) are almost immediate. We prove (*c*) and (*d*). In order to prove (*c*) it is sufficient to prove

$$|\rho_\varepsilon * u-u|_{\mathscr{E}^m(K)}\to 0 \quad (\varepsilon\to 0),$$

where K is an arbitrary compact set. On the other hand, let $\alpha\in\mathscr{D}$ be a function identically equal to 1 in a neighbourhood of K. For $x\in K$ and a sufficiently small ε, we have

$$\rho_\varepsilon * u=\rho_\varepsilon *(\alpha u),\quad \mathrm{D}^\nu(\rho_\varepsilon * u)=\rho_\varepsilon * \mathrm{D}^\nu(\alpha u) \quad (|\nu|\leqslant m).$$

It is therefore sufficient to show $|\rho_\varepsilon * u-u|_0\to 0$ $(\varepsilon\to 0)$ for a continuous u having compact support. In this case, using $\int \rho_\varepsilon(y)\,\mathrm{d}y=1$, we have

$$\begin{aligned}\rho_\varepsilon * u-u&=\int \rho_\varepsilon(y)\,u(x-y)\,\mathrm{d}y-u(x)\\ &=\int \rho_\varepsilon(y)\,[u(x-y)-u(x)]\,\mathrm{d}y. \qquad (1.41)\end{aligned}$$

† [...] is inserted (translator).

Therefore

$$|(\rho_\varepsilon * u - u)(x)| \leqslant \max_{x \in R^n,\, |h| \leqslant \varepsilon} |u(x+h) - u(x)| \to 0 \quad (\varepsilon \to 0).$$

Next we prove (*d*). We put $\rho_\varepsilon * u = u_\varepsilon$. From (1.41), using Hölder's inequality when $p > 1$, we have

$$|u_\varepsilon(x) - u(x)|^p \leqslant \int_{|y| \leqslant \varepsilon} \rho_\varepsilon(y)\, |u(x-y) - u(x)|^p \,\mathrm{d}y .$$

We put

$$\delta(h) = \sup_{|y| \leqslant h} \int |u(x+y) - u(x)|^p \,\mathrm{d}x .$$

Because $u \in L^p$ we can claim $\delta(h) \to 0$ as $h \to 0$ (the n-dimensional case of lemma 1.1). Hence,

$$\int |u_\varepsilon(x) - u(x)|^p \,\mathrm{d}x \leqslant \int_{|y| \leqslant \varepsilon} \rho_\varepsilon(y)\,\mathrm{d}y \int |u(x-y) - u(x)|^p \,\mathrm{d}x \leqslant \delta(\varepsilon) \to 0 .$$

Q.E.D.

From this lemma we observe that $\rho_\varepsilon *$ plays the rôle of the approximation by the smooth function in various function spaces. This convolution was first observed by Friedrichs. He termed this operator $\rho_\varepsilon *$ '*mollifier*'.

8 *Multiple Fourier series*

Hitherto we have followed the line of historical development not only for its own significance but also for the importance of the meaning of various expressions obtained on the way. Before describing the multiple Fourier series we should mention a complex Fourier series.

Complex Fourier series

It is obvious that the system

$$(2\pi)^{-\frac{1}{2}}\, \mathrm{e}^{\mathrm{i}nx} \qquad (n = 0, \pm 1, \pm 2, \ldots)$$

is orthonormal in $L^2(0, 2\pi)$. Now, we consider a Fourier series defined as

$$f(x) \sim \sum_{n=-\infty}^{+\infty} c_n\, \mathrm{e}^{\mathrm{i}nx}, \quad c_n = \frac{1}{2\pi} \int_0^{2\pi} f(x)\, \mathrm{e}^{-\mathrm{i}nx} \,\mathrm{d}x . \tag{1.42}$$

This is essentially the same Fourier expansion we have encountered before. In fact, from (1.2) and (1.42),

$$\left.\begin{aligned}&c_p=\tfrac{1}{2}(a_p-ib_p),\\ &a_{-p}=a_p,\quad b_{-p}=-b_p,\quad b_0=0,\qquad p=0,\pm1,\pm2,\ldots\end{aligned}\right\}\tag{1.43}$$

Therefore,

$$\sum_{p=-n}^{+n} c_p\,\mathrm{e}^{\mathrm{i}px}=\sum_{p=-n}^{+n}\tfrac{1}{2}(a_p-ib_p)\,(\cos px+\mathrm{i}\sin px)\,.$$

Taking account of (1.43), this sum is equal to

$$\tfrac{1}{2}a_0+\sum_{p=1}^{n}(a_p\cos px+b_p\sin px)$$

i.e. it coincides with the Fourier series. In this case, therefore, theorem 1.4 (Jordan–Lebesgue) is also valid.

Multiple Fourier series

It is readily seen that the system of functions

$$\{(2\pi)^{-\frac{1}{2}n}\exp\left[\mathrm{i}(m_1x_1+m_2x_2+\cdots+m_nx_n)\right]\}_{m_i=0,\,\pm1,\,\pm2,\,\ldots}$$

is orthonormal in $L^2(\Omega)$ where the n-tuple (x_i) belongs to $\Omega=(0,2\pi)\times(0,2\pi)\times\cdots\times(0,2\pi)$, an open set in R^n. The Fourier expansion obtained by the above orthonormal system is called a *multiple Fourier series*. In other words, for an arbitrary $f\in L^2(\Omega)$, we write a multiple Fourier series

$$\begin{aligned}f(x)\sim\sum_{(m)}c_m\,\mathrm{e}^{imx},\quad m&=(m_1,m_2,\ldots,m_n),\\ mx&=m_1x_1+\cdots+m_nx_n\,;\end{aligned}\tag{1.44}$$

where (m) under $\sum$ indicates the whole combination of n-tuple integers.

In this case, the Fourier coefficient is naturally

$$c_{m_1\ldots m_n}=(2\pi)^{-n}\int\ldots\int f(x_1,\ldots,x_n)\exp\left[-\mathrm{i}(m_1x_1+\cdots+m_nx_n)\right]\mathrm{d}x_1\ldots\mathrm{d}x_n\,,$$

or, in abbreviated form

$$c_m=(2\pi)^{-n}\int_{\Omega}f(x)\,\mathrm{e}^{-imx}\,\mathrm{d}x\,.$$

We note that by Bessel's inequality, the right hand series of (1.44) is convergent in $L^2(\Omega)$. With regard to the pointwise convergence, because this series is a multiple Fourier series it is, in general, only meaningful when the series is absolutely convergent.

THEOREM 1.10. The orthogonal system $\{e^{imx}\}$ is complete in $L^2(\Omega)$.

Proof. From theorem 1.5, it is enough to show that Parseval's equality

$$\int_\Omega |f(x)|^2 \, dx = (2\pi)^{-n} \sum_{(m)} |c_m|^2 \tag{1.46}$$

is true for any $f \in L^2(\Omega)$. For this to hold it is sufficient to show that it is true for any function in a dense linear subspace, E, of $L^2(\Omega)$. In fact, for an arbitrary $f \in L^2(\Omega)$, there exists $f_\nu \in E$, and $f_\nu \to f$ in $L^2(\Omega)$. For this f_ν, Parseval's equality (1.46) is true. We then write

$$f_\nu(x) \sim \sum c_m^{(\nu)} e^{imx}; \quad \|f_\nu(x)\|_{L^2}^2 = (2\pi)^{-n} \sum_{(m)} |c_m^{(\nu)}|^2.$$

On the other hand, $\|f_\nu\|_{L^2} \to \|f\|_{L^2}$ is obvious. Hence

$$\sum_{(m)} |c_m^{(\nu)}|^2 \to \sum_{(m)} |c_m|^2 \quad (\nu \to +\infty).$$

In fact, $c_m^{(\nu)} \to c_m \, (\nu \to \infty)$ is true, and f_ν is a Cauchy sequence in $L^2(\Omega)$. So that, for an arbitrary ε, there exists an N, such that

$$\sum_{|m| \geqslant N} |c_m^{(\nu)}|^2 < \varepsilon \quad (\nu = 1, 2, \ldots).$$

Hence, for an arbitrary $f \in L^2(\Omega)$, (1.46) is true.

Now, we must first find E. We shall proceed step by step proving the following lemmas.

(1) For a basis of E, we select a set of functions of the 'separate variable' type:

$$\psi_1(x_1), \psi_2(x_2), \ldots, \psi_n(x_n), \tag{1.47}$$

where $\psi_i(\xi)$ is continuously differentiable of order 2 defined for $\xi \in [0, 2\pi]$ and its support lies within $(0, 2\pi)$.

LEMMA 1.4. We consider the whole set of finite linear combinations of the form (1.47). We write E for this linear space. Then, E is dense in $L^2(\Omega)$.

Proof. Let $f \in L^2(\Omega)$. $f(x)$ is represented by a limit in measure of a sequence of step functions. Therefore, for an arbitrary $\varepsilon(>0)$, there exists a certain step function φ satisfying

$$\int_{\Omega} |f(x)-\varphi(x)|^2 \, dx < \varepsilon. \tag{1.48}$$

Here $\varphi(x)=\sum c_i x_i(x)$, where x_i is a characteristic function defined on a certain interval I_i in Ω, and c_i is a constant. We can write

$$x_i(x)=\psi_1(x_1)\dots\psi_n(x_n), \quad x_i \in (0, 2\pi),$$

where ψ_i is a characteristic function of a certain interval in $(0, 2\pi)$. Now, we can redefine $\psi_i(x_i)$ as $\tilde{\psi}_i(x_i)$ in order to make the new function satisfy (1.47). Then, we pick up $\tilde{\varphi}(x)$ corresponding to these $\tilde{\psi}_i(x)$. We see that $\tilde{\varphi}(x)$ satisfies (1.48).

(2) We now prove:

LEMMA 1.5. We consider the subspace E defined in the previous lemma 1.4. The multiple Fourier series of $\psi(x)$ $(\psi \in E)$ is normally convergent and the limit of the series represents $\psi(x)$.

Note. We say that the series of functions

$$u_1(x)+u_2(x)+\cdots+u_n(x)+\cdots$$

is *normally convergent* in Ω if and only if there exist constants $\bar{U}_i$ satisfying $|u_i(x)| \leqslant \bar{U}_i (i=1, 2, \dots)$ and $\sum \bar{U}_i < +\infty$.

Proof. It suffices to prove the lemma for the function of the form (1.47). For a function such as

$$\psi(x)=\psi_1(x_1)\dots\psi_n(x_n)$$

we have

$$c_{m_1 \dots m_n}(\psi)=(2\pi)^{-n} \int_0^{2\pi} \psi_1(x_1) \exp(-im_1x_1)\, dx_1 \dots \int_0^{2\pi} \psi_n(x_n) \exp(-im_nx_n)\, dx_n$$
$$= c_{m_1}(\psi_1)\, c_{m_2}(\psi_2) \dots c_{m_n}(\psi_n).$$

On the other hand, by the definition of $\psi_i(x_i)$,

$$|c_{m_i}(\psi_i)| \leqslant \frac{1}{m_i^2} |\psi_i(x_i)|_2 . \dagger$$

† When $m_i=0$, the right hand term is $|\psi_i(x_i)|_2$.

From this we can write

$$\sum_{(m)} c_m \,\mathrm{e}^{\mathrm{i}mx} = \prod_{i=1}^{n} \left(\sum_{n=-\infty}^{+\infty} c_n(\psi_i)\,\mathrm{e}^{\mathrm{i}nx} \right) = \prod_{i=1}^{n} \psi_i(x_i).$$

Here the last equality holds according to the argument of a complex Fourier series stated at the beginning of this section. The first equality holds because $\sum_{(m)} |c_m| < +\infty$. This is obvious from

$$\sum_{-N \leqslant m_i \leqslant N} |c_{m_1 \dots m_n}| = \prod_{i=1}^{n} \sum_{m_i=-N}^{+N} |c_{m_i}| \leqslant \prod_{i=1}^{n} 2\,|\psi_i(x_i)|_2 \sum_{m_i=0}^{\infty} \frac{1}{m_i^{\,2}} < +\infty .$$

After examining this inequality we conclude that the multiple Fourier series in question is in fact normally convergent.

Q.E.D.

(3) From lemma 1.5 and the fundamental relationship (1.18), section 4, it it is now obvious that the Parseval's equality is true for any function picked from E. Hence, theorem 1.10 is proved.

Q.E.D.

We have now established the completeness of $\{e^{imx}\}$. From this and the argument in section 4 it follows that, given a function, the condition to be satisfied for having Fourier expansion in an intrinsic sense is absolute and uniform convergence of its Fourier series. The result which follows is still too rough for practical purposes. Nevertheless, we prove:

THEOREM 1.11. Let $f(x)=f(x_1, \dots, x_n)$ where $f(x)$ is continuously differentiable to order $2n$. Also, assume that $f(x)$ has a period 2π for each of its variables, then the multiple Fourier series of $f(x)$ is normally convergent and represents $f(x)$.

Proof.

$$c_{m_1 \dots m_n}(f) = (2\pi)^{-n} \int \dots \int f(x_1, \dots, x_n) \exp[-\mathrm{i}(m_1x_1 + \dots + m_nx_n)]\, \mathrm{d}x_1 \dots \mathrm{d}x_n .$$

We integrate the left-hand side by parts twice with respect to x_1, to give

$$(2\pi)^{-n} \left(\frac{\mathrm{i}}{m_1} \right)^2 \int \dots \int \frac{\partial^2}{\partial x_1^{\,2}} f(x) \exp[-\mathrm{i}(m_1x_1 + \dots + m_nx_n)]\, \mathrm{d}x_1 \dots \mathrm{d}x_n .$$

Similarly with respect to $x_2, \ldots, x_n$, we have

$$(2\pi)^{-n} \frac{(-1)^n}{m_1^{\,2} \ldots m_n^{\,2}} \int \ldots \int \frac{\partial^{2n}}{\partial x_1^{\,2} \ldots \partial x_n^{\,2}} f(x)\, \mathrm{e}^{-imx}\, \mathrm{d}x_1 \ldots \mathrm{d}x_n$$

$$= \frac{(-1)^n}{m_1^{\,2} \ldots m_n^{\,2}} c_{m_1 \ldots m_n} \left(\frac{\partial^{2n}}{\partial x_1^{\,2} \ldots \partial x_n^{\,2}} f \right).$$

(Here we do not integrate by parts when $m_i = 0$.) Therefore,

$$|c_m(f)| \leqslant c \frac{1}{(m_1^{\,2}+1) \ldots (m_n^{\,2}+1)} |f(x)|_{2n},$$

where c is a constant depending only upon n, the dimension of the space of f. Hence,

$$\sum_{(m)} |c_m| \leqslant c\, |f(x)|_{2n} \sum_{(m)} \frac{1}{(m_1^{\,2}+1) \ldots (m_n^{\,2}+1)} < +\infty .$$

Q.E.D.

Note. From the proof we see that if f has period 2π with respect to each variable, and is infinitely differentiable, then

$$|c_m(f)| \leqslant c \frac{1}{(|m_1|+1)^N \ldots (|m_n|+1)^N} |f(x)|_{Nn} \tag{1.49}$$

is true where c is dependent only upon n, but N is arbitrary. In other words, the Fourier series decreases rapidly. Hence, the multiple Fourier series is 'term by term' differentiable any number of times.

We shall state a theorem which is an application of a multiple Fourier series.

THEOREM 1.12. Let f be a function which is continuously differentiable to order p with compact support. In this case, f can be approximated by the finite sum of the functions of the following form

$$\varphi_1(x_1), \varphi_2(x_2), \ldots, \varphi_n(x_n) \quad (\varphi_i \in \mathscr{D}(R^1)) \tag{1.50}$$

to any degree, with respect to the norm in B^p, i.e., for an arbitrary ε, we can make

$$|f(x) - \sum \psi_i(x)|_p < \varepsilon,$$

where each $\psi_i(x)$ satisfies the conditions of (1.50).

Proof. Without loss of generality, we may assume that the support of $f(x)$

lies within the basic domain Ω of the multiple Fourier series. For an arbitrary $\varepsilon(>0)$, when we use a Friedrichs' mollifier ρ_δ in $f^*=\rho_\delta * f$ for a sufficiently small δ, we observe that f^* has the same property as f, i.e. the property which we have assumed. Therefore, $f^* \in \mathscr{D}(\Omega)$, and

$$|f^*(x)-f(x)|_p<\tfrac{1}{3}\varepsilon \quad \text{(from lemma 1.3)}. \tag{1.51}$$

On the other hand, the coefficients of the multiple Fourier series of $f^*(x)$ decrease rapidly (see note after theorem 1.11). We see that for an arbitrary $\varepsilon'(>0)$, there exists a sufficiently large N, such that

$$|f^*(x)-\sum_{|m|\leqslant N} c_m \, \mathrm{e}^{imx}|_{\mathscr{E}^p(\Omega)}<\varepsilon'. \tag{1.52}$$

Next, we consider $\alpha(\xi)$ where $\xi \in [0, 2\pi]$. We assume α is infinitely differentiable with support in $(0, 2\pi)$. Let

$$\beta(x)=\alpha(x_1)\,\alpha(x_2)\ldots\alpha(x_n).$$

We can then take the value of β as identically 1 in a neighbourhood of the support of f^*. We write the trigonometrical polynomial of (1.52) as $\varphi(x)$. Consider

$$|f^*(x)-\beta(x)\,\varphi(x)|_p\leqslant|f^*(x)-\varphi(x)|_p+|(\beta(x)-1)\,\varphi(x)|_p.$$

$\beta(x)$ in the second term of the right hand side of the inequality is identically equal to 1 on the support of f^*.

Therefore, from (1.52), the value of the term does not exceed $c(n,p)|\beta(x)|_p\cdot\varepsilon'$. From this observation, if we choose $\varepsilon'<\frac{1}{3}\varepsilon$ and $c(n,p)|\beta(x)|_p\cdot\varepsilon'<\frac{1}{3}\varepsilon$, then we have $|f^*(x)-\beta(x)\varphi(x)|_p<\frac{2}{3}\varepsilon$. Thus, from (1.51), we see

$$|f(x)-\beta(x)\,\varphi(x)|_p<\varepsilon.$$

We can therefore write

$$\beta(x)\,\varphi(x)=\sum_{|m|\leqslant N} c_m\alpha(x_1)\exp(im_1x_1)\ldots\alpha(x_n)\exp(im_nx_n)$$

and the theorem is proved.

Q.E.D.

THEOREM 1.13 *(Weierstrass polynomial approximation)*. Let $f \in \mathscr{E}^p$. We can approximate the value of f by polynomials with the topology of $\mathscr{E}^p$, i.e. for an arbitrary $\varepsilon\,(>0)$ and an arbitrary compact set K, there exists a polynomial $P(x)$ satisfying

$$|f(x)-P(x)|_{\mathscr{E}^P(K)}<\varepsilon.$$

Proof. We proceed as with the previous theorem. Without loss of generality, we assume that K lies within Ω. Choose $\alpha\in\mathscr{D}(\Omega)$ as $\alpha(x)=1$ in a neighbourhood of K. As we did earlier with (αf), we choose $f^*(x)$ satisfying

$$|(\alpha f)(x)-f^*(x)|_p<\tfrac{1}{3}\varepsilon.$$

For $f^*(x)$ we put $\varepsilon'=\frac{1}{3}\varepsilon$, and we choose $\sum_{|m|\leq N} c_m\,e^{imx}$ to satisfy (1.52). We note that for $x\in K$, $\alpha f=f$, therefore

$$|f(x)-\sum_{|m|\leqslant N} c_m\,e^{imx}|_{\mathscr{E}^P(K)}<\tfrac{2}{3}\varepsilon.$$

We now consider the Taylor expansion of e^{imx}. As we can approximate this uniformly in K to its derivative up to pth degree, so

$$|\sum_{|m|\leqslant N} c_m\,e^{imx}-P(x)|_{\mathscr{E}^P(K)}<\tfrac{1}{3}\varepsilon,$$

where $P(x)$ is a polynomial. Using these two inequalities, we then obtain the result claimed.

Q.E.D.

9 *Fourier transform of several variables*

Let $f\in L^1$. We set

$$\hat{f}(\xi_1,\ldots,\xi_n)=\int\ldots\int\exp[-2\pi i(x_1\xi_1+\cdots+x_n\xi_n)]\,f(x_1,\ldots,x_n)\,dx_1\ldots dx_n \tag{1.53}$$

or, simply

$$\hat{f}(\xi)=\int e^{-2\pi ix\xi}f(x)\,dx \tag{1.54}$$

as the Fourier transform of $f(x)$. In this case we write $\hat{f}(\xi)=\mathscr{F}[f]$. If $g\in L^1$, then

$$(\bar{\mathscr{F}}g)(x)=\int e^{2\pi ix\xi}\,g(\xi)\,d\xi. \tag{1.55}$$

We call this the *inverse Fourier transform* of $g(\xi)$.

First we state some fundamental properties:

(1) $f(x)=f_1(x_1)\,f_2(x_2)\ldots f_n(x_n)$, where $f_i(x_i)\in L^1$ $(i=1,2,\ldots,n)$

implies

$$\hat{f}(\xi)=\hat{f}_1(\xi_1)\hat{f}_2(\xi_2)\ldots\hat{f}_n(\xi_n). \tag{1.56}$$

In fact

$$\exp(-2\pi i x\xi)=\prod_{j=1}^{n}\exp(-2\pi i x_j\xi_j)$$

therefore

$$\int\exp(-2\pi i x\xi)f(x)\,dx=\prod_{j=1}^{n}\int\exp(-2\pi i x_j\xi_j)f_j(x_j)\,dx_j=\prod_{j=1}^{n}\hat{f}_j(\xi_j).$$

(2) Let $f\in\mathscr{D}^m$, i.e. f has a compact support and is m-times continuously differentiable, or, more generally, when $|x|^{n-1}|D^\alpha f(x)|\to 0$ (uniformly convergent), $|\alpha|\leqslant m-1$, as $|x|\to\infty$, and $D^\alpha f\in L^1$, $|\alpha|\leqslant m$, we have

$$(2\pi i\xi)^\alpha\hat{f}(\xi)=\int e^{-2\pi i x\xi}D^\alpha f(x)\,dx, \tag{1.57}$$

where

$$(2\pi i\xi)^\alpha=(2\pi i\xi_1)^{\alpha_1}\ldots(2\pi i\xi_n)^{\alpha_n}.$$

In fact, if we take the integration by parts of (1.53) with respect to x_j, we obtain

$$\frac{1}{2\pi i\xi_j}\int e^{-2\pi i x\xi}\frac{\partial}{\partial x_j}f(x)\,dx.$$

We repeat the same operation until we reach (1.57). In other words,

$$\mathscr{F}[D^\alpha f(x)]=(2\pi i\xi)^\alpha\mathscr{F}[f(x)]\quad(|\alpha|\leqslant m). \tag{1.58}$$

$$|(2\pi\xi)^\alpha\hat{f}(\xi)|\leqslant\|D^\alpha f(x)\|_{L^1}\quad(|\alpha|\leqslant m). \tag{1.59}$$

(3) Let $(1+|x|)^m f\in L^1\,(m\geqslant 0)$. We differentiate (1.53) under the integration sign

$$D^\alpha\hat{f}(\xi)=\mathscr{F}[(-2\pi i x)^\alpha f(x)]\quad(|\alpha|\leqslant m), \tag{1.60}$$

$$|D^\alpha\hat{f}(\xi)|\leqslant\|(2\pi x)^\alpha f(x)\|_{L^1}\quad(|\alpha|\leqslant m). \tag{1.61}$$

(4) $\mathscr{F}[f(x-h)]=e^{-2\pi i h x}\hat{f}(\xi)$.

(5) $$\bar{\mathscr{F}}f=\overline{\mathscr{F}\bar{f}}. \tag{1.62}$$

Next, we state a direct extension of theorem 1.9 to the n-variable case:

THEOREM 1.14. For $f\in\mathscr{D}^{2n}$, the following Fourier inversion formulae can be established;

$$\bar{\mathscr{F}}\mathscr{F}f=f,\quad \mathscr{F}\bar{\mathscr{F}}f=f, \tag{1.63}$$

and

$$\|\mathscr{F}f\|_{L^2}=\|\bar{\mathscr{F}}f\|_{L^2}=\|f\|_{L^2} \quad \text{(Parseval–Plancherel)}. \tag{1.64}$$

Proof. We separate our proof as follows.

(1) From the property (1.59), we have

$$\prod_{j=1}^{n}(1+\xi_j^2)\,|\hat{f}(\xi)|\leqslant c(n)\sum_{|\alpha|\leqslant 2n}\|\mathrm{D}^\alpha f(x)\|_{L^1} \tag{1.65}$$

where $c(n)$ is a constant dependent solely upon the dimension n. From this, we conclude $\hat{f}\in L^1$.

(2) From theorem 1.12 (putting $p=2n$) and the note after it, there exists $\{f_j\}$ and $f_j\in\mathscr{D}$ satisfying the condition of theorem 1.12, and $f_j(x)\to f(x)$ in $\mathscr{B}^{2n}$, where we note that the support of each f_j is contained in a fixed compact set K.

For each f_j, on the other hand, we see from (1) and the result of the Fourier transform of the function of one variable (theorem 1.9), that the Fourier inversion formula is true, i.e.

$$f_j(x)=\int \mathrm{e}^{2\pi i x\xi}\hat{f}_j(\xi)\,\mathrm{d}\xi\,.$$

(3) From (1.65), $f_j(x)\to\int \mathrm{e}^{2\pi i x\xi}\hat{f}(\xi)\,\mathrm{d}\xi$ $(j\to\infty)$ uniformly with respect to x. In fact, from (2)

$$\begin{aligned}|f_j(x)-\int \mathrm{e}^{2\pi i x\xi}\hat{f}(\xi)\,\mathrm{d}\xi| &\leqslant \int |\hat{f}_j(\xi)-\hat{f}(\xi)|\,\mathrm{d}\xi\\ &\leqslant \sup_\xi\left(\prod_{i=1}^{n}(1+\xi_i^2)\,|\hat{f}_j(\xi)-\hat{f}(\xi)|\right)\cdot\int\prod_{i=1}^{n}(1+\xi_i^2)^{-1}\,\mathrm{d}\xi\,.\end{aligned}$$

From (1.65), the value of the last term is less than or equal to

$$c(n)\,\pi^n\sum_{|\alpha|\leqslant 2n}\|\mathrm{D}^\alpha(f_j(x)-f(x))\|_{L^1},$$

where, if we note that $f_j(x)\to f(x)$ in $\mathscr{D}^{2n}(K)$,† the value tends to 0.

(4) From the foregoing results $f(x)=\int \mathrm{e}^{2\pi i x\xi}\hat{f}(\xi)\,\mathrm{d}\xi$ is clear; $\bar{\mathscr{F}}\mathscr{F}=I$ follows.

† I.e. all supports of f_j are contained in K, and $f_j\to f$ is uniformly convergent including derivatives of $2n$ degree.

(5) Parseval's equality holds as in the case of one variable.

$$\int f(x)\overline{f(x)}\,\mathrm{d}x = \int f(x)\,\mathrm{d}x\,\overline{\int \mathrm{e}^{2\pi \mathrm{i}x\xi}\hat{f}(\xi)\,\mathrm{d}\xi}$$
$$= \int \hat{f}(\xi)\,\mathrm{d}\xi \int f(x)\,\mathrm{e}^{-2\pi \mathrm{i}x\xi}\,\mathrm{d}x = \int \hat{f}(\xi)\,\hat{f}(\xi)\,\mathrm{d}\xi\,.$$

Q.E.D.

10 *Plancherel's theorem*

Up to this point Fourier transforms have been defined only for functions belonging to L^1 and an inversion formula (theorem 1.14) has been established for the functions which belong to $\mathscr{D}^{2n}$. We will generalize these facts to the entire space of functions L^2.

More precisely, if $f \in L^2 \cap L^1$, i.e. f is summable and square-integrable, the new generalized Fourier transform coincides with the old one, and, for an arbitrary $f \in L^2$, $\mathscr{F}f \in L^2$ and an inversion formula can also be established.

In detail, we note first that the function space $\mathscr{D}^{2n}$, which appeared in theorem 1.14 in the previous paragraph, is dense in L^2. In fact, for $f \in L^2$ we define

$$f_A(x) = \begin{cases} f(x) & (|x| \leqslant A) \\ 0 & (|x| > A)\,. \end{cases} \tag{1.66}$$

Then, if A is sufficiently large, the value of $\|f(x) - f_A(x)\|_{L^2}$ can be made small enough, and we then employ the Friedrichs mollifier to get

$$\|\rho_\varepsilon * f_A - f_A\|_{L^2} \to 0 \quad (\varepsilon \to 0) \quad \text{(lemma 1.3)}$$

where $\rho_\varepsilon * f_A \in \mathscr{D}$. We now see that $\mathscr{D}$ is dense in L^2.

Now, for $f \in L^2$, we can choose $\{f_j\}$ where $f_j \in \mathscr{D}^{2n}$ and $f_j(x) \to f(x)$ in L^2. We observe that $\{f_j\}$ is a Cauchy sequence in L^2. Therefore, from (1.64) in the previous theorem 1.14, we can see that $\{\hat{f}_j(\xi)\}$ is also a Cauchy sequence in $L^2(R_\xi^n)$. Hence, $\{\hat{f}_j(\xi)\}$ has a unique limit because L^2-space is complete. We denote the limit as $\hat{f}(\xi)$;

$$\hat{f}_j(\xi) \to \hat{f}(\xi) \quad \text{in} \quad L^2$$

and define

$$\hat{f}(\xi) = \mathscr{F}[f(x)]\,.$$

That $\hat{f}(\xi)$ does not depend on the choice of $f_j(x)$ is readily ascertained. Thus, our Fourier transform is an isometric operator from $L^2(R_x^n)$ into

$L^2(R_\xi^n)$:

$$\|\mathscr{F}f\|_{L^2}=\|f\|_{L^2}.$$

or, equivalently, for arbitrary $f, g\in L^2$,

$$(f(x), g(x))_{L^2}=(\hat{f}(\xi), \hat{g}(\xi))_{L^2}. \tag{1.67}$$

Similarly, we can generalize the inverse Fourier transform to L^2. We note that

$$\begin{aligned}\bar{\mathscr{F}}\mathscr{F}f(x)=f(x) \quad (f\in L^2),\\ \mathscr{F}\bar{\mathscr{F}}g(\xi)=g(\xi) \quad (g\in L^2).\end{aligned} \tag{1.68}$$

We prove the first equality. Let $f_j\in\mathscr{D}^{2n}$, and $f_j(x)\to f(x)$ in L^2. For $f_j(x)$ the equality holds. (See (1.63).) On the other hand, $\mathscr{F}[f_j(x)]\to\mathscr{F}[f(x)]$ in $L^2(R_\xi^n)$, and $\bar{\mathscr{F}}$ is a continuous operator from $L^2(R_\xi^n)$ into $L^2(R_x^n)$. Therefore,

$$\bar{\mathscr{F}}\mathscr{F}[f_j(x)]\to\bar{\mathscr{F}}\mathscr{F}[f(x)] \quad \text{in} \quad L^2(R_x^n).$$

Since left hand side is $f_j(x)$ itself, therefore $f(x)=\bar{\mathscr{F}}\mathscr{F}[f(x)]$. (1.68) means that $\mathscr{F}$ is a surjective mapping which maps $L^2(R_x^n)$ onto $L^2(R_\xi^n)$, and $\bar{\mathscr{F}}$ is the inverse mapping of $\mathscr{F}$.

Next, we prove that, for $f\in L^2\cap L^1$, the newly defined Fourier transform is the same as that used hitherto. To this end, we first note the fact that we can make $f_j\in\mathscr{D}$ convergent as $f_j(x)\to f(x)$ in L^2 and, at the same time, in L^1. Therefore,

$$\hat{f}_j(\xi)\to\int e^{-2\pi ix\xi}f(x)\,dx \quad \text{in } \mathscr{B}^0$$

From our definition, $\hat{f}_j(\xi)\to\hat{f}(\xi)$ in L^2, so that

$$\hat{f}(\xi)=\int e^{-2\pi ix\xi}f(x)\,dx$$

is valid almost everywhere.†

From this, we see that $f_A(x)\to f(x)$ in L^2 $(A\to\infty)$, and $f_A\in L^2\cap L^1$. Hence,

$$\hat{f}(\xi)=\underset{A\to\infty}{\text{l.i.m.}}\int_{|x|\leqslant A}e^{-2\pi ix\xi}f(x)\,dx, \tag{1.69}$$

† In fact, for an arbitrary compact set K in R_ξ^n,

$$\|\hat{f}_j(\xi)-\int e^{-2\pi ix\xi}f(x)\,dx\|_{L^2(K)}\to 0 \quad (j\to\infty).$$

From

$$\|\hat{f}_j(\xi)-\hat{f}(\xi)\|_{L^2(K)}\to 0$$

we see that two functions coincide almost everywhere on K.

where l.i.m. means 'limit in mean', i.e. limit in L^2. In other words,

$$\|\hat{f}(\xi) - \int_{|x| \leqslant A} e^{-2\pi i x \xi} f(x)\, dx\|_{L^2} \to 0 \quad (A \to \infty). \tag{1.70}$$

To sum up, we have:

THEOREM 1.15 *(Plancherel's theorem)*. For $f \in L^2$, the Fourier transform $\mathscr{F}$, and the Fourier inverse transform $\bar{\mathscr{F}}$ can be defined.

$\mathscr{F}$ is an isometric operator from $L^2(R_x{}^n)$ onto $L^2(R_\xi{}^n)$. Hence, the equality (1.67) holds. $\bar{\mathscr{F}}$ is the inverse operator of $\mathscr{F}$:

$$\bar{\mathscr{F}}\mathscr{F} = I \quad \text{in} \quad L^2(R_x{}^n),$$
$$\mathscr{F}\bar{\mathscr{F}} = I \quad \text{in} \quad L^2(R_\xi{}^n).$$

Thus the newly defined $\mathscr{F}$ coincides with the previously defined $\mathscr{F}$ when $f \in L^2 \cap L^1$. (1.69) is a defining equation for an arbitrary function belonging to L^2.

We now state a fact following directly from the theorem.

The Fourier transform of a convolution

Let

$$f \in L^1, g \in L^p, p \geqslant 1.$$

We define $h(x)$ as

$$h(x) = \int_{R^n} f(x-t)\, g(t)\, dt \tag{1.71}$$

and denote

$$h = f * g \tag{1.72}$$

calling this operation $*$ a *convolution*. Note that (1.71) is meaningful for x almost everywhere. To prove this, we first assume $p > 1$. From Hölder's inequality and Fubini's theorem, we have

$$\int |h(x)|^p\, dx \leqslant \left(\int |f(x-t)|\, dt\right)^{p/p'} \iint |f(x-t)|\, |g(t)|^p\, dx\, dt.$$

where $(1/p) + (1/p') = 1$. The right hand side of the inequality can be written as

$$\|f(x)\|_{L^1}^{(p/p')+1} \|g(x)\|_{L^p}^p$$

so that $h \in L^p$, and

$$\|h\|_{L^p} \leqslant \|f\|_{L^1} \cdot \|g\|_{L^p} \quad \text{(Hausdorff–Young inequality)}. \tag{1.73}$$

(1.73) is obvious for $p = 1$, therefore:

LEMMA 1.6. If $f \in L^1$ and $g \in L^p (p \geqslant 1)$, then, $h = f * g \in L^p$ and the inequality (1.73) holds. Also, $f * g = g * f$.

Q.E.D.

For the Fourier transform of a convolution, we have the following property;

THEOREM 1.16. For $f \in L^1$, $g \in L^2$

$$\mathscr{F}[f*g] = \mathscr{F}[f]\,\mathscr{F}[g]. \tag{1.74}$$

Proof. The argument is similar to (1.66). First let $g_A(x) \to g(x)$ in L^2 as $A \to \infty$. Accordingly, we define

$$h_A(x) = \int f(x-t)\, g_A(t)\, dt.$$

From this and (1.73) we have $h_A(x) \to h(x)$ in L^2. On the other hand, $h_A \in L^2 \cap L^1$ and

$$\hat{h}_A(\xi) = \hat{f}(\xi)\, \hat{g}_A(\xi).$$

In fact,

$$\int e^{-2\pi i x \xi}\, h_A(x)\, dx = \int e^{-2\pi i x \xi}\, dx \int f(x-t)\, g_A(t)\, dt,$$
$$= \int g_A(t)\, dt \int e^{-2\pi i x \xi} f(x-t)\, dx,$$

from Fubini's theorem,

$$= \int e^{-2\pi i t \xi}\, g_A(t)\, dt \int e^{-2\pi i x' \xi} f(x')\, dx'$$
$$= \hat{f}(\xi)\, \hat{g}_A(\xi)$$

putting $x - t = x'$. At this point we note that, from Plancherel's theorem, we have $\hat{g}_A(\xi) \to \hat{g}(\xi)$ in L^2. So that

$$\hat{f}(\xi)\, \hat{g}_A(\xi) \to \hat{f}(\xi)\, \hat{g}(\xi) \quad \text{in} \quad L^2$$

follows from the fact that $\hat{f}(\xi)$ is bounded. Let $A \to \infty$, and we establish the desired property.

Q.E.D.

11 *Plancherel's theorem reconsidered*

In the previous paragraph we obtained theorem 1.15 (Plancherel's theorem) by extending the Fourier transform to L^2-space. This is significant by itself, but for practical purposes the form of the theorem as stated is not convenient. It seems, therefore, to be natural for us to consider whether we can establish $\bar{\mathscr{F}}\mathscr{F} = I$ for a larger domain by sacrificing the condition (1.67). One difficulty is that, if $f \in L^1$, its Fourier image $\hat{f}(\xi)$ is not generally summable. This also means that the inverse Fourier transform of $\hat{f}(\xi)$

$$\int e^{2\pi i x\xi} \hat{f}(\xi)\, d\xi \tag{1.75}$$

is meaningless.

However, following Plancherel's theorem, let us define the inverse transform

$$\lim_{A\to\infty} \int_{|x|\leq A} e^{2\pi i x\xi} \hat{f}(\xi)\, d\xi \tag{1.76}$$

instead of (1.75). We wish to show that it is in fact equal to $f(x)$. Obviously this involves other problems and we show in the next chapter that our argument here can be justified. On attempting the argument it is necessary to revise our old definition of limit.

Let us slightly modify the problem by considering that when f is bounded and measurable, $f_A(x)$ is defined as (1.66), and

$$\hat{f}_A(\xi) = \int e^{-2\pi i x\xi} f_A(x)\, dx = \int_{|x|\leq A} e^{-2\pi i x\xi} f(x)\, dx\,.$$

We can thus regard the limit of $\hat{f}_A(\xi)$ $(A\to\infty)$ as the Fourier transform of $f(x)$.

Let us explain this point more precisely. We regard $\hat{f}_A$ as a linear functional on $\mathscr{D}_\xi$. Taking a test function $\varphi \in \mathscr{D}_\xi$, we consider $\lim_{A\to\infty} \langle \hat{f}_A(\xi), \varphi(\xi)\rangle$ where

$$\langle \hat{f}_A(\xi), \varphi(\xi)\rangle = \int \hat{f}_A(\xi)\, \varphi(\xi)\, d\xi$$

and see that this limit definitely exists. In fact, from the definition of $\hat{f}_A(\xi)$ and Fubini's theorem

$$\langle \hat{f}_A(\xi), \varphi(\xi)\rangle = \langle f_A(x), \int e^{-2\pi i x\xi}\, \varphi(\xi)\, d\xi\rangle = \langle f_A(x), \mathscr{F}\varphi\rangle$$

can be established where $\mathscr{F}\varphi$ decreases rapidly as $|x|\to\infty$ (by (1.59)). We let $A\to\infty$, and see that

$$\lim_{A\to\infty} \langle \hat{f}_A(\xi), \varphi(\xi)\rangle = \langle f(x), (\mathscr{F}\varphi)(x)\rangle\,.$$

When, for an arbitrary $\varphi \in \mathscr{D}_\xi$, there exists

$$\lim_{A\to\infty} \langle \hat{f}_A(\xi), \varphi(\xi)\rangle$$

as a finite definite limit, Schwartz gave 'the status of citizenship' to $\lim_{A\to\infty} \hat{f}_A(\xi)$ and defined it thus

$$\mathscr{F}[f(x)] = \lim_{A\to\infty} \hat{f}_A(\xi)\,.$$

Note that in this case the right hand side no longer represents a function. It represents a 'distribution' (a continuous linear functional on $\mathscr{D}$), and it is of course necessary to regard this convergence also as being on a topology inserted in the distribution space.

We shall give a detailed discussion on 'distribution' in the following chapter. Here we content ourselves with a glance at the power of this notion. Let us first see what the Fourier image of $f(x) \equiv 1$ is. We see

$$\langle \mathscr{F}[1], \varphi(\xi) \rangle = \langle 1, (\mathscr{F}\varphi)(x) \rangle = \int (\mathscr{F}\varphi)(x) \mathrm{d}x = \int \phi(x) \mathrm{d}x = \varphi(0).$$

The last equality holds because we can put $\xi = 0$ in

$$\varphi(\xi) = \int \mathrm{e}^{2\pi i x \xi} \phi(x) \, \mathrm{d}x.$$

However, for $\varphi \in \mathscr{D}_\xi$, the functional $\varphi(\xi) \to \varphi(0)$ is precisely the δ-function introduced by Dirac. Therefore,

$$\mathscr{F}[1] = \delta.$$

Returning to the main theme, we claim that (1.76) is in fact $f(x)$. To see this, we define (1.76) as $\mathscr{F}[\hat{f}(\xi)]$. For an arbitrary $\varphi \in \mathscr{D}$,

$$\begin{aligned} \langle \mathscr{F}[\hat{f}(\xi)], \overline{\varphi(x)} \rangle = \lim_{A \to \infty} \langle \mathscr{F}\hat{f}_A(\xi), \overline{\varphi(x)} \rangle \\ = \lim_{A \to \infty} \langle \hat{f}_A(\xi), \overline{\mathscr{F}\varphi} \rangle = \langle \hat{f}(\xi), \overline{\phi(\xi)} \rangle. \end{aligned}$$

Here we made use of the fact that $\hat{f}(\xi)$ is bounded. The last term is equal to

$$\langle \int \mathrm{e}^{-2\pi i x \xi} f(x) \, \mathrm{d}x, \overline{\phi(\xi)} \rangle$$

From Fubini's theorem this equals

$$\langle f(x), \overline{\int \mathrm{e}^{2\pi i x \xi} \phi(\xi) \, \mathrm{d}\xi} \rangle = \langle f(x), \overline{\varphi(x)} \rangle,$$

where $\varphi \in \mathscr{D}$ is taken arbitrarily, so that $\mathscr{F}[\hat{f}(\xi)] = f(x)$. The foregoing explanations will enable the reader to appreciate how essential the notion of 'distribution' is to the treatment of Fourier transforms.

CHAPTER 2

Distributions

1 *Definition of distribution, convergence of sequence of distributions*

For all $\varphi\in\mathscr{D}$ (infinitely differentiable functions with compact support), we consider a mapping T which gives a uniquely determined finite complex value to φ. T is defined as follows:

A. T is a linear functional on $\mathscr{D}$, i.e., for an arbitrary complex number λ,

(1) $T(\lambda\varphi)=\lambda T(\varphi)$

(2) for $\varphi_1, \varphi_2\in\mathscr{D}$ $\quad T(\varphi_1+\varphi_2)=T(\varphi_1)+T(\varphi_2)$.

B. By a given sequence of functions $\{\varphi_j\}$, φ_j tends to 0 where $\varphi_j\in\mathscr{D}$, we mean that the support of each φ_j is contained in a certain compact set, and the derivatives of each degree of each $\varphi_j(x)$ tend to 0 uniformly. (2.1)

Using our agreed notation,

$\varphi_i\to 0$ means supp $[\varphi_j]\subset K$ (K is a suitably chosen compact set), and for an arbitrary α, $D^\alpha\varphi_j(x)\to 0$ uniformly. (2.2)

We therefore require that

$$\varphi_j\to 0 \text{ implies } T(\varphi_j)\to 0 \qquad (2.3)$$

Schwartz called T satisfying the conditions A and B 'distribution'.

Following Schwartz we shall introduce some new notation and terminology for future usage. We say *linear form* instead of *linear functional*, and write

$$\langle T, \varphi\rangle \qquad (2.4)$$

instead of $T(\varphi)$. We also write $\mathscr{D}'$ for the space of all distributions ($T\in\mathscr{D}'$). Obviously $\mathscr{D}'$ is a vector space.

We give some examples.

(1) $f\in L^1_{\text{loc}}$; $\langle f(x), \varphi(x)\rangle=\int_{R^n} f(x)\,\varphi(x)\,dx$.

(2) δ (Dirac's δ-measure); $\langle\delta, \varphi(x)\rangle=\varphi(0)$.

(3) $Y(x)$ (Heaviside's function); $\langle Y(x), \varphi(x)\rangle=\int_0^{+\infty}\varphi(x)\,dx \quad (n=1)$.

These are all straightforwardly derived by setting $\alpha=0$ in (2.2), i.e. $\varphi_i(x)\to 0$ uniformly, and $\operatorname{supp}[\varphi_j]\subset K$ means $\langle T, \varphi_j\rangle\to 0$.

Note. Example (1) is a typical case which we often encounter. Indeed, in physics, point functions do not usually appear in direct measurements, but they appear as the integral of product functions, where the product comprises a certain function and the original point function.

We discuss this situation in more concrete terms. Let $f(x)$ be a density, then

$$\int_V f(x)\,\mathrm{d}x$$

is the mass of V. We can write this as

$$\int f(x)\,\alpha_V(x)\,\mathrm{d}x$$

where α_V is the characteristic function of V. In other words, for $\varphi(x)$, we choose the set of characteristic functions of bounded sets (measurable sets). (Historically, this is in fact the origin of the term 'distribution').

However, this naive way of viewing $\varphi(x)$ is not practical for more general cases. Our way of defining $\varphi(x)$, shown above, is considered to be the most natural. We call φ a *test function.* We note that the continuity of T is defined on a set of functions where these functions have derivatives of any order which are uniformly convergent. On the other hand, to see the continuity of T defined as above, we may simply check, for each φ_j, the uniform convergence of its derivatives up to a certain order, i.e.,

LEMMA 2.1. We write $\mathscr{D}_K$ for the set of $\varphi\in\mathscr{D}$ which has $\operatorname{supp}[\varphi]\subset K$. T is a distribution. Then, for an arbitrary compact set K, there exists p, such that, if

$$\mathrm{D}^\nu\varphi_j(x)\to 0, \quad |\nu|\leqslant p; \quad \varphi_j\in\mathscr{D}_K$$

uniformly, then $\langle T, \varphi_j\rangle\to 0$. In other words, for an arbitrary compact set K, C and p are such that for all $\varphi\in\mathscr{D}_K$,

$$|\langle T, \varphi\rangle|\leqslant C\,|\varphi|_p, \tag{2.5}$$

where C, p are constants and dependent on T and K.

Proof. (By contradiction). If (2.5) does not hold, there exists a certain K and a series of functions $\{\varphi_p(x)\}_{p=1,2,\ldots}$, where $\varphi_p \in \mathscr{D}_K$ with

$$|\varphi_p|_p \leqslant 1, \quad \text{hence,} \quad \sum_{|\nu| \leqslant p} \sup |D^\nu \varphi_p(x)| \leqslant 1.$$

Also, $|\langle T, \varphi_p \rangle| \geqslant p$. We consider $\{\varphi_p(x)/p\}_{p=1,2,\ldots}$, whose derivatives of any degree are uniformly convergent, but

$$|\langle T, \varphi_p(x)/p \rangle| \geqslant 1 \quad (p=1, 2, \ldots).$$

This contradicts the assumption that T is a distribution.

Q.E.D.

DEFINITION 2.1 *(Functional space $\mathscr{D}_K$)*. Let $\mathscr{D}_K$ be a subset of $\mathscr{D}$ comprising all functions whose supports are contained in a compact set K. Obviously $\mathscr{D}_K$ is a linear subspace of $\mathscr{D}$. The topology of $\mathscr{D}_K$ given by countable semi-norms which are defined as

$$p_m(f) = |f|_m = \sum_{|\alpha| \leqslant m} \sup_x |D^\alpha f(x)| \quad (m=0, 1, 2, \ldots).$$

In other words, a fundamental system of neighbourhoods is given as

$$V_{mn} = \{f; p_m(f) < 1/n\} \quad (m=0, 1, 2, \ldots)$$

where $n=1, 2, \ldots$

$f_j \to 0$ in $\mathscr{D}_K$ means that for an arbitrary V_{mn}, there exists N, and whenever $j > N$, then $f_j \in V_{mn}$. $f_j \to f$ in $\mathscr{D}_K$ means $(f_j - f) \to 0$ in $\mathscr{D}_K$, i.e. we choose the fundamental system of neighbourhoods as $f + V_{mn}$.

Note. In general, we say that a function p defined on a vector space E is a *semi-norm* if and only if p satisfies the following conditions.

(1) For every $x \in E$, $p(x) \geqslant 0$ and $p(x)$ is finite.
(2) For an arbitrary complex number λ, $p(\lambda x) = |\lambda| p(x)$.
(3) For arbitrary $x, y, \in E$, $p(x+y) \leqslant p(x) + p(y)$.

We now rephrase the definition of distribution as follows.

DEFINITION 2.2. T is a distribution if and only if T is a linear form on $\mathscr{D}$, and is a continuous linear form on $\mathscr{D}_K$ if we restrict the domain of T to an arbitrary $\mathscr{D}_K$.

Definition 2.3 *(Derivatives of distribution).* For $T \in \mathscr{D}'$, $\partial T/\partial x_i$ is defined as

$$\left\langle \frac{\partial T}{\partial x_i}, \varphi \right\rangle = -\left\langle T, \frac{\partial \varphi}{\partial x_i} \right\rangle. \tag{2.6}$$

Then $\partial T/\partial x_i$ is also a distribution.

In fact, we observe that the right hand side of (2.6) is a linear form on $\mathscr{D}$, and when $\varphi_j \to 0$ in $\mathscr{D}_K$, we see $\partial \varphi_j/\partial x_i \to 0$ in $\mathscr{D}_K$. So that,

$$\left\langle T, \frac{\partial \varphi_j}{\partial x_i} \right\rangle \to 0 \quad (j \to \infty).$$

Incidentally, the definition of $\partial T/\partial x_i$ given above coincides with the usual definition of derivative when T is a once differentiable function, i.e. $T = f \in \mathscr{E}^1$, because we see that

$$\frac{\partial T}{\partial x_i} = \frac{\partial f}{\partial x_i}.$$

In fact, the right hand side is equal to

$$-\int \cdots \int f(x) \frac{\partial \varphi}{\partial x_i}(x)\, \mathrm{d}x_1 \ldots \mathrm{d}x_n .$$

Integration by parts gives

$$\int \cdots \int \frac{\partial f}{\partial x_i}(x)\, \varphi(x)\, \mathrm{d}x_1 \ldots \mathrm{d}x_n .$$

Another way of stating the definition is to define $\mathrm{D}^\alpha T$ for $T \in \mathscr{D}'$ as

$$\langle \mathrm{D}^\alpha T, \varphi \rangle = (-1)^{|\alpha|} \langle T, \mathrm{D}^\alpha \varphi \rangle, \quad |\alpha| = \alpha_1 + \alpha_2 + \cdots + \alpha_n .$$

We give some examples:

Example 1. $Y'(x) = \delta$. In fact,

$$\langle Y'(x), \varphi(x) \rangle = -\langle Y(x), \varphi'(x) \rangle = -\int_0^\infty \varphi'(x)\, \mathrm{d}x = \varphi(0) = \langle \delta, \varphi(x) \rangle .$$

Example 2.

$$(\log |x|)' = \mathrm{v.p.}\, \frac{1}{x} \quad (n=1).$$

Here $\log|\cdot|\in L^1{}_{\text{loc}}$, so that it is the same case as example 1 at the beginning of this paragraph. v.p. on the right hand side is the abbreviation of *valeur principale de Cauchy* (Cauchy principal value).

Formally we define

$$\langle \text{v.p.}\, \frac{1}{x}, \varphi(x)\rangle = \lim_{\varepsilon\to 0} \int_{|x|\geqslant\varepsilon} \frac{\varphi(x)}{x}\, dx\,.$$

We write

$$\langle (\log|x|)', \varphi(x)\rangle = -\langle \log|x|, \varphi'(x)\rangle = -\int_{-\infty}^{+\infty} \log|x|\, \varphi'(x)\, dx$$

$$= -\lim_{\varepsilon\to 0}\left[\int_{-\infty}^{-\varepsilon} \log|x|\, \varphi'(x)\, dx + \int_{\varepsilon}^{+\infty} \log|x|\, \varphi'(x)\, dx\right]$$

$$= \lim_{\varepsilon\to 0}\left\{\log\varepsilon[\varphi(\varepsilon)-\varphi(-\varepsilon)] + \int_{\varepsilon}^{\infty} \frac{\varphi(x)}{x}\, dx + \int_{-\infty}^{-\varepsilon} \frac{\varphi(x)}{x}\, dx\right\},$$

by integrating by parts. As first term tends to 0 as $\varepsilon\to 0$, the final expression is

$$\text{v.p.} \int \frac{\varphi(x)}{x}\, dx\,.$$

The convergence of a sequence of distributions in the topology of $\mathscr{D}'$

DEFINITION 2.4. We say that a sequence of distributions $\{T_j\}$ tends to a distribution T when for an arbitrary $\varphi\in\mathscr{D}$, we have

$$\langle T_j, \varphi\rangle \to \langle T, \varphi\rangle \quad (j\to\infty).$$

In this case we simply write $T_j\to T$ in $\mathscr{D}'$.

Note. The topology we are interested in here is the so-called 'simple topology',† and the convergence of T_j to T is a 'simple convergence'. We cannot define a strong topology for $\mathscr{D}'$ unless the topology of $\mathscr{D}$ is already given, but it is convenient in practice to use the simple topology of $\mathscr{D}'$ defined without considering the topology of $\mathscr{D}$.

In fact, we can prove that for a sequence of T_j, the convergence of the simple topology coincides with the convergence of the strong topology of $\mathscr{D}'$.

† 'Simple topology' is more often called 'weak topology'.

(There is a detailed discussion on the topology of $\mathscr{D}$ in the original book by Schwartz.)

The topology we have given to $\mathscr{D}'$ is in fact a very weak topology compared with ordinary topologies in classical analysis. The following examples illustrate this.

EXAMPLE 1. If

$$f_j \in L^1_{\text{loc}}, \quad f_j(x) \to f(x) \quad \text{in} \quad L^1_{\text{loc}},$$

i.e. for an arbitrary compact set K, if

$$\int_K |f_j(x) - f(x)| \, dx \to 0 \qquad (j \to \infty),$$

it follows $f_j(x) \to f(x)$ in $\mathscr{D}'$. In fact, for $\varphi \in \mathscr{D}$, supp $[\varphi(x)] = K$, we have

$$|\langle f_j(x) - f(x), \varphi \rangle| = |\int_K [f_j(x) - f(x)] \, \varphi(x) \, dx|$$
$$\leqslant \max |\varphi(x)| \int_K |f_j(x) - f(x)| \, dx \to 0.$$

EXAMPLE 2. $\sin nx \to 0$ in $\mathscr{D}'$ $(n \to \infty)$. This means

$$\int_{-\infty}^{+\infty} \varphi(x) \sin nx \, dx \to 0 \qquad (n \to \infty) \qquad \text{(Riemann–Lebesgue theorem)}.$$

EXAMPLE 3. In one-dimensional linear space,

$$\frac{\sin nx}{x} \to \pi\delta \quad \text{in} \quad \mathscr{D}' \qquad (n \to \infty).$$

This follows easily from theorem 1.6 (Fourier's formula).

EXAMPLE 4. For $\varepsilon (>0) \to 0$,

$$\frac{\delta_{(\varepsilon)} - \delta_{(-\varepsilon)}}{2\varepsilon} \to -\delta' \quad \text{in} \quad \mathscr{D}'.\dagger$$

† Translator's note: Where $\delta_{(\alpha)}$ is the Dirac distribution at the point α of R^n (see L. Schwartz [2]).

In fact,

$$\left\langle \frac{\delta_{(\varepsilon)}-\delta_{(-\varepsilon)}}{2\varepsilon},\ \varphi \right\rangle = \frac{\varphi(\varepsilon)-\varphi(-\varepsilon)}{2\varepsilon} \to \varphi'(0) = -\langle \delta', \varphi \rangle .$$

EXAMPLE 5. In n-dimensional linear space, we consider the limit of the series of functions

$$\frac{1}{|x|^{n-\alpha}},\quad \alpha>0,\quad |x| = \sqrt{(x_1^2+\cdots+x_n^2)}.$$

Let $\alpha \to 0$. We assume that supp $[\varphi(x)]$ is contained in $|x| \leqslant A$, then we have

$$\left\langle \frac{1}{|x|^{n-\alpha}},\ \varphi(x) \right\rangle = \int_{|x|\leqslant A} \frac{\varphi(x)}{|x|^{n-\alpha}}\,\mathrm{d}x\,;$$

$$\varphi(x) = \varphi(0) + \sum_{i=1}^{n} x_i \varphi_i(x) \qquad (\varphi_i \in \mathscr{B}),$$

so that,

$$\left\langle \frac{1}{|x|^{n-\alpha}},\ \varphi(x) \right\rangle = \varphi(0) \int_{|x|\leqslant A} \frac{\mathrm{d}x}{|x|^{n-\alpha}} + \sum_{i=1}^{n} \int_{|x|\leqslant A} \frac{x_i}{|x|^{n-\alpha}}\,\varphi_i(x)\,\mathrm{d}x\,.$$

The second term of the last part of the estimation is meaningful in the neighbourhood of $\alpha=0$. More precisely, this means that it is a holomorphic function of α in $\mathrm{Re}\,\alpha > -1$. By using the polar coordinate, the first term becomes $\varphi(0)\, S_n \cdot (A^{\alpha}/\alpha)$, where S_n is the surface area of the unit ball:

$$S_n = \frac{2\pi^{\frac{1}{2}n}}{\Gamma(\frac{1}{2}n)}.$$

From this,

$$\lim_{\alpha\to+0} \left\langle \alpha \cdot \frac{1}{|x|^{n-\alpha}},\ \varphi(x) \right\rangle = S_n \cdot \varphi(0),$$

i.e.

$$\alpha \cdot \frac{1}{|x|^{n-\alpha}} \to S_n \delta \ \text{ in } \ \mathscr{D}', \quad (\alpha\to+0)\,; \qquad S_n = \frac{2\pi^{\frac{1}{2}n}}{\Gamma(\frac{1}{2}n)}. \tag{2.7}$$

In fact,

$$\left\langle \frac{1}{r^{n-\alpha}},\ \varphi(x) \right\rangle$$

is defined in $\operatorname{Re}\alpha>0$ and can be extended to the whole plane of α as a meromorphic function. It has poles of order one at $\alpha=0, -2, -4,\ldots$ (The idea of the *extension* is Hadamard's.)

The next theorem is useful.

THEOREM 2.1. Let $\{T_j\}$ be a sequence of distributions. If $\{\langle T_j, \varphi\rangle\}$ has a finite and definite limit for all $\varphi\in\mathscr{D}$ when $j\to\infty$, there exists a unique distribution T satisfying

$$\lim_{j\to\infty}\langle T_j, \varphi\rangle=\langle T, \varphi\rangle.$$

Note. This can be considered as a special case of the Banach–Steinhaus theorem. In fact, if we put $\lim_{j\to\infty}\langle T_j, \varphi\rangle=L(\varphi)$, we see that L is obviously a linear form defined on $\mathscr{D}$. It is therefore sufficient to prove that $L(\varphi)$ has continuity. Speaking more precisely, we must prove that $L(\varphi)$ is a continuous linear form on an arbitrary $\mathscr{D}_K$.

Because the Banach–Steinhaus theorem has important applications and the method of proving the theorem is instructive in itself, we give a proof in the next section.

COROLLARY OF THE THEOREM 2.1 *(Series of distributions)*. Let $\{T_n\}$ be distributions,

For an arbitrary $\varphi\in\mathscr{D}$, suppose $\sum_{n=1}^{\infty}\langle T_n, \varphi\rangle$ is a convergent series, then $S=\sum_{n=1}^{\infty} T_n$ defined as

$$\langle S, \varphi\rangle=\sum_{n=1}^{\infty}\langle T_n, \varphi\rangle$$

is a distribution.

Proof. If $S_j=\sum_{n=1}^{j} T_n$, we have

$$\lim_{j\to\infty}\langle S_j, \varphi\rangle=\sum_{n=1}^{\infty}\langle T_n, \varphi\rangle.$$

Therefore, the right hand term is in fact a linear form and a distribution according to the previous theorem 2.1.

PROPOSITION 2.1. Let $T_j \to T$ in $\mathscr{D}'$ $(j \to \infty)$ then

$$\frac{\partial T_j}{\partial x_i} \to \frac{\partial T}{\partial x_i} \quad \text{in} \quad \mathscr{D}'$$

If $a \in \mathscr{E}$, we have $a(x)T_j \to a(x)T$.

Note. $\langle a(x)T, \varphi \rangle = \langle T, a(x)\varphi(x) \rangle$ follows immediately from the definition. From this $a(x)T \in \mathscr{D}'$ is obvious.

Proof of proposition 2.1. From our hypothesis, for an arbitrary $\varphi \in \mathscr{D}$, we have $\langle T_j, \varphi \rangle \to \langle T, \varphi \rangle$. Therefore,

$$\left\langle \frac{\partial T_j}{\partial x_i}, \varphi \right\rangle = -\left\langle T_j, \frac{\partial \varphi}{\partial x_i} \right\rangle \underset{j \to \infty}{\to} -\left\langle T, \frac{\partial \varphi}{\partial x_i} \right\rangle = \left\langle \frac{\partial T}{\partial x_i}, \varphi \right\rangle.$$

The remainder of the proof is easy.

Q.E.D.

Combining proposition 2.1 with the corollary of theorem 2.1 we have:

PROPOSITION 2.2 *(Theorem of term by term differentiation)*. Let $S = \sum_{n=1}^{\infty} T_n$ be a convergent series in the sense of the corollary of theorem 2.1. Then, for an arbitrary α,

$$D^{\alpha}S = \sum_{n=1}^{\infty} D^{\alpha}T_n.$$

We now give some basic notions used in the next section.

DEFINITION 2.5 *(Fréchet Space, F-space)*. Suppose that in a vector space E, an at most countable set of semi-norms $p_m(x)$, $m = 1, 2, \dots$ is given, and, for an arbitrary x,

$$p_1(x) \leqslant p_2(x) \leqslant \cdots \leqslant p_m(x) \leqslant \cdots \quad \text{(increasing sequence)}.$$

We then define the fundamental neighbourhoods of 0 as

$$V_{mn}=\{x;\, p_m(x)<1/n\}\,.\ ^{\dagger}$$

Furthermore, the following two conditions are satisfied:

(1) Condition of separation: for an arbitrary $x \neq 0$, there exists a certain p_m such that $p_m(x)>0$.

(2) Condition of completeness: let $\{x_n\}$, $n=1, 2, \ldots$, be a Cauchy sequence, i.e. for an arbitrary neighbourhood V of x, there exists an N, such that $x_p - x_q \in V$, for $p, q > N$.

Then we claim there exists $x_0 \in E$ with $x_n \to x_0$.

The space E satisfying all the above conditions is called a *Fréchet-space* or F-space.

Note 1. In general, the completeness of an F-space is defined in terms of a Cauchy filter instead of Cauchy sequence. In our case, the base space E has a metric so that a Cauchy sequence may be used rather than a Cauchy filter.

Note 2. By the definition of semi-norm, for $x, y \in V_{mn}$ and for two arbitrary complex numbers α, β, if $|\alpha|+|\beta| \leqslant 1$, then $\alpha x + \beta y \in V_{mn}$.

Note 3. Every Banach space is an F-space. In this case we put $p=\|x\|$ as p_m.

LEMMA 2.2. Every $\mathscr{D}_K$ is an F-space.

Proof. Observe that

$$p_m(f)=|f|_m=\sum_{|\alpha|\leqslant m} \sup|D^\alpha f(x)|\,.$$

p_m is obviously a semi-norm. The separation of $\mathscr{D}_K$ follows from the fact that if $p_0(f)=0$ then $f=0$.

† In general, $(p_n(x))$ is not an increasing sequence. In that case, we pick an arbitrary finite set of semi-norms, and form a fundamental system of neighbourhoods as

$$\{x;\, p_{i_1}(x)<1/n,\, p_{i_2}(x)<1/n, \ldots, p_{i_s}(x)<1/n\}.$$

In this way we obtain an equivalent topology as in the case of increasing sequence, if we take $q_m(x)=\sup_{k=1,2\ldots,m} p_k(x)$ as a semi-norm sequence.

We did not mention earlier what happens if K does not have any interior point. In this case, $\mathscr{D}_K$ is reduced to a *trivial* space which consists of only 0. Therefore the assumption of the existence of interior points of K is reasonable.

Next, we show the completeness of $\mathscr{D}_K$. Let $\{f_j(x)\}$ be a Cauchy sequence. For an arbitrary α, $\{D^\alpha f_j(x)\}$ is also a Cauchy sequence in the sense of maximum norm. Therefore, $D^\alpha f_j(x) \to f_\alpha(x)$ uniformly. Then $f_\alpha(x)$ becomes continuous, and its support is contained in K.

It is sufficient to show $D^\alpha f(x) = f_\alpha(x)$. We put $\alpha = (\alpha_1, \dots, \alpha_n)$ and for convenience assume $\alpha_{p+1} = \alpha_{p+2} = \dots = \alpha_n = 0$. (There may be some of $\alpha_1, \dots, \alpha_n$ which are 0.) For $p = 1, 2, \dots, n$,

$$f_j(x_1, \dots, x_n) = \int_{-\infty}^{x_1} \cdots \int_{-\infty}^{x_p} \frac{(x_1 - y_1)^{\alpha_1 - 1} \dots (x_p - y_p)^{\alpha_p - 1}}{(\alpha_1 - 1)! \dots (\alpha_p - 1)!} \times D^\alpha f_j(y_1, \dots, y_p, x_{p+1}, \dots, x_n)\, dy_1 \dots dy_p .$$

supp $[D^\alpha f_j] \subset K$, and $D^\alpha f_j$ converges uniformly. So that

$$f(x_1, \dots, x_n) = \int_{-\infty}^{x_1} \cdots \int_{-\infty}^{x_p} \frac{(x_1 - y_1)^{\alpha_1 - 1} \dots (x_p - y_p)^{\alpha_p - 1}}{(\alpha_1 - 1)! \dots (\alpha_p - 1)!} \times f_\alpha(y_1, \dots, y_p, x_{p+1}, \dots, x_n)\, dy_1 \dots dy_p .$$

Hence, $D^\alpha f(x) = f_\alpha(x) \qquad (x \in R^n)$.

Q.E.D.

We mention one more example of an F-space.

EXAMPLE. *(H(D)-space)*. Let D be a domain in the n-complex number space $z = (z_1, \dots, z_n)$ i.e. $D \subset C^n$. We form $H(D)$ as a whole set of holomorphic functions on D. Then, we pick a sequence of compact sets

$$K_1 \Subset K_2 \Subset \cdots \Subset K_m \Subset \cdots \to D .^\dagger$$

We note that, for an arbitrary compact set K in D, there exists N with $K \subset K_N$. We now define semi-norms for $H(D)$:

$$p_m(f) = \max_{z \in K_m} |f(z)| , \qquad f \in H(D) \qquad (m = 1, 2, \dots).$$

† $K_m \Subset K_{m+1}$ means K_m belongs to the interior of K_{m+1}.

$H(D)$ is obviously an F-space. The detailed proof is left as an exercise.

Bounded set in F-space

Let E be a F-space. By saying that a set B in E is bounded, we mean that for an arbitrary fundamental system of neighbourhoods V_{mn} of 0, there exist $\lambda > 0$ and $\lambda B \subset V_{mn}$ †

LEMMA 2.3. The necessary and sufficient condition for the boundedness of B in $\mathscr{D}_K$ is that, for an arbitrary m, $\sup_{f \in B} p_m(f)$ is finite, i.e.

$$\max |D^\alpha f(x)| \leqslant M_\alpha < +\infty \quad (f \in B, |\alpha| \geqslant 0).$$

2 *Fundamental properties of Fréchet spaces*

The principal aim of this section is to establish theorem 2.1. The line of proof follows mainly the study of complete metric spaces by Banach. His original work was concerned with extending Baire's ideas to abstract spaces, and made a great contribution to the progress of classical analysis in the twentieth century. It has served as one of the driving forces in the development of modern analysis.

By saying that a space E is a metric space we mean that, for two arbitrary points $x, y \in E$, a distance $(x, y) \in R$ is defined, where distance $(\ , \ )$ is a function, satisfying

$$\begin{aligned}
&0 \leqslant (x, y) < +\infty, \\
&(x, y) = (y, x), \\
&(x, y) = 0 \Leftrightarrow x = y, \\
&(x, y) + (y, z) \geqslant (x, z).
\end{aligned}$$

We can apply most notions and methods of elementary point-set theory in an Euclidean space to a metric space thus defined, and we therefore do not propose to discuss this aspect further, except to give a warning to the reader. That is, the notion of a bounded set in an Euclidean space is not valid for metric spaces in general. Therefore, the Bolzano–Weierstrass theorem, i.e. that in R^n a closed bounded set is compact, is not applicable. We give an alternative definition.

† Translator's note: λ depends on m and n.

By saying that a set A of a complete metric space E is *compact*, we mean either†

(1) A is closed. For an arbitrary ε, there exists a finite set of distinct points $x_1, x_2, \ldots, x_p \in A$ and every point of A lies within the ε-neighbourhood of one of these points.

or

(2) For an arbitrarily chosen sequence $\{x_n\}$ in A there exists a subsequence of $\{x_n\}$ with $x_{n_p} \to x_0 \in A$, i.e. $\{x_{n_p}\}$ is a convergent sequence.‡

These conditions are equivalent.

DEFINITION 2.6. Let E be a metric space. $G(\subset E)$ is said to be a set of the first category, if and only if, it is the union of a countable family of nowhere dense (non-dense, Fr. rare) subsets of G. G is of the second category if, and only if, it is not of the first category.

We recall the definition of a nowhere-dense set: F is nowhere-dense if and only if the closure (Fr. adhérence) $\bar{F}$ of F has no interior points.

EXAMPLE. The whole set of the points in a plane having their coordinates in rational numbers is of the first category. We note that the set is also countable. The union of countable sets of lines is also of the first category because a line has no interior points. From this example we see that even if G is of the first category, in some cases $\bar{G}$ may be the whole space.

THEOREM 2.2. A complete metric space is not of the first category, i.e. it is of the second category.*

Theorem 2.2. is often used in the following form:

COROLLARY. Let E be a complete metric space. Assume that there exists a countable sequence of closed sets $\{F_n\}$ and E is covered by the union of $\{F_n\}$: $E=\bigcup_{n=1}^{\infty} F_n$.

† In this text, we use 'compact' and 'relatively compact' in different senses. By saying that a set A of a complete space E is relatively compact, we mean the closure $\bar{A}$ of A is compact.

‡ The necessary and sufficient conditions for $\bar{A}$ to be relatively compact are (1) and (2) above, but without the conditions 'A is closed' and '$x_0 \in A$'.

* This is known as Baire's category theorem.

Then, at least one of $\{F_n\}$ has interior points, i.e. there exist F_n, $x_0 \in F_n$, and $\varepsilon(>0)$, such that the open ball of radius ε: $|x-x_0|<\varepsilon$, is contained in F_n.

Proof of theorem 2.2.

(1) We note that the following fact can be established. If there exists a sequence of closed balls $K_1 \supset K_2 \supset K_3 \supset \cdots$ in a complete metric space E, and the radius Y_n of K_n tends to 0 as $n \to \infty$, then, there is a point $x_0 \in \bigcap_n K_n$. To see this, let x_p be the centre of K_p; then it is obvious that $\{x_p\}$ is a Cauchy sequence. From the completeness of E, for a certain $x_0 \in E$, $x_p \to x_0$. Thus, x_0 belongs to every ball K_p. In fact, for an arbitrary p,

$$(x_p, x_0) = \lim_{q\to\infty} (x_p, x_q) \leqslant r_p.$$

(2) Now we are ready to prove the theorem by contradiction. We assume that there exists a sequence of nowhere-dense subsets $\{G_n\}$ and

$$E = \bigcup_{n=1}^{\infty} G_n. \tag{2.8}$$

Because G_1 is nowhere-dense, we see that there is some point which does not belong to $\bar{G}_1$. We pick such a point x_1. Let us consider a ball

$$K_1 : \{x;\ |x-x_1| \leqslant r_1\}$$

which has a sufficiently small radius r_1 with centre x_1. We see that $K_1 \cdot G_1 = \emptyset$. That is K_1 does not intersect G_1. Next, we consider a ball of radius $\frac{1}{2}r_1$ with centre x_1.

It is impossible for the intersection of this ball and G_2 to coincide in the same ball. Therefore, there exists x_2 which does not belong to $\bar{G}_2$ but belongs to the ball.

We then consider a ball having sufficiently small radius r_2, centre x_2:

$$K_2 : \{x;\ |x-x_2| \leqslant r_2\}.$$

We can take $K_2 \subset K_1$ and $K_2 \cdot G_2 = \emptyset$. We repeat this process until we have $K_1 \supset K_2 \supset K_3 \supset \cdots$ (a sequence of closed balls) satisfying the following conditions:

(*a*) r_n (the radius of K_n) tends to 0 as $n \to \infty$,

(*b*) $K_n \cdot G_n = \emptyset$ $(n=1, 2, \ldots)$.

From (1) we see that there exists $x_0 \in \bigcap_n K_n$, but, from (*b*), $x_0 \notin G_n$ for all n. This is contrary to our hypothesis (2.8).

Q.E.D.

DEFINITION 2.7 *(F-type space)*. By saying that a vector space E is an F-type space, we mean that E is a complete metric space and its distance (x, y) satisfies:

(1) $(x, y) = (x-y, 0)$.

(2) For an arbitrary $x \in E$, if $\lim_{n\to\infty} \lambda_n = 0$, then $\lim_{n\to\infty} \lambda_n x = 0$.

(3) For an arbitrary λ, if $\lim_{n\to\infty} x_n = 0$, then $\lim_{n\to\infty} \lambda x_n = 0$.

We write $|x| = (x, 0)$.

LEMMA 2.4. Every F-space can be regarded as an F-type space. This is equivalent to saying that an F-space is metrizable. More precisely we can introduce a distance into an F-space and make the F-space a metric space. In this case we require that the metric space thus obtained has an equivalent topology to that already given for the F-space, and that at the same time it is an F-type space.

Proof. We use the same idea of an increasing sequence of semi-norms as in the definition of an F-space. For arbitrary $x, y \in E$, we define

$$(x, y) = \sum_{n=1}^{\infty} \frac{1}{2^n} \cdot \frac{p_n(x-y)}{1+p_n(x-y)}.$$

It is obvious that this is a metric. We check the condition of triangular inequality. Because

$$p_n(x+y) \leqslant p_n(x) + p_n(y),$$

we see that

$$\frac{p_n(x+y)}{1+p_n(x+y)} \leqslant \frac{p_n(x)+p_n(y)}{1+p_n(x)+p_n(y)} \leqslant \frac{p_n(x)}{1+p_n(x)} + \frac{p_n(y)}{1+p_n(y)}.$$

Also, we have $(x-y, 0) = (x, y)$. We note that the equivalence relation exists between the metric topology of these semi-norms and the original topology of the F-space if and only if:

(*a*) for arbitrary p_n and ε, there exists δ such that, if $|x|<\delta$, then $p_n(x)<\varepsilon$;

(*b*) for an arbitrary ε, there exists p_n and δ, such that if $p_n(x)<\delta$ then $|x|<\varepsilon$.

(*a*) is easy to prove, so we leave it as an exercise for the reader. We prove (*b*). First given ε, we pick a sufficiently large N so that

$$\sum_{n=N+1}^{\infty} \frac{1}{2^n} < \tfrac{1}{2}\varepsilon,$$

and we see that for an arbitrary x satisfying $p_N(x)<\frac{1}{2}\varepsilon$,

$$p_j(x)<\tfrac{1}{2}\varepsilon \quad (j=1, 2, \ldots, N).$$

Therefore

$$|x| \leqslant \tfrac{1}{2}\varepsilon \sum_{n=1}^{N} \frac{1}{2^n} + \tfrac{1}{2}\varepsilon < \varepsilon.$$

Finally, the conditions (2), (3) for an F-type space are satisfied because these are built-in properties of the topology of an F-space.

Q.E.D.

Topology of a Fréchet space and its dual space

Let E be an F-space and E' be the dual space consisting of all continuous linear forms defined on E. E' is a vector space. There are usually three different topologies defined on E'. We give them arranged in order from the weakest to the strongest. (We use the property of the vector space E'.)

(*a*) E_s' (simple topology): $u_j \to 0$ in E_s' if and only if, for an arbitrary $x \in E$, $\langle u_j, x\rangle \to 0$.

(*b*) E_c' (compact topology): for an arbitrary compact set A in E, $u_j \to 0$ in A uniformly, i.e. $u_j \to 0$ in E_c' if and only if $\sup_{x\in A}|\langle u_j, x\rangle| \to 0$.

(*c*) E_b' (strong topology): $u_j \to 0$ in E_b' if and only if, for an arbitrary bounded set B,

$$\sup_{x\in B} |\langle u_j, x\rangle| \to 0 .^{\dagger}$$

We have written u_j; by this we mean that if $\{u_j\}$ is a sequence, then $j\to\infty$. If u_j is a single object, the convergence is understood to be filter-convergence.

† In this text, by the topology of E' we mean E_b' unless the contrary is stated.

We are only considering the former case in order to prove the Banach–Steinhaus theorem.

DEFINITION 2.8. When we say that a set H of E' is *bounded in the simple topology*, we mean that for an arbitrary $x \in E$

$$\sup_{u \in H} |\langle u, x \rangle| < +\infty .$$

Note. If $\{u_j\} \in E'$, then the above definition means that the sequence $\langle u_j, x \rangle$ is bounded whenever x is chosen and fixed. If $\{u_j\}$ is a Cauchy sequence in the simple topology, i.e. for an arbitrary $x \in E$, $\{\langle u_j, x \rangle\}$ is a Cauchy sequence, then $\{u_j\}$ is bounded in the simple topology.

The boundedness which we have just explained does not seem to be very restrictive. However, we see some of the power of the boundedness in the following lemma. (We use Baire's theorem (2.2) to prove this.)

LEMMA 2.5 *(The fundamental lemma of Fréchet space)*. Let E be an F-space. A set bounded in the simple topology of E' is an equi-continuous set. By saying that a set $H \subset E'$ is equi-continuous, we mean that for a certain neighbourhood V of 0 in E,

$$\sup_{u \in H, x \in V} |\langle u, x \rangle| < +\infty .$$

In other words, for an arbitrary $\varepsilon (>0)$, there exists a neighbourhood V of 0, such that

$$\sup_{u \in H, x \in V} |\langle u, x \rangle| < \varepsilon .$$

Note. Historically, the origin of the idea 'equi-continuity' came from the Ascoli–Arzela theorem.

In general, let f_j be a continuous function, then a sequence $\{f_j\}$ is equi-continuous at a point x_0 if and only if for an arbitrary $\varepsilon (>0)$, there exists $\delta (>0)$, and if $(x - x_0) < \delta$, then $|f_j(x) - f_j(x_0)| < \varepsilon$ is valid at the same time for all j. In our case, we use the same idea which appeared in the classical theorem in a more generalized form (considering the linearity of $f_j = u_j \in E'$).

Proof of lemma 2.5. Let H be bounded in the simple topology. We put

$f(x)=\sup_{u\in H}|\langle u, x\rangle|$, then $f(x)$ is a function taking only finite values. Next, we put $G_n=\{x; f(x)\leqslant n\}$, then we see $E=\bigcup_{n=1}^{\infty} G_n$ where each G_n is a closed subset. In fact, if we fix u, $\{x; |\langle u, x\rangle|\leqslant n\}$ is closed because $|\langle u, x\rangle|$ is a continuous function of x. G_n is the union of these closed sets when u runs through H. Therefore G_n is closed.

We recall the fact that any F-space is an F-type space (lemma 2.4). Therefore, by Baire's category theorem, (the corollary of theorem 2.2), we see that at least one of G_n, G_{n_0} contains an open set of E, i.e. there exists x_0 and a neighbourhood V of 0, with $x_0+V\subset G_{n_0}$. G_{n_0} is symmetric with respect to 0, and a convex set. Therefore, $\frac{1}{2}V\subset G_{n_0}$. Hence, $U\subset G_{n_0}$ (U is a neighbourhood of 0). Thus for $u\in H$, $\sup_{x\in U}|\langle u, x\rangle|\leqslant n_0$.

Q.E.D.

We examine the proof of the Ascoli–Arzela theorem in order to prove:

THEOREM 2.3. Let E be an F-space, and E' be the dual space of E. A Cauchy sequence in the simple topology of E' is a Cauchy sequence in the topology of *compact convergence* (=uniform convergence on compacta).

Proof. Let $\{u_j\}$ be a Cauchy sequence in E_s'. $\{u_j\}$ is naturally a bounded set in E_s'. Therefore $\{u_j\}$ is an equi-continuous set and, for an arbitrary ε, there exists a neighbourhood V of 0 in E satisfying

$$|\langle u_j, y\rangle|<\tfrac{1}{2}\varepsilon \quad (y\in V, j=1, 2, \ldots). \tag{2.9}$$

Next, we consider an arbitrary compact set A in E. From the definition of compact set, we see that there exists a finite set $x_1, x_2, \ldots, x_p$ $(x_i\in A)$ and A is covered by $\{x_i+V\}_{i=1,2,\ldots,p}$. From our hypothesis on $\{u_j\}$, for a sufficiently large N, we have

$$|\langle u_p-u_q, x_i\rangle|<\tfrac{1}{2}\varepsilon \quad (i=1, 2, \ldots, p) \text{ for } p, q>N. \tag{2.10}$$

Therefore, for an arbitrary $x\in A$, we establish

$$|\langle u_p-u_q, x\rangle|<\varepsilon \text{ for } p, q>N, \tag{2.11}$$

and in fact, there exists x_i, with $x=x_i+y$ $(y\in V)$. We then have a decomposition

$$\langle u_p-u_q, x\rangle=\langle u_p-u_q, x_i\rangle+\langle u_p-u_q, y\rangle.$$

From (2.9) and (2.10), we see that (2.11) is valid. Hence, $\{u_j\}$ is a Cauchy sequence in the topology E_c'.

Q.E.D.

We establish one more lemma relating to the theorem.

LEMMA 2.6. Let u be a linear form in an F-space E. If u is bounded on all campact sets in E, then u is a continuous linear form: $u \in E'$.

Proof. Let $\{P_n(x)\}$ be an increasing sequence of semi-norms (see the definition of a Fréchet space). If u is not continuous, then there exists a sequence $\{x_n\}$ in E satisfying

$$p_n(x_n) \leqslant 1, \, |\langle u, x_n \rangle| \geqslant n \qquad (n=1, 2, \ldots).$$

On the other hand, $\{x_n/\sqrt{n}\}$ is obviously a sequence tending to 0. Therefore, adding 0 to the sequence, we have a compact set in E. However, we see that

$$\left|\left\langle u, \frac{1}{\sqrt{n}} x_n \right\rangle\right| \geqslant \sqrt{n} \to +\infty \qquad (n \to \infty).$$

This is a contradiction and u is therefore continuous.

Q.E.D.

THEOREM 2.4 *(Banach–Steinhaus)*. Let $\{u_j\}_{j=1,2,\cdots}$ be a sequence of continuous linear forms in F-space E, which form a Cauchy sequence in the sense of the simple topology of E'. Then there exists a unique continuous linear form u_0, with

$$\langle u_j, x \rangle \to \langle u_0, x \rangle \qquad (j \to \infty),$$

where the converge is uniform on an arbitrary compact set in E.

Proof. We see first, by theorem 2.3, that $\{u_j\}$ is a Cauchy sequence in E_c'-topology. Therefore, if we denote the linear form of their limit as u_0, then u_0 satisfies the conditions of lemma 2.6. The reason is this. Let A be an arbitrary compact set in E. For any ε, there exists n, and if $p, q > n$, for all $x \in A$, $|\langle u_p - u_q, x \rangle| < \varepsilon$.

Therefore, letting $q \to +\infty$, we have

$$|\langle u_p, x\rangle - \langle u_0, x\rangle| \leqslant \varepsilon \quad \text{for} \quad p > n, x \in A, \tag{2.12}$$

i.e. $|\langle u_0, x\rangle| \leqslant \varepsilon + |\langle u_p, x\rangle|$ is valid for $x \in A$. For this the continuity of u_p, and the compactness of A, we see that $\sup_{x \in A}|\langle u_p, x\rangle| < +\infty$. Hence the linear form u_0 is bounded in A. Hence, from lemma 2.6, $u_0 \in E'$. The rest of the theorem follows directly from (2.12).

Q.E.D.

Note on theorem 2.1. While proving theorem 2.4, we complete the proof of theorem 2.1. We have some remarks:

(1) $\{T_j\}$ in theorem 2.1 is uniformly convergent to T on an arbitrary compact set A in $\mathscr{D}_K$. However, the closure B of an arbitrary bounded set B in $\mathscr{D}_K$ is, from lemma 2.3, a compact set in $\mathscr{D}_K$ (by the Ascoli–Arzela theorem). Therefore, $\{T_j\}$ is in fact uniformly convergent on an arbitrary bounded set in $\mathscr{D}_K$.

(2) Schwartz introduced a so-called 'locally convex topology' into $\mathscr{D}$; he defined the topology by which a linear form coincides with a linear form as defined in definition 2.2. Moreover, for an arbitrary bounded set B in $\mathscr{D}$, there exists a compact set K, $B \subset \mathscr{D}_K$ (K depends on B), and B is a bounded set by the topology $\mathscr{D}_K$. (The converse is also true.) From this and the result of (1), if a sequence $\{T_j\} \to T$ by a simple topology, then $\{T_j\}$ is convergent uniformly on B.

The converse is of course true. Therefore, we have the following result: the convergence of $\{T_j\}$ in the simple topology and the convergence of $\{T_j\}$ in the strong topology of $\mathscr{D}'$ (a topology of uniform convergence on an arbitrary bounded set in $\mathscr{D}$) are completely identical.

One of the important results based on Baire's category is Banach's closed graph theorem which we shall prove after some preparation. The proof given here is based on Banach's work found in *Theorie des Opérations Linéaires*.

LEMMA 2.7 *(Fundamental property of an F-type space)*. Let E and E_1 be F-type spaces, and u be a continuous linear operator from E to E_1. If the image set $u(E)$ of E by u is not of the first category in E_1, then, for an arbitrary $\varepsilon(>0)$, a sufficiently small $\eta(>0)$ can be taken and the image of the ball $|x| < \varepsilon$ under u covers the whole open ball $|y| < \eta$ in E_1.

First we prove this in the following weaker form.

(1) For an arbitrary $\bar{\varepsilon}(>0)$, a sufficiently small $\eta(>0)$ can be taken and the image of the ball $|x|<\bar{\varepsilon}$ under u is dense in the ball $|y|<\eta$ in E_1.

The proof proceeds as follows. Let G be the ball $|x|<\frac{1}{2}\bar{\varepsilon}$, and $2G_1, 3G_1, \ldots, nG_1$ be respectively written as $G_2, G_3 \ldots, G_n$, where nG_1 denotes the set $\{nx\}$, $x \in G_2$. Then an arbitrary point in E belongs to G_n.

In fact, $\lim_{n\to\infty} (x/n)=0$, so that for a sufficiently large n, $(x/n) \in G_1$. Therefore, $x \in nG_1 = G_n$. Hence, $E=\bigcup_{n=1}^{\infty} G_n$

We put $u(G_n)=H_n$. Then $u(E)=\bigcup_{n=1}^{\infty} H_n$. By our hypothesis, $u(E)$ is of the second category (because the union of sets of the first category is again of the first category). Therefore, the open ball $K_1 \subset H_{n_0}$. Furthermore, from the fact that an F-type space is a vector space, we see that H_{n_0}' (the derived set of H_{n_0}) $\supset K_1$.

Let $K_1 : |y-y_0|<\eta_1$ From this

$$H_1' \supset K_2 : \left|y-\frac{1}{n_0}y_0\right| < \frac{1}{n_0}\eta_1 .$$

In fact, for $y \in K_2$,

$$|n_0 y - y_0| = \left|n_0\left(y-\frac{1}{n_0}y_0\right)\right| \leqslant n_0\left|y-\frac{1}{n_0}y_0\right| < \eta_1 ,$$

so that $n_0 y \in K_1 \ (\subset H_{n_0}')$. There exists

$$y_j \in H_{n_0} \to n_0 y \ (j\to\infty) \quad \text{and} \quad \frac{1}{n_0} y_j \in H_1 \to y ,$$

i.e. $y \in H_1'$ follows from $y \in K_2$.

Now, let y_3 be a point $H_1 \subset K_2$. Construct an open ball $K_3 \subset K_2$ of radius η centre y_3. (If the centre of K_2 belongs to H_1, we put $K_3 = K_2$.) H_1 is, of course, dense in K_3. Therefore $H_1 \setminus y_3$ is dense in $|y|<\eta$. From $y_3 \in u(G_1)$ we have $H_1 \setminus y_3 \subset u(G_1 \setminus G_1)$. $G \setminus G_1$ is contained in the open ball $|x|<\bar{\varepsilon}$ and hence, the image of the open ball $|x|<\bar{\varepsilon}$ is dense in $|y|<\eta$.

(2) From this fact, the completeness of E, and the continuity of u, we prove the lemma as follows. Let ε be an arbitrarily given positive number. Put

$$\varepsilon_i = \frac{\varepsilon}{2^i} \quad (i=1, 2, \ldots).$$

We regard each ε_i as $\bar{\varepsilon}$ as before, and choose η_i for ε_i. Now, we see that the image of the ball $|x|<\varepsilon_i$ is dense in $|y|<\eta_i$. Without loss of generality, we let $\eta_2 \to 0$. We show that if $\eta=\eta_1$, then for an arbitrary point y in the ball $|y|<\eta$, there exists $x \in E$, $y=u(x)$ and $|x|<\varepsilon$. To see this, we choose y_1, sufficiently close to y, with $|y-y_1|<\eta_2$, $y_1=u(x_1)$, $|x_1|<\varepsilon$. Next, for $y-y_1$, we pick y_2 sufficiently close to $y-y_1$, such that

$$|(y-y_1)-y_2|<\eta_3, \quad y_2=u(x_2), \quad |x_2|<\varepsilon_2.$$

We repeat the same process until

$$\left|y-\sum_{i=1}^{n} y_i\right|<\eta_{n+1}, \quad y_i=u(x_i), \quad |x_i|<\varepsilon_i \quad (i=1, 2, \ldots, n),$$

so that

$$\sum_{i=1}^{n} y_i \to y \quad (n \to \infty).$$

From the completeness of E,

$$\sum_{i=1}^{\infty} x_i = x \in E, \quad \sum_{i=1}^{\infty} |x_i|<\varepsilon.$$

Hence $|x|<\varepsilon$. Hence $u(x)=y$ because u is continuous.

Q.E.D.

LEMMA 2.8. Let E and E_1 be F-types spaces, and u be a continuous linear operator from E to E_1. Then, either the image set $u(E)$ of E is of the first category in E_1, or $u(E)$ coincides with E_1.

Proof. If $u(E)$ is not of the first category then, from the previous lemma, for an arbitrary $y \in E_1$

$$\lim_{n \to \infty} (y/n) = 0.$$

We can pick a sufficiently large n_0 satisfying

$$|y/n_0|<\eta.$$

Therefore, for an arbitrary $x \in E$, $u(x)=y/n_0$,
i.e. $u(n_0 x)=y$. Hence, $u(E)=E_1$.

LEMMA 2.9. Let E, E_1, and u be the same as in the previous lemma 2.8 and assume that $u(E)=E_1$. Then, for an arbitrary convergent sequence $y_n \to y_0$ in E_1, we can find $\{x_n\}$ and x_0 such that $u(x_n)=y_n$, $u(x_0)=y_0$, with $x_n \to x_0$ in E.

Proof. This follows directly from the fundamental lemma 2.7. Choose a sequence $\varepsilon_1 > \varepsilon_2 > \cdots > \varepsilon_n > \cdots \to 0$. From this sequence $\eta_1, \eta_2, \ldots, \eta_n, \ldots$ can be determined by lemma 2.7. That is to say that the image of the ball $(x-x_0)<\varepsilon_n$ under u contains the ball $(y-y_0)<\eta_n$.

For $y_m \in E_1$:

(*a*) If $y_m = y_0$, then take $x_m = x_0$.

(*b*) If $y_m \neq y_0$, then choose x_m in the following way. If there exists η_n such that $|y_m - y_0| < \eta_n$, then pick n_m, the maximum of such n. Take x_m satisfying $|x_m - x_0| < \varepsilon_{n_m}$ and $u(x_m) = y_m$. We note that $n_m \to \infty$ as $m \to \infty$.

(*c*) $y_m \neq y_0$, but when there is no η_n as in (*b*) for any $n=1, 2, \ldots$, take x_m as an arbitrary inverse image of y_m of u. Such m are all finite. We see that $\{x_m\}$ thus constructed meets our requirements.

THEOREM 2.5. Let E and E_1, be F-type spaces. If a continuous linear operator is a bijection from E onto E_1, then the inverse operator of u is also continuous, i.e. u is bicontinuous. The proof is obvious from the previous lemma.

From the theorem we have:

THEOREM 2.6 *(Banach's closed graph theorem)*. Let E and F be F-type spaces, and u be a linear operator from E to F. Then if the graph of u is closed, this is equivalent to saying:

$$\text{If } \lim_{n\to\infty} x_n = x_0 \text{ and } \lim_{n\to\infty} u(x_n) = y_0, \text{ then } u(x_0) = y_0. \qquad \text{(C)}$$

Then u is a continuous linear operator from E to F.

Note. Considering the linearity of u, we see that the condition (C) is equivalent to the same condition (C) when $x_0 = 0$, i.e. (C) is equivalent to the condition:

$$\text{If } \lim_{n\to\infty} x_n = 0, \lim_{n\to\infty} u(x_n) = y_0 \text{ then } y_0 = 0. \qquad (\text{C}_0)$$

In fact, if (C_0) is true, then the conditions of (C) are written as $\lim_{n\to\infty} u(x_n - x_0)$

$=y_0-u(x_0)$ and $\lim_{n\to\infty}(x_n-x_0)=0$. Therefore, from ($C_0$), $u(x_0)=y_0$ follows.

Proof of theorem 2.6. Let E_1 be the same space as E but with a distance defined by $(x, x')_1=(x, x')_E+(u(x), u(x'))_F$.
We see that E_1 is also an F-type space. In fact, a distance thus defined satisfies the necessary conditions for an F-type space. We need only prove the completeness of E_1. To do this, we pick a Cauchy sequence $\{x_n\}$ in E_1

$$|x_p-x_q|_1\to 0 \quad (p, q\to+\infty).$$

From this we have $x_p\to x_0$, $u(x_p)\to y_0$ $(p\to\infty)$, by virtue of the completeness of both E and F.

From (C), we see that $y_0=u(x_0)$, so that

$$|x_p-x_0|_1=|x_p-x_0|_E+|u(x_p)-u(x_0)|_F\to 0 \quad (p\to\infty).$$

Hence, E_1 is complete. Also, $(x, x')_E\leqslant(x, x')_{E_1}$, therefore, the identity map from E_1 to E is continuous. Applying theorem 2.5, we see that the inverse operator, like the identity map from E to E_1, is also continuous.

Then, as $x_n\to 0$ in E, we conclude that $u(x_n)\to 0$ in F.

Q.E.D.

3 *Function spaces* $\mathscr{E}_{L^p}{}^m(\Omega)$, $\mathscr{D}_{L^p}{}^m(\Omega)$

In the first section of this chapter we defined distributions and gave a rather strong topology to the subspace of test functions through which we were able to define very general linear functionals. The power of this type of topology can be clearly seen in theorem 2.1. This, roughly speaking, is analogous to a fact in the theory of Lebesgue integrals where we see that, after defining measurable functions, we can derive the measurability of the limit function of a sequence of measurable functions. However, we need to consider more restricted classes of distributions for the purposes of practical application. In the next paragraph we examine such classes.

DEFINITION 2.9. By a space $\mathscr{E}_{L^p}{}^m(\Omega)$ $(1\leqslant p<+\infty;\ m=0, 1, 2, \ldots)$ where Ω is an open set in R^n, we mean the set of functions f such that $f\in L^p(\Omega)$, and for all α, $|\alpha|\leqslant m$, $D^\alpha f\in L^p(\Omega)$ where the derivatives up to the mth order are taken as distributions.

For the norm of the space, we define

$$\|f(x)\|_{m,\, L^p(\Omega)} = \sum_{|\alpha| \leqslant m} \|D^\alpha f(x)\|_{L^p(\Omega)}. \tag{2.13}$$

When $p=2$, we take essentially the same topology as above, but in accordance with its Hilbert space structure, we define

$$\|f(x)\|^2_{m,\, L^2(\Omega)} = \sum_{|\alpha| \leqslant m} \|D^\alpha f(x)\|^2_{L^2(\Omega)}. \tag{2.14}$$

In particular, if $\Omega = R^n$, we sometimes write simply $\mathscr{E}_{L^p}{}^m$

Note. When $\Omega \neq R^n$, we may consider the distributions $\mathscr{D}'(\Omega)$ – distributions on Ω – as those, in the definition of $\mathscr{D}'$, replaced by compact sets of Ω rather than those on compact sets K of R^n. Therefore, in this case, we have a test function $\varphi \in \mathscr{D}(\Omega)$, i.e. $\varphi \in C^\infty$ whose support is a compact set of Ω. Note that it is not generally possible to extend a distribution on Ω to a distribution on the whole space.

PROPOSITION 2.3. $\mathscr{E}_{L^p}{}^m(\Omega)$ is complete.

Proof. Let $\{f_j(x)\}$ be a Cauchy sequence. From the completeness of $L^p(\Omega)$ the following limit exists

$$D^\alpha f_j(x) \to f_\alpha(x) \quad \text{in} \quad L^p(\Omega),\ |\alpha| \leqslant m.$$

Write $f_j(x) \to f_0(x)$. It is enough to prove $f_\alpha(x) = D^\alpha f_0(x)$, but this is obvious from proposition 2.1., i.e. $D^\alpha f_j(x) \to D^\alpha f_0(x)$ in $\mathscr{D}'(\Omega)$. Therefore, $D^\alpha f_0(x) = f_\alpha(x)$. In detail, for $\varphi\,(\in \mathscr{D}(\Omega))$,

$$\begin{aligned}\langle D^\alpha f_j(x), \varphi(x)\rangle &= (-1)^{|\alpha|}\langle f_j(x), D^\alpha \varphi(x)\rangle \underset{j\to\infty}{\to} (-1)^{|\alpha|}\langle f_0(x), D^\alpha\varphi(x)\rangle\\ &= \langle D^\alpha f_0(x), \varphi(x)\rangle.\end{aligned}$$

On the other hand,

$$\langle D^\alpha f_j(x), \varphi(x)\rangle \underset{j\to\infty}{\to} \langle f_\alpha(x), \varphi(x)\rangle.$$

Hence,

$$D^\alpha f_0(x) = f_\alpha(x)$$

Q.E.D.

Note. In this proof we note that the completeness was naturally derived because $D^\alpha f$ was defined in the sense of a distribution. If we did not use the notion of a distribution, our situation would be different and we would have had to consider the process of completion. It might not be always obvious to us to recognize the properties of the added elements by the completion. Nevertheless, it is necessary for the reader to know the significance of the idea of a distribution when we introduce it to derivatives in $L^p(\Omega)$-space.

We touch upon this argument again in the following paragraph.

Derivatives of a function

When we defined derivatives in section 1, if $f \in \mathscr{E}^1$, we noted that the ordinary derivative of f and the newly defined 'distribution derivative' coincide.

We now discuss the detailed consequences of this fact.

LEMMA 2.10. Let $f \in L^1_{\text{loc}}$. The necessary and sufficient condition for $\partial f/\partial x_1 = 0$ in the sense of distribution is that there exists a certain $g(x_2, \ldots, x_n)$, $g \in L^1_{\text{loc}}(R^{n-1})$, and $f(x_1, \ldots, x_n) = g(x_2, \ldots, x_n)$ is true in R^n almost everywhere.

Proof. We show that under this condition, for an arbitrary h

$$f(x_1 + h, x_2, \ldots, x_n) = f(x_1, x_2, \ldots, x_n) \tag{2.15}$$

is true in the sense of distribution.

For an arbitrary $\varphi \in \mathscr{D}$ we put

$$\langle f(x_1 + h, x_2, \ldots, x_n), \varphi(x) \rangle = \psi(h).$$

We see

$$\psi(h) = \langle f(x), \varphi(x_1 - h, \ldots, x_n) \rangle,$$

therefore $\psi'(h) \equiv 0$. Hence $\psi(h)$ is a constant. Because φ is arbitrary, (2.15) holds. From this fact we can establish the following result: for an arbitrary h and an arbitrary characteristic function ψ which is defined on a bounded and measurable set,

$$\int f(x)\,\psi(x)\,dx = \int f(x)\,\psi(x - h_1)\,dx; \quad h_1 = (h, 0, \ldots, 0). \tag{2.16}$$

Next, we apply the following Lebesgue theorem to $f(x)$:

$$\frac{1}{m(e)} \int_e f(x)\,dx \to f(x_0) \quad (e \to x_0) \tag{2.17}$$

which is valid for $x_0 \in R^n$ almost everywhere e can be taken as the family of balls with centre x_0.

Let us call the point which satisfies (2.17) a *regular point.* Using this new terminology, we can say that there are regular points almost everywhere on the straight lines which are parallel to the x_1-axis in R^n (as one-dimensional sets) except measure 0 sets in the space $(x_2, \ldots, x_n)$. Now, we assume that $f(x)$ is not essentially a constant on one of such straight lines, that is, $f(x)$ never becomes a constant even if we alter the value of points in certain measure sets. Under this assumption we conclude there exist two regular points on such a straight line and the values for f of these points are different. This is contrary to the facts, (2.16) and (2.17.) The converse is obvious, and is therefore omitted.

Q.E.D.

Using the lemma, we state:

THEOREM 2.7 *(Nikodym).* Let $f \in L^1{}_{\text{loc}}$. We avoid a possible confusion by writing $[\partial f/\partial x_1]$ when the derivatives are derivatives in the ordinary sense (if they exist), and $\partial f/\partial x_1$ when they are derivatives in the sense of distribution.

(1) Let f be absolutely continuous on straight lines except for measure 0 sets in the $(x_2, \ldots, x_n)$-space (we call this almost everywhere parallel lines to the x_1-axis). Accordingly, there exists $[\partial f/\partial x_1]$ almost everywhere in R^n as Lebesgue's theorem guarentees. If we assume furthermore $[\partial f/\partial x_1] \in L^1{}_{\text{loc}}$, then $\partial f/\partial x_1 = [\partial f/\partial x_1]$.

(2) Conversely, if $\partial f/\partial x_1 \in L^1{}_{\text{loc}}$, $f(x)$ has the property (1). More precisely, we can alter the value of f on measure 0 sets if necessary, and define a new function f (we use the same notation). We then see that f is absolutely continuous on 'a line parallel with the x_1-axis', and $[\partial f/\partial x_1] = \partial f/\partial x_1$.

Proof.

(1) First we see

$$\left\langle \frac{\partial f}{\partial x_1}, \varphi \right\rangle = -\left\langle f, \frac{\partial \varphi}{\partial x_1} \right\rangle$$

$$= -\int \cdots \int \mathrm{d}x_2 \ldots \mathrm{d}x_n \int_{-\infty}^{+\infty} f(x_1, x_2, \ldots, x_n) \frac{\partial \varphi}{\partial x_1}(x_1, \ldots, x_n)\, \mathrm{d}x_1 .$$

By our hypothesis, we can perform integration by parts in the $(x_2, x_3, \ldots, x_n)$-

space almost everywhere;

$$\left\langle \frac{\partial f}{\partial x_1}, \varphi \right\rangle = \int \cdots \int dx_2 \ldots dx_n \int_{-\infty}^{+\infty} \left[\frac{\partial f}{\partial x_1}(x_1, \ldots, x_n) \right] \varphi(x_1, \ldots, x_n)\, dx_1$$
$$= \left\langle \left[\frac{\partial f}{\partial x_1} \right], \varphi \right\rangle.$$

(2) Put

$$g(x_1, \ldots, x_n) = \int_0^{x_1} \frac{\partial f}{\partial x_1}(t_1, x_2, \ldots, x_n)\, dt_1 .$$

Then we see by Lebesgue's theorem, $[\partial g/\partial x_1] = \partial f/\partial x_1$, where $g(x)$ satisfies the condition of (1). Therefore $\partial g/\partial x_1 = [\partial g/\partial x_1]$ and $\partial f/\partial x_1 = \partial g/\partial x_1$. We use the previous lemma here to derive

$$f(x) - g(x) = \psi(x_2, \ldots, x_n); \quad \psi \in L^1_{\text{loc}}(R^{n-1}).$$

Hence,

$$\left[\frac{\partial f}{\partial x_1} \right] = \left[\frac{\partial g}{\partial x_1} \right] = \frac{\partial f}{\partial x_1} .$$

Q.E.D.

EXAMPLE. Let $f(x) = 1/|x|^\lambda, \lambda < n$ so $f \in L^1_{\text{loc}}$. Consider the following equation

$$\frac{\partial}{\partial x_i} \frac{1}{|x|^\lambda} = -\lambda \frac{x_i}{|x|^{\lambda+2}} . \tag{2.18}$$

This is true by the previous theorem (see (1)) when $\lambda < n-1$. To see this we observe $f \in C^\infty$ in $R^n \setminus \{0\}$, and the function on the right hand side of (2.18) belongs to L^1_{loc}. Similarly,

$$\frac{\partial^2}{\partial x_i^2} \frac{1}{|x|^\lambda} = -\frac{\lambda}{|x|^{\lambda+2}} + \lambda(\lambda+2) \frac{x_i^2}{|x|^{\lambda+4}} \quad (\lambda < n-2).$$

Therefore,

$$\Delta \left(\frac{1}{|x|^\lambda} \right) = \lambda(\lambda+2-n) \frac{1}{|x|^{\lambda+2}} \quad (\lambda < n-2).$$

Then we put $\lambda = n-2-\alpha$, $\alpha > 0$ to have

$$\Delta \left(\frac{1}{|x|^{n-2-\alpha}} \right) = -\alpha(n-2-\alpha) \frac{1}{|x|^{n-\alpha}} \quad (\alpha > 0). \tag{2.19}$$

Now, if we let $\alpha \to +0$, the left hand side of (2.19) tends to $\Delta(1/|x|^{n-2})$ by lemma 2.1. For the right hand side of (2.19), we see that

$$-\alpha(n-2-\alpha)\frac{1}{|x|^{n-\alpha}} \to -(n-2)\,S_n\delta \quad \text{in} \quad \mathscr{D}' \quad \text{as} \quad \alpha \to +0.$$

Hence,

$$\Delta\left(\frac{1}{|x|^{n-2}}\right) = -(n-2)\,S_n\delta \quad (n>2); \quad S_n = \frac{2\pi^{n/2}}{\Gamma(n/2)}. \tag{2.20}$$

The space $\mathscr{E}_{L^p}{}^m(\Omega)$ is important because it will be considered in the following chapter in relation to boundary value problems (the Dirichlet problem, the Neumann problem, etc.). We give some elementary properties of the space here. The reader will find a more detailed account of the space in the following chapter.

PROPOSITION 2.4. Let $f \in \mathscr{E}_{L^p}{}^m(\Omega)$. Then the following statements are true.

(1) For $a \in \mathscr{B}^m(\Omega)$, consider a map $(a(x), f) \to a(x)\,f$. This is a continuous function from $\mathscr{B}^m(\Omega) \times \mathscr{E}_{L^p}{}^m(\Omega)$ into $\mathscr{E}_{L^p}{}^m(\Omega)$, i.e.

$$\|a(x)\,f\|_{m,\,L^p(\Omega)} \leqslant C|a(x)|_m \cdot \|f\|_{m,\,L^p(\Omega)}, \tag{2.21}$$

where C is a constant.

(2) Consider a map

$$f(x) \to \frac{\partial f}{\partial x_i}(x).$$

This is a continuous function from $\mathscr{E}_{L^p}{}^m(\Omega)$ to $\mathscr{E}_{L^p}{}^{m-1}(\Omega)$.

(3) Let $\rho_\varepsilon(x)$ be a Friedrichs mollifier (see the end of Section 7, Chapter 1). Consider $\rho_\varepsilon * f \in \mathscr{E}^\infty$. As $\varepsilon \to 0$ this tends to $f(x)$ on an arbitrary compact set in Ω (including its derivates of up to the mth order) in the topology given to L^p.

Proof.

(1) We see

$$\frac{\partial}{\partial x_i}(a(x)\,f) = \frac{\partial a}{\partial x_i} f + a(x)\frac{\partial f}{\partial x_i}.$$

In fact, if we put $\varphi\in\mathscr{D}(\Omega)$, then

$$\left\langle\frac{\partial}{\partial x_i}(af),\varphi\right\rangle=-\left\langle af,\frac{\partial}{\partial x_i}\varphi\right\rangle=-\left\langle f,a\frac{\partial}{\partial x_i}\varphi\right\rangle$$

$$=-\left\langle f,\frac{\partial}{\partial x_i}(a\varphi)-\frac{\partial a}{\partial x_i}\varphi\right\rangle=\left\langle a(x)\frac{\partial f}{\partial x_i}+\frac{\partial a}{\partial x_i}f,\varphi\right\rangle.$$

Therefore, Leibniz formula is valid

$$\mathrm{D}^\alpha(af)=\sum_{\beta\leqslant\alpha}C_\beta{}^\alpha\ \mathrm{D}^{\alpha-\beta}a(x)\cdot\mathrm{D}^\beta f,\quad(|\alpha|\leqslant m).$$

From this

$$\|\mathrm{D}^\alpha(af)\|_{L^p(\Omega)}\leqslant C|a(x)|_{\mathscr{B}^m(\Omega)}\cdot\|f\|_{m,L^p(\Omega)}.$$

Hence (2.21) is valid.

(2) is obvious.

(3) Let

$$f_\varepsilon(x)=\int_\Omega\rho_\varepsilon(x-y)f(y)\,\mathrm{d}y.$$

If K is a compact set in Ω, then for a sufficiently small ε,

$$\mathrm{D}^\alpha f_i(x)=\int_\Omega\rho_\varepsilon(x-y)\,\mathrm{D}^\alpha f(y)\,\mathrm{d}y\quad(|\alpha|\leqslant m)\tag{2.22}$$

for all $x\in K$. In fact, if we put $\varphi\in\mathscr{D}(\Omega)$, then

$$\langle\mathrm{D}^\alpha f_\varepsilon(x),\varphi(x)\rangle=(-1)^{|\alpha|}\langle f_\varepsilon(x),\mathrm{D}^\alpha\varphi(x)\rangle$$

$$=(-1)^{|\alpha|}\int\mathrm{D}^\alpha\varphi(x)\,\mathrm{d}x\int_\Omega f(x-y)\,\rho_\varepsilon(y)\,\mathrm{d}y;$$

by Fubini's theorem, this is equal to

$$(-1)^{|\alpha|}\int\rho_\varepsilon(y)\,\langle f(x-y),\mathrm{D}^\alpha\varphi(x)\rangle\,\mathrm{d}y.$$

On the other hand, if $\varepsilon_0(>0)$ is small enough, for $|y|<\varepsilon_0$, then

$$\langle f(x-y),\mathrm{D}^\alpha\varphi(x)\rangle=(-1)^{|\alpha|}\langle\mathrm{D}^\alpha f(x-y),\varphi(x)\rangle.$$

Note that, in order to get the equality we have used the definition of the derivative of f. We have written $f(x-y)$ which means that, putting $f(x)=0$ outside Ω, we extend $f(x)$ to $\hat{f}(x)$, then operate a parallel transformation on it, so that, more precisely, the right hand side term should be written as

$$(-1)^{|\alpha|}\langle D^\alpha \tilde{f}(x-y), \varphi(x)\rangle.$$

However, because $\varphi \in \mathscr{D}(\Omega)$, for a sufficiently small $\varepsilon_0(>0)$,

$$\langle D^\alpha \tilde{f}(x-y), \varphi(x)\rangle = \langle D^\alpha f(x-y), \varphi(x)\rangle$$

in $|y|<\varepsilon_0$. Then, applying Fubini's theorem again, we have

$$\langle D^\alpha f_\varepsilon(x), \varphi(x)\rangle = \langle \rho_\varepsilon * D^\alpha f, \varphi\rangle$$

for $\varepsilon<\varepsilon_0$.

Let $\tilde{K}$ be a compact set in Ω containing K, and let $\varphi \in \mathscr{D}(\tilde{K})$ be an arbitrary test function. We see now that (2.22) is true. Also,

$$\|D^\alpha f_\varepsilon(x) - D^\alpha f(x)\|_{L^p(K)} \leqslant \|\rho_\varepsilon * D^\alpha f - D\, f\|_{L^p(\Omega)}.$$

Hence, by lemma 1.3, (3) is valid.

Q.E.D.

DEFINITION 2.10 (*$\mathscr{D}_{L^p}{}^m(\Omega)$-space*). We write $\mathscr{D}_{L^p}{}^m(\Omega)$ as the closure of $\mathscr{D}(\Omega)$ in $\mathscr{E}_{L^p}{}^m(\Omega)$. More precisely, $f \in \mathscr{D}_{L^p}{}^m(\Omega)$ means that there exists a sequence $\varphi_j(x)$ of $\mathscr{D}(\Omega)$, with $\varphi_j \to f$ in $\mathscr{E}_{L^p}{}^m(\Omega)$. $\mathscr{D}_{L^p}{}^m(\Omega)$ is a closed linear subspace of $\mathscr{E}_{L^p}{}^m(\Omega)$.

Note. In the following chapter, we shall see that in general $\mathscr{D}_{L^p}{}^m(\Omega) \neq \mathscr{E}_{L^p}{}^m(\Omega)$, but equality holds when $\Omega = R^n$. In fact, we can prove this as follows. We choose a sequence $\alpha_j \in \mathscr{D}$ satisfying the following condition. For an arbitrary compact set K, there exists j_0, and for $j>j_0$, $\alpha_j(x)\equiv 1$, for all $x \in K$, and then each derivative of $\alpha_j(x)$ is bounded:

$$|\alpha_j(x)|_{\mathscr{B}^m} \leqslant M \qquad (j=1, 2, \ldots).$$

Then, obviously, for $f \in \mathscr{E}_{L^p}{}^m(R^n)$, we see that $\alpha_j f \to f$ in $\mathscr{E}_{L^p}{}^m(R^n)$. On the other hand, each $\alpha_j f$ has a compact support so that, by using Friedrichs' mollifier, we can pick an element of $\mathscr{D}$ which is sufficiently close to $\alpha_j f$ under the topology of $\mathscr{E}_{L^p}{}^m(R^n)$. Therefore, $\varphi_j \in \mathscr{D} \to f$ in $\mathscr{E}_{L^p}{}^m(R^n)$.

Next, we state the results thus obtained in the language of L^2-space.

THEOREM 2.8. Let $f(x),\ g\in\mathscr{D}_{L^2}{}^m(=\mathscr{E}_{L^2}{}^m(R^n))$. Then the following statements are true.

(1) $\mathscr{D}_{L^2}{}^m$ is a Hilbert space, and its inner product is defined as

$$(f, g)_{m, L^2}=\sum_{|\alpha\leqslant m|}(\mathrm{D}^\alpha f(x), \mathrm{D}^\alpha g(x))_{L^2}. \tag{2.23}$$

(2) Let $\rho_\varepsilon(x)$ be a mollifier. We have

$$\rho_\varepsilon * f \to f \quad \text{in} \quad \mathscr{D}_{L^2}{}^m.$$

(3) $\mathscr{D}$ is dense in $\mathscr{D}_{L^2}{}^m$.

(4) (A special case of Sobolev's lemma)

$$\mathscr{B}^0 \supset \mathscr{D}_{L^2}{}^{[\frac{1}{2}n]+1}.$$

That is $f\in\mathscr{D}_{L^2}{}^{[\frac{1}{2}n]+1}$ is bounded and continuous and

$$|f(x)|_0 \equiv \sup|f(x)| \leqslant c(n)\|f(x)\|_{[\frac{1}{2}n]+1, L^2}. \tag{2.24}$$

More generally,

$$\mathscr{B}^p \supset \mathscr{D}_{L^2}{}^{[\frac{1}{2}n]+1+p} \qquad (p=0, 1, 2, \ldots). \tag{2.25}$$

(5) $(\mathrm{D}^\alpha f)\hat{}\,(\xi)=(2\pi\mathrm{i}\xi)^\alpha \hat{f}(\xi) \quad (|\alpha|\leqslant m)$.

(6) $(a(x), f)\to a(x)f$ is a continuous function from $\mathscr{B}^m\times\mathscr{D}_{L^2}{}^m$ to $\mathscr{D}_{L^2}{}^m$. Also, $f(x)\to \mathrm{D}^\alpha f(x)$ is a continuous function from $\mathscr{D}_{L^2}{}^m$ to $\mathscr{D}_{L^2}{}^{m-|\alpha|}$.

Proof We have already proved (1) (3) and (6). For an arbitrary ε, and an arbitrary $K(\supset R^n)$, (2.22) is true, so that for an arbitrary ε, (2.22) is true for all $x\in R^n$. Hence (2) is true. We prove (4).

Considering (3) first we prove (4) for an arbitrary $f\in\mathscr{D}$. From theorem 1.14

$$f(x)=\int \mathrm{e}^{2\pi\mathrm{i}x\xi}\hat{f}(\xi)\,\mathrm{d}\xi; \quad \mathscr{F}[\mathrm{D}^\alpha f(x)]=(2\pi\mathrm{i}\xi)\hat{f}(\xi);$$

$$\|f\|^2_{m, L^2}=\sum_{|\alpha|\leqslant m}\|(2\pi\xi)^\alpha\hat{f}(\xi)\|_{L^2}{}^2. \tag{2.26}$$

So that

$$c_0\|f\|_{m, L^2}\leqslant\|(1+|\xi|)^m\hat{f}(\xi)\|_{L^2}\leqslant c_1\|f\|_{m, L^2}, \tag{2.27}$$

where c_0 and c_1 are positive constant numbers which are fixed by (n, m). Therefore, for $f\in\mathscr{D}$, we consider

$$|f(x)|\leqslant\int|\hat{f}(\xi)|\,\mathrm{d}\xi=\int(1+|\xi|)^{[\frac{1}{2}n]+1}|\hat{f}(\xi)|\cdot(1+|\xi|)^{-[\frac{1}{2}n]-1}\mathrm{d}\xi.$$

By Schwartz inequality, and from (2.27), we see that

$$|f(x)|_0 \leqslant c_1 \left(\int (1+|\xi|)^{-2([\frac{1}{2}n]+1)} \, d\xi\right)^{\frac{1}{2}} \|f\|_{[\frac{1}{2}n]+1, L^2}.$$

Also,

$$2([\tfrac{1}{2}n]+1) \geqslant n+1.$$

Hence, (2.24) is true.

Next, we prove (2.24) is also true when $f \in \mathscr{D}_{L^2}{}^{[\frac{1}{2}n]+1}$. To do this, we note that in the sense of distribution we identify all $f \in L^1{}_{\text{loc}}$ when their values are equal except on sets which have measure 0. We do the same when $f(x) \in \mathscr{D}_{L^2}{}^m$ Therefore when we say $\mathscr{B}^0 \supset \mathscr{D}_{L^2}{}^{[\frac{1}{2}n]+1}$ it means actually that an arbitrary $f \in \mathscr{D}_{L^2}{}^{[\frac{1}{2}n]+1}$ can be made into a bounded continuous function if we alter its values on measure 0 sets.

By (3), for $f(x)$, there exists a sequence $f_j \in \mathscr{D}$, and $f_j(x) \to f(x)$ in $\mathscr{D}_{L^2}{}^{[\frac{1}{2}n]+1}$. By (2.24), $f_j(x)$ is a Cauchy sequence in $\mathscr{B}^0$. Therefore, there exists a limit function $g \in \mathscr{B}^0$ with $f_j(x) \to g(x)$ in $\mathscr{B}^0$. Hence, $f(x)$ and $g(x)$ are equal except on measure 0 sets, i.e. $g = f$ in $\mathscr{D}_{L^2}{}^{[\frac{1}{2}n]+1}$. Hence, if we choose $g(x)$ instead of $f(x)$, and re-write it as $f(x)$, we have

$$|f_j(x) - f(x)|_0 \to 0, \quad \|f_j(x) - f(x)\|_{[\frac{1}{2}n]+1, L^2} \to 0 \quad (j \to \infty).$$

So that, we see that (2.24) is valid for $f(x)$ by applying a limiting process $j \to \infty$ to $f_j(x)$ in (2.24). (2.25) is obvious.

As far as (5) is concerned, we note that it is true for $f \in \mathscr{D}$. As we have done before, we take $f_j \to f$ in $\mathscr{D}_{L^2}{}^m$ for $f_j \in \mathscr{D}$. Then, by Plancherel's theorem (theorem 1.15), we see that $(2\pi i\xi)^\alpha \hat{f}_j(\xi) \to (D^\alpha f)^\wedge(\xi)$ in L^2 $(|\alpha| \leqslant m)$. On the other hand, from $\hat{f}_j(\xi) \to \hat{f}(\xi)$ in L^2, on an arbitrary compact set, we have $(2\pi i\xi)^\alpha \hat{f}_j(\xi) \to (2\pi i\xi)^\alpha \hat{f}(\xi)$ in L^2. Hence, $(D^\alpha f)^\wedge(\xi) = (2\pi i\xi)^\alpha \hat{f}(\xi)$.

Q.E.D.

The converse of the theorem is also true. That is

THEOREM 2.9 *(The converse of theorem 2.8).*

(1) Let $f \in L^2$. If $\rho_\varepsilon * f$ is a convergent sequence by the topology of $\mathscr{D}_{L^2}{}^m$, then $f \in \mathscr{D}_{L^2}{}^m$.

(2) Let $f \in L^2$. If $(1+|\xi|)^m \hat{f} \in L^2$, then $f \in \mathscr{D}_{L^2}{}^m$.

Proof.

(1) We consider a sequence of mollified functions $(\rho_\varepsilon * f)$ in L^2-space. From lemma 1.3, we see that $(\rho_\varepsilon * f) \to F$ as $\varepsilon \to 0$. On the other hand, if we con-

sider the sequence in $\mathscr{D}_{L^2}{}^m$ a subspace of L^2, the sequence converges by the topology of $\mathscr{D}_{L^2}{}^m$. Therefore, from the completeness of $\mathscr{D}_{L^2}{}^m$, the limit function of the sequence also belongs to $\mathscr{D}_{L^2}{}^m$.

(2) $D^\alpha(\rho_\varepsilon * f) = D^\alpha \rho_\varepsilon(x) * f(x)$ is obvious. Taking the Fourier image and considering theorem 1.16, we have

$$\mathscr{F}[(D^\alpha \rho_\varepsilon) * f] = (2\pi i\xi)^\alpha \hat{\rho}(\varepsilon\xi) \hat{f}(\xi).$$

In fact,

$$\hat{\rho}_\varepsilon(\xi) = \left(\frac{1}{\varepsilon}\right)^n \int e^{-2\pi i x\xi} \rho\left(\frac{x}{\varepsilon}\right) dx = \hat{\rho}(\varepsilon\xi).$$

Then, from

$$|\hat{\rho}(\xi)| \leqslant \int \rho(x)\, dx = 1, \quad \hat{\rho}(0) = \int \rho(x)\, dx = 1$$

we have

$$\hat{\rho}(\varepsilon\xi)(2\pi i\xi)^\alpha \hat{f}(\xi) \to (2\pi i\xi)^\alpha \hat{f}(\xi) \quad \text{in} \quad L^2 \quad (|\alpha| \leqslant m),$$

as $\varepsilon \to 0$, i.e. $\rho_\varepsilon * f$ is a convergent sequence in $\mathscr{D}_{L^2}{}^m$. By (1) we see that $f \in \mathscr{D}_{L^2}{}^m$.

Q.E.D.

4 *Structures of* $\mathscr{D}'_{L^2}{}^m(\Omega)$ and $\mathscr{E}'$

In this section we shall examine the structures of the dual spaces of $\mathscr{D}_{L^2}{}^m$ (Ω) and $\mathscr{E}$. Before doing this, we state Riesz theorem in a Hilbert space which we use quite often in what follows. First, we remind the reader of the definition of a Hilbert space.

DEFINITION 2.11 (*Hilbert space*). Let H be a vector space in which a positive Hermitian form, a so-called inner product, is defined. The ***inner product*** satisfies the following conditions A and B:

$$\left.\begin{array}{l}(1)\ (f, g) = \overline{(g, f)},\\ (2)\ (f_1 + f_2, g) = (f_1, g) + (f_2, g);\ (\lambda f, g) = \lambda(f, g),\\ (3)\ (f, f) \geqslant 0. \text{ The quality holds if and only if } f = 0.\end{array}\right\} \qquad \text{(A)}$$

Furthermore,

Let $\|f\| = (f, f)^{\frac{1}{2}}$. Then, $\|f\|$ is a ***norm*** in H. Also, H is complete in regard to the norm now defined. (B)

Then we call H a *Hilbert space*.

Note 1. Obviously, $(f, g_1 + g_2) = (f, g_1) + (f, g_2)$, $(f, \lambda g) = \bar{\lambda}(f, g)$.

Note 2. When we speak about a Hilbert space H, we consider that a certain inner product has been already given. Let us denote another positive Hermitian form as $(f, g)_1$ satisfying the condition A.

By $(f, f)_1$, we denote $\| f \|_1$, a norm which satisfies B. We assume that there exist positive constants c_0 and c_1, and that $c_0\| f \| \leqslant \| f \|_1 \leqslant c_1\| f \|$, is true for $\| f \|$ and $\| f \|_1$, i.e. $\| f \|_1$ is equivalent to $\| f \|$.

Then, because H is also complete in regard to the topology defined by $\| \ \|_1$, the underlying set and its topology by $\| \ \|_1$ are identical with those defined by $\| \ \|$. However, we regard this Hilbert space defined by $\| \ \|_1$ as different from the Hilbert space defined by $\| \ \|$ because of the difference of their inner products. This is significant if we compare the situation when the given spaces are simply Banach spaces. To understand the necessity of distinction, we remind the reader of the definitions of orthogonal functions or symmetric operators in a Hilbert space.

EXAMPLE. An inner product of $\mathscr{D}_{L^2}{}^1(\Omega)$ is given by

$$(f, g)_{1, L^2(\Omega)} = \sum_{i=1}^{n} \left(\frac{\partial f}{\partial x_i}, \frac{\partial g}{\partial x_i}\right)_{L^2(\Omega)} + (f, g)_{L^2(\Omega)}.$$

Also, if a is a bounded real-valued function and $a(x) \geqslant c > 0$, we take

$$(f, g)_1 = \sum_{i=1}^{n} \left(\frac{\partial f}{\partial x_i}, \frac{\partial g}{\partial x_i}\right)_{L^2(\Omega)} + (a(x) f, g)_{L^2(\Omega)} \tag{2.28}$$

as another norm. These norms are equivalent. Note that in the next chapter, when we consider the Dirichlet problem for

$$-\Delta u + a(x)\, u = f(x) \quad \left(\Delta = \sum_{i=1} \frac{\partial^2}{\partial x_i{}^2}\right),$$

we take the norm in $\mathscr{D}_{L^2}{}^1(\Omega)$ as defined in (2.28).

THEOREM 2.10 (*Riesz theorem*). Let H be a Hilbert space. For an arbitrary continuous linear form $u \in H'$, there exists a unique $g \in H$, and for all $f \in H$,

$$\langle u, f \rangle = (f, g). \tag{2.29}$$

Also,

$$\| u \|_{H'} = \| g \|; \quad \| u \|_{H'} = \sup_{\| f \| \leqslant 1} |\langle u, f \rangle|. \tag{2.30}$$

Note. For a fixed $g \in H$, there corresponds a continuous linear form $f \to (f, g)$. The theorem claims that this correspondence exhausts all continuous linear forms which belong to H'.

Proof of theorem 2.10. Let M be a set of all f such that $\langle u, f \rangle = 0$. Then M is a closed subspace of H. Let N be an ortho-complement of M. Then, N is a linear space. In fact, from

$$H = M \oplus N$$

we see that u is a bijection from N into the space of complex numbers when we restrict $\langle u, f \rangle$ to N. Therefore we choose an element $g \in N$ for which (2.29) is valid in case we restrict f to N. To do this, we pick g_0 such that $\|g_0\| = 1$. Put $g = cg_0$, and $f = g_0$. Then from

$$\langle u, g_0 \rangle = (g_0, cg_0) = \bar{c},$$

i.e. $c = \overline{\langle u, g_0 \rangle}$. We see that $g = \overline{\langle u, g_0 \rangle} g_0$ satisfies our requirements. That is, for an arbitrary $f \in N$, (2.29) is valid. The uniqueness of g and (2.30) are obvious.

Q.E.D.

THEOREM 2.11. We write the dual space of $\mathscr{D}_{L^2}{}^m(\Omega)$ as $\mathscr{D}'_{L^2}{}^m(\Omega)$. For $T \in \mathscr{D}'_{L^2}(\Omega)$,

$$T = \sum_{|\alpha| \leq m} D^\alpha f_\alpha(x) \quad (f_\alpha \in L^2(\Omega)). \tag{2.31}$$

Conversely, a distribution T such as (2.31) belongs to $\mathscr{D}'_{L^2}{}^m(\Omega)$.

Proof. By Riesz theorem, for $T \in \mathscr{D}'_{L^2}{}^m(\Omega)$ there exists $g \in \mathscr{D}_{L^2}{}^m(\Omega)$ with

$$\langle T, f \rangle = (f, g)_{m, L^2(\Omega)} = \sum_{|\alpha| \leq m} (D^\alpha f, D^\alpha g)_{L^2(\Omega)}.$$

In particular, if $f \in \mathscr{D}(\Omega)$, then

$$\langle T, f \rangle = \sum_{|\alpha| \leq m} (-1)^{|\alpha|} \langle f(x), D^{2\alpha} \overline{g(x)} \rangle.$$

On the other hand, $\mathscr{D}(\Omega)$ is dense in $\mathscr{D}_{L^2}{}^m(\Omega)$. So that, from

$$T = \sum_{|\alpha| \leq m} (-1)^{|\alpha|} D^{2\alpha} \overline{g(x)}; \quad g \in \mathscr{D}_{L^2}{}^m(\Omega), \tag{2.32}$$

we have

$$(-1)^{|\alpha|} D^\alpha \overline{g(x)} = f_\alpha(x) \in L^2(\Omega).$$

Hence, (2.31) is established.

Conversely, we consider a distribution T in the form of (2.31). Let $\varphi \in \mathscr{D}(\Omega)$. Then,

$$\langle T, \varphi \rangle = \sum_{|\alpha| \leqslant m} \langle D^\alpha f_\alpha(x), \varphi(x) \rangle = \sum_{|\alpha| \leqslant m} (-1)^{|\alpha|} \langle f_\alpha(x), D^\alpha \varphi(x) \rangle.$$

However, for $\varphi \in \mathscr{D}_{L^2}{}^m(\Omega)$, $\mathscr{D}(\Omega)$ is dense in $\mathscr{D}_{L^2}{}^m(\Omega)$, so that we have

$$\langle T, \varphi \rangle = \sum (-1)^{|\alpha|} \langle f_\alpha(x), D^\alpha \varphi(x) \rangle \quad \text{for} \quad \varphi \in \mathscr{D}_{L^2}{}^m(\Omega).$$

It is obvious that this linear form is continuous.

Q.E.D.

Note. (1) $f_\alpha(x)$ in (2.31) is *not* uniquely determined for a given T. We see that, in general, different $f_\alpha(x)$ appear if we calculate T by taking an inner product, while the same $\mathscr{D}'_{L^2}{}^m(\Omega)$ appears if we define a new structure of the given Hilbert space which gives $\mathscr{D}_{L^2}{}^m(\Omega)$ a 'norm' equivalent to the old one. We call (f_α) in the theorem *a normal representation.*

From (2.30) and (2.32), we see that the normal representation $T \to \{f_\alpha(x)\}_{|\alpha| \leqslant m}$ is linear, and also that

$$\|T\|'_{m, L^2(\Omega)} = \left(\sum_{|\alpha| \leqslant m} \|f_\alpha(x)\|^2_{L^2(\Omega)} \right)^{\frac{1}{2}} \tag{2.33}$$

where $\|T\|_{\mathscr{D}'_{L^2}{}^m(\Omega)} \equiv \|T\|'_{m, L^2(\Omega)}$ (we adopt this notation in what follows).

(2) In the proof of theorem 2.11 we saw that the denseness of $\mathscr{D}(\Omega)$ in $\mathscr{D}_{L^2}{}^m(\Omega)$ played an important rôle. In general, we assume that if linear forms T_1 and T_2 each defined on a dense subspace E_1 in a topological vector space E are identical on E_1 and also continuous, then $T_1 = T_2$. The reason is that they have a uniquely extended continuous form $\tilde{T}$. Also, we note that if we replace $\mathscr{D}_{L^2}{}^m(\Omega)$ by $\mathscr{E}_{L^2}{}^m(\Omega)$ then theorem 2.11 is no longer valid. To see this we observe:

EXAMPLE. For simplicity, we write Ω for the interior of a unit circle in a two-dimensional space. Then, we consider a continuous physical distribution $\mu(s)$ on the boundary $\dot{\Omega}$ (circle) of Ω. On the other hand, we write $\varphi(s)$ for the boundary value (its existence will be shown in the next chapter) of

$\varphi \in \mathscr{E}_{L^2}{}^1(\Omega)$. Then, we see that

$$\varphi \in L^2(\dot{\Omega}); \quad \|\varphi(s)\|_{L^2(\dot{\Omega})} \leqslant c\|\varphi(x)\|_{1, L^2(\Omega)}.$$

So that a linear form

$$\varphi \in \mathscr{E}_{L^2}{}^1(\Omega) \to \int_{\dot{\Omega}} \varphi(s)\, \mu(s)\, ds$$

is continuous, but it is not representable in the form which we have established in theorem 2.11. In fact, the linear form is 0 on $\mathscr{D}(\Omega)$.

Relationship between the convergence of a sequence of distributions and of a weakly convergent sequence in $\mathscr{D}_{L^2}{}^m$

We have given a rather detailed account of $(\mathscr{D}'_{L^2}{}^m)$ or $\mathscr{D}'_{L^2}{}^m(\Omega)$. One of the purposes of doing this is so that when we consider the Fourier transform of a distribution, we can regard it as a natural generalization of Plancherel's theorem starting from a concrete representation. In fact, historically, Bochner first initiated research in this direction; he extended the notion of function itself. In relation to this problem a natural question arose; how do we represent a distribution in analytical terms?

We shall examine this problem. Furthermore, we shall discuss the convergence of a sequence of distributions. (The convergence is defined by a simple topology such as we have given in definition 2.4). Historically, a weak convergence was already regarded as an important notion by Hilbert himself when he recognized a so-called Hilbert space; it is a very powerful notion when we study a Hilbert space analytically (we see this in the following theorem 2.12).

DEFINITION 2.12. By saying that a point-sequence $\{f_j\}$ in a Hilbert space H is a *weakly convergent sequence*, we mean that for an arbitrary $g \in H$ there exists $\lim_{j\to\infty}(f_j, g)$. From Riesz theorem, this is nothing but a convergent sequence by the simple topology in F-space because a Hilbert space is a Fréchet space. Therefore, the results obtained in section 2 are immediately available.

We state some of the consequences.

PROPOSITION 2.5. Let $\{f_j\}$ be a weakly convergent sequence in a Hilbert space H. Then the following statements are true.

(1) $\|f_j\|$ is bounded.

(2) $\{f_j\}$ is uniformly convergent on an arbitrary compact set A of H, i.e.

$$\sup_{g\in A}|(f_p-f_q, g)|\to 0 \quad (p, q\to\infty).$$

(3) There exists a uniquely determined f in H, and for an arbitrary $g\in H$,

$$\lim_{j\to\infty}(f_j, g)=(f, g).$$

(4)

$$\|f\|\leqslant\underline{\lim}_{j\to\infty}\|f_j\|. \tag{2.34}$$

Proof. By Riesz theorem (2.10), there exists an anti-linear isomorphism between H and H'. It is anti-linear because, if for $u\in H'$, there is a corresponding $g\in H$, then λu corresponds to $\bar{\lambda}g$. Therefore, $f_j\in H$ corresponds to $u_j\in H'$, and

$$\left.\begin{aligned}&\text{For all } g\in H, (g, f_j)=\langle u_j, g\rangle;\\&\|u_j\|_{H'}=\|f_j\|_H.\end{aligned}\right\} \tag{2.35}$$

We see that (1) is lemma 2.5 itself, and (2) and (3) are just the Banach–Steinhaus theorem (2.4). We prove (4). Let

$$\underline{\lim}_{j\to\infty}\|f_j\|=\alpha.$$

If necessary we pick a subsequence $\{f_{j_p}\}$ from $\{f_j\}$ and put $\lim_{p\to\infty}\|f_{j_p}\|=\alpha$ where $\{f_{j_p}\}$ is of course a weakly convergent sequence which tends to f. We rewrite f_{j_p} as f_j. Now, we put $\lim_{j\to\infty}\|f_j\|=\alpha$. Also, we write $u\in H'$, which corresponds to $f\in H$. We take $g\in H$ as satisfying $\|g\|<1$; then

$$|\langle u, g\rangle|=\overline{\lim_{j\to\infty}}|\langle u_j, g\rangle|\leqslant\lim_{j\to\infty}\|u_j\|_{H'}=\overline{\lim_{j\to\infty}}\|f_j\|=a$$

Because $\|g\|<1$, $\|u\|_{H'}\leqslant\alpha$, hence, $\|f\|=\|u\|_{H'}\leqslant\alpha$.

Q.E.D.

THEOREM 2.12 *(Weakly compact in a Hilbert space).* Let B be a bounded set in a Hilbert space H. There exists an appropriate subsequence of an arbitrary sequence in B which is weakly convergent.

Proof. This is an obvious extension of the Ascoli–Arzela theorem to a Hilbert space. Nevertheless we shall give a proof. Let M be a closed subspace which is determined by $\{f_j\}$, i.e. M is a closure of a subspace which is spanned by every finite linear combination of elements taken from the set $\{f_j\}$. Then, obviously, M is separable. Therefore, if we pick an appropriate denumerable set of points $\{g_k\}$ becomes a dense set in M.

We consider M (a Hilbert space) instead of H. We choose an appropriate subsequence of $\{f_j\}$ as follows.

First, we choose a subsequence

$$f_{11}, f_{12}, \ldots, f_{1j}, \ldots$$

such that each $\{f_j, g_1\}$ becomes a convergent sequence. Secondly, we choose a subsequence $\{f_{2j}\}\, j=1,2,\ldots$ of $\{f_{1j}\}$ such that each $\{f_j, g_2\}_{j=1,2,\ldots}$ becomes a convergent sequence, and so on. Then we pick their diagonal elements $f_{11}, f_{22}, \ldots, f_{jj}, \ldots$. We see now that this sequence of the diagonal elements is convergent on an arbitrary g_k, i.e. $\lim_{j\to\infty} \{f_{jj}, g_k\}$ exists. On the other hand, the set $\{f_{jj}\}$ is bounded. Therefore, it is an equi-continuous set. (See lemma 2.5 and the following note). Also, the set $\{g_k\}$ is dense in M. Therefore, for an arbitrary $g \in M$, there exists $\lim_{j\to\infty}\{f_{jj}, g\}$. Now, for an arbitrary $g \in H$, $g = g_1 + g_2$, $g_1 \in M$, $g_2 \in N$ (the ortho-complementary space of M), and

$$\{f_{jj}, g\} = \{f_{jj}, g_1\} + \{f_{jj}, g_2\} = \{f_{jj}, g_1\} \quad (g_1 \in M).$$

So that $\{f_{jj}\}\, j=1, 2,\ldots$ is weakly convergent as $j \to \infty$.

Q.E.D.

Note. The notion 'weakly convergent sequence' is invariant with respect to the structure of a Hilbert space. In other words, a weakly convergent sequence with respect to $\{f, g\}_1$, a positive Hermitian form which defines an equivalent norm (see note 2 of definition 2.11), and the original weakly convergent sequence coincide.

We affirm the following proposition 2.6 about a weakly convergent sequence in $\mathscr{E}_{L^2}{}^m(\Omega)$. Of course, it is also true in $\mathscr{D}_{L^2}{}^m(\Omega)$.

PROPOSITION 2.6. If a sequence $\{f_j(x)\}$ is weakly convergent to $f(x)$ in $\mathscr{E}_{L^2}{}^m(\Omega)$ by its weak topology, then, for an arbitrary $\alpha \in L^2(\Omega)$,

$$\lim_{j\to\infty} (D^\alpha f_j(x), \varphi(x))_{L^2(\Omega)} = (D^\alpha f(x), \varphi(x))_{L^2(\Omega)} \quad (|\alpha| \leqslant m). \tag{2.36}$$

Conversely, if for an arbitrary $\varphi \in L^2(\Omega)$

$$\lim_{j\to\infty} (D^\alpha f_j(x), \varphi(x))_{L^2(\Omega)} = (f_\alpha(x), \varphi(x))_{L^2(\Omega)} \quad (|\alpha| \leqslant m), \tag{2.37}$$

then $f_\alpha(x) = D^\alpha f(x) (|\alpha| \leqslant m)$ and $\{f_j(x)\}$ is weakly convergent to $f(x)$ in the weak topology of $\mathscr{E}_{L^2}{}^m(\Omega)$.

Proof. Let us prove (2.36). For an arbitrary $\varphi \in L^2(\Omega)$ we see that the mapping $f \in \mathscr{E}_{L^2}{}^m(\Omega) \to (D^\alpha f(x), \varphi(x))_{L^2(\Omega)}$ is a continuous linear form. Therefore, by Riesz theorem, there exists $\psi \in \mathscr{E}_{L^2}{}^m(\Omega)$ with

$$(D^\alpha f(x), \varphi(x))_{L^2(\Omega)} = (f(x), \psi(x))_{m, L^2(\Omega)}. \tag{2.38}$$

Hence,

$$\lim_{j\to\infty} (D^\alpha f_j(x), \varphi(x))_{L^2(\Omega)} = \lim_{j\to\infty} (f_j(x), \psi(x))_{m, L^2(\Omega)} = (f(x), \psi(x))_{m, L^2(\Omega)}$$
$$= (D^\alpha f(x), \varphi(x))_{L^2(\Omega)}.$$

Conversely, let (2.37) be true. First we note that $(f_j(x), \varphi(x))_{m, L^2(\Omega)} = \sum_{|\alpha| \leqslant m} (D^\alpha f_j(x), D^\alpha \varphi(x))_{L^2(\Omega)}$ so that $\{f_j\}$ is a weakly convergent sequence in $\mathscr{E}_{L^2}{}^m(\Omega)$. We write the limit of $\{f_j\}$ as $f \in \mathscr{E}_{L^2}{}^m(\Omega)$ and see that (2.36) is true. With regard to (2.37), we have $f_\alpha(x) = D^\alpha f(x)$.

Q.E.D.

The following proposition 2.7 is obtained from the fact that $\mathscr{D}(\Omega)$ is dense in $\mathscr{D}_{L^2}{}^m(\Omega)$.

PROPOSITION 2.7 *(Weakly convergent sequence in $\mathscr{D}_{L^2}{}^m(\Omega)$).* The necessary and sufficient conditions for $\{f_j(x)\}$ to converge to 0 are:

(1) $\|f_j(x)\|_{m, L^2(\Omega)}$ is bounded, and

(2) For an arbitrary $\varphi \in \mathscr{D}(\Omega)$, $\lim_{j\to\infty} \langle f_j(x), \varphi(x) \rangle = 0$.

Proof (Necessity). This follows immediately after proposition 2.5(1) and (2.36).

(Sufficiency). We note that $\mathscr{D}(\Omega)$ is dense in $\mathscr{D}_{L^2}{}^m(\Omega)$.

From (1) we see that $\{f_j\}$ is an equi-continuous set. Using the last half of proposition 2.6, we assert the proposition.

Q.E.D.

Note. We can paraphrase the condition of the proposition as 'f_j is bounded in $\mathscr{D}_{L^2}{}^m(\Omega)$ and tends to 0 as $j\to\infty$ when we regard it as a sequence of distributions.'

Incidentally, if we drop the condition (1), then, in general, $\{f_j\}$ is no longer a weakly convergent sequence in $\mathscr{D}_{L^2}{}^m(\Omega)$ by the only remaining condition (2).

EXAMPLE. Let $\Omega=(-\infty,+\infty)$ and $f_j(x)=j\alpha(x)\sin jx (j=1,2,\ldots)$, where $\alpha\in\mathscr{D}$ and $\alpha(x)$ is 1 throughout the interval $I=[0,\pi]$. Then $\{f_j(x)\}\to 0$ in $\mathscr{D}(\Omega)$ but not in $L^2(\Omega)$. In fact, for $\varphi(\in\mathscr{D}(\Omega))$,

$$(f_j,\varphi)_{L^2}=j\int_{-\infty}^{+\infty}\alpha(x)\overline{\varphi(x)}\sin jx\, dx$$
$$=+\int_{-\infty}^{+\infty}\cos jx\cdot[\alpha(x)\overline{\varphi(x)}]'\, dx\to 0 \quad (j\to\infty).$$

If, $\psi_I(x)$ is the characteristic function on I, then

$$(f_j,\psi_I)_{L^2}=j\int_0^{\pi}\sin jx\, dx=1-\cos j\pi \quad (j=1,2,\ldots).$$

So that this will not converge to 0.

We have so far considered weakly convergent sequences in $\mathscr{D}_{L^2}{}^m(\Omega)$ and $\mathscr{E}_{L^2}{}^m(\Omega)$. The basic idea of our study lies in the fact that, for example, in $\mathscr{D}_{L^2}{}^m(\Omega)$.

$$\left.\begin{array}{l}\text{The condition that } \{u_j\}\to 0 \text{ by a simple topology of } \mathscr{D}'_{L^2}{}^m(\Omega),\\ \text{and the condition that } \{f_j(x)\}\to 0 \text{ by a weak topology } \mathscr{D}_{L^2}{}^m(\Omega)\\ \text{via Riesz condition } \langle u_j,\varphi\rangle=(\varphi,f_j)_{m,L^2(\Omega)} \text{ are equivalent.}\end{array}\right\} \quad (2.39)$$

We return briefly to $\mathscr{D}'_{L^2}{}^m(\Omega)$ to state some facts about this space.

PROPOSITION 2.8A. Let $\{u_j\}$ be a convergent sequence under a simple topology of $\mathscr{D}'_{L^2}{}^m(\Omega)$.

(1) $\|u_j\|'_{m,L^2(\Omega)}$ is bounded. There exists a unique $u\in\mathscr{D}'_{L^2}{}^m(\Omega)$, and $\{u_j\}\to u$ by a simple topology of $\mathscr{D}'_{L^2}{}^m(\Omega)$.

(2) $\{u_j\}$ converges uniformly on an arbitrary compact A in $\mathscr{D}_{L^2}{}^m(\Omega)$ i.e.

$$\sup_{f\in A}|\langle u_j-u,f\rangle|\to 0 \quad (j\to\infty).$$

PROPOSITION 2.8B. The necessary and sufficient conditions for $\{u_j\}$ to be a convergent sequence in the simple topology of $\mathscr{D}'_{L^2}{}^m(\Omega)$ are

$$\left.\begin{aligned} &\|u_j\|'_{m,L^2(\Omega)} \text{ is bounded, and for an arbitrary} \\ &\varphi\in\mathscr{D},\ \lim_{j\to\infty}\langle u_j, \varphi\rangle \text{ exists.} \end{aligned}\right\} \quad (2.40)$$

(The proof is left to the reader).

$\mathscr{E}'$-space

Let us generalize the idea 'support of a function' to 'support of a distribution'. Suppose T, a distribution, is given. Let x_0 be a point in R^n, and let V be a neighbourhood of x_0. We consider the whole set of x_0 such that, for an arbitrary small neighbourhood of x_0, there exists a function which has its support in V and $\langle T, \varphi\rangle \neq 0$.

We call this set the *support of T*. We write supp$[T]$ in this case. If T is a function this definition coincides with the previously given definition (section 7, chapter 1). It is obvious that supp$[T]$ is not empty if $T\neq 0$ and also the support of T is a closed set. If the support of φ $(\in\mathscr{D})$ and the support of T have no point in common, then $\langle T, \varphi\rangle = 0$. We write a set of distributions whose supports are compact as $\mathscr{E}'$. In the next paragraph we see that this notation is justified when we define a topological vector space $\mathscr{E}$ (we write some-times $\mathscr{E}^\infty$). The dual space of $\mathscr{E}$ coincides with the space of distributions whose supports are compact.

$\mathscr{E}$-space

We write $\mathscr{E}$ for the whole set of functions φ which are differentiable an infinite number of times. For this set we obtain a denumerable set of semi-norms as follows. Let $K_1\subset K_2\subset K_3\subset\cdots\to R^n$ be a sequence of compact sets such that there exists a K_j in the sequence which contains any compact set in R^n. Then we define

$$p_{n,m}(\varphi)=\sum_{|\alpha|\leqslant m}\sup_{x\in K_n}|D^\alpha\varphi(x)| \qquad (n=1,2,\ldots;\ m=0,1,2,\ldots).$$

A topology thus obtained by semi-norms has completeness. Therefore $\mathscr{E}$ is in fact a Fréchet space. In other words, the topology given to $\mathscr{E}$ is defined as

$$\left.\begin{aligned} &\varphi_j(x)\to 0 \text{ in } \mathscr{E} \text{ if and only if on arbitrary compact set in } R^n \\ &\{\varphi_i(x)\}\to 0 \text{ uniformly, including all their derivatives.} \end{aligned}\right\} \quad (2.41)$$

Note. Concerning the notation: C^∞ is widely used to represent the set of C^∞-

functions. The difference between $\mathscr{E}$ and C^∞ is that C^∞ usually has no assigned topologies in contrast to $\mathscr{E}$. However, for an individual function, we can write either a C^∞-function or a $\mathscr{E}$-function; both have the same meaning.

LEMMA 2.11. A continuous linear form on $\mathscr{E}$ is a distribution having a compact support. The converse of the statement is also true.

Proof. The meaning of this lemma is as follows. Because $\mathscr{D}$ is dense in $\mathscr{E}$ (the proof was essentially given at the note after definition 2.10) a linear form on $\mathscr{D}$ which is continuous by the topology of $\mathscr{E}$ (given above) and a continuous linear form on $\mathscr{E}$ can be regarded as the same. From this point of view we see that $\mathscr{E}' \subset \mathscr{D}'$. On the other hand, if the support of $T(\in \mathscr{D}')$ is not bounded, then there exists a sequence $\{x_\nu\}$ and $x_\nu \in \operatorname{supp}[T]$, $|x_\nu| \geqslant \nu$. Also, there exists $\varphi_\nu(\in \mathscr{D})$ such that the support of φ_ν is contained in the ball having unit radius centre x_ν, and $|\langle T, \varphi_\nu \rangle| \geqslant 1$.

Obviously, $\varphi_\nu(x) \to 0$ in $\mathscr{E}$. Therefore, T is *not* continuous by the topology of $\mathscr{E}$. Hence $\mathscr{E}'$ must be a space of distributions having compact supports. Conversely, such a distribution is a continuous linear form by the topology of $\mathscr{E}$.

Q.E.D.

The extension of the domain of distributions

Consider $\alpha(\in \mathscr{D})$ which is identically equal to 1 in the neighbourhood of the support of $T(\in \mathscr{E}')$. For $\varphi \in \mathscr{E}$, we put

$$\langle T, \varphi \rangle = \langle T, \alpha\varphi \rangle$$

to define the left hand term by the right hand term. This is essentially the same extension of the domain of a distribution which we made during the proof of the previous lemma. In fact, if we consider β satisfying the same conditions as α, then $\langle T, (\alpha-\beta)\varphi \rangle = 0$ because,

$$\operatorname{supp}[T] \cap \operatorname{supp}[(\alpha-\beta)\varphi] = \emptyset .$$

In general, for an arbitrary distribution T, we can extend the domain of the definition according to the nature of the support of T. Let A be a closed set which is not necessarily compact. On the other hand, we assume that $K \equiv \operatorname{supp}[T] \cap A$ is a compact set. Then, we pick an arbitrary $\alpha(\in \mathscr{D})$ which is identically equal to 1 in a neighbourhood of K. We denote by $\mathscr{D}_A$ a space which is defined as the set of C^∞-functions having their supports contained in A. For an arbitrary $\varphi \in \mathscr{D}_A$, we define

$$\langle T, \varphi\rangle = \langle T, \alpha\varphi\rangle .$$

We see that this is the definition of a continuous linear form on $\mathscr{D}_A$, a subspace of $\mathscr{E}$, and a unique extension of the domain of T which is $\mathscr{D} \cap \mathscr{D}_A$.

The following proposition establishes some relations between $\mathscr{E}'$ and $\mathscr{D}'_{L^2}{}^m$.

PROPOSITION 2.9

(1) Let B be a bounded set of $\mathscr{E}'$. For a certain m, B is bounded in $\mathscr{D}'_{L^2}{}^m$, i.e. $\sup_{T\in B} \|T\|'_{m,L^2} < +\infty$.

(2) Let $T_j \to T$ in $\mathscr{E}'$. For a certain m, $T_j, T \in \mathscr{D}'_{L^2}{}^m$ and $T_j \to T (j \to \infty)$ under a simple topology of $\mathscr{D}'_{L^2}{}^m$.

Proof.

(1) $\mathscr{E}$ is a Fréchet space, so that B is an equi-continuous set of $\mathscr{E}'$ (lemma 2.5). Therefore, for a certain V, a neighbourhood of 0 in $\mathscr{E}$,

$$\sup_{T\in B,\, \varphi\in V} |\langle T, \varphi\rangle| \leqslant 1 ,$$

where

$$V = \{\varphi;\ \sum_{|\alpha|\leqslant p}\ \sup_{x\in K} |D^\alpha \varphi(x)| < \eta\} .$$

By Sobolev's lemma (2.25), for a sufficiently small η' (>0), if

$$\|\varphi\|_{[\frac{1}{2}n]+1+p,\, L^2} < \eta'$$

then $\varphi \in V$. Therefore, for all $T \in B$, if $\|\varphi\|_{[\frac{1}{2}n]+1+p,\,L^2} < 1$, then $|\langle T, \varphi\rangle| < C$, i.e.

$$T \in \mathscr{D}'_{L^2}{}^m,\ \sup_{T\in B} \|T\|'_{m,L^2} \leqslant C \qquad (m = [\tfrac{1}{2}n] + 1 + p) .$$

(2) T_j is a bounded set of $\mathscr{E}'$ (lemma 2.5) so that from the result which we have now obtained T_j is a bounded set of $\mathscr{D}'_{L^2}{}^m$.

Therefore, by part B of 2.8, we obtain the proposition. We note that because a bounded closed set of $\mathscr{E}$ is compact for an arbitrary sequence, convergence by the simple topology of $\mathscr{E}'$ and convergence by the strong topology of $\mathscr{E}'$ coincide (theorem 2.3).

Q.E.D.

Note. The converse of each part (1) and (2) of the above proposition is *not* true as it was stated. However, if we impose a condition on (1) (or(2)):

$$\text{there exists a fixed compact set } K \text{ and for all } T \in B, \operatorname{supp}[T] \subset K \qquad (2.42)$$

then its converse is also true. The fact that for a bounded set of $\mathscr{E}'$, the supports of distributions belonging to it are uniformly bounded can be similarly proved as we have done in lemma 2.11. In fact, if we assume that there is no such K, we see that B does not become an equi-continuous set of $\mathscr{E}'$.

THEOREM 2.13 *(Representation of $\mathscr{E}'$).*

(1) Let B be a bounded set of $\mathscr{E}'$. For a certain m, and an arbitrary $T(\in B)$, we have

$$T = \sum_{|\alpha| \leqslant m} D^\alpha f_\alpha(x) \quad (f_\alpha \in L^2).$$

When T runs through B, we can make $\sum_{|\alpha| \leqslant m} \|f_\alpha(x)\|_{L^2}^2$ bounded. At the same time, $\{f_\alpha(x)\}_{|\alpha| \leqslant m}$ have their supports in an arbitrarily fixed neighbourhood Ω of the compact set K which we have mentioned in the condition (2.42).

(2) Let $T_j \to T$ in $\mathscr{E}'$. For a certain m,

$$T_j = \sum_{|\alpha| \leqslant m} D^\alpha f_\alpha^{(j)}(x); \quad T = \sum_{|\alpha| \leqslant m} D^\alpha f_\alpha(x).$$

Also, for each α, we have $f_\alpha^{(j)}(x) \to f_\alpha(x)$ $(j \to \infty)$, which is valid in the weak topology of L^2 where

$$\operatorname{supp}[f_\alpha^{(j)}(x)], \operatorname{supp}[f_\alpha(x)] \subset \Omega.$$

Note.

(1) The representations of T and T_j are not unique, they are dependent on m and Ω.

(2) We can regard the above representation theorem of $\mathscr{E}'$ as a local representation theorem of distributions in general. We see the reason for this in what follows. Let ω be an open set such that $\bar{\omega}$ is compact. Consider T which is restricted to ω. Let $\alpha \in \mathscr{D}$ such that it is identically 1 in the neighbourhood of $\bar{\omega}$. Then, for $\varphi(\in \mathscr{D}(\omega))$,

$$\langle T, \varphi \rangle = \langle \alpha T, \varphi \rangle$$

and $\alpha T \in \mathscr{E}'$.

Proof of theorem 2.13. The theorem is essentially the summary of results already obtained. Nevertheless, we set out the proof in detail for the convenience of the reader.

(1) By proposition 2.9, B is a bounded set of a certain $\mathscr{D}'_{L^2}{}^m$. By Riesz's theorem, there is a bijection:

$$T \in \mathscr{D}'_{L^2}{}^m \leftrightarrow g \in \mathscr{D}_{L^2}{}^m.$$

Also

$$\|T\|'_{m, L^2} = \|g(x)\|_{m, L^2};$$

we note that

$$\langle T, \varphi \rangle = (\varphi, g)_{m, L^2} = \sum_{|\alpha| \leqslant m} (D^\alpha \varphi, D^\alpha g)_{L^2}.$$

Because $\mathscr{D}$ is dense in $\mathscr{D}_{L^2}{}^m$, we have that this is equal to

$$\sum_{|\alpha| \leqslant m} \langle (-1)^{|\alpha|} D^\alpha \cdot D^\alpha \overline{g(x)}, \varphi(x) \rangle.$$

Therefore, if we put

$$f_\alpha(x) = (-1)^{|\alpha|} D^\alpha \overline{g(x)} \in L^2 \tag{2.43}$$

(See (2.32)), then

$$T = \sum_{|\alpha| \leqslant m} D^\alpha f_\alpha(x).$$

From $\|T\|'_{m, L^2} = \|g(x)\|_{m, L^2}$ and (2.43), we see that the boundedness of B is equivalent to the boundedness of $\|f_\alpha(x)\|_{L^2}$ for each α when T runs through the whole B.

We shall derive an expression for T with a restricted support. Let Ω be a fixed neighbourhood of K, and let $\beta(\in \mathscr{D}(\Omega))$ be a C^∞-function which is identically equal to 1 in the neighbourhood of K. For $T(\in B)$, we have $\langle T, \varphi \rangle = \langle T, \beta\varphi \rangle$.
Therefore

$$\langle T, \varphi \rangle = \langle T, \beta\varphi \rangle = \sum_{|\alpha| \leqslant m} (-1)^{|\alpha|} \langle f_\alpha(x), D^\alpha(\beta\varphi) \rangle. \tag{2.44}$$

By Leibniz' formula we have

$$D^\alpha(\beta\varphi) = \sum_\gamma C_\gamma^\alpha \, D^{\alpha-\gamma}\beta \cdot D^\gamma \varphi.$$

Therefore

$$(-1)^{|\alpha|}\langle f_\alpha(x), \mathrm{D}^\alpha(\beta\varphi)\rangle=(-1)^{|\alpha|}\sum_\gamma \langle \mathrm{D}^{\alpha-\gamma}\beta(x)f_\alpha(x), \mathrm{D}^\gamma\varphi(x)\rangle$$
$$=\sum_\gamma (-1)^{|\alpha|+|\gamma|}C_\gamma^\alpha\langle \mathrm{D}^\gamma(\mathrm{D}^{\alpha-\gamma}\beta\cdot f_\alpha), \varphi\rangle.$$

We put

$$\tilde{f}_\gamma(x)=\sum_\alpha (-1)^{|\alpha|+|\gamma|}C_\gamma^\alpha\, \mathrm{D}^{\alpha-\gamma}\beta(x)f_\alpha(x)\in L^2, \tag{2.45}$$

so that

$$T=\sum_{|\gamma|\leqslant m} \mathrm{D}^\gamma\tilde{f}_\gamma(x), \tag{2.46}$$

where the support of $\tilde{f}_\gamma$ is contained in Ω. It is obious that $\|\tilde{f}_\gamma(x)\|_{L^2}$ is bounded for each γ when T runs through B (see (2.45)).

(2) By (2) of the previous proposition, $\{T_j\}\to T$ is true in the simple topology of $\mathscr{D}'_{L^2}{}^m$. By Riesz theorem, we have

$$\langle T_j, \varphi\rangle=(\varphi, g_j)_{m, L^2}\to(\varphi, g)_{m, L^2} \quad (j\to\infty).$$

Therefore, in the weak topology of $\mathscr{D}_{L^2}{}^m$ $\{g_j(x)\}\to g(x)$. On the other hand, by proposition 2.6, $\mathrm{D}^\alpha g_j(x)\to \mathrm{D}^\alpha g(x)$ is valid for $|\alpha|\leqslant m$ in the weak topology of L^2. Therefore, $f_\alpha^{(j)}(x)$ defined as (2.46) is convergent as $j\to 0$ in the weak topology of L^2. So that, observing the representation of T_j defined by (2.46), $\tilde{f}_\gamma^{(j)}(x)$ is also convergent to $\tilde{f}_\gamma(x)$ as $j\to\infty$ in the weak topology of L^2.

Q.E.D.

We give some miscellaneous results and a remark before embarking on the next section.

COROLLARY

(1) $$\mathscr{D}\subset\mathscr{D}_{L^2}{}^\infty\subset\cdots\subset\mathscr{D}_{L^2}{}^{m+1}\subset\mathscr{D}_{L^2}{}^m\subset\cdots\subset\mathscr{D}_{L^2}{}^1\subset\mathscr{D}_{L^2}{}^0(=L^2)$$
$$\subset\mathscr{D}'_{L^2}{}^1\subset\cdots\subset\mathscr{D}'_{L^2}{}^m\subset\mathscr{D}'_{L^2}{}^{m+1}\subset\cdots\subset\mathscr{D}'_{L^2}\subset\mathscr{D}'$$

where

$$\mathscr{D}_{L^2}{}^\infty=\bigcap_{m=1}^\infty \mathscr{D}_{L^2}{}^m; \quad \mathscr{D}'_{L^2}=\bigcup_{m=1}^\infty \mathscr{D}'_{L^2}{}^m.$$

$\mathscr{D}$ is dense in any of these spaces which contain $\mathscr{D}$.

(2) $\mathscr{E}'\subset\mathscr{D}'_{L^2}$ (this follows from lemma 2.9).

(3) The Dirac measure $\delta\in\mathscr{D}'_{L^2}{}^{[\frac{1}{2}n]+1}$ (this follows from Sobolev's lemma (2.14)).

(4) $\mathscr{D}'_{L^2}{}^m$ contains neither the function which is identically 1 nor any polynomial $P(x)$.

Proof of (4) *(By contradiction)*. Assume that there exists m for which $1\in\mathscr{D}'_{L^2}{}^m$. Let $\{\alpha_j\}(\alpha_j(x)\geqslant 0)$ be a sequence of C^∞-functions satisfying

$$\alpha_j(x)=\begin{cases}1 & (|x|\leqslant j)\\ 0 & (|x|\geqslant j+1);\end{cases} \qquad |\mathrm{D}^\alpha\alpha_j(x)|\leqslant M_\alpha \quad (j=1,2,\ldots). \tag{2.47}$$

Under this condition, it is easy to see that if $T\in\mathscr{D}'_{L^2}{}^m$, then $\alpha_jT\to T$ under the simple topology of $\mathscr{D}'_{L^2}{}^m$. On the other hand, if we put

$$\varphi(x)=\frac{1}{(1+|x|^2)^{(\frac{1}{4}n+\varepsilon)}} \qquad (0<\varepsilon<1),$$

then, $\varphi\in\mathscr{D}_{L^2}{}^\infty$, i.e. $\mathrm{D}^\alpha\varphi\in L^2$, $|\alpha|\geqslant 0$. Therefore,

$$\langle 1,\varphi\rangle=\lim_{j\to\infty}\langle\alpha_j(x)1,\varphi\rangle=\lim_{j\to\infty}\int\frac{\alpha_j(x)}{(1+r^2)^{\frac{1}{4}(n+\varepsilon)}}\,\mathrm{d}x\,.$$

The right hand side is obviously divergent. In fact, $\alpha_j\geqslant 0$, $\varphi(x)\geqslant 0$, $\varphi\notin L^1$, so that we reach a contradiction. Hence, $P\notin\mathscr{D}'_{L^2}{}^m$, because if we consider $P(x)\not\equiv 0$, $P\in\mathscr{D}'_{L^2}{}^m$ $\deg P(x)=p$; and its pth order derivative of $\mathrm{D}^\alpha P\in\mathscr{D}'_{L^2}{}^{m+p}$ then there exists α, $|\alpha|=p$ and $\mathrm{D}^\alpha P(x)=c_\alpha\neq 0$, so that $1\in\mathscr{D}'_{L^2}{}^{m+p}$, which is a contradiction.

Q.E.D.

The above argument shows that $\mathscr{D}'_{L^2}{}^m$ is too small to apply Fourier transforms.

Note.[†] In the previous corollary we have considered $\mathscr{D}_{L^2}{}^\infty$ (we write also $\mathscr{D}_{L^2}$). The topology considered on $\mathscr{D}_{L^2}{}^\infty$ was given by semi-norms:

$$p_m(f)=\sum_{|\alpha|\leqslant m}\|\mathrm{D}^\alpha f(x)\|_{L^2} \qquad (m=1,2,\ldots).$$

Then the completeness of the space is clear. Therefore, $\mathscr{D}_{L^2}{}^\infty$ is a Fréchet space.

It is also obvious that $\mathscr{D}'_{L^2}$ is the dual space of $\mathscr{D}_{L^2}{}^\infty$. Next, we shall show that $\mathscr{D}$ is dense in $\mathscr{D}'_{L^2}{}^m$. To see this, we consider the representation of

[†] This note is not directly connected with the following discussion and can therefore be disregarded.

$T\in\mathscr{D}'_{L^2}{}^m$ following theorem 2.13. If we substitute $f_\alpha(\in L^2)$ by $\varphi_\alpha(\in\mathscr{D})$, we have

$$T'=\sum_{|\alpha|\leqslant m} D^\alpha\varphi_\alpha\in\mathscr{D}.$$

From this it follows that

$$\|T-T'\|^{2\prime}_{m,L^2}\leqslant\sum_{|\alpha|\leqslant m}\|f_\alpha(x)-\varphi_\alpha(x)\|_{L^2}{}^2.$$

We see that the expression on the right hand side can be taken as small as we wish. Therefore, $\mathscr{D}$ is dense in $\mathscr{D}'_{L^2}{}^m$.

Finally, we show $\mathscr{D}$ *is dense in* $\mathscr{D}'$. *To do this we first define a Friedrichs* mollifier $\rho_\varepsilon(x)*$ by generalizing the idea of the convolution of two functions to that of two distributions: for an arbitrary $T\in\mathscr{D}'$,

$$(\rho_\varepsilon * T)(x)=\langle T_t, \rho_\varepsilon(x-t)\rangle.$$

So that we write the right hand term as $\theta_\varepsilon(x)$. Then $\theta_\varepsilon(x)$ is obviously a C^∞-function. Now, let $\varphi\in\mathscr{D}$. We see that

$$\langle\theta_\varepsilon(x),\varphi(x)\rangle=\int\langle T_t,\rho_\varepsilon(x-t)\rangle\,\varphi(x)\,dx=\langle T_t,\int\rho_\varepsilon(x-t)\,\varphi(x)\,dx\rangle.$$

Consider the family of functions $\int\rho_\varepsilon(x-t)\varphi(x)dx$, depending on a parameter $\varepsilon(>0)$. It has a support which is contained in an ε-neighbourhood of supp $[\varphi]$. Therefore, if we write K for the ε_0-neighbourhood of supp $[\varphi]$, then we have

$$\int\rho_\varepsilon(x-t)\,\varphi(x)\,dx\to\varphi(t)\quad\text{in}\quad\mathscr{D}_K$$

for $\varepsilon<\varepsilon_0$ as $\varepsilon\to0$. Hence, $\langle\theta_\varepsilon(x),\varphi(x)\rangle\to\langle T,\varphi(x)\rangle$, i.e.

$$(\rho_\varepsilon * T)(x)\to T\quad\text{in}\quad\mathscr{D}'. \tag{2.48}$$

We choose $\varepsilon_1>\varepsilon_2>\cdots\to0$ and let $\{\alpha_j(x)\}$ be a sequence satisfying (2.47). We put $\theta_j(x)=\rho_{\varepsilon_j}(x)*T\in\mathscr{E}$ and we see that

$$\alpha_j(x)\,\theta_j\in\mathscr{D}\to T\quad\text{in}\quad\mathscr{D}'\quad(j\to\infty). \tag{2.49}$$

5 *Fourier transforms of distributions*

As we remarked at the end of chapter 1, if we wish to generalize the notions of Fourier's inverse formula, etc., beyond L^2-spaces, then we have to introduce the idea distribution. In this section we shall show that the required properties of Fourier transforms can be established in the function space $\mathscr{S}'$ which was introduced by Schwartz.

$\mathscr{S}$-space.

We define $\varphi \in \mathscr{S}$ if and only if $\varphi \in C^\infty$, and $\varphi(x)$ and its all derivatives, which we are allowed to multiply by any polynomial of $|x|$, tend to 0 uniformly as $|x| \to +\infty$.

We shall give the following topology $\mathscr{S}$: $\varphi_j(\in \mathscr{S})$ converges to 0 if and only if for arbitrary k and α,

$$(1+|x|^2)^k \, D^\alpha \varphi_j(x) \tag{2.50}$$

tends to 0 uniformly in R^n. This condition is equivalent to saying that, as a fundamental system of neighbourhoods of $\mathscr{S}$, we take

$$V(m; k; \varepsilon) = \{\varphi; \, (1+|x|^2)^k |D^\alpha \varphi(x)| < \varepsilon, \, |\alpha| \leqslant m\} \tag{2.51}$$

where $m = 0, 1, 2, \ldots$; $k = 0, 1, 2, \ldots$

Also, these conditions are equivalent to the condition that the topology on $\mathscr{S}$ is given by a sequence of semi-norms:

$$p_{m,k}(\varphi) = \sum_{|\alpha| \leqslant m} \sup_x (1+|x|^2)^k |D^\alpha \varphi(x)| .$$

Some properties of $\mathscr{S}$-space.

First we note that $\mathscr{S}$ is complete. Therefore, it is an F-space, $\mathscr{D}$ is dense in $\mathscr{S}$. Moreover, as is the case for $\mathscr{D}_K$, a closed and bounded set of $\mathscr{S}$ is compact. Therefore, for $\{T_j\}$ a sequence of continuous linear forms the two convergences in the simple topology and in the strong topology coincide. We shall outline the proofs of these properties.

(1) Completeness: Let $\{\varphi_j\}$ be a Cauchy sequence of $\mathscr{S}$. We see that $\{\varphi_j\}$ is uniformly convergent. We therefore write its limit function as φ. Then, φ is continuous. $\{D^\alpha \varphi_j\}$ are also uniformly continuous. Therefore, we write the limit functions of the latter as φ_α.

For a fixed k, there exists N, such that

$$(1+r^2)^k |\varphi_j(x) - \varphi_{j'}(x)| < \varepsilon \quad \text{for} \quad j, j' > N .$$

Therefore, if $j' \to \infty$, then $(1+r^2)^k |\varphi_j(x) - \varphi(x)| \leqslant \varepsilon$. For $\{D^\alpha \varphi_j\}$ a similar property holds. Hence, for an arbitrary k, and $x \in R^n$,

$$(1+r^2)^k |D^\alpha \varphi_j(x) - \varphi_\alpha(x)| \to 0 \quad (j \to \infty)$$

uniformly.

Finally, we have to show that $D^\alpha \varphi(x) = \varphi_\alpha(x)$. We do not need to prove this because we can use an essentially similar argument to that of the proof of lemma 2.2 (where we proved the completeness of $\mathscr{D}_K$).

(2) Denseness. Let $\{\alpha_j\}$ be a sequence satisfying (2.47) for $\varphi \in \mathscr{S}$. We see immediately that $\alpha_j \varphi \to \varphi$ in $\mathscr{S}$.

(3) The necessary and sufficient condition for B to be a bounded set of $\mathscr{S}$ is that, for arbitrary k and α,

$$\sup_x (1+|x|^2)^k |D^\alpha \varphi(x)| < M_{k,\alpha} < +\infty \quad (\varphi \in B).$$

From this, (1) and the Ascoli–Arzela theorem, we see that if B is bounded and closed then it is compact. (See the definition of *compactness in a metric space* at the beginning of section 2.) The rest of the properties of an $\mathscr{S}$-space follow immediately from theorem 2.3.

DEFINITION 2.13 *($\mathscr{S}'$-space)*. We call a continuous linear form on $\mathscr{S}$ a *tempered distribution* (Fr. distribution tempérée). In this case we write $\mathscr{S}'$ as the whole set of tempered distributions.

We should justify the terminology 'distribution'. We see that by the topology given to $\mathscr{S}$, in fact, the elements of $\mathscr{S}'$ become distributions when we restrict the domain of the definition to $\mathscr{D}$. On the other hand, $\mathscr{D}$ is dense in $\mathscr{S}$. Therefore, we can identify continuous distributions with respect to the topology on $\mathscr{S}$ with continuous linear forms on $\mathscr{S}$. Next, we examine some examples of tempered distribution.

(1) If $f \in L^p(1 \leqslant p < +\infty)$, then $f \in \mathscr{S}'$. Furthermore, suppose that α is a function being at most of the same order of infinity as a polynomial i.e. $|\alpha(x)| \leqslant C(1+|x|)^l$ then the product of these functions $g: x \to \alpha(x) f(x)$ belongs to $\mathscr{S}'$. In fact,

$$\langle \alpha f, \varphi \rangle = \int f(x)\alpha(x)\varphi(x)\,dx; \quad |\langle \alpha f, \varphi \rangle| \leqslant \|f\|_{L^p} \cdot \|\alpha\varphi\|_{L^{p'}} \quad (p' = p/(p-1)).$$

On the other hand, from $(1+|x|)^{n+l}|\varphi(x)| < \varepsilon$ it follows that $\|\alpha\varphi\|_{L^{p'}} \leqslant C'\varepsilon$

(2) Let $f \in L^1{}_{\text{loc}}$, and $|x| \to +\infty$.

For a certain $l>0$, if $|f(x)|\leqslant A\cdot|x|^l$, then $f\in\mathscr{S}'$. In fact

$$|\langle f,\varphi\rangle|\leqslant\int_{|x|\leqslant R}|f(x)|\,|\varphi(x)|\,\mathrm{d}x\leqslant A\int_{|x|\geqslant R}|x|^l\,|\varphi(x)|\mathrm{d}x.$$

We can be more precise about this if $f\in L^1_{\mathrm{loc}}$, and when $|x|\to+\infty$ f remains of the same sign. For example, if $f(x)\geqslant 0$, then the necessary and sufficient condition for $f\in\mathscr{S}'$ is that for suitable positive numbers C and l

$$\int_{|x|\leqslant A}|f(x)|\,\mathrm{d}x\leqslant CA^l \tag{2.52}$$

as $A\to+\infty$. Note that for the one-dimensional case $\mathrm{e}^x\notin\mathscr{S}'$.

Proof of (2.52). It is obvious that the condition is sufficient. In fact, let $(1+|x|)^{l+2}|\varphi(x)|<\varepsilon$, then if (2.52) is valid when $A\geqslant N$, we have

$$\int_{|x|\geqslant N}|f(x)|\,|\varphi(x)|\,\mathrm{d}x=\sum_{n=N+1}^{\infty}\int_{n-1\leqslant|x|\leqslant n}|f(x)|\,|\varphi(x)|\,\mathrm{d}x\leqslant\varepsilon C\sum\frac{1}{n^2}.$$

We prove the necessity. Without loss of generality, we assume that, when $f(x)\geqslant 0$, (2.52) is valid irrespective of A. Let us choose $\{\alpha_j\}$ satisfying (2.47). From $f\in\mathscr{S}'$, we have $\alpha_j f\to f$ in $\mathscr{S}'$. In fact, $\langle(\alpha_j-1)f,\varphi\rangle=\langle f,(\alpha_j-1)\varphi\rangle$ and $(\alpha_j-1)\varphi\to 0$ in $\mathscr{S}$. Therefore, $\{\alpha_j f\}$ form a equi-continuous set (lemma 2.5). Hence, in a neighbourhood of 0 of $\mathscr{S}$, if $|\mathrm{D}^\alpha\varphi(x)|\leqslant\eta/(1+|x|)^k$ $(|\alpha|\leqslant m)$ then $|\langle\alpha_j f,\varphi\rangle|\leqslant 1$ $(j=1,2,\ldots)$. On the other hand, if we take ε sufficiently small and l sufficiently large, $\varphi(x)=\varepsilon/(1+|x|^2)^l$ satisfies the condition which we have just mentioned. Therefore,

$$\varepsilon\int\frac{\alpha_j(x)\,f(x)}{(1+|x|^2)^l}\,\mathrm{d}x\leqslant 1.$$

This is nothing but the inequality (2.52).

Q.E.D.

(3) Let $T\in\mathscr{S}'$. Then, $\mathrm{D}^\alpha T\in\mathscr{S}'$. Also $\mathscr{E}'\subset\mathscr{D}'_{L^2}\subset\mathscr{S}'$. Now we see that

$$\langle\mathrm{D}^\alpha T,\varphi\rangle=(-1)^{|\alpha|}\langle T,\mathrm{D}^\alpha\varphi\rangle$$

so that $\varphi_j\to 0$ in $\mathscr{S}$. Hence $(\mathrm{D}^\alpha\varphi_j)\to 0$. From this, $\mathrm{D}^\alpha T\in\mathscr{S}'$. The rest can be proved by considering the representation in $\mathscr{D}'_{L^2}{}^m$ (theorem 2.13) and (1) above).

(4) let $\alpha \in C^\infty$ and let its all derivatives be of the same order at infinity as a polynomial, i.e.

$$|D^\nu \alpha(x)| \leqslant M_\nu (1+|x|)^{k_\nu} \quad (|\nu| \geqslant 0).$$

Then from $T \in \mathscr{S}'$ we see that $\alpha T \in \mathscr{S}'$. In fact, $\langle \alpha(x) T, \varphi \rangle = \langle T, \alpha\varphi \rangle$. From $\varphi_j \to 0$ in $\mathscr{S}$. it follows that $\alpha\varphi_j \to 0$ in $\mathscr{S}$.

PROPOSITION 2.10

(1) Let B be a bounded set of $\mathscr{S}'$. For certain $k(>0)$ and $m\,(>0)$, $T/(1+r^2)^k$ is a bounded set of $\mathscr{D}'_{L^2}{}^m$ where T is allowed to run through B.

(2) If we let $T_j \to T$ in $\mathscr{S}'$, for certain k (>0) and m (>0), $T_j/(1+r^2)^k$ tends to $T/(1+r^2)^k$ under the simple topology of $\mathscr{D}'_{L^2}{}^m$.

Proof.

(1) B is an equi-continuous set of $\mathscr{S}'$ (lemma 2.5). Therefore, for certain $k(>0)$ and p, if

$$(1+r^2)^k\, |D^\alpha \varphi(x)| < \varepsilon \quad (|\alpha| \leqslant p) \tag{2.53}$$

then,

$$|\langle T, \varphi \rangle| \leqslant 1 \quad (T \in B).$$

On the other hand, if we take η sufficiently small, then (2.53) is valid for all φ satisfying

$$\|D^\alpha[(1+r^2)^k\, \varphi(x)]\|_{L^2} \leqslant \eta, \quad |\alpha| \leqslant [\tfrac{1}{2}n]+1+p \equiv m. \tag{2.54}$$

In fact, from Sobolev's lemma (2.25) we have

$$|D^\alpha[(1+r^2)^k\, \varphi(x)]\,| < c\eta \quad (|\alpha| \leqslant p),$$

where c is a positive constant which depends only upon the dimension of the space. Therefore, if we put

$$\langle T, \varphi \rangle = \langle T/(1+r^2)^k, (1+r^2)^k\, \varphi(x) \rangle$$

then, from (2.54), $\langle T/(1+r^2)^k, \psi \rangle$ is uniformly bounded for $T \in B$ when $\|\psi(x)\|_{m,L^2} \equiv \|(1+r^2)^k \varphi(x)\|_{m,L^2}$ is bounded. Hence, $\|T/(1+r^2)^k\|_{m,L^2}$ is bounded for $T \in B$.

(2) Because T_j and T are bounded in S', by (1). we see that for certain

$k(>0)$ and $m(>0)$ $T_j/(1+r^2)^k$ is a bounded set of $\mathscr{D}'_{L^2}{}^m$. By our hypothesis, it is a convergent sequence at $\varphi(\in\mathscr{D})$ as $j\to\infty$. For all $\varphi(\in\mathscr{D}_{L^2}{}^m)$, we have

$$\left\langle\frac{T_j}{(1+r^2)^k},\varphi\right\rangle\to\left\langle\frac{T}{(1+r^2)^k},\varphi\right\rangle \quad (j\to\infty),$$

since $\mathscr{D}$ is dense in $\mathscr{D}_{L^2}{}^m$.

Q.E.D.

From the proposition and theorem 2.13 we have:

THEOREM 2.14 *(Representation theorem for $\mathscr{S}'$)*

(1) Let B be a bounded set of $\mathscr{S}'$. There are some $k(\geqslant 0)$ and $m(\geqslant 0)$ such that any $T(\in B)$ can be represented as

$$T=(1+|x|^2)^k \sum_{|\alpha|\leqslant m} D^\alpha f_\alpha(x) \quad (f_\alpha\in L^2). \tag{2.55}$$

When T runs through B, each set of f_α which appears as a component in the representation of T (2.55) forms a bounded set of L^2.

(2) If $T_j\to T$ in $\mathscr{S}'$, then for some $k\geqslant 0$ and $m\geqslant 0$, T_j and T have representations in the form of (2.55). Moreover, if $T_j=(1+|x|^2)^k \sum_{|\alpha|\leqslant m} D^\alpha f_\alpha{}^{(j)}(x)$, then $f_\alpha{}^{(j)}\to f_\alpha$ in the weak topology on L^2 as $j\to\infty$.

Note. The above discussion guarantees that an arbitrary tempered distribution may have its representation expressed by a product of a polynomial and an element of $\mathscr{D}'_{L^2}{}^m$. The converse of this statement is also true. This fact can be conveniently remembered as $(\mathscr{S}')=(\text{polynomial})\times(\mathscr{D}'_{L^2})$, although this is not a precise notation.

Fourier transform in $\mathscr{S}'$

In Chapter 1 we defined the Fourier transform $\mathscr{F}f$ for $f\in L^1$.

We note that in this case for an arbitrary $\varphi\in\mathscr{D}$,

$$\langle\mathscr{F}f,\varphi(\xi)\rangle=\langle f(x),\mathscr{F}\varphi\rangle, \tag{2.56}$$

$$\langle\overline{\mathscr{F}}f,\varphi(\xi)\rangle=\langle f(x),\overline{\mathscr{F}}\varphi\rangle. \tag{2.57}$$

In fact,

$$\langle\mathscr{F}f,\varphi(\xi)\rangle=\int\varphi(\xi)\,d\xi\int e^{-2\pi i x\xi}f(x)\,dx;$$

from Fubini's theorem, this is equal to

$$\int f(x)\,\mathrm{d}x \int \mathrm{e}^{-2\pi i x\xi}\varphi(\xi)\,\mathrm{d}\xi = \langle f(x), \mathscr{F}\varphi\rangle .$$

We obtain similar results for $\overline{\mathscr{F}}$. From this we may conceive an idea how to extend Fourier transforms to distributions; replacing f by a distribution T and defining $\mathscr{F}T$ by (2.56) etc.

Unfortunately, in this case, $\mathscr{F}\varphi \notin \mathscr{D}$ and supp $[\mathscr{F}\varphi]$ covers the whole space. Therefore, in general, the right hand term of (2.56) is meaningless. For this reason, we must give up the idea of defining the Fourier transform for every distribution; we satisfy ourselves with defining $\mathscr{F}U$ for $U(\in\mathscr{S}')$ by (2.56). This is possible by virtue of the following:

LEMMA 2.12. A Fourier transform $\mathscr{F}$ is a bijective and bi-continuous linear mapping from $\mathscr{S}_x$ onto $\mathscr{S}_\xi$. (This is the same as saying that $\mathscr{F}$ is a topological isomorphism between $\mathscr{S}_x$ and $\mathscr{S}_\xi$.) The inverse of the mapping is $\overline{\mathscr{F}}$.

Proof. This is immediate from what we have already proved in section 9, chapter 1. In fact, we apply (1.60) and (1.58) to $\varphi(\in\mathscr{S}_x)$, then we have

$$\begin{aligned}(1+|\xi|^2)^k\,\mathrm{D}^\alpha\phi(\xi) &= (1+|\xi|^2)^k\,\mathscr{F}[(-2\pi i x)^\alpha\varphi(x)]\\ &= \mathscr{F}\left[\left(1-\frac{\Delta}{4\pi^2}\right)^k\{(-2\pi i x)^\alpha\varphi(x)\}\right]; \quad \Delta=\sum_{i=1}^n\frac{\partial^2}{\partial x_i^2}.\end{aligned}$$

Therefore,

$$|(1+|\xi|^2)^k\,\mathrm{D}^\alpha\phi(\xi)| \leqslant \left\|\left(1-\frac{\Delta}{4\pi^2}\right)^k\{(-2\pi i x)^\alpha\varphi(x)\}\right\|_{L^1} \tag{2.58}$$

for arbitrary α and k. From this we see that $\varphi\in\mathscr{S}_\xi$. Furthermore, we see also that $\mathscr{F}$ is a continuous mapping from $\mathscr{S}_x$ to $\mathscr{S}_\xi$†. The proof for $\overline{\mathscr{F}}$ is similar.

Q.E.D.

DEFINITION 2.14. Let $U_x\in\mathscr{S}_x'$. We define $\mathscr{F}U_x\in\mathscr{S}_\xi'$ and $\overline{\mathscr{F}}U_x\in\mathscr{S}_\xi'$ respectively as

$$\langle\mathscr{F}U_x, \varphi(\xi)\rangle = \langle U_x, (\mathscr{F}\varphi)(x)\rangle \quad (\varphi(\xi)\in\mathscr{S}_\xi), \tag{2.59}$$

$$\langle\overline{\mathscr{F}}U_x, \varphi(\xi)\rangle = \langle U_x, (\overline{\mathscr{F}}\varphi)(x)\rangle \quad (\varphi(\xi)\in\mathscr{S}_\xi). \tag{2.60}$$

† A linear mapping $\mathscr{F}$ is continuous if and only if for an arbitrary neighbourhood V of $0(\in\mathscr{S}_\xi)$, there exists a fundamental neighbourhood U of $0(\in\mathscr{S}_x)$ such that the image of U by $\mathscr{F}$ is contained in V.

Note.

(1) We give some explanations concerning the above definition. We see that a linear form on $\mathscr{S}_\xi$:

$$\varphi(\xi) \to \langle U_x, (\mathscr{F}\varphi)(x) \rangle$$

is continuous. In fact, if we know $\varphi_j(\xi) \to 0$ in $\mathscr{S}_\xi$ then, from lemma 2.12, we have $(\mathscr{F}\varphi_j)(x) \to 0$ in $\mathscr{S}_x$. Also, from $U_x \in \mathscr{S}_x'$, we have $\langle U_x, (\mathscr{F}\varphi_j)(x) \rangle \to 0$. From this we can write $\langle S_\xi, \varphi(\xi) \rangle = \langle U_x, (\mathscr{F}\varphi)(x) \rangle$ for a uniquely determined element $S_\xi (\in \mathscr{S}_\xi')$.

In the above definition we wrote S_ξ as $\mathscr{F} U_x$.

(2) In chapter 1 we defined Fourier transforms for $f \in L^1$ or $f \in L^2$. It is easy to see that the above new definition coincides with these old definitions. For instance, if we write $\mathscr{F}_P f$ as the Fourier transform defined by Plancherel's theorem, then for $\varphi \in \mathscr{S}$ we have $\mathscr{F}_P \varphi = \mathscr{F}\varphi$. Therefore, for $f \in L^2$, we can re-write Parseval's equality (1.67) as $\langle \mathscr{F}_P f, \overline{\varphi(\xi)} \rangle = \langle f(x), (\mathscr{F}\overline{\varphi})(x) \rangle$. We put $\varphi = \overline{\varphi}$. Then this is nothing but (2.59).

We note that in the case of the Fourier transform on $\mathscr{S}'$. precise results such as Plancherel's theorem (1.15) are no longer available. Nevertheless, we can obtain a similar result as follows:

THEOREM 2.15. Let $\mathscr{F}$ be a bi-continuous bijection from $\mathscr{S}_x'$ onto $\mathscr{S}_\xi'$. $\overline{\mathscr{F}}$ is the inverse of $\mathscr{F}$. Thus,

$$\overline{\mathscr{F}}\mathscr{F} U = U, \quad \mathscr{F}\overline{\mathscr{F}} U = U \quad (U \in \mathscr{S}'). \tag{2.61}$$

Proof. From $\langle \mathscr{F}U, \varphi(\xi) \rangle = \langle U, (\mathscr{F}\varphi)(x) \rangle$ and $U_j \to 0$ in $\mathscr{S}_x'$ we have $\langle U_j, (\mathscr{F}\varphi)(x) \rangle \to 0$. Therefore, $\langle \mathscr{F}U_j, \varphi(\xi) \rangle \to 0$. In this case the $\{U_j\}$ of $\langle \mathscr{F}U_j, \varphi(\xi) \rangle$ can be replaced by converging filters. In fact, the set of the Fourier images of bounded sets which belong to $\mathscr{S}$ is also a bounded set.

Therefore, $\mathscr{F}$ is a continuous mapping from $\mathscr{S}_x'$ onto $\mathscr{S}_\xi'$. A similar argument holds for $\overline{\mathscr{F}}$. In $\mathscr{S}$, $\overline{\mathscr{F}}\mathscr{F}\varphi = \mathscr{F}\overline{\mathscr{F}}\varphi = \varphi$ therefore

$$\langle \overline{\mathscr{F}}\mathscr{F} U, \varphi \rangle = \langle \mathscr{F} U, \overline{\mathscr{F}}\varphi \rangle = \langle U, \mathscr{F}\overline{\mathscr{F}}\varphi \rangle = \langle U, \varphi \rangle,$$

i.e. $\overline{\mathscr{F}}\mathscr{F} = I$ in $\mathscr{S}'$ and similarly $\mathscr{F}\overline{\mathscr{F}} = I$ in $\mathscr{S}'$.

Q.E.D.

PROPOSITION 2.11 *(Fundamental property of the Fourier transform)*. Let

$U \in \mathscr{S}'$. Then the following equalities are true

$$\mathscr{F}[D^\alpha U] = (2\pi i\xi)^\alpha \mathscr{F}[U], \tag{2.62}$$

$$\mathscr{F}[(-2\pi ix)^\alpha U] = D_\xi^\alpha \mathscr{F}[U], \tag{2.63}$$

$$\mathscr{F}[\tau_h U] = e^{-2\pi ih\xi} \mathscr{F}[U], \tag{2.64}$$

$$\mathscr{F}[e^{2\pi ihx} U] = \tau_h \mathscr{F}[U], \tag{2.65}$$

where τ_h in (2.64) and (2.65) is a distribution which is obtained from the distribution U by a parallel translation h. satisfying $\langle \tau_h U, \varphi(x)\rangle = \langle U, \varphi(x+h)\rangle$. (When U is a function, $(\tau_h f)(x) = f(x-h)$.)

Proof. (2.62) and (2.63) have already been proved under certain conditions (see (1.58) and 1.60)). Of course, they are true for a function φ of $\mathscr{S}$. We show that (2.62) is true.

Let $\varphi \in \mathscr{S}$. Then, $\langle \mathscr{F}[D^\alpha U], \varphi(\xi)\rangle = \langle D^\alpha U, (\mathscr{F}\varphi)(x)\rangle = (-1)^{|\alpha|}\langle U, D^\alpha(\mathscr{F}\varphi)\rangle$. From (1.60) this is equal to $(-1)^{|\alpha|}\langle U, \mathscr{F}[(-2\pi i\xi)^\alpha \varphi(\xi)]\rangle$. From the definition of the Fourier transform, we can write this as

$$(-1)^{|\alpha|}\langle \mathscr{F}U, (-2\pi i\xi)^\alpha \varphi(\xi)\rangle = \langle (2\pi i\xi)^\alpha \mathscr{F}U, \varphi(\xi)\rangle.$$

We show that (2.64) is true. It is obvious that for $\varphi(\in \mathscr{S})$, (2.64) and (2.65) are valid, i.e.

$$\mathscr{F}[\varphi(\xi-h)] = e^{-2\pi ihx}\mathscr{F}[\varphi(\xi)]; \qquad \mathscr{F}[e^{2\pi ih\xi}\varphi(\xi)] = \mathscr{F}[\varphi](x-h).$$

From this we have

$$\begin{aligned}\langle e^{2\pi ih\xi}\mathscr{F}[\tau_h U], \varphi(\xi)\rangle &= \langle \mathscr{F}[\tau_h U], e^{2\pi ih\xi}\varphi(\xi)\rangle \\ &= \langle \tau_h U, \mathscr{F}[e^{2\pi ih\xi}\varphi(\xi)]\rangle = \langle \tau_h U, \mathscr{F}[\varphi](x-h)\rangle.\end{aligned}$$

By the definition of T_h, this is equal to

$$\langle U, (\mathscr{F}\varphi)(x)\rangle,$$

(2.64) is true.

Q.E.D.

Concrete representation of the Fourier transform

(1) Let $U \in \mathscr{D}'_{L^2}{}^m$. We can write $U = \sum_{|\alpha| \leqslant m} D^\alpha f_\alpha(x) (f_\alpha \in L^2)$. So that for f_α there exists $\hat{f}_\alpha \in L^2$ by Plancherel's theorem. As we explained in the note following definition 2.14, this is exactly the Fourier image of an element of $\mathscr{S}'$. By (2.62) we have

$$\mathscr{F}[U] = \sum_{|\alpha| \leqslant m} (2\pi i\xi)^\alpha \hat{f}_\alpha(\xi) \quad (\hat{f}_\alpha \in L^2). \tag{2.66}$$

Therefore, the Fourier image of any element of $\mathscr{D}'_{L^2}$ is in fact always a function.

(2) By the representation theorem of $\mathscr{S}'$ (theorem 2.14), we have for U $(\in \mathscr{S}')$

$$U=(1+|x|^2)^k \sum_{\alpha} \mathrm{D}^\alpha f_\alpha(x) \quad (f_\alpha \in L^2),$$

so that by (2.62) and (2.63).

$$\mathscr{F}U=\left(1-\frac{\Delta}{4\pi^2}\right)^k \sum_{\alpha} (2\pi\mathrm{i}\xi)^\alpha \hat{f}_\alpha(\xi). \tag{2.67}$$

The reader should note that this $\mathscr{F}U$ is also in the form of (2.55) which was obtained by the representation theorem of $\mathscr{S}'$ which we have just mentioned

(3) If $U_x \in \mathscr{E}'$, then we can write

$$\mathscr{F}[U_x]=\langle U_x, \mathrm{e}^{-2\pi\mathrm{i}x\xi}\rangle. \tag{2.68}$$

The right hand term is not only a C^∞-function but also an entire function of ξ, i.e. since it is a regular function we can extend ξ to a complex variable $\zeta=(\zeta_1, \dots, \zeta_n) \in C^n$ by an analytic continuation.

Proof of (3). It is easy to prove this by the representation theorem of U_x. In fact, by theorem 2.13 (the fact that supp $[f_\alpha] \subset K$ where K is a certain compact set), we have $U_x=\sum_{\alpha} \mathrm{D}^\alpha f_\alpha(x)$ $(f_\alpha \in L^2 \cap L^1)$.

Therefore,

$$\mathscr{F}[U_x]=\sum_{\alpha} (2\pi\mathrm{i}\xi)^\alpha \hat{f}_\alpha(\xi)=\sum_{\alpha} \int_K \mathrm{e}^{-2\pi\mathrm{i}x\xi}(2\pi\mathrm{i}\xi)^\alpha f_\alpha(x)\,\mathrm{d}x, \tag{2.69}$$

We can rewrite this as follows

$$\sum_{\alpha} \langle \mathrm{D}^\alpha f_\alpha(x), \mathrm{e}^{-2\pi\mathrm{i}x\xi}\rangle=\langle U_x, \mathrm{e}^{-2\pi\mathrm{i}x\xi}\rangle.$$

From (2.69) $\mathscr{F}U_x$ must be a entire function. Actually, we do not need to use the representation theorem to prove (2.68); it can be proved directly as follows

We note that (2.68) is valid when $U_x \in \mathscr{D}$. On the other hand, if we take $\{\varphi_j\}$, a sequence of $\mathscr{D}$, such that $\varphi_j(x) \to U_x$ in $\mathscr{E}'$, then this convergence of $\{\varphi_j\}$ can be interpreted as in $\mathscr{S}_x'$. Therefore, $\mathscr{F}[\varphi_j] \to \mathscr{F}[U_x]$ in $\mathscr{S}_\xi'$ (see theorem 2.15). Also, $\mathrm{e}^{-2\pi\mathrm{i}x\xi} \in \mathscr{E}_x$ is infinitely differentiable with respect to the parameter ξ.

Such an element forms a bounded set of $\mathscr{E}_x$ when ξ runs through a compact set of R^n. Therefore, $\theta_j(\xi)=\langle\varphi_j(x), \mathrm{e}^{-2\pi i x\xi}\rangle$ is uniformly convergent to $\theta(\xi)=\langle U_x, \mathrm{e}^{-2\pi i x\xi}\rangle$ with respect to ξ on the compact set. Hence, $\mathscr{F}[U_x]=\theta(\xi)$.

Q.E.D.

(4)

$$\mathscr{F}[\delta]=1, \quad \mathscr{F}[1]=\delta. \tag{2.70}$$

$$\overline{\mathscr{F}}[\delta]=1, \quad \overline{\mathscr{F}}[1]=\delta. \tag{2.71}$$

In fact, from (2.68) we have $\mathscr{F}[\delta]=\langle\delta, \mathrm{e}^{-2\pi i x\xi}\rangle_x=1$. Similarly, $\overline{\mathscr{F}}[\delta]=1$. Operating with $\overline{\mathscr{F}}$ on both sides of the equation $\mathscr{F}(\delta)=1$ from the left, we have

$$\overline{\mathscr{F}}\mathscr{F}[\delta]=\overline{\mathscr{F}}[1], \quad \overline{\mathscr{F}}\mathscr{F}=I \text{ in } \mathscr{S}'.$$

Similarly $\delta=\overline{\mathscr{F}}[1]$, $\delta=\mathscr{F}[1]$.

In section 6 we shall demonstrate some concrete examples of the Fourier transforms of distributions. The following facts are useful for understanding these examples.

When the dimension of the base space is 1, we readily see that

$$\mathscr{F}[\exp(-\pi r^2)]=\exp(-\pi\xi^2). \tag{2.72}$$

From this and (1.56), in the case of n-dimensions, we have

$$\mathscr{F}[\exp[-\pi(x_1{}^2+\cdots x_n{}^2)]]=\exp[-\pi(\xi_1{}^2+\cdots\xi_n{}^2)]. \tag{2.73}$$

Next, let $f\in\mathscr{S}'$ and let $f, \mathscr{F}f\in L^1{}_{\mathrm{loc}}$, i.e. f and $\mathscr{F}f$ are both functions. For a real number $\lambda(\neq 0)$, we have

$$\mathscr{F}[f(\lambda x)]=\frac{1}{|\lambda|^n}\hat{f}(\xi/\lambda), \tag{2.74}$$

where n is the dimension of the base space. Similar results can be obtained for $\overline{\mathscr{F}}$.

To see this we note that for $f\in\mathscr{S}$, this is nothing but a change of the variable of integration. Let $\varphi\in\mathscr{S}$. Then we have

$$\langle\mathscr{F}[f(\lambda x)], \varphi(\xi/\lambda)\rangle=\langle f(\lambda x), \mathscr{F}[\varphi(\xi/\lambda)]\rangle=|\lambda|^n\langle f(\lambda x), (\mathscr{F}\varphi)(\lambda x)\rangle.$$

Changing the variable of integration we obtain

$$\langle f(x), (\mathscr{F}\varphi)(x)\rangle=\langle(\mathscr{F}f)(\xi), \varphi(\xi)\rangle,$$

which is the same as in the above equation. We now put $\varphi(\xi/\lambda)=\psi(\xi)$. Then we have

$$\langle \mathscr{F}[f(\lambda x)], \psi(\xi)\rangle = \langle (\mathscr{F}f)(\xi), \psi(\lambda\xi)\rangle = \frac{1}{|\lambda|^n}\langle (\mathscr{F}f)(\xi/\lambda), \psi(\xi)\rangle,$$

and we see that this is just (2.74).

Finally, let $f\in\mathscr{S}'$, $f\in L^1{}_{\text{loc}}$, and $\mathscr{F}f\in L^1{}_{\text{loc}}$, and let f be a function depending on r only. Then, first we see that $\mathscr{F}f=\hat{f}(\xi)$ is a function depending on $|\xi|\equiv\rho$ only, i.e.

$$\text{if} \quad f(x)=\Phi(r), \quad \text{then} \quad \hat{f}(\xi)=\Psi(\rho). \tag{2.75}$$

To prove this, we shall begin with the case when $f\in L^1$. Let S be an arbitrary rotation around the origin. Let $\langle x, S\xi\rangle=\langle S^{-1}x, \xi\rangle$ and $S^{-1}x=x'$. From this $\mathrm{d}x=\mathrm{d}x'$, and

$$\begin{aligned}\hat{f}(S\xi)=\int\exp(-2\pi\mathrm{i}x\cdot S\xi)\,f(x)\,\mathrm{d}x&=\int\exp(-2\pi\mathrm{i}S^{-1}x\cdot\xi)\,f(x)\,\mathrm{d}x\\&=\int\exp(-2\pi\mathrm{i}x\cdot\xi)\,f(Sx)\,\mathrm{d}(Sx).\end{aligned}$$

Also, $f(Sx)=f(x)$ so that

$$f(S\xi)=\int \mathrm{e}^{-2\pi\mathrm{i}x\xi}f(x)\,\mathrm{d}x=\hat{f}(\xi).$$

Now we have $\hat{f}(\xi)=\Psi(\rho)$.

For f in general, we prove the property (2.75) as follows. Let $\alpha_j(x)$ satisfy (2.47). Furthermore, if we take $\alpha_j(x)$ is a function depending only on $|x|$, and put

$$\hat{f}_j(\xi)=\int \mathrm{e}^{-2\pi\mathrm{i}x\xi}\alpha_j(r)\,\Phi(r)\,\mathrm{d}x,$$

we see that $\hat{f}_j(\xi)$ is a function depending only on ρ. Also, $\hat{f}_j(\xi)\to\hat{f}(\xi)$ under the topology of $\mathscr{S}'$. Therefore, $\hat{f}(\xi)$ is a function depending only on ρ.

6 *Concrete examples of Fourier transforms*

The reason why we did not demonstrate concrete examples of Fourier transforms in chapter 1 is that the functions in which we shall be interested in the theory of partial differential equations do not usually belong to L^1 or L^2-space. Even if they did belong to L^1 or L^2-space, we did not know some fundamental relations for dealing with the relation between Fourier transforms and differential operators which we have later obtained as (2.62) and (2.63). We are now fully prepared to do this.

EXAMPLE 1.

$$\mathscr{F}\left[\text{v.p.}\,\frac{1}{x}\right]=\begin{cases}-\pi i & (\xi>0)\\ \pi i & (\xi<0),\end{cases} \tag{2.76}$$

and $f\in\mathscr{D}'_{L^2}$ where $f(x)=\text{v.p.}(1/x)$. In fact,

$$\text{v.p.}\,\frac{1}{x}=\alpha(x)\,\text{v.p.}\,\frac{1}{x}+(1-\alpha(x))\,\frac{1}{x},$$

where α is a function taking the value 1 in the neighbourhood of the origin and $\alpha\in\mathscr{D}$. The first term of the above decomposition $\in\mathscr{E}'$, and the second term $\in L^2$. Therefore, the sum $\text{v.p.}(1/x)\in\mathscr{D}'_{L^2}$. Hence,

$$\mathscr{F}[\text{v.p.}\,(1/x)],$$

is a function by (2.66).

Next, we assume $\varphi\in\mathscr{S}$. Then, we have

$$\left\langle \text{v.p.}\,\frac{1}{x},\,\varphi\right\rangle=\lim_{\substack{\varepsilon\to 0\\ A\to+\infty}}\int_{\varepsilon\leqslant|x|\leqslant A}\frac{\varphi(x)}{x}\,dx$$

so that if we put

$$f_{\varepsilon,A}(x)=\begin{cases}1/x & (\varepsilon\leqslant|x|\leqslant A)\\ 0 & (\text{otherwise})\end{cases}$$

we have $\text{v.p.}(1/x)=\lim f_{\varepsilon,A}(x)$, by the topology on $\mathscr{S}'$. Therefore, from the continuity of $\mathscr{F}$ (theorem 2.15), we have

$$\mathscr{F}[\text{v.p.}(1/x)]=\lim\mathscr{F}[f_{\varepsilon,A}(x)],$$

by the toplogy on $\mathscr{S}'_{\xi}$.

On the other hand,

$$\int_{\varepsilon\leqslant|x|\leqslant A} e^{-2\pi i x\xi}\,\frac{dx}{x}=\int_{\varepsilon\leqslant|x|\leqslant A}\frac{-i\sin 2\pi x\xi}{x}\,dx=-2i\int_{\varepsilon}^{A}\frac{\sin 2\pi x\xi}{x}\,dx.$$

Therefore, if we assume $\xi\neq 0$, we have

$$f(\xi)=-2i\lim_{A\to+\infty}\int_0^A\frac{\sin 2\pi x\xi}{x}\,dx=-2i\lim_{A\to\infty}\int_0^{2\pi A\xi}\frac{\sin x}{x}\,dx=-2i\int_0^{\to\pm\infty}\frac{\sin x}{x}\,dx$$

By (1.6)

$$f(\xi)=\begin{cases}-\pi i & (\xi>0)\\ \pi i & (\xi<0).\end{cases}$$

Note that the above limiting process occurs in the sense of compact uniform convergence in $R \setminus \{0\}$. Therefore, it is a convergence in $\mathscr{S}'$.

EXAMPLE 2. Let $Y(x)$ be a Heaviside function, i.e. $\langle Y, \varphi \rangle = \int_0^\infty \varphi(x)\, dx$. Let us calculate $\mathscr{F}[Y(x)]$. It is obvious that

$$\bar{\mathscr{F}}\left[\text{v.p.}\ \frac{1}{x}\right] = \begin{cases} \pi i & (\xi > 0) \\ -\pi i & (\xi < 0). \end{cases}$$

Therefore, $\bar{\mathscr{F}}[\delta] = 1$. By (2.71)

$$\bar{\mathscr{F}}\left[\frac{1}{\pi i}\ \text{v.p.}\ \frac{1}{x} + \delta\right] = 2Y(\xi).$$

$\mathscr{F}\bar{\mathscr{F}} = I$ gives

$$\mathscr{F}[Y(x)] = \frac{1}{2\pi i}\ \text{v.p.}\ \frac{1}{\xi} + \frac{1}{2}\delta. \tag{2.77}$$

EXAMPLE 3. Let the base space be n-dimensional. Put $|x| = r$.

$$\mathscr{F}\left[\frac{1}{r^k}\right] = \pi^{k - \frac{1}{2}n} \frac{\Gamma\left(\dfrac{n-k}{2}\right)}{\Gamma\left(\dfrac{k}{2}\right)} \frac{1}{\rho^{n-k}} \quad (0 < k < n,\ \rho = |\xi|). \tag{2.78}$$

Also,

$$\bar{\mathscr{F}}\left[\frac{1}{r^k}\right] = \mathscr{F}\left[\frac{1}{r^k}\right].$$

To see this, let $\frac{1}{2}n < k < n$. As we have done in example 1 we decompose $1/r^k$ as

$$\frac{1}{r^k} = \alpha(x)\frac{1}{r^k} + (1 - \alpha(x)) \cdot \frac{1}{r^k}.$$

The second term of this decomposition $\in L^2$, so that $1/r^k \in \mathscr{D}'_{L^2}$. Therefore $\mathscr{F}[1/r^k]$ is a function.

Next, we put

$$\mathscr{F}[1/r^k] = f(\xi).$$

From (2.75), we see that $\hat{f}$ is a function depending only on $\rho=|\xi|$. Furthermore, by (2.74) we have

$$\hat{f}(\lambda\xi)=\frac{1}{\lambda^{n-k}}\hat{f}(\xi)$$

for $\lambda>0$. Therefore, we see that $\hat{f}$ is a homogeneous function of degree $k-n$. So that

$$\mathscr{F}\left[\frac{1}{r^k}\right]=c_k\cdot\frac{1}{\rho^{n-k}} \qquad (c_k=\text{constant}).$$

This is sufficient for practical purposes, but the following is a simple way of finding c_k.

As the Fourier transform (2.59) we put $\varphi(\xi)=\exp(-\pi|\xi|^2)$. By (2.73) we have $\mathscr{F}\varphi=\exp(-\pi|x|^2)$ so that

$$c_k\int\exp(-\pi|\xi|^2)\frac{\mathrm{d}\xi}{|\xi|^{n-k}}=\int\exp(-\pi|x|^2)\frac{\mathrm{d}x}{|x|^k}.$$

On the other hand, for $0<m<n$ we have

$$\int\exp(-\pi r^2)\frac{\mathrm{d}x}{r^m}=\pi^{\frac{1}{2}(m-n)}\,\Gamma\left(\frac{n-m}{2}\right)S_n,$$

where S_n is the surface area of the unit hall of R^n. Hence we have (2.78). Next, we see that

$$\bar{\mathscr{F}}[1/r^k]=\mathscr{F}[1/r^k]$$

as $1/r^k$ and $\hat{f}(\xi)$ are both real. Also, when $k=\frac{1}{2}n$, by a limiting process we conclude that our equality is true. When $0<k\leqslant\frac{1}{2}n$, we multiply both sides of (2.78) by $\bar{\mathscr{F}}$, or by an analytical continuation† of (2.78), the equality is true for $0<k<n$ (Because $1/\lambda$ is an analytic function of λ as an element of $\mathscr{S}'$, and $\mathscr{F}$ is continuous.).

EXAMPLE 4. From the previous example 3, we have the following immediate

† In general, by $U_x(\lambda)$ $(\in\mathscr{S}')$ is *analytical* in $\lambda(\in\mathscr{D})$, we mean that for an arbitrary $\varphi\in\mathscr{S}$ $F(\lambda)=\langle U_x(\lambda),\varphi(x)\rangle$ is a regular function in $\mathscr{D}$. From this it follows that if $U_x(\lambda)$ is analytic on $\mathscr{D}$, $\mathscr{F}[U_x(\lambda)]\in\mathscr{S}_\xi'$ is also analytic in $\mathscr{D}$ because $\langle\mathscr{F}[U_x(\lambda)],\varphi(\xi)\rangle=\langle U_x(\lambda),(\mathscr{F}\varphi)(x)\rangle$.

result:

$$\mathscr{F}\left[\text{v.p.}\ \frac{x_j}{|x|^{n+1}}\right] = -\mathrm{i}\ \frac{\pi^{\frac{1}{2}(n+1)}}{\Gamma\left(\dfrac{n+1}{2}\right)}\ \frac{\xi_j}{|\xi|} \tag{2.79}$$

where v.p. means Cauchy's principal value (Fr. valeur principale) in the n-dimensional case. We shall touch upon this notion later in example 5. In this example we use the following property:

$$\frac{x_j}{|x|^{n+1-\varepsilon}} \to \text{v.p.}\ \frac{x_j}{|x|^{n+1}} \quad \text{in} \quad \mathscr{S}' \qquad (\varepsilon \to +0).$$

(To see this we consider an integral over the surface of the sphere $|x|=r$:

$$\int_{|x|=r} \frac{x_j}{|x|^{n+1}}\, \mathrm{d}S = 0.$$

Note that $x_j/|x|^{n+1}$ is not a function belonging to $L^1{}_{\text{loc}}$ because it is a homogeneous function of the nth degree.)

Now, we shall prove (2.79). From (2.18)

$$\frac{\partial}{\partial x_j}\left(\frac{1}{|x|^{n-1-\varepsilon}}\right) = -\frac{(n-1-\varepsilon)\, x_j}{|x|^{n+1-\varepsilon}} \qquad (\varepsilon > 0).$$

We make $\varepsilon \to +0$, then we have

$$\frac{\partial}{\partial x_j}\left(\frac{1}{|x|^{n-1}}\right) = -(n-1)\cdot \text{v.p.}\ \frac{x_j}{|x|^{n+1}}.$$

in the topology on $\mathscr{S}'$. Let us consider the Fourier transform of both sides of the above equation:

$$2\pi\mathrm{i}\xi_j \mathscr{F}\left[\frac{1}{|x|^{n-1}}\right] = -(n-1)\, \mathscr{F}\left[\text{v.p.}\ \frac{x_j}{|x|^{n+1}}\right].$$

We put $k=n-1$ in (2.78). Then, we use this result to obtain

$$\mathscr{F}\left[\text{v.p.}\ \frac{x_j}{|x|^{n+1}}\right] = -\frac{1}{(n-1)}\, 2\pi\mathrm{i}\xi_j \cdot \pi^{n-1-\frac{1}{2}n}\ \frac{\Gamma(\frac{1}{2})}{\Gamma(\frac{1}{2}(n-1))} \cdot \frac{1}{|\xi|}.$$

By using the property $\Gamma(\frac{1}{2}) = \pi^{\frac{1}{2}}$ we have (2.79).

EXAMPLE 5 *(Fourier transforms of v.p.$(k(\omega)\,/\,r^n)$ and $K(\omega)$).* Let us extend v.p.$(1/x)$ to the n-dimensional case. Assume f is homogeneous of degree n. From this it follows immediately that $f(x)|x|^n$ is a homogeneous function of degree 0, and also a function on the unit sphere. Write

$$f(x)=\frac{k(\omega)}{|x|^n} \quad (\omega\in\Omega_0). \tag{2.80}$$

where f is continuous on $R^n\setminus\{0\}$. We assume

$$\int_{\Omega_0} k(\omega)\,d\omega=0. \tag{2.81}$$

We define for $\varphi\in\mathscr{D}$

$$\langle \text{v.p. } f(x), \varphi(x)\rangle = \lim_{\varepsilon\to+0} \int_{\varepsilon\leqslant|x|\leqslant A} f(x)\,\varphi(x)\,dx.$$

as it is defined where A satisfies the condition: the sphere $|x|\leqslant A$ contains the support of $\varphi(x)$. We decompose $\varphi(x)$ as

$$\varphi(x)=\varphi(0)+\sum_{i=1}^{n} x_i\varphi_i(x).$$

From this we have

$$\varphi(0)\int_{\varepsilon\leqslant|x|\leqslant A} f(x)\,dx=0.$$

Then,

$$\begin{aligned}\langle \text{v.p.} f(x), \varphi(x)\rangle &= \int_{|x|\leqslant A} [\varphi(x)-\varphi(0)]\,\frac{k(\omega)}{r^n}\,dx \\ &=\sum_i \int_{|x|\leqslant A} \frac{x_i}{r^n}\,k\left(\frac{x}{|x|}\right)\varphi_i(x)\,dx\end{aligned} \tag{2.82}$$

so that we have v.p.$f\in\mathscr{D}'_{L^2}$. In fact, let α be a function in $\mathscr{D}$ which is identically equal to 1 in the neighbourhood of the origin. Then, we have a decomposition

$$\text{v.p. } f(x)=\alpha(x)\,\text{v.p.} f(x)+(1-\alpha(x))f(x), \tag{2.83}$$

where the first term belongs to $\mathscr{E}'$ and the second to L^2. Therefore, $\mathscr{F}[\text{v.p.} f(x)]$ is a function.

For $\delta>0$, we put

$$f_\delta(x)=\frac{k(\omega)}{|x|^{n-\delta}}.$$

From (2.82) we have immediately

$$\alpha(x) f_\delta(x) \to \alpha(x) \text{ v.p.} f(x) \quad \text{in} \quad \mathscr{D}'_{L^2} \quad (\delta \to +0).$$

We know

$$(1-\alpha(x)) f_\delta(x) \to (1-\alpha(x)) f(x) \quad \text{in} \quad L^2,$$

so that

$$f_\delta(x) \to \text{v.p.} f(x) \quad \text{in} \quad \mathscr{D}'_{L^2}.$$

On the other hand the Fourier transform $\hat{f}_\delta(\xi)$ of $f_\delta(x)$ is a homogeneous function of $(-\delta)$-degree by (2.74) so that the function $\mathscr{F}[\text{v.p.} f(x)](\xi)$ is homogeneous of degree 0 as the limit of $\hat{f}_\delta(\xi)$.

Let $\mathscr{F}[\text{v.p.} f(x)] = K(\xi)$. Then, $K(\xi) = K(\xi/|\xi|) = K(\omega)$. Now we let $\varphi(\xi) = \Phi(\rho) \in \mathscr{S}$. Its Fourier transform is $\mathscr{F}\varphi = \mathscr{F}[\Phi(\rho)] = \Psi(r)$ by (2.75), so that we now establish the relationship for the Fourier transform

$$\langle K(\omega), \Phi(\rho) \rangle = \langle \text{v.p.} f(x), \Psi(r) \rangle.$$

We see that the right hand term of the above equation is equal to 0 irrespective of Φ (If we regard v.p. $f(x)$ as the limit of $f_\delta(x)$, this is clear.) However $\Phi(\rho)$ is arbitrary. Hence

$$\int_{\Omega_0} K(\omega)\, d\omega = 0.$$

Next, we show that the smoothness of $f(x)$ implies the smoothness of $K(\xi)$. To see this, let f be continuously differentiable to the first degree on $R^n \setminus \{0\}$, and let α be a function taken from $\mathscr{D}$ such that its value is identically 1 in the neighbourhood of the origin. We have a decomposition

$$K(\xi) = \mathscr{F}[\alpha(x) \text{ v.p.} f(x)] + \mathscr{F}[(1-\alpha(x)) f(x)]$$

where the first term of the right hand side is a C^∞-function according to (2.68). So that it is enough to know what the second term is. Now, as $|x| \to +\infty$, we have

$$\left| \frac{\partial}{\partial x_j} \{(1-\alpha(x)) f(x)\} \right| = O\left(\frac{1}{|x|^{n+1}} \right).$$

In fact, $\partial f(x)/\partial x_j$ is a homogeneous function of degree $(-n-1)$, so that the left hand partial derivative belongs to L^1. From this we see that $\mathscr{F}[\partial \{(1-\alpha(x)) f(x)\}/\partial x_j]$ is bounded and continuous,† i.e.

† Because $\hat{f}(\xi) = \int \exp(-2\pi i x \xi) f(x) dx$, $|\hat{f}(\xi)| \leqslant \int |f(x)| dx$.

$2\pi i\xi_j\mathscr{F}[(1-\alpha(x))f(x)]$ is bounded and continuous. Therefore, K is continuous on $R^n\setminus\{0\}$.

In general, if $f\in\mathscr{E}^{m+1}(R_x{}^m\setminus\{0\})$, then $K\in\mathscr{E}^m(R_\xi{}^n\setminus\{0\})$. In fact, from

$$D_\xi{}^\alpha\mathscr{F}[(1-\alpha)f]=\mathscr{F}[(-2\pi ix)^\alpha(1-\alpha(x))f(x)],$$

we see that $x^\alpha f(x)$ is a homogeneous function of degree $(|\alpha|-n)$, so that

$$D^\beta\{(1-\alpha(x))\,x^\alpha f(x)\}=(1-\alpha(x))\,D^\beta(x^\alpha f(x))+\gamma(x)\in L^1,$$

where β satisfies $|\beta|\geqslant|\alpha|+1$. Therefore,

$$(2\pi i\xi)^\beta\,D_\xi{}^\alpha\mathscr{F}[(1-\alpha(x))\,f(x)]$$

is bounded and continuous. Hence, we may take α as satisfying $|\alpha|\leqslant m$.

Summing up we have:

PROPOSITION 2.12. Let f be a homogeneous function of degree $-n$ satisfying $\int_{|x|=1}f(x)dS_x=0$, where $f\in\mathscr{E}^{m+1}(R^n\setminus\{0\})$. Then $\mathscr{F}[\text{v.p.}f(x)]=K(\xi)$ is a homogeneous function of degree 0 and $\int_{|\xi|=1}K(\xi)dS_\xi=0$ where $K\in\mathscr{E}^m(R^n\setminus\{0\})$ $(m=0,1,2,\ldots)$. In what follows, the converse of this proposition 2.12 is more important.

We investigate some basic properties in order to prove the converse.

PROPOSITION 2.13. Let f be a homogeneous function of degree k, and let $0<k<n$ and $f\in\mathscr{E}^{n+1}(R^n\setminus\{0\})$. Then, $\mathscr{F}[f(x)]=\hat{f}(\xi)$ is a homogeneous function of degree $(k-n)$ and $\hat{f}\in\mathscr{E}^0(R^n\setminus\{0\})$. If $\int_{|x|=1}f(x)\,dS_x=0$, then $\int_{|\xi|=1}\hat{f}(\xi)\,dS_\xi=0$.

Proof. Let $\alpha\in\mathscr{D}$ where α is a function such that its value is identically 1 in the neighbourhood of the origin. We decompose $f(x)$ as follows:

$$f(x)=\alpha(x)\,f(x)+[1-\alpha(x)]\,f(x).$$

We note that $\mathscr{F}[\alpha f]\in C^\infty$. Because

$$D^\alpha\{(1-\alpha(x))\,f(x)\}=(1-\alpha(x))\,D^\alpha f(x)+\gamma(x)$$

we have $(1-\alpha)D^{\alpha}f \in L^1$ and $\gamma \in L^1$ if we put $|\alpha|=n+1$. Therefore,

$$(2\pi i\xi)^{\alpha}\,\mathscr{F}[(1-\alpha(x))\,f(x)]$$

is bounded and continuous. So that if we restrict the distribution $\mathscr{F}f$ to the open set $R^n{}_{\xi}\setminus\{0\}$, then this is a continuous function. Also $f \in L^1{}_{\text{loc}}$ and $\mathscr{F}f \in \mathscr{E}^0(R^n\setminus\{0\})$.

On the other hand, if we take the test function $\varphi(\xi)$ which we used at the proof of (2.74) as an element of $\mathscr{D}(R^n\setminus\{0\})$, then we see that $\mathscr{F}f$ is a homogeneous function of degree $(k-n)$ defined on $R^n\setminus\{0\}$. We write this function as $\kappa(\xi)$. We prove $\mathscr{F}[f]=\kappa(\xi)$ defined on $R_{\xi}{}^n$. In fact, $\mathscr{F}[f]-\kappa(\xi)$ has its support only at the origin, so that we can write†

$$\mathscr{F}[f]=\kappa(\xi)+\sum c_{\alpha}\, D^{\alpha}\delta\,.$$

We see that the last term of this expression is 0. To see this, first we note that if $\frac{1}{2}n<k<n$, then $f \in \mathscr{D}'_{L^2}$. Therefore, $\mathscr{F}[f]$ is a function, so that the coefficient of the distribution which appears under the summation sign is 0, i.e. $c_{\alpha}=0$.

Next, if $0<k<\frac{1}{2}n$, then we operate with $\overline{\mathscr{F}}$ on the expression on the left to obtain the relation

$$f(x)=\overline{\mathscr{F}}[\kappa(\xi)]+\sum c_{\alpha}(-2\pi i x)^{\alpha}\,.$$

Now, we see that $\overline{\mathscr{F}}[\kappa(\xi)]$ is a function because $\kappa \in \mathscr{D}'_{L^2}$. By (2.74), it is a

---

† Let T be a distribution which has its support only on the origin. We see that $T \in \mathscr{E}'$. So that there exists a certain m, and if $|\varphi_j|_m \to 0$ $(j\to\infty)$ then $\langle T, \varphi_j(x)\rangle \to 0$. Next, we pick a function α which belongs to $\mathscr{D}$ and let the value of α be identically 1 at the origin. We put $|\nu|=m+1$. For an arbitrary $\varphi \in \mathscr{E}$, we see that

$$\left| x^{\nu}\alpha\!\left(\frac{x}{\varepsilon}\right)\varphi(x)\right|_m \to 0 \qquad (\varepsilon \to +0).$$

By our hypothesis, we have

$$\langle x^{\nu}T, \varphi(x)\rangle = \langle x^{\nu}T, \alpha\!\left(\frac{x}{\varepsilon}\right)\varphi(x)\rangle = \langle T, x^{\nu}\alpha\!\left(\frac{x}{\varepsilon}\right)\varphi(x)\rangle \to 0 \qquad (\varepsilon\to +0).$$

Therefore, $x^{\nu}T=0$ $(|\nu|=m+1)$.
From

$$\varphi(x)=\sum_{|\nu|\leqslant m}\frac{D^{\nu}\varphi(0)x^{\nu}}{\nu!}+\sum_{|\nu|=m+1} x^{\nu}\varphi_{\nu}(x) \qquad (\varphi_{\nu}(x)\in\mathscr{E})$$

we have

$$\langle T, \varphi\rangle=\sum_{|\nu|\leqslant m}\frac{D^{\nu}\varphi(0)}{\nu!}\langle T, x^{\nu}\rangle=\sum_{|\nu|\leqslant m}\langle c_{\nu}D^{\nu}\delta, \varphi(x)\rangle.$$

homogeneous function of degree k. Therefore $f(x)-\mathscr{F}[\kappa(\xi)]$ is the same type of function. Hence, it is necessary that the terms appearing under $\sum$ are 0. Finally $\int_{|\xi|=1} \kappa(\xi)\,dS_\xi=0$. This is obvious if we recall the previous case v.p.f just before proposition 2.12.

From Proposition 2.13 we have the following:

THEOREM 2.16. Let $K(x)\equiv K(\omega)$ be a homogeneous function of degree 0, and let $\int_{\Omega_0} K(\omega)\,d\omega=0$, and $K\in\mathscr{E}^{n+1+m}(R^n\setminus\{0\})$ $(m=0, 1, 2, \ldots)$. In this case there exists a uniquely determined homogeneous function f of degree $-n(\in\mathscr{E}^m(R^n\setminus\{0\})$, and $\mathscr{F}[K(x)]=\text{v.p.}f(\xi)$; $\int_{|\xi|=1} f(\xi)\,dS_\xi=0$. Under the same conditions, the inequality

$$\sum_{|\alpha|\leqslant m}\sup_{|\xi|\geqslant 1}|D_\xi^\alpha f(\xi)|\leqslant c(n, m)\sum_{|\alpha|\leqslant n+1+m}\sup_{|x|\geqslant 1}|D_x^\alpha K(x)| \tag{2.84}$$

holds with respect to the correspondence $K(x)\to f(\xi)$.

Proof. Let ε be a parameter satisfying $0\leqslant\varepsilon\leqslant\varepsilon_0(<1)$. Consider $K(x)r^{-\varepsilon}\in\mathscr{S}'$. This is continuous with respect to ε including the case $\varepsilon=0$, i.e. $K(x)r^{-\varepsilon}\to K(x)$ in $\mathscr{S}'(\varepsilon\to 0)$. Therefore, $\mathscr{F}[K(x)r^{-\varepsilon}]\to\mathscr{F}[K(x)]$ in $\mathscr{S}_\xi'$ $(\varepsilon\to 0)$.

Now, let α be a function which is identically 1 in the neighbourhood of the origin and which belongs to $\mathscr{D}$. We decompose $K(x)r^{-\varepsilon}$ as $K(x)r^{-\varepsilon}$ $=\alpha(x)K(x)r^{-\varepsilon}+(1-\alpha(x))K(x)r^{-\varepsilon}$. We now see that $\mathscr{F}[\alpha(x)K(x)r^{-\varepsilon}]$ $\to\mathscr{F}[\alpha(x)K(x)]$ in $\mathscr{E}_\xi^\infty$ $(\varepsilon\to 0)$. Next let $|\nu|=n+1$, and

$$D^\nu\{(1-\alpha(x))\,K(x)\,r^{-\varepsilon}\}=(1-\alpha(x))\,D^\nu(K(x)\,r^{-\varepsilon})+\gamma_\varepsilon(x).$$

This function belongs to L^1, and if we write $\theta^\varepsilon(\xi)$ as its Fourier image, then $\theta^{(\varepsilon)}(\xi)$ is bounded and continuous, satisfying the following properties:

(i) $|\theta^{(\varepsilon)}(\xi)|\leqslant c(n)\sum_{|\nu|\leqslant n+1}\sup_{|x|\geqslant 1}|D^\nu K(x)| \quad (0\leqslant\varepsilon\leqslant\varepsilon_0)$,

(ii) $\theta^{(\varepsilon)}(\xi)\to\theta^{(0)}(\xi)$ uniformly on R_ξ^n. (The property (i) follows from $|\theta^{(\varepsilon)}(\xi)|\leqslant\|(1-\alpha(x))D^\nu[K(x)r^{-\varepsilon}]\|_{L^1}+\|\gamma_\varepsilon(x)\|_{L^1}$.

From the previous proposition 2.13, we put

$$\mathscr{F}[K(x)r^{-\varepsilon}]=f_\varepsilon(\xi)\equiv\frac{k_\varepsilon(\omega)}{|\xi|^{n-\varepsilon}}.$$

For $\varepsilon>0$ and for an arbitrarily chosen $|\nu|=n+1$, whereupon we see that f_ε tends to f_0 uniformly on an arbitrary compact set of $R^n\backslash\{0\}$. Also, we have

$$\int_{|\xi|=1} f_0(\xi)\,\mathrm{d}S_\xi=0.$$

Now, we see that f_0 is a homogeneous function of degree $-n$. So that, from the representation formula of v.p. (2.82), we have $f_\varepsilon(\xi)\to$ v.p. $f_0(\xi)$ in $\mathscr{S}'$ $(\varepsilon\to+0)$.
From this, $\mathscr{F}[K(x)]=$ v.p. $f_0(\xi)$.

We have yet to prove (2.84). It is not difficult to do this if we recall the arguments already put forward, i.e. for $|\gamma|\leqslant m$, we have

$$\mathrm{D}_\xi{}^\gamma\mathscr{F}[(1-\alpha(x))K(x)]=\mathscr{F}[(1-\alpha(x))(-2\pi\mathrm{i}x)^\gamma K(x)]$$

and for $|\beta|=n+1+m$, by Leibniz's formula, we have a decomposition

$$\mathrm{D}_x{}^\beta\{(1-\alpha(x))x^\gamma K(x)\}=(1-\alpha(x))\,\mathrm{D}^\beta\{x^\gamma K(x)\}+\eta(x).$$

Now, K is a homogeneous function of degree 0, so that it is obvious that the L^1-norm of the function which appears on the right-hand side of the above equation can be estimated by the expression which appears on the right hand side of the inequality (2.84).

Q.E.D.

7 *The relationship of Fourier transforms and convolutions*

We have defined 'convolution' in section 10, chapter 1 and we have mentioned some relations with Fourier transforms (theorem 1.16), i.e. for $f\in L^1$ and $g\in L^2$,

$$\mathscr{F}[f*g]=\mathscr{F}[f]\mathscr{F}[g]. \tag{1.74}$$

On the other hand, after we extended Fourier transforms to the case of distribution, we saw, for an arbitrary $U\in\mathscr{S}'$, that

$$\mathscr{F}[\mathrm{D}^\alpha U]=(2\pi\mathrm{i}\xi)^\alpha\mathscr{F}[U]. \tag{2.62}$$

However, (2.62) alone is not very useful. We require a relation such as (1.74) in a more general case. To establish such a relation we must have a necessary tool, a new 'convolution' which is defined not only on functions but also on distributions. To illustrate the situation we examine the following example.

EXAMPLE. Historically, an operator $f \to g$ defined as

$$g(x) = \text{v.p.} \int_{-\infty}^{+\infty} \frac{f(t)}{x-t} \, dt = \lim_{\varepsilon \to 0} \int_{|x-t| \geqslant \varepsilon} \frac{f(t)}{x-t} \, dt$$

for $f \in \mathscr{D}$ (the dimension of the base space $n=1$) was often encountered in the classical study of analysis. Roughly speaking, according to the definition of convolution (see chapter 1) this seems to be $g(x) = (1/x) * f(x)$, but this is not correct because $1/x$ is no longer a function, i.e. it does not belong to L^1_{loc}. To be precise, we should have written $g(x)$ as

$$g(x) = \left(\text{v.p.} \frac{1}{x}\right) * f(x).$$

Then, following example 1 of the previous section, we see that v.p.$(1/x) \times$ $\times \in \mathscr{D}'_{L^2}$.

Let S and T be elements of $\mathscr{D}'_{L^2}$. Consider the definition of $S*T$. To do this, we assume $S \in \mathscr{D}'_{L^2}{}^l$ and $T \in \mathscr{D}'_{L^2}{}^m$. Then, we pick sequences $\{f_j(x)\}$ and $\{g_j(x)\}$ such that they tend to S and T respectively as $j \to \infty$ by the topology on $\mathscr{D}'_{L^2}{}^l$ and on $\mathscr{D}'_{L^2}{}^m$ respectively. In this case, we let

$$h_j(x) = \int f_j(x-t) \, g_j(t) \, dt = \int f_j(t) \, g_j(x-t) \, dt \in \mathscr{D}.$$

Now, can we define $S*T$ as the limit (if any) of $\{h_j(x)\}$ under a topology on $\mathscr{D}'$ for example? The answer is yes; this idea of defining $*$ in fact works. We are going to show this in what follows. Before doing so we establish a lemma which we need to justify our claim, and to define convolution explicitly.

LEMMA 2.13. Let $S \in \mathscr{D}'_{L^2}{}^l$, and $T \in \mathscr{D}'_{L^2}{}^m$. For $\varphi \in \mathscr{D}$, we have

$$|\langle S_x, \langle T_y, \varphi(x+y) \rangle \rangle| \leqslant C \|S\|'_{m,L^2} \|T\|'_{l,L^2} \|\varphi(x)\|_{l+m,L^1}, \tag{2.85}$$

where C is a positive constant depending only on l, m and the dimension n of the base space.

Proof. Concerning $\mathscr{D}_{L^1}{}^{l+m}$ the reader should recall the note after definition 2.10. Now, $\theta(x) = \langle T_y, \varphi(x+y) \rangle$ is a C^∞-function, and obviously $D^\alpha\theta(x)$

$=\langle T_y, D^\alpha\varphi(x+y)\rangle$. We apply the normal representation of $\mathscr{D}'_{L^2}{}^m$.

$$T=\sum_{|\beta|\leqslant m} D^\beta f_\beta(x);\quad \sum \|f_\beta(x)\|_{L^2}{}^2=\|T\|^{2\prime}{}_{m,L^2}\quad \text{(by (2.33))}.$$

Also,

$$D^\alpha\theta(x)=\sum_{|\beta|\leqslant m}(-1)^{|\beta|}\langle f_\beta(y), D^{\alpha+\beta}\varphi(x+y)\rangle.$$

By the Hausdorff–Young inequality (1.73)

$$\|\langle f_\beta(y), D^{\alpha+\beta}\varphi(x+y)\rangle\|_{L^2(R_x{}^n)}\leqslant\|f_\beta(y)\|_{L^2}\cdot\|D^{\alpha+\beta}\varphi(x)\|_{L^1}.$$

Therefore, for $|\alpha|\leqslant l$,

$$\|D^\alpha\theta(x)\|_{L^2}\leqslant C(\sum_{|\beta|\leqslant m}\|f_\beta(y)\|_{L^2}{}^2)^{\frac{1}{2}}\|\varphi(x)\|_{l+m,L^1}\leqslant C\|T\|'_{m,L^2}\|\varphi\|_{l+m,L^1}.$$

From this, we have (2.85) immediately.

Q.E.D.

Note that by (2.85) we see that the linear form:

$$\varphi(x)\to\langle S_x, \langle T_y, \varphi(x+y)\rangle\rangle,$$

is in fact a continuous linear form on $\mathscr{D}_{L^1}{}^{l+m}$; we write this as $S*T$, i.e., we define convolution as

$$\langle S*T, \varphi(x)\rangle=\langle S_x, \langle T_y, \varphi(x+y)\rangle\rangle,\qquad S*T\in(\mathscr{D}_{L^1}{}^{l+m})'\subset\mathscr{S}'. \tag{2.86}$$

We shall establish the following properties of convolution thus defined.

PROPOSITION 2.14. Let S and $T\in\mathscr{D}'_{L^2}$.

(1) $S*T$ is linear with respect to S and T. For example, let S_1 and $S_2\in\mathscr{D}'_{L^2}$. Then,

$$(aS_1+bS_2)*T=a(S_1*T)+b(S_2*T)$$

where a and b are complex numbers.

(2) Let $S\in\mathscr{D}'_{L^2}{}^l$, and let $T\in\mathscr{D}'_{L^2}{}^m$. We pick sequences $\{f_j(x)\}$ and $\{g_j(x)\}$ from $\mathscr{D}$ which are convergent to S and T respectively by the topology of $\mathscr{D}'_{L^2}{}^m$ and $\mathscr{D}'_{L^2}{}^l$. As $j\to\infty$, we have $f_j(x)*g_j(x)\to S*T$ in $(\mathscr{D}_{L^1}{}^{l+m})'$.

(3) $S*T=T*S$; $D^\alpha(S*T)=(D^\alpha S)*T=S*(D^\alpha T)$.

(4)
$$\mathscr{F}[S*T]=\mathscr{F}[S]\mathscr{F}[T], \tag{2.87}$$

where $\mathscr{F}[S]$ and $\mathscr{F}[T]\in L^2{}_{\text{loc}}$. (Therefore the term on either side of (2.87) $\in L^1{}_{\text{loc}}$.)

Proof. (1) is obvious from the definition of convolution. We prove (2); from

$$f_j*g_j-S*T=(f_j-S)*g_j+S*(g_j-T),$$

we have

$$\|(f_j*g_j)-(S*T)\|'_{(l+m),L^1}\leqslant C[\|f_j-S\|'_{l,L^2}\|g_j\|'_{m,L^2}+ \\ +\|S\|'_{l,L^2}\|g_j-T\|'_{m,L^2}].$$

Now, we see that the right hand side of this inequality tends to 0 as $j\to\infty$.

We now prove (3). The last half of (3) is obvious, therefore, we prove only the first half. To do this, it is sufficient to prove

$$\langle S_x, \langle T_y, \varphi(x+y)\rangle\rangle=\langle T_x, \langle S_y, \varphi(x+y)\rangle\rangle.$$

By Fubini's theorem the equality is guaranteed if S and $T\in\mathscr{D}$. In general, we can also ensure this is true by replacing S and T with $\{f_j(x)\}$ and $\{g_j(x)\}$ respectively (see (2)).

To prove (4), we pick the same sequences which we mentioned in the above argument and (2). By (1.74) obviously $\mathscr{F}[f_j(x)*g_j(x)]=\mathscr{F}[f_j]\mathscr{F}[g_j]$. We see that as $j\to\infty$, the left hand side of the above equation tends to $\mathscr{F}(S*T)$ in the topology on $\mathscr{S}'_\xi$. Using the normal representation of S and T,

$$\mathscr{F}[f_j](\xi)\to\mathscr{F}[S](\xi) \quad \text{in} \quad L^2{}_{\text{loc}} \quad \text{by (2.66)}$$

Therefore,

$$\mathscr{F}[f_j]\mathscr{F}[g_j]\to\mathscr{F}[S]\mathscr{F}[T] \quad \text{in} \quad L^1{}_{\text{loc}} \quad (j\to\infty).$$

Hence, we have (2.87)

Q.E.D.

Proposition 2.14 plays a fundamental rôle in applications; we give the following simple example.

OPERATOR $S*$. Let $S\in\mathscr{S}'$, and let the Fourier image of S be $s(\xi)$. We assume that $s(\xi)$ is a function having at most the same order of increment as polynomials, i.e. for certain C and M,

$$|s(\xi)| \leqslant C(1+|\xi|)^M. \tag{2.88}$$

Under this condition, we see that $S \in \mathscr{D}'_{L^2}$. In fact, if we pick a sufficiently large N,

$$s(\xi) = (1+|\xi|^2)^N \frac{s(\xi)}{(1+|\xi|^2)^N}.$$

Now, we see that $S(\xi)/(1+|\xi|^2)^N \in L^2$. Therefore, we can write

$$S = \left(1 - \frac{\Delta}{4\pi^2}\right)^N \psi(x) \quad (\psi \in L^2); \quad \hat{\psi}(\xi) = \frac{s(\xi)}{(1+|\xi|^2)^N}.$$

So that, for $T \in \mathscr{D}'_{L^2}$, we can define $S * T$ which satisfies (2.87). In particular, for $f \in L^2$, $(S * f)\hat{}(\xi) = s(\xi)\hat{f}(\xi)$.
If we use Plancherel's theorem when $s(\xi)$ is bounded, then we have

$$\|S * f\|_{L^2} = \|s(\xi)\hat{f}(\xi)\|_{L^2} \leqslant (\sup_\xi |s(\xi)|) \cdot \|\hat{f}(\xi)\|_{L^2}.$$

We now obtain the following:

Proposition 2.15. Let S be a tempered distribution (definition 2.13) and let the Fourier image of S satisfy (2.88). Then, $S \in \mathscr{D}'_{L^2}$.

The necessary and sufficient condition for the operator $S * f$ to be bounded in L^2 is that $s(\xi)$ is bounded. In this case we have

$$\|S * f\|_{L^2} \leqslant (\sup_\xi |s(\xi)|) \|f(x)\|_{L^2}. \tag{2.89}$$

This is the best possible estimate for $S * f$ because

$$\|S * \|_{\mathscr{L}(L^2, L^2)} = \sup_\xi |s(\xi)|.$$

Example. We see that

$$\left\| \left(\text{v.p. } \frac{1}{x}\right) * f(x) \right\|_{L^2} = \pi \|f(x)\|_{L^2}.$$

In general, from proposition 2.12, the operator

$$\text{v.p.}\,\frac{k(\omega)}{\gamma^n}\,*$$

(see example 5, section 6) is bounded in $L^2(R^n)$.

CONVOLUTION $(\mathscr{E}')*(\mathscr{S}')$. The operand of this operator is $\mathscr{S}'$. Let $S \in \mathscr{E}'$ and $T \in \mathscr{S}'$. Then,

$$\varphi(x) \to \langle S_x, \langle T_y, \varphi(x+y)\rangle\rangle$$

defines a continuous linear form on $\mathscr{S}$. In fact, let $\varphi_j(x) \to 0$ in $\mathscr{S}$. Then

$$\theta_j(x) = \langle T_y, \varphi_j(x+y)\rangle$$

tends uniformly to 0 on any compact set. The same argument holds for an arbitrary $D^\alpha \theta_j(x)$, so that $\langle Sx, \theta_j(x)\rangle \to 0$. Therefore, $S * T \in \mathscr{S}'$.

We have the following proposition:

PROPOSITION 2.16. Let $S \in \mathscr{E}'$ and $T \in \mathscr{S}'$. We consider a bilinear transformation $(S, T) \to S * T$.

This transformation from $\mathscr{E}' \times \mathscr{S}'$ to $\mathscr{S}'$ is in fact continuous in the following sense. Let S run through a bounded set of $\mathscr{E}'$, and $T_j \to 0$ in $\mathscr{S}'$. Then, $S * T_j \to 0$ in $\mathscr{S}'$. Also, let T run through a bounded set of $\mathscr{S}'$, and $S_j \to 0$ in $\mathscr{E}'$. Then, $S_j * T \to 0$ in $\mathscr{S}'$. In this case we call this type of continuity of a bilinear transformation *hypocontinuity*.

Proof. We prove only the first half of the proposition.

We write $\theta_j(x) = \langle T_j, \varphi(x+y)\rangle$. If φ runs through a bounded set B of $\mathscr{S}$, and x takes values such that for an arbitrary $A, |x| \leqslant A$, then the value set $\varphi\,(x+y)$ is a bounded set of $\mathscr{S}_y$.

Therefore, $\theta_j(x) \to 0$ uniformly by the topology on $\mathscr{E}_x$ as φ runs through B. So that, if S runs through a certain bounded set of $\mathscr{E}'$, then by theorem 2.13, $\langle S, \theta_j(x)\rangle$ tends to 0 uniformly for $\varphi(\in B)$ and $S(\in B')$.

Properties similar to proposition 2.14 hold for the convolution which is defined on $(\mathscr{E}') \times (\mathscr{S}')$. The proof of this fact is left to the reader. The following proposition can be easily established.

PROPOSITION 2.17. Let $S \in \mathscr{E}'$ and let $S, T \in \mathscr{D}'_{L^2}$.

(1) The associativity law holds for $S * T * U$ that is $(S * T) * U = S * (T * U)$.
Also, the commutativity law holds for convolutions.

(2) The equality $\mathscr{F}[S * T * U] = \mathscr{F}[S]\mathscr{F}[T]\mathscr{F}[U]$ is true for any S, T, and U.

8 *Laplace transforms of functions*

In this text we shall not use the Laplace transform as often as the more traditional treatises on partial differential equations. Nevertheless, we discuss briefly simple Laplace transforms (one-dimensional case) for the reader's convenience. Let $f \in L^1{}_{\text{loc}}$, and let $f(t)$ (t is the variable of f) be identically 0 if $t < 0$. In this section we impose these conditions on our argument unless we state to the contrary. In this case, we define the *Laplace transform* of $f(t)$ as

$$F(p) = \int_0^\infty e^{-pt} f(t)\,dt \tag{2.90}$$

and we write

$$f(t) \sqsupset F(p), \quad \text{or} \quad F(p) = \mathscr{L}[f(t)]. \tag{2.91}$$

We see that the Laplace transform can be defined if and only if the right hand side of the expression (2.90) is meaningful, i.e. it is defined for a complex parameter p such that the expression under the integral sign (integrand) is summable. We examine this more closely. Let $p = \xi + i\eta$. It is clear that

$$|e^{-pt} f(t)| = e^{-\xi t} |f(t)|$$

so that the following properties hold.

(1) If (2.90) is meaningful for $p = \xi + i\eta$, then, it is meaningful for an arbitrary $p = \xi + i\eta'$, i.e. it is dependent only on the real part of p.

(2) If (2.90) is meaningful for $\operatorname{Re} p_0 = \xi_0$, then for an arbitrary p_0 satisfying $\xi = \operatorname{Re} p \geqslant \operatorname{Re} p_0 = \xi_0$ it is meaningful. In fact, we can write

$$\exp(-\xi t)|f(t)| = \exp[-(\xi - \xi_0)t] \exp(-\xi_0 t)|f(t)|$$

where $\exp[-(\xi - \xi_0)t]$ is less than 1 at $t \geqslant 0$.

From these facts we see that an arbitrary real number ξ can be classified into two mutually disjoint classes, namely, an 'upper' class consisting of all ξ satisfying $e^{-\xi t}|f(t)| \in L^1$, and the rest, the 'lower' class consisting of all ξ

not satisfying the relation. We see therefore that we can define a Dedekind cut in this way because if ξ belongs to the upper class, then all $\xi'(\geqslant\xi)$ belong to the upper class, and the corresponding statement holds for the lower class.

By the abscissa of absolute convergence of a Laplace transform we mean the Dedekind cut defined as above. Let a be a Dedekind cut thus defined. If $e^{-\xi t}|f(t)|$ is summable for any ξ, we put $a=-\infty$, and if there exists no such ξ, we put $a=+\infty$.

PROPOSITION 2.18. *$F(p)$ is a holomorphic function in* $\operatorname{Re} p>a$.

Proof. First, we note that F is continuous. In fact, let p_0 be a complex number such that $\operatorname{Re} p_0>a$. Let us choose a p which lies in a neighbourhood p_0 and satisfies $\operatorname{Re} p>a+\delta(\delta>0)$. For p thus chosen, we have uniformly

$$|e^{-pt}f(t)|\leqslant e^{-(\alpha+\delta)t}|f(t)|\in L^1,$$

also, $\exp(-pt)f(t)\to\exp(-p_0t)f(t)$ as $p\to p_0$. Therefore, by Lebesgue's theorem, we have $F(p)F(p_0)$. Similarly,

$$\frac{\partial}{\partial\xi}F(\xi+i\eta)=\frac{1}{i}\frac{\partial}{\partial\eta}F(\xi+i\eta)=-\int_0^\infty e^{-(\xi+i\eta)t}tf(t)\,dt\qquad(\xi>a)$$

can be established where the right hand side is continuous. Hence, $F(p)$ is holomorphic. From this,

$$(-1)^m t^m f(t)\sqsupset F^{(m)}(p)\quad(\xi>a,\ m=1,2,\ldots).\tag{2.92}$$

Q.E.D.

PROPOSITION 2.19. *Let f and g be elements of $L^1{}_{\text{loc}}$. We assume that*

$$f(t)\sqsupset F(p)\ (\xi>a)\quad\text{and}\quad g(t)\sqsupset G(p)\ (\xi>b).$$

We have

$$h(t)=\int_0^t (ft-s)\,g(s)\,ds\equiv f*g\sqsupset F(p)\,G(p)\quad(\xi>\max(a,b)).$$

Proof. We see that $h\in L^1{}_{\text{loc}}$. In fact, let us pick $T(>0)$ and keep it fixed.

First we see that

$$\int_0^T\int_0^t |f(t-s)|\,|g(s)|\,\mathrm{d}t\,\mathrm{d}s \leqslant \int_0^T\int_0^T |f(t)|\,|g(s)|\,\mathrm{d}t\,\mathrm{d}s < +\infty,$$

so that, from Fubini's theorem, $h(t)$ is meaningful almost everywhere for $t \in [0, T]$. It follows from this that $h \in L^1{}_{\mathrm{loc}}$.

Next, let us assume that $\mathrm{Re}\,p = \xi > \max(a, b)$. It is clear that

$$(\mathrm{e}^{-\xi t}f(t)) * (\mathrm{e}^{-\xi t}g(t)) = \mathrm{e}^{-\xi t}(f(t) * g(t)).$$

From this and by the Hausdorff–Young inequality (1.73) we have

$$\|\mathrm{e}^{-\xi t}(f * g)\|_{L^1} \leqslant \|\mathrm{e}^{-\xi t} f\|_{L^1} \|\mathrm{e}^{-\xi t} g\|_{L^2}.$$

The following calculation can be easily carried out:

$$\begin{aligned}\int_0^\infty \mathrm{e}^{-pt}(f * g)(t)\,\mathrm{d}t &= \int_0^\infty \mathrm{e}^{-\mathrm{i}\eta t}\,\mathrm{e}^{-\xi t}(f * g)(t)\,\mathrm{d}t \\ &= \int_0^\infty \mathrm{e}^{-\mathrm{i}\eta t}\{(\mathrm{e}^{-\xi t} f) * (\mathrm{e}^{-\xi t} g)\}\,\mathrm{d}t \\ &= \mathscr{F}_\eta[(\mathrm{e}^{-\xi t} f) * (\mathrm{e}^{-\xi t} g)].^\dagger\end{aligned}$$

On the other hand,

$$L^1 \subset \mathscr{D}'_{L^2}{}^{[\frac{1}{2}n]+1},$$

in fact, by Sobolev's lemma, for $f \in L^1$ and $\varphi \in \mathscr{D}$, we have

$$|\langle f, \varphi\rangle| \leqslant \|f\|_{L^1} \cdot \sup|\varphi| \leqslant C\|f\|_{L^1}\|\varphi\|_{[\frac{1}{2}n]+1}.$$

Therefore, by (4) of proposition 2.14, we conclude that

$$\mathscr{F}_\eta[(\mathrm{e}^{-\xi t} f) * (\mathrm{e}^{-\xi t} g)] = \mathscr{F}_\eta[\mathrm{e}^{-\xi t} f]\mathscr{F}_\eta[\mathrm{e}^{-\xi t} g] = F(p)\,G(p).$$

Q.E.D.

We deal with the Laplace transforms of derivatives in what follows.

† The Fourier transform is defined as $\int \mathrm{e}^{-\mathrm{i}\eta t}\,\psi(t)\,\mathrm{d}t = \mathscr{F}_\eta[\psi]$. There is no essential difference between this new definition and the usual one so we employ the same notation. In this section we shall always use Fourier transforms in this sense.

PROPOSITION 2.20. Let $f(t) \overline{\mathscr{F}} F(p)$ $(\xi > a)$. Then, we have

$$f^{(m)}(t) \sqsupset p^m F(p) \qquad (\xi > a) \tag{2.93}$$

where $f^{(m)}(t)$ is a derivative in the sense of distribution.

Proof. We rewrite

$$e^{-\xi t} f'(t) = (e^{-\xi t} f(t))' + \xi e^{-\xi t} f(t).$$

From $\mathscr{F}_\eta[(e^{-\xi t} f(t))'] = i\eta \mathscr{F}_\eta[e^{-\xi t} f(t)]$,

we have $\mathscr{F}_\eta[e^{-\xi t} f'(t)] = (\xi + i\eta)\mathscr{F}_\eta[e^{-\xi t} f(t)]$, i.e. (2.93) is established when $m = 1$. Subsequently we apply this equation to establish (2.93) for $m > 1$.

Q.E.D.

Note that obviously $Y(t) \sqsupset 1/p$ $(\xi > 0)$, so that, $Y'(t) = \delta$.† Therefore,

$$\begin{cases} \delta \sqsupset 1 \\ \delta^{(m)} \sqsupset p^m \qquad (m = 1, 2, \ldots). \end{cases} \tag{2.94}$$

The argument we have presented here also holds in L^2. We shall examine this in more detail. Let $e^{-\xi t} f(t) \in L^2$ $(\xi > a')$. In this case, we pick ξ' satisfying $a' < \xi' < \xi$. We write

$$\exp(-\xi t) f(t) = \exp[-(\xi - \xi')t] \exp(-\xi' t) f(t).$$

We see that $e^{-\xi t} f(t) \in L^1$ so that $F(p)$ is a holomorphic function for $\xi > a'$.
Also, from the fact that the Fourier image of $e^{-\xi t} f(t)$ is $F(\xi + i\eta)$, and by Plancherel's theorem, we can establish

$$\begin{aligned} f(t) &= \underset{A \to +\infty}{\text{l.i.m.}} \frac{1}{2\pi} \int_{\xi - iA}^{\xi + iA} e^{(\xi + i\eta)} F(\xi + i\eta)\, d\eta \\ &= \underset{A \to \infty}{\text{l.i.m.}} \frac{1}{2\pi i} \int_{\xi - iA}^{\xi + iA} e^{pt} F(p)\, dp. \end{aligned} \tag{2.95}$$

Here we encounter a difficulty. That is, when $F(p)$ is a holomorphic function for $\xi > a$, what is the condition for $F(p)$ to be the image of a Laplace transform? In this case we need to know whether the domain of a Laplace transform can be uniquely determined. The reader should note that this problem

† Translator's note: $Y(t)$ is not explicitly defined here. See Example 2 p. 109.

can be solved only after observing the Laplace transforms of distributions. This is discussed in the following section.

9 *Laplace transforms of distributions*

We are extending what we have observed in the previous section to the case of distribution. To do this we shall use the results which we have already obtained when we studied Fourier transforms in the space $\mathscr{S}'$.

DEFINITION 2.15. Let us write $D_+{}'$ for the set of all distributions which are defined on $R^1=(-\infty, +\infty)$ with support $t\geqslant 0$ where t is real. Let $T\in D_+{}'$. We further assume that $\exp(-\xi t)T\in\mathscr{S}'$. We claim that, for an arbitrary $\xi'(\geqslant\xi)$, $\exp(-\xi' t)T\in\mathscr{S}'$.

In fact, let $\alpha(t)$ be a C^∞-function taking the value 1 for $t\geqslant -\frac{1}{2}$, and 0 for $t\leqslant -1$. Then we see that the support of T lies on $t\geqslant 0$. We can write

$$\exp(-\xi' t)T=\alpha(t)\exp[-(\xi'-\xi)t]\exp(-\xi t)T.$$

For $\xi'>\xi$, $\alpha(t)\exp[-(\xi'-\xi)t]\in\mathscr{S}$ is true. Therefore, from the argument at the beginning of section 5, we have $(\mathscr{S})\times(\mathscr{S}')\subset(\mathscr{S}')$ so that $\exp(-\xi' t)T\in\mathscr{S}'$.

From this, we can assign a Dedekind cut if we classify the whole field of real numbers ξ by the conditions $\exp(-\xi t)T\in\mathscr{S}'$ and $\exp(-\xi t)T\notin\mathscr{S}'$. We write b as the real number thus obtained as the Dedekind cut. Let us call *b the abscissa of convergence by a Laplace transform with respect to a distribution.*

When $T\in L^1{}_{\text{loc}}$, we note that the relation $b\leqslant a$ holds where a is the abscissa of absolute convergence of the same Laplace transform (see the previous section). Note that $a=b$ is not generally true. In our case, we only consider distributions so that we write a instead of b. We have described the representation of $\mathscr{S}'$ in (2.55) of theorem 2.14. Considering this we use the following form of representation for our future arguments.

LEMMA 2.14. Let $T\in\mathscr{S}'$. In this case

$$T=(1+|x|^2)^k\sum D^\alpha f_\alpha(x) \quad (f_\alpha\in\mathscr{B}^0) \tag{2.96}$$

where k is a positive integer and $\mathscr{B}^0$ is the space of bounded continuous functions.

Proof. Observe (2.55). We see that it is sufficient to prove the following. If $f \in L^2$, then $f(x) = (1-\Delta)^p \varphi(x)$. where $\varphi \in \mathscr{B}^0$ and p is a positive integer. In fact, from

$$\|\hat{f}(\xi)(1+4\pi^2|\xi|^2)^{-p}\|_{L^1} \leqslant \|\hat{f}(\xi)\|_{L^2} \cdot \|(1+4\pi^2|\xi|^2)^{-2p}\|_{L^2},$$

we can choose p $(4p>n)$ with

$$\varphi(x) = \bar{\mathscr{F}}[(1+4\pi^2|\xi|^2)^{-p}\hat{f}(\xi)].$$

Then, φ satisfies our condition.

Q.E.D.

Let $\exp(-\xi t)T \in \mathscr{S}'$ $(\xi > a)$. Of course, in this section, we assume that $T \in D_+'$. Let us choose ξ_0 so that it satisfies $a < \xi_0 < \xi$, and let α be the same function as we defined at the beginning of this section.

Then, we decompose as follows:

$$\exp(-\xi t)T = \alpha(t)\exp[-(\xi-\xi_0)t]\exp(-\xi_0 t)T.$$

From $\exp(-\xi_0 t)T \in \mathscr{S}'$ and lemma 2.14, which we have just obtained, we have

$$\exp(-\xi t)T = \exp[-(\xi-\xi_0)t]\beta(t)\sum_j \left(\frac{\mathrm{d}}{\mathrm{d}t}\right)^j f_j(t)$$

where $\beta(t)$ is the product of $\alpha(t)$ and a polynomial, and $f_j(t) \in \mathscr{B}^0$. From this

$$\mathscr{F}_\eta[\mathrm{e}^{-\xi t}T] = \sum_j \left\langle \exp(-\mathrm{i}\eta t), \exp[-(\xi-\xi_0)t]\beta(t)\left(\frac{\mathrm{d}}{\mathrm{d}t}\right)^j f_j(t) \right\rangle$$

$$= \sum_j (-1)^j \left\langle \left(\frac{\mathrm{d}}{\mathrm{d}t}\right)^j [\beta(t)\exp(-(\xi+\mathrm{i}\eta-\xi_0)t)],\ f_j(t) \right\rangle.$$

This last expression can be re-written in an integrated form which represents a holomorphic function of $p = \xi + \mathrm{i}\eta$ for $\xi > \xi_0$ where ξ_0 can be taken in an arbitrary small neighbourhood of a.

Proposition 2.21. $F(p) \equiv F(\xi+\mathrm{i}\eta) = \mathscr{F}_\eta[\mathrm{e}^{-\xi t}T]$ is a holomorphic function for $\xi > a$. Furthermore, if we choose ξ_0 as satisfying $a < \xi_0 < \xi$, we can write $F(p)$ as

$$F(p) = \langle \exp[-(p-\xi_0)t], \exp(-\xi_0 t)T \rangle \quad (\operatorname{Re} p > \xi_0). \tag{2.97}$$

We call the $F(p)$ which satisfies the conditions of proposition 2.21 the *Laplace transform* of T, or the *Laplacian image*, and we write

$$T \sqsupset F(p) \quad (\xi > a). \tag{2.98}$$

Related to proposition 2.19, we have the following:

PROPOSITION 2.22. Let $S \times F(p)$ $(\xi > a)$ and let $T \times G(p)$ $(\xi > b)$. Then,

$$S * T \sqsupset F(p)G(p) \quad (\xi > \max(a, b)). \tag{2.99}$$

Proof. Let $\xi > \max(a, b)$. We first show that

$$(e^{-\xi t}S) * (e^{-\xi t}T) = e^{-\xi t}(S * T). \tag{2.100}$$

To see this, we observe $e^{-\xi t}S$, $e^{-\xi t}T \in \mathscr{D}'_{L^2}$ (see the proof of the previous proposition). From the definition of convolution (see the previous paragraph), if $\varphi \in \mathscr{D}$, then

$$\begin{aligned}\langle (e^{-\xi t}S) * (e^{-\xi t}T), \varphi(t)\rangle &= \langle e^{-\xi s}S_s, \langle e^{-\xi t}T_t, \varphi(s+t)\rangle\rangle \\ &= \langle S_s, \langle e^{-\xi(s+t)}T_t, \varphi(s+t)\rangle\rangle = \langle S * T, e^{-\xi t}\varphi(t)\rangle \\ &= \langle e^{-\xi t}(S * T), \varphi(t)\rangle,\end{aligned}$$

i.e. (2.100) is true. From this

$$\mathscr{F}_\eta[e^{-\xi t}(S * T)] = \mathscr{F}_\eta[(e^{-\xi t}S) * (e^{-\xi t}T)].$$

Therefore, from proposition 2.14, (4), we see that this is exactly equal to

$$\mathscr{F}_\eta[e^{-\xi t}S]\mathscr{F}_\eta[e^{-\xi t}T] = F(p)G(p).$$

Q.E.D.

In relation to proposition 2.20 we have the following:

PROPOSITION 2.23. Let $T \times F(p)$ $(\xi > a)$. Then,

$$\left(\frac{d}{dt}\right)^m T \sqsupset p^m F(p) \quad (\xi > a, m = 1, 2, \ldots). \tag{2.101}$$

Proof. It is sufficient to show the statement is true when $m = 1$.

To see this, we observe (2.97). We see this means

$$\exp(-\xi_0 t)T \in \mathscr{S}' \quad \text{and} \quad \alpha(t)\exp[-(p-\xi_0)t] \in \mathscr{S},$$

where $\alpha(t)$ has been defined at the beginning of this section, so that this is a dual form. On the other hand, from

$$\exp(-\xi_0 t)T' = (\exp(-\xi_0 t)T)' + \xi_0 \exp(-\xi_0 t)T,$$

we have

$$\begin{aligned}\langle \exp[-(p-\xi_0)t], (\exp(-\xi_0 t)T)' \rangle = \\ &= -\langle (\exp[-(p-\xi_0)t])', \exp(-\xi_0 t)T \rangle \\ &= (p-\xi_0)\langle \exp[-(p-\xi_0)t], \exp(-\xi_0 t)T \rangle = \\ &= (p-\xi_0)F(p),\end{aligned}$$

for $p > \xi_0$, i.e. for $\xi > \xi_0$ (2.101) is true when $m=1$. Because ξ_0 can be taken in any small neighbourhood of a, we see that (2.101) is true for $\xi > a$.

Q.E.D.

THEOREM 2.17. Let $T \in D_+'$ and let $T \sqsupset F(p)$ $(\xi > a)$. If $F(p)=0$, then $T=0$, i.e. the correspondence of T and $F(p)$ is bijective.

Proof. We know $F(\xi + i\eta) = \mathscr{F}_\eta[e^{-\xi t}T]$ $(\xi > a)$. Also, the Fourier transform between $(\mathscr{S}')_t$ and $(\mathscr{S}')_\eta$ is bijective. Therefore, $e^{-\xi t}T=0$ is true from our assumptions, i.e. $T=0$.

Q.E.D.

Characterization of Laplacian images

In the previous section we have mentioned a problem. What is the necessary and sufficient condition for $F(p)$, (a holomorphic function defined for $\mathrm{Re}\, p > a$) to be the Laplace image of $T(\in D_+')$? The answer to this question is as follows:†

THEOREM 2.18.

(1) Let the abscissa of convergence of the Laplace transform of $T(\in D_+')$

† L. Schwartz: *Méthodes mathématiques pour les sciences physiques*, Paris, (1961), p. 249.

be a $(<+\infty)$. In this case, for an arbitrary $\varepsilon(>0)$, there exist positive real numbers A and k, and for $T \sqsupset F(p)$,

$$|F(p)| \leqslant A(1+|p|)^k \quad (\operatorname{Re} p \geqslant a+\varepsilon) \tag{2.102}$$

i.e. $F(p)$ is at most of the same increasing degree as polynomials.

(2) Conversely, for a holomorphic function $F(p)$ satisfying (2.102), there exists a unique $T(\in D_+')$ which satisfies the condition: $T \sqsupset F(p)$ $(\xi > a)$.

Proof.

(1) This part of the theorem seems to be already proved by using the expression

$$F(\xi+i\eta)=\sum_j (-1)^j \left\langle \left(\frac{d}{dt}\right)^j [\beta(t)\exp[-(\xi+i\eta-\xi_0)t]],\ f_j(t) \right\rangle \quad (f_j \in \mathscr{B}^0),$$

which has been obtained previously. It is not; the reason is that we must prove the following fact. The support of $f_j(t)$ is $t \geqslant 0$.

So far we have proved that for an arbitrary $\varepsilon(>0)$, the support is $t \geqslant -\varepsilon$. To see this, we use the fact that if the support of a distribution T is contained in $[a, b]$, then, there is an expression

$$\langle T, \varphi \rangle = \sum_{j=0}^{m} \int_a^b \varphi^{(j)}(t)\, dg_j(t)$$

where g_j $(j=0, 1, \ldots, m)$ are functions of bounded variations. The fact we have just stated is not immediately clear, so that we prove this later. We assume it for the time being.

Now, let $\alpha(t)$ be a function belonging to $\mathscr{D}$ which is identically 1 in the neighbourhood of $t=0$. Let us decompose $T=\alpha T+(1-\alpha)T \equiv T_1+T_2$. Obviously, the abscissae of convergence of T and T_2 coincide $(=a)$. From the fact which we have assumed

$$\langle e^{-pt}, T_1 \rangle = \sum_{j=0}^{m} \int_a^b (e^{-pt})^{(j)}\, dg_j(t).$$

(1) is therefore true for T_1.

On the other hand, for T_2, the support is $t>0$. So that, for example, we take ξ_0 as $a+\frac{1}{2}\varepsilon$ in the first expression. Then, the support of f_j is $t \geqslant 0$, i.e. the Laplace image of T_2 satisfies (2.102). Therefore, part (1) is proved.

(2) To see this, let

$$\xi \geqslant a+\varepsilon(\varepsilon>0) \quad \text{and} \quad |F(p)| \leqslant A(1+|p|)^{-2}.$$

Then,

$$f_\xi(t)=\frac{1}{2\pi i}\int_{\xi-i\infty}^{\xi+i\infty} e^{pt}F(p)\,dp \tag{2.103}$$

is meaningful, i.e. $e^{-\xi t}f_\xi(t)$ is a bounded and continuous function with respect to t. In fact, the value of $f_\xi(t)$ is independent of ξ. The reason is that for ξ we take ξ_1, ξ_2 satisfying $a+\varepsilon\leqslant\xi_1<\xi_2$, then

$$\int_\Gamma e^{pt}F(p)\,dp=0$$

because $F(p)$ is holomorphic where Γ is the circumference of the rectangle which is formed by $\xi=\xi_1$, $\xi=\xi_2$, and $\eta=\pm R$. We put $R\to+\infty$. We see that $f_{\xi_1}(t)-f_{\xi_2}(t)=0$. So that we write $f(t)$ instead of $f_\xi(t)$.

Next, let us prove that $F(p)$ is the Laplace transform of $f(t)$. To do this, we use the inversion formula of Fourier transforms in distribution. From

$$f(t)=\frac{1}{2\pi}e^{\xi t}\int_{-\infty}^{+\infty} e^{+i\eta t}F(\xi+i\eta)\,d\eta \tag{2.104}$$

it follows that

$$e^{-\xi t}f(t)=\frac{1}{2\pi}\mathscr{F}_\eta[F(\xi+i\eta)].$$

That is

$$F(\xi+i\eta)=\mathscr{F}_\eta[e^{-\xi t}f(t)]$$

where $\mathscr{F}_\eta$ is a Fourier transform in the sense of distribution. Therefore,

$$f(t)\sqsupset F(p)\qquad(\xi\geqslant a+\varepsilon).$$

We note that $f(t)=0$ at $t<0$. In fact, from (2.104) we have

$$|f(t)|\leqslant\frac{e^{\xi t}}{2\pi}A\int_{-\infty}^{+\infty}\frac{d\eta}{1+\xi^2+\eta^2}\leqslant\frac{A}{2}e^{\xi t}.$$

If we make $\xi>a+\varepsilon$ and let $\xi\to+\infty$, then $|f(t)|=0$ when $t<0$.

Let us consider the general case. Let $a\neq-\infty$ in (2.102) and consider $F(p)/(p-a)^{k+2}$. This satisfies our condition in $\xi\geqslant a+\varepsilon$. Therefore, there exists a continuous function $f\in\mathscr{D}_+'$ satisfying

$$f(t)\sqsupset\frac{F(p)}{(p-a)^{k+2}}\qquad(\xi\geqslant a+\varepsilon).$$

Therefore, referring to (2.101) we have[†]

$$T=(\underbrace{(\delta'-a\delta)*\cdots*(\delta'-a\delta)}_{k+2})*f(t)\sqsupset F(p).$$

Hence $T\in\mathscr{D}_+'$. The uniqueness of T has already been proved (theorem 2.17). The case when $a=-\infty$ is obvious.

Q.E.D.

Next, we prove the fact which we left unproved in our previous discussion. That is:

LEMMA 2.15. If the support of a distribution T is contained in $[a, b]$, then for $\varphi\in\mathscr{E}$, we have a representation

$$\langle T,\varphi\rangle=\sum_{j=0}^{m}\int_a^b \varphi^{(j)}(t)\,\mathrm{d}g_j(t)$$

where $g_j(t)$ is of bounded variation defined in $[a, b]$.

Proof. We separate the proof into two steps.

(A) Let the order of T be m, i.e. if $\varphi_j\in\mathscr{E}(R^1)$ satisfies $|\varphi_j|_m\to 0$, then $\langle T,\varphi_j\rangle\to 0$. In this case T can be uniquely extended to a continuous linear form on $\mathscr{E}^m(R^1)$, and if $\varphi\in\mathscr{E}^m$ is identically 0 in $[a, b]$, then $\langle T,\varphi\rangle=0$. In fact let us take $[a, b]$ as $(-\infty, 0]$ to simplify the argument.

Also, let α be a C^∞-function, $\alpha(t)\equiv 1$ in $t\leqslant\frac{1}{2}$, and 0 in $t\geqslant 1$, satisfying $0\leqslant\alpha(t)\leqslant 1$ otherwise, and let $\varphi\in\mathscr{E}^m$ be a function which is identically 0 in $(-\infty, 0]$. Under these assumptions, we see that

$$|\alpha(t/\varepsilon)\,\varphi(t)|_m\to 0\quad(\varepsilon\to+0).$$

On the other hand, $\alpha(t/\varepsilon)$ is identically 1 in the neighbourhood of $(-\infty, 0]$. Therefore,

$$\langle T,\varphi\rangle=\langle T,\alpha(t/\varepsilon)\varphi(t)\rangle.$$

Hence, as $\varepsilon\to 0$, this tends to 0. Hence, $\langle T,\varphi\rangle=0$. In general, we see our argument is valid for any $[a, b]$.

† Otherwise expressed: $T=((\mathrm{d}/\mathrm{d}t)-a)^{k+2}f(t)$.

(B) For an arbitrary $\varphi \in \mathscr{E}^m(R^1)$, let us define $\psi(t)$ as follows: $\psi(t)$ is equal to $\varphi(t)$ on $[a, b]$ including mth derivatives satifyings

$$\psi(t)=\begin{cases}\varphi(t) & (a\leqslant t\leqslant b),\\ \beta(t)\sum_{\nu=0}^{m} a_\nu\varphi(a+\nu(a-t)) & (t\leqslant a),\\ \gamma(t)\sum_{\nu=0}^{m} a_\nu\varphi(b-\nu(t-b)) & (t\geqslant b),\end{cases}$$

where

(1) a_ν is defined such that $\psi(t)$ is equal to $\varphi(t)$ at $t=a$ and $t=b$, i.e. it is a constant satisfying $\sum_{\nu=0}^{m} a_\nu(-\nu)^j=1, j=0, 1,\ldots, m$. This system of equations has a unique solution $(a_0, a_1,\ldots, a_m)$.

(2) $\beta(t)$ is a C^∞-function such that for example it is identically equal to 1 for $a-(b-a)/2m\leqslant t\leqslant a$ and 0 for $t\leqslant a-(b-a)/m$. $\gamma(t)$ is also similarly defined. We can now establish the following two facts:

(*a*) $\psi \in \mathscr{D}^m$, and $\psi(t)=\varphi(t)$ on $[a, b]$. Therefore, using the result which has been previously obtained, we have $\langle T, \varphi\rangle=\langle T, \psi\rangle$.

(*b*) The linear mapping from $\psi \in \mathscr{E}^m[a, b]$† to $\psi \in \mathscr{D}^m$ is continuous. Therefore, T can be regarded as a continuous linear form defined in a product space‡ $(\psi(t), \psi'(t),\ldots, \psi^{(m)}(t))\subset\prod \mathscr{E}^0[a, b]$.

Now, by the Hahn–Banach theorem* we see that there exists $\tilde{T}$, a continuous extension of T, and $\tilde{T}$ is a continuous linear form on $\prod \mathscr{E}^0[a, b]$. By the representation theorem of continuous linear forms on $\mathscr{E}^0[a, b]$, for $\psi \in \mathscr{E}^m[a, b]$, there exists a function g_j of bounded variation satisfying

$$\langle T, \psi\rangle=\sum_{j=0}^{m}\int_a^b \psi^{(j)}(t)\, dg_j(t).$$

Therefore,

$$\langle T, \varphi\rangle=\langle T, \psi\rangle=\sum_{j=0}^{m}\int_a^b \varphi^{(j)}(t)\, dg_j(t).$$

Q.E.D.

† The meaning of this is that ψ is defined only on $[a, b]$.

‡ Because $\psi(t)\in\mathscr{E}^m[a, b]\to\langle T, \psi(t)\rangle_{\mathscr{D}^m}$ is continuous, there exists a unique $U\in\mathscr{E}'^m[a, b]$ and $\langle T, \psi(t)\rangle_{\mathscr{D}^m}=\langle U, \psi(t)\rangle_{\mathscr{E}^m[a,b]}$. We replace U by T.

* For example see Lang: *Analysis II* (Addison-Wesley, 1969), or Yosida: *Functiona Analysis I* (Springer-Verlag, 1968).

10 *Laplace transforms of vector-valued functions*

Let E be a Banach space. We consider the extension of the idea of the Laplace transform of scalar-valued functions to that of vector-valued functions, which take their values in E. More generally, this extension can be achieved even in the case of distributions, but here we satisfy ourselves with dealing with the case of functions only.

In this chapter, to avoid possible confusion, we denote the vector-value of a function as $\vec{u}(t)$, etc. Also, for simplicity, we assume

(1) $\vec{u}(t)=0$ at $t<0$, and $\vec{u}$ is integrable in an arbitrary finite interval of t in the sense of Riemann. For example, it can be continuous, or of bounded variation. Note that by saying u is of *bounded variation* in $[a, b]$ we mean that for an arbitrary partition $\Delta: a=t_0<t_1<\cdots<t_n=b$, we contend that

$$\sum_{i=1}^{n} \|\vec{u}(t_i)-\vec{u}(t_{i-1})\|$$

is *not* greater than a constant M.

(2) With respect to the increasing order when $t \to +\infty$, we assume that there exist constants $C(>0)$, and β such that

$$\|\vec{u}(t)\| \leqslant Ce^{\beta t} \tag{2.105}$$

is always true. In this case it is easy to see that

$$\vec{U}(p)=\int_0^{\infty} e^{-pt}\vec{u}(t)dt \quad (\operatorname{Re} p=\xi>\beta) \tag{2.106}$$

is a holomorphic function of p having values in E.

We call this $\vec{U}(p)$ the *Laplace transform* of $\vec{u}(t)$. Now, we consider the inversion formula of the Laplace transform. To do this, we show the inversion formula of Fourier, (theorem 1.6, chapter 1) is also valid in our vector-valued case.

LEMMA 2.16. Let $\vec{f}$ be defined in $(-\infty, +\infty)$ satisfying

$$\int_{-\infty}^{+\infty} \|\vec{f}(\theta)\| d\theta < +\infty .$$

Also, let $\vec{f}$ be of bounded variation in an arbitrary finite interval. Then, we can establish

$$\lim_{A\to+\infty}\frac{1}{\pi}\int_{-\infty}^{+\infty}\vec{f}(\theta)\frac{\sin A(t-\theta)}{t-\theta}\,d\theta=\frac{1}{2}[\vec{f}(t+0)+\vec{f}(t-0)]$$

where it is to be understood that the convergence takes place in the topology of E. Moreover, we note that this is uniform convergence in any interval which lies entirely in the interior of an arbitrary finite interval where $\vec{f}$ is continuous.

Proof. Let us show that we can reduce this case to the case of a scalar-valued function. To do this, we first redefine $\vec{f}$ as

$$\vec{f}(t)=\tfrac{1}{2}[\vec{f}(t+0)+\vec{f}(t-0)]$$

at its points of discontinuity. Then, we take

$$\vec{\varphi}(\theta)=\vec{f}(t+\theta)+\vec{f}(t-\theta)$$

and show

$$\lim_{A\to+\infty}\int_{0}^{+\infty}\varphi(\theta)\frac{\sin A\theta}{\theta}\,d\theta=\pi\vec{f}(t).$$

Now, let us take $\delta(>0)$ sufficiently small and N sufficiently large. Then, we decompose the above integral as follows:

$$\int_{0}^{\delta}\varphi(\theta)\frac{\sin A\theta}{\theta}\,d\theta+\int_{\delta}^{N}\dots d\theta+\int_{N}^{\infty}\dots d\theta\equiv\vec{I}_1+\vec{I}_2+\vec{I}_3.$$

Because N is large enough we can see that, irrespective of the magnitude of A, $\|\vec{I}_3\|<\varepsilon$.

Next, we consider $\vec{I}_2$. Pick $\vec{e}'\in E'$ (where E' is the dual space of E) satisfying $\|\vec{e}'\|=1$. Then, we have

$$\langle\vec{I}_2,\vec{e}'\rangle=\int_{\delta}^{N}\langle\vec{\varphi}(\theta),\vec{e}'\rangle\frac{\sin A\theta}{\theta}\,d\theta.$$

Following the same argument as the proof of theorem 1.3 (Riemann–Lebesgue), we have to prove that

$$\int_{\delta}^{N}\left|\left\langle\frac{1}{\theta+h}\vec{\varphi}(\theta+h)-\frac{1}{\theta}\vec{\varphi}(\theta),\vec{e}'\right\rangle\right|d\theta \tag{2.107}$$

tends to 0 uniformly as $h \to 0$ when $\vec{e}'$ moves around the unit sphere in E'. Then, it can be proved that $\vec{I}_2 \to 0$ in E $(A \to \infty)$. We can show this as follows: (2.107) is estimated by

$$\frac{2|h|}{\delta^2} \int_{\delta}^{N} \|\varphi(\theta)\| \mathrm{d}\theta + \frac{2}{\delta} \int_{\delta}^{N} \|\varphi(\theta+h) - \vec{\varphi}(\theta)\| \mathrm{d}\theta$$

because $\|\vec{e}'\| = 1$. In fact this expression tends to 0 as $h \to 0$. We note that the second integral is estimated by

$$\int_{-N}^{N} \|\vec{f}(\theta+h) - \vec{f}(\theta)\| \mathrm{d}\theta$$

so that we need only estimate $\vec{I}_1$. It is obvious that

$$\vec{I}_1 - \pi \vec{f}(t) = \int_0^{\delta} \vec{\psi}(\theta) \frac{\sin A\theta}{\theta} \mathrm{d}\theta - 2\vec{f}(t) \int_{\delta A}^{\infty} \frac{\sin \theta}{\theta} \mathrm{d}\theta,$$

and

$$\vec{\psi}(\theta) = \vec{\varphi}(\theta) - 2\vec{f}(t) = \vec{f}(t+\theta) + \vec{f}(t-\theta) - 2\vec{f}(t).$$

The second integral tends to 0 as $A \to 0$, therefore we consider the first integral.

Together with $\vec{e}' \in E'$ and $\|\vec{e}'\| = 1$, we consider

$$\left\langle \int_0^{\delta} \vec{\psi}(\theta) \frac{\sin A\theta}{\theta} \mathrm{d}\theta, \vec{e}' \right\rangle = \int_0^{\delta} \langle \vec{\psi}(\theta), \vec{e}' \rangle \frac{\sin A\theta}{\theta} \mathrm{d}\theta.$$

Let $v(\delta) = \underset{0 \leqslant \theta < \delta}{\text{Total variation}} [\vec{\varphi}(\theta)]$. Then, we see

$$\left| \int_0^{\delta} \langle \vec{\psi}(\theta), \vec{e}' \rangle \frac{\sin A\theta}{\theta} \mathrm{d}\theta \right| \leqslant 2\sqrt{2}\, C v(\delta),$$

where

$$C = \sup_{\varepsilon \geqslant 0} \left| \int_0^{\xi} \frac{\sin \theta}{\theta} \mathrm{d}\theta \right|.$$

By our assumption $v(\delta)$ can be taken as small as we like, if we make δ small enough. Therefore

$$\|\vec{I}_1 - \pi \vec{f}(t)\| < \varepsilon.$$

The proof of the uniformity of this convergence is left as an (easy) exercise for the reader.

Q.E.D.

An immediate result from this lemma is:

THEOREM 2.19.[†] Let $\vec{u}(t)=0$ for $t<0$, and let $\vec{u}(t)$ satisfy (2.105) for $t\geqslant 0$. If $\vec{u}$ is of bounded variation in an arbitrary finite interval, then we have a Laplace's inversion formula in the following sense:

$$\tfrac{1}{2}[\vec{u}(t+0)+\vec{u}(t-0)]=\lim_{A\to+\infty}\frac{1}{2\pi i}\int_{\xi-iA}^{\xi+iA} e^{pt}\vec{U}(p)dp \qquad (\xi>\beta),$$

where the convergence is to be understood in the sense of the topology of E. Furthermore, we see that the convergence is uniform in an arbitrary finite interval which lies completely in the interior of an interval where $\vec{u}$ is continuous.

Proof. By (2.106) and Fubini's theorem, we see

$$\begin{aligned}&\lim_{A\to+\infty}\int_0^\infty \vec{u}(\theta)\,d\theta\,\frac{1}{2\pi i}\int_{\xi-iA}^{\xi+iA} e^{p(t-\theta)}\,dp\\ &=\lim_{A\to+\infty}\frac{1}{\pi}\int_0^\infty \vec{u}(\theta)\,e^{\xi(t-\theta)}\,\frac{\sin A(t-\theta)}{t-\theta}\,d\theta.\end{aligned}$$

Then, applying the previous lemma to this expression we have our desired result

Q.E.D.

11 *Fourier transform of a spherically symmetric function*

First, we note that when we wish to calculate the Fourier transform of a certain function we can consider this as the calculation of a Hankel transform providing the given function is dependent only on the distance r from the origin. We see this as follows.

Let

$$f(x_1, x_2, \dots, x_n)=\Phi(r), \quad r=|x|=(x_1{}^2+\cdots+x_n{}^2)^{\frac{1}{2}}.$$

† See E. Hille & R. S. Phillips: *Functional analysis and semi-groups*, (*Amer. Math. Soc. Coll. Publ.* 31 1957), where the reader can find a detailed study of the same type of relation.

To simplify our argument we assume $f \in L^1$. Recall the discussion at the end of section 5. Then, we see that in this case, $\hat{f}(\xi)$, the Fourier transform of $f(x)$, is in fact a function dependent on $p=|\xi|$. Let us write

$$\Psi(\rho)=\int e^{-2\pi i x\xi} f(x)\, dx \quad (\rho=|\xi|).$$

Then we have

$$\Psi(\rho)=\int_0^\infty I(r)\, dr$$

$$I(r)=\int_{|x|=r} e^{-2\pi i x\xi}\Phi(r)\, dS=\Phi(r)\int_{|x|=r} e^{-2\pi i x\xi}\, dS.$$

We write

$$\xi=(\xi_1, 0, \dots, 0)=(\rho, 0, \dots, 0)$$

by which we introduce a polar coordinate system.

Let the angle between $\overrightarrow{Ox}$ and the x_1-axis be θ. Now, we see that the intersection between the sphere $|x|=r$ and the hyperplane $X_1=r\cos\theta$ is in fact the $(n-2)$-dimensional sphere with the diameter $r\sin\theta$ and its surface area is

$$2\frac{\pi^{\frac{1}{2}(n-1)}}{\Gamma(\frac{1}{2}(n-1))}(r\sin\theta)^{n-2}.$$

Therefore we can write

$$dS=2\frac{\pi^{\frac{1}{2}(n-1)}}{\Gamma(\frac{1}{2}(n-1))}(r\sin\theta)^{n-2} r\, d\theta.$$

Hence,

$$I(r)=2\frac{\pi^{\frac{1}{2}(n-1)}}{\Gamma(\frac{1}{2}(n-1))}\Phi(r)r^{n-1}\int_0^\pi e^{-2\pi i r\rho\cos\theta}(\sin\theta)^{n-2}\, d\theta,$$

where the right hand integral is equal to an integral representation of a Bessel function.

$$J_\nu(x)=\frac{(\frac{1}{2}x)^\nu}{\pi^{\frac{1}{2}}\Gamma(\nu+\frac{1}{2})}\int_0^\pi e^{\pm ix\cos\theta}\sin^{2\nu}\theta\, d\theta.$$

If we put

$$\nu=\tfrac{1}{2}(n-2), \quad x=2\pi\rho r,$$

we have

$$I(r)=2\frac{\pi^{\frac{1}{2}(n-1)}}{\Gamma(\frac{1}{2}(n-1))}\cdot\frac{\pi^{\frac{1}{2}}\Gamma(\frac{1}{2}(n-1))}{(\pi r\rho)^{\frac{1}{2}(n-2)}}\Phi(r)r^{n-1}J_{\frac{1}{2}(n-2)}(2\pi\rho r)$$

$$=\frac{2\pi}{\rho^{\frac{1}{2}(n-2)}}\Phi(r)r^{\frac{1}{2}n}J_{\frac{1}{2}(n-2)}(2\pi\rho r).$$

Therefore, we obtain

$$\Psi(\rho)=\frac{2\pi}{\rho^{\frac{1}{2}(n-2)}}\int_0^{+\infty}\Phi(r)\,r^{\frac{1}{2}n}J_{\frac{1}{2}(n-2)}(2\pi\rho r)\,\mathrm{d}r. \tag{2.108}$$

Next, we assume that $\Phi(r)$ has at most the increasing order of polynomials.† We have already seen that $f(x)=\Phi(r)\in\mathscr{S}'$ at the beginning of section 5. By referring to the argument there we conclude that $\Phi_N(r)\to\Phi(r)$ in $\mathscr{S}_x'$ where $\Phi_N(r)$ is defined as

$$\Phi_N(r)=\begin{cases}\Phi(r) & (r\leqslant N),\\ 0 & (r>N).\end{cases}$$

If we take $\varepsilon(>0)$, we have

$$\Phi(r)\,\mathrm{e}^{-\varepsilon r}\to\Phi(r)\quad\text{in}\quad\mathscr{S}_x'\quad(\varepsilon\to+0)$$

Therefore, the following formula is true:

$$\begin{aligned}\Psi(\rho)&=\lim_{A\to+\infty}\frac{2\pi}{\rho^{\frac{1}{2}(n-2)}}\int_0^{A}\Phi(r)\,r^{\frac{1}{2}n}J_{\frac{1}{2}(n-2)}(2\pi\rho r)\,\mathrm{d}r\\ &=\lim_{\varepsilon\to+0}\frac{2\pi}{\rho^{\frac{1}{2}(n-2)}}\int_0^{+\infty}e^{-\varepsilon r}\Phi(r)\,r^{\frac{1}{2}n}J_{\frac{1}{2}(n-2)}(2\pi\rho r)\,\mathrm{d}r,\end{aligned} \tag{2.109}$$

where the convergence is of course in the sense of the topology given to $\mathscr{S}_\xi'$. Therefore, $\Psi(\rho)$ is, in general, a distribution.

Example. In the two-dimensional case, the Fourier image of $\Phi(r)=(\sin ar)/r\,(a>0)$ is

$$\Psi(\rho)=\begin{cases}0 & (\rho\geqslant a/2\pi),\\ \dfrac{2\pi}{\sqrt{[a^2-(2\pi\rho)^2]}} & (\rho<a/2\pi).\end{cases} \tag{2.110}$$

To see this, we consider (2.109). We have

$$\Psi_\varepsilon(\rho)=2\pi\int_0^{+\infty}\mathrm{e}^{-\varepsilon r}\sin ar J_0(2\pi\rho r)\,\mathrm{d}r=2\pi\,\mathrm{Im}\int_0^{+\infty}\mathrm{e}^{(-\varepsilon+ia)r}J_0(2\pi\rho r)\,\mathrm{d}r.$$

† I.e. there exist C and l satisfying $|\Phi(r)|\leqslant Cr^l\quad(r\to+\infty)$.

On the other hand, as is well known,

$$\int_0^{+\infty} e^{-pr} J_0(r)\,dr = \frac{1}{\sqrt{(1+p^2)}} \qquad (\mathrm{Re}\, p>0), \tag{2.111}$$

where the value of the root is taken as positive when p is positive. Therefore,

$$\int_0^{+\infty} e^{(-\varepsilon+ia)r} J_0(2\pi\rho r)\,dr = \frac{1}{\sqrt{[(2\pi\rho)^2-(a+i\varepsilon)^2]}}$$

where the value of the root is taken as $\sim(2\pi\rho)^{-1}$ as $\rho\to+\infty$. From this, we have the above representation of $\Psi(\rho)$.

Note. So far we have discussed Fourier images. The formula for obtaining inverse Fourier images should be exactly the same as (2.109). This is obvious if we consider the procedure which we employed to obtain (2.109), but, nevertheless, we shall describe the situation in what follows.

Let us consider $\psi(\xi)$, a function dependent only on $|\xi|=\rho$ which has at most the increasing order of polynomials as $\rho\to+\infty$. Let the Fourier inverse image of $\psi(\xi)=\Psi(\rho)$ be $f(x)=\Phi(r)$. Then, we have

$$\begin{aligned}\Phi(r) &= \lim_{A\to+\infty} \frac{2\pi}{r^{\frac{1}{2}(n-2)}} \int_0^A \Psi(\rho)\,\rho^{\frac{1}{2}n} J_{\frac{1}{2}(n-2)}(2\pi r\rho)\,d\rho \\ &= \lim_{\varepsilon\to+0} \frac{2\pi}{r^{\frac{1}{2}(n-2)}} \int_0^{\varepsilon} e^{-\varepsilon\rho}\Psi(\rho)\,\rho^{\frac{1}{2}n} J_{\frac{1}{2}(n-2)}(2\pi r\rho)\,d\rho.\end{aligned} \tag{2.112}$$

From

$$J_{-\frac{1}{2}}(x) = \sqrt{\left(\frac{2}{\pi x}\right)}\cos x, \qquad J_{\frac{1}{2}}(x) = \sqrt{\left(\frac{2}{\pi x}\right)}\sin x,$$

when $n=1, 2, 3$ we can calculate (2.112) as

$$\Phi(r) = 2\int_0^{+\infty} \Psi(\rho)\cos(2\pi r\rho)\,d\rho \quad (n=1),^{\dagger} \tag{2.112$_1$}$$

$$\Phi(r) = 2\pi\int_0^{+\infty} \Psi(\rho)\,\rho J_0(2\pi r\rho)\,d\rho \quad (n=2), \tag{2.112$_2$}$$

$$\Phi(r) = \frac{2}{r}\int_0^{+\infty} \Psi(\rho)\,\rho\sin(2\pi r\rho)\,d\rho \quad (n=3), \tag{2.112$_3$}$$

where the integrals are taken in the same way as in (2.112).

† When $n=1$, this is true by the representation formula of Fourier transforms.

On the other hand, we know

$$\frac{1}{x}\frac{\mathrm{d}}{\mathrm{d}x}\left(\frac{J_\nu(\lambda x)}{x^\nu}\right)=-\lambda\frac{J_{\nu+1}(\lambda x)}{x^{\nu+1}}.$$

By using this, when $n=2p$, the right hand side of (2.112) becomes

$$(-2\pi)^{-(p-1)}\left(\frac{1}{r}\frac{\partial}{\partial r}\right)^{p-1}2\pi\int_0^{+\infty}\mathrm{e}^{-\varepsilon\rho}\Psi(\rho)\,\rho J_0(2\pi r\rho)\,\mathrm{d}\rho,$$

and when $n=2p+1$

$$(-2\pi)^{-(p-1)}\left(\frac{1}{r}\frac{\partial}{\partial r}\right)^{p-1}\left[\frac{2\pi}{r^{\frac{1}{2}}}\int_0^{+\infty}\mathrm{e}^{-\varepsilon\rho}\Psi(\rho)\,\rho^{\frac{1}{2}}J_{\frac{1}{2}}(2\pi r\rho)\,\mathrm{d}\rho\right].$$

Therefore, when $n=2p$, (2.112) becomes

$$\Phi(r)=(-2\pi)^{-(p-1)}\lim_{\varepsilon\to+0}\left(\frac{1}{r}\frac{\partial}{\partial r}\right)^{p-1}\left[2\pi\int_0^{+\infty}\mathrm{e}^{-\varepsilon\rho}\Psi(\rho)\,\rho J_0(2\pi r\rho)\,\mathrm{d}\rho\right],\tag{2.113}$$

and when $n=2p+1$,

$$\Phi(r)=(-2\pi)^{-(p-1)}\lim_{\varepsilon\to+0}\left(\frac{1}{r}\frac{\partial}{\partial r}\right)^{p-1}\left[\frac{2}{r}\int_0^{+\infty}\mathrm{e}^{-\varepsilon\rho}\Psi(\rho)\,\rho\sin(2\pi r\rho)\,\mathrm{d}\rho\right].\tag{2.114}$$

Note that the distribution within [ ] is in fact the Fourier inverse image (respectively when $n=2$ and $n=3$) as $\varepsilon\to+0$.

12 *Elementary solutions for elliptic operators with constant coefficients*

In this section we discuss an application of Fourier transforms to show how we can obtain an elementary solution for an elliptic operator $P(\partial/\partial x)$.

By saying that $P(\partial/\partial x)$ is an *elliptic operator* we mean that, assuming P is a differential operator of mth order, for the homogenous part P_m of P of mth degree, there exists a certain $\delta\,(>0)$, and

$$|P_m(\xi)|\geqslant\delta|\xi|^m\quad(\delta>0)\tag{2.115}$$

is true for all real vectors ξ. It is easy to see from (2.115) that if $|\xi|$ is sufficiently large, then $|P(2\pi\mathrm{i}\xi)|\geqslant\delta'|\xi|^m$ $(\delta'>0)$. Therefore, if $P(2\pi\mathrm{i}\xi)\neq0$, then $1/P(2\pi\mathrm{i}\xi)$ is bounded and continuous in the whole space. Hence, it belongs to $\mathscr{S}'$. If we write the Fourier inverse image of $P(2\pi i\xi)$ as $E(x)$, i.e.

$$E(x)=\int e^{2\pi i x\xi}\frac{1}{P(\xi)}\,d\xi, \tag{2.116}$$

then the Fourier image of $P(\partial/\partial x)\,E(x)$ is, by (2.62),

$$P(2\pi i\xi)\frac{1}{P(2\pi i\xi)}\equiv 1.$$

Therefore,

$$P\left(\frac{\partial}{\partial x}\right)E(x)=\delta, \tag{2.117}$$

elementary solution for $P(\partial/\partial x)$.

How can we cope with the situation when $P(2\pi i\xi)=0$? Hörmander has devised the following method.†

Let us first consider C such that for $|\xi|\geqslant C, |P(2\pi i\xi)|\geqslant 1$. Next, we consider the case $|\xi|\leqslant C$. Let $(\xi_2,\dots,\xi_n)$ satisfy $\sqrt{(\xi_2{}^2+\dots+\xi_n{}^2)}\leqslant C$. In this case, there exists C_0 and a solution ζ_1 for $P(2\pi i\zeta_1, 2\pi i\xi_2,\dots,2\pi i\xi_n)=0$ which satisfies $|\zeta_1|\leqslant C_0$. In fact, from the fact that P is elliptic (by our assumption), it follows that P can be expressed by an equation in ζ_1, in which the coefficient of $\zeta_1{}^m$ is a constant $\neq 0$ independent of $(\xi_2,\dots,\xi_n)$. Therefore, if we choose C' as a constant which is sufficiently large, then

$$|P(2\pi i\zeta_1, 2\pi i\xi_2,\dots,2\pi i\xi_n)|\geqslant 1$$

for $|\zeta_1|\geqslant C'$.

We put $R=\sqrt{(C^2+C'^2)}$. Then, we rewrite the part

$$\int_{|\xi|\leqslant R}\frac{e^{2\pi i x\xi}}{P(2\pi i\xi)}\,d\xi$$

in (2.116) as

$$E_1(x)=\int\cdots\int_{\sqrt{(\xi_2{}^2+\dots+\xi_n{}^2)}\leqslant R} d\xi_2\cdots d\xi_n\int_\Gamma\frac{\exp[2\pi i(x_1\zeta_1+x_2\xi_2+\dots+x_n\xi_n)]}{P(2\pi i\zeta_1, 2\pi i\xi_2,\dots,2\pi i\xi_n)}\,d\zeta_1, \tag{2.118}$$

where Γ is defined as:

(1) In the case when $\sqrt{(\xi_2{}^2+\dots+\xi_n{}^2)}\leqslant C$ the curve from

† L. Hörmander: 'On the theory of general partial differential operators', *Acta Math.* 94 (1955), 161–248.

$-\sqrt{[R^2-(\xi_2{}^2+\cdots+\xi_n{}^2)]}$ to $\sqrt{[R^2-(\xi_2{}^2+\cdots+\xi_n{}^2)]}$ replacing the part $[-C',+C']$ by the lower semi circle centre the origin of the ζ_1-plane.

(2) In the case when $\sqrt{(\xi_2{}^2+\cdots+\xi_n{}^2)}>C$, this can be taken to be on the real axis without need of replacement. Then, obviously, $|P|\geqslant 1$ along Γ. Hence, if we define

$$E(x)=E_1(x)+E_2(x), \quad E_2(x)=\int_{|\xi|\geqslant R} e^{2\pi ix\xi}\frac{d\xi}{P(\xi)}, \tag{2.119}$$

then $E(x)$ should be an elementary solution. In fact,

$$P\left(\frac{\partial}{\partial x}\right)E_1(x)=\int\cdots\int d\xi_2\cdots d\xi_n\int_\Gamma \exp[2\pi i(x_1\rho_1+\cdots+x_n\xi_n)]\, d\xi_1 .$$

In this case this value will remain unchanged if we replace the integral along Γ by the integral along the real axis. Hence, the right hand side term of this equation becomes

$$\int_{|\xi|\leqslant R} e^{2\pi ix\xi}\, d\xi .$$

As for $E_2(x)$, we have similarly,

$$P\left(\frac{\partial}{\partial x}\right)E_2(x)=\int_{|\xi|\geqslant R} e^{2\pi ix\xi}\, d\xi$$

Hence,

$$P\left(\frac{\partial}{\partial x}\right)E(x)=\mathscr{F}[1]=\delta .$$

Estimation of the derivative of $E(x)$ in the neighbourhood of the origin
We note that $E(x)$ is a C^∞-function throughout except at the origin. This is clear for $E_1(x)$ because it is analytic over the whole space. So that we consider only $E_2(x)$. Let $\alpha(\xi)=\alpha(|\xi|)\in\mathscr{D}$ and let $\alpha(\xi)=0$ for $|\xi|\leqslant R/2$ and $\alpha(\xi)=1$ for $|\xi|\geqslant 2R$. Then we observe

$$\tilde{E}(x)=\int \alpha(\xi)\frac{e^{2\pi ix\xi}}{P(2\pi i\xi)}\, d\xi$$

where $P\neq 0$. In this case we have

$$(x_1{}^2+\cdots+x_n{}^2)^p\,\tilde{E}(x)\equiv|x|^{2p}\tilde{E}(x)=\text{const}\int \Delta_\xi{}^p\left[\alpha(\xi)\frac{1}{P(2\pi i\xi)}\right]e^{2\pi ix\xi}\, d\xi$$

where the function $\Delta_\xi{}^p[\cdots]$ tends to 0 like $|\xi|^{-m-2p}$ as $|\xi|\to\infty$.

Therefore, $|x|^{2p}\tilde{E}(x)$ is continuous including its derivatives up to the appropriate order. Hence, for an arbitrary p, $\tilde{E}(x)$ is a C^∞-function except at the origin if we redefine

$$\tilde{E}(x)=\frac{1}{|x|^{2p}}\,[|x|^{2p}\tilde{E}(x)]\,.$$

The following estimation is valid in the neighbourhood of $x=0$:

$$\left|\left(\frac{\partial}{\partial x}\right)^{\alpha}E(x)\right|\leqslant c_0+c_1|x|^{m-n-|\alpha|}\quad(|\alpha|=0,1,2,\dots,m-1),\qquad(2.120)$$

where we replace the right hand side of the inequality by $c_0+c_1\log(1/|x|)$ if $m-n-|\alpha|=0$. We establish this inequality as follows:

LEMMA 2.17. Let

$$F(x)=\int_{|\xi|\geqslant 1}\mathrm{e}^{2\pi i x\xi}Q(\xi)\,\mathrm{d}\xi$$

where $Q(\xi)$ is a homogeneous C^∞-function of $-k$ degree except at the origin. Then, for an appropriate positive constant c_0, we have the following estimation of $F(x)$ in the neighbourhood of $x=0$:†

$$|F(x)|\leqslant\begin{cases}c_0|x|^{k-n} & (k=1,2,\dots,n-1),\\ c_0\log(1/|x|) & (k=n),\\ c_0 & (k>n).\end{cases}\qquad(2.121)$$

Proof. The first inequality of (2.121) can be shown as follows. Let

$$F(x)=\int_{R^n}\mathrm{e}^{2\pi i x\xi}Q(\xi)\,\mathrm{d}\xi-\int_{|\xi|<1}\mathrm{e}^{2\pi i x\xi}Q(\xi)\,\mathrm{d}\xi\,.$$

Then, we apply proposition 2.13 to the first term on the right hand side. The second term is bounded and continuous.

The second inequality of (2.121) can be shown as follows. Let

$$Q(\xi)=\frac{k(\omega)}{|\xi|^n}\,.$$

† We have conveniently defined the lower bound of integration as 1, but in fact we can fix the lower bound as any R.

We introduce polar coordinates. Put $|x|=r$ and $|\xi|=p$. Also, we write θ as the angle between x and ξ. Then, we have

$$\int_{1\leqslant|\xi|\leqslant A} e^{2\pi i x\xi}\frac{k(\omega)}{|\xi|^n}\,d\xi=\int_1^A\frac{d\rho}{\rho}\int_\Omega \exp(2\pi i r\rho\cos\theta)\,k(\omega)\,d\Omega$$

$$=\int_\Omega k(\omega)\left[\int_{r\cos\theta}^{Ar\cos\theta}\frac{e^{2\pi i\delta}}{s}\,ds\right]d\Omega.$$

We consider the effect of letting $A\to+\infty$.

$$\left|\int_{r\cos\theta}^{Ar\cos\theta}\frac{e^{2\pi is}}{s}\,ds\right|\leqslant c+\int_{r|\cos\theta|}\frac{\cos 2\pi s}{s}\,ds\leqslant c+\int_{r|\cos\theta|}\frac{ds}{s}$$

$$\leqslant c+\log\frac{1}{r}+\log\frac{1}{|\cos\theta|}$$

where c is a constant independent of A, r and θ. Obviously,

$$\int_\Omega |k(\omega)|\log\frac{1}{|\cos\theta|}\,d\Omega<+\infty,$$

so that the second inequality is true.

The last inequality is obvious.

Q.E.D.

We return to the proof of (2.120) (proposition 2.17). To show this, we should estimate $E_2(x)$. Let

$$P(2\pi i\xi)=P_m(2\pi i\xi)+Q(2\pi i\xi)$$

where Q is a polynomial of degree $(m-1)$.

On the other hand we can decompose

$$\frac{1}{P}=\frac{1}{P_m}-\frac{Q}{P_m{}^2}+\cdots+(-1)^s\frac{Q^s}{P_m{}^{s+1}}+(-1)^{s+1}\frac{1}{P}\cdot\frac{Q^{s+1}}{P_m{}^{s+1}}$$

when $|\xi|\geqslant R$. In this decomposition we take s such that the last term tends to 0 with a higher order of convergence than the degree $-(n+m)$ as $|\xi|\to\infty$. We observe the Fourier integral of the right hand expression together with the Fourier integral of the product obtained by multiplication of polynomials of degree not greater than $(m-1)$ (as the Fourier image of derivatives). By the previous lemma, we see that (2.120) is true.

Q.E.D.

Chapter 3

Elliptic equations (fundamental theory)

1 *Introduction*

In this chapter we mainly consider simple partial differential equations of the elliptic type:

$$L_\lambda[u] \equiv -\Delta u + c(x)u + \lambda u = f(x), \quad \left(\Delta(\text{Laplacian}) = \sum_{i=1}^{n} \frac{\partial^2}{\partial x_i{}^2}\right). \tag{3.1}$$

We are particularly concerned with the Dirichlet and the Neumann problems related to this type of equation. In (3.1) we assume that λ is a complex parameter and $c(x)$ is a real-valued function of a real variable. Let us recall the history of the Dirichlet problem ignoring the Neumann problem. Historically speaking, the origin of the problem can be traced back to the following problem which puzzled mathematicians during the later half of the nineteenth century.

Dirichlet problem

In the Euclidean space R^n, we consider a certain interior domain Ω which is surrounded by a smooth surface S.† Then, we observe a solution $u(x)$ of

$$\Delta u(x) = 0 \qquad (x \in \Omega)$$

which takes the value of a given function $\varphi(s)$ $(s \in S)$ on S. How do we obtain $u(x)$? An answer to this problem was partly given by Poisson who discovered the (unique) existence of the solution when S was a sphere.

Nevertheless, the general case of this problem, when S is an arbitrary smooth surface, remained intact despite the efforts of mathematicians who tried to apply the results of the theory of analytic functions, which was the centrepiece of mathematical activity at that period. Then, people started to realize that the hard core of the problem could be solved only by a global method, not by a traditional local method. At the beginning of the twentieth

† Strictly speaking, this is a hypersurface.

century the final solution was presented by Fredholm in his monumental work on integral equations. Using a potential function he succeeded in representing the solution $u(x)$.

In this chapter we shall not follow Fredholm's classical approach. Instead, we shall treat the problem within the framework of the L^2-theory.† Related to this problem, we note that there is another well-known problem, the so-called mixed problem (or the initial boundary value problem). For example, we consider the generalized wave equation:

$$\frac{\partial^2 u}{\partial t^2} = \Delta u - c(x)u. \tag{3.2}$$

We look for the solution(s) to the equation satisfying the following two conditions. Namely:

(*a*) Dirichlet's condition, i.e. $u(x, t)$ is identically 0 on S.

(*b*) Initial condition, i.e.

$$\left.\begin{aligned} u(x, 0) &= f_0(x), \\ \frac{\partial u}{\partial t}(x, 0) &= f_1(x). \end{aligned}\right\} \tag{3.3}$$

Another example in the same vein is the problem, for a generalized heat-equation

$$\frac{\partial u}{\partial t} = \Delta u - c(x)u, \tag{3.4}$$

of finding the solution(s) satisfying the following conditions:

$$\left.\begin{aligned} u(x, t)|_{x \in S} &= 0, \\ u(x, 0) &= f_0(x). \end{aligned}\right\} \tag{3.5}$$

We can observe the connection between the above-mentioned examples and the Dirichlet problem as follows. Let us try to find the solution of the generalized wave equation (the first example) by expansions in eigenfunctions. First, we are looking for a solution in the form

$$u(x, t) = u(x)v(t).$$

† See chapter 8 where we explain Fredholm's method using the classical theory of integral equations.

From this relation we must have $u(x)\,|_{x\in S}=0$. On the other hand, (3.2) can be re-written as

$$u(x)v''(t)=v(t)\,(\Delta-c(x))u(x).$$

If we put

$$\frac{v''(t)}{v(t)}=\frac{(\Delta-c(x))u(x)}{u(x)}=-\lambda$$

we shall find

$$\left.\begin{aligned} &v''(t)=-\lambda v(t)\\ &\lambda u(x)=(-\Delta+c(x))u(x), \end{aligned}\right\} \tag{3.6}$$

where the second relation can be written as

$$(\lambda I-L)\,u(x)=0 \tag{3.7}$$

where $-\Delta+c(x)=L$.

In (3.7), if λ is taken as satisfying $u(x)\not\equiv 0$, we shall call λ an *eigenvalue* of L. The terminology is analogous to the theory of linear equations. Also, in this case, we call $u(x)$ an *eigenfunction* for the eigenvalue λ. Later we shall see that such eigenvalues λ are all in fact real, and the only point of accumulation is at $+\infty$. $\{u_\nu(x)\}$ is a complete system of $L^2(\Omega)$ where $u_1(x), u_2(x), u_3(x),\ldots$ are the eigenfunctions of $\lambda_1, \lambda_2, \lambda_3,\ldots$ respectively, and $\lambda_1\leqslant\lambda_2\leqslant\lambda_3\leqslant\cdots\to+\infty$. So that, without loss of generality, we can assume that $\{u_\nu(x)\}$ is an orthonormal system.

Let

$$f_0(x)=\sum_{\nu=1}^{\infty} a_\nu u_\nu(x)$$

$$f_1(x)=\sum_{\nu=1}^{\infty} b_\nu u_\nu(x)$$

be the expansions of the elements of $L^2(\Omega)$. We can easily see that we can formally write

$$u(x,t)=\sum_{\nu=1}^{\infty}\cos\lambda_\nu^{\frac{1}{2}}t a_\nu u_\nu(x)+\sum_{\nu=1}^{\infty}\frac{\sin\lambda_\nu^{\frac{1}{2}}t}{\lambda_\nu^{\frac{1}{2}}}\,b_\nu u_\nu(x) \tag{3.8}$$

where we replace $(\cos\lambda_\nu^{\frac{1}{2}}t, \sin\lambda_\nu^{\frac{1}{2}}t/\lambda_\nu^{\frac{1}{2}})$ by $(\cosh(-\lambda_\nu)^{\frac{1}{2}}t, \sinh(-\lambda_\nu)^{\frac{1}{2}}t/(-\lambda_\nu)^{\frac{1}{2}})$ in the terms under $\sum$ when $\lambda_\nu<0$, and we replace $\sin\lambda_\nu^{\frac{1}{2}}t/\lambda_\nu^{\frac{1}{2}}$ by t when $\lambda_\nu=0$. We shall later examine the condition under which (3.8) is valid.†

† See section 5, chapter 5, and chapter 6.

Similarly, we see that the solution of (3.4) and (3.5) can be formally written as

$$u(x, t) = \sum_{\nu=1}^{\infty} \exp(-\lambda_\nu t) a_\nu u_\nu(x). \tag{3.9}$$

2 *The solution of the Dirichlet problem (Green's operator)*

As we stated in the previous section, in this section we consider the Dirichlet problem within L^2. First, let

$$L_\lambda[u(x)] \equiv -\Delta u(x) + c(x)u(x) + \lambda u(x) = f. \tag{3.1}$$

We assume further that $c(x)$ is a real-valued function which is bounded and measurable. (3.10)

This assumption is not always necessary. In fact, when we deal with the spectral theory of the Schrödinger operator, it is too strong so that we shall relax it.†

Now, we define the Dirichlet problem for (3.1) as follows. Let Ω be an arbitrary open set in R^n.

We wish to obtain $u \in \mathscr{D}_{L^2}{}^1(\Omega)$ such that, for $f \in \mathscr{D}'_{L^2}{}^1(\Omega)$, (3.1) is satisfied in the sense of the distribution of Ω. (P.1)

Note. Roughly speaking, $u \in \mathscr{D}_{L^2}{}^1(\Omega)$ can be regarded as a function which becomes 0 on the boundary of Ω where Ω is not necessarily bounded. In other words Ω can be the whole of R^n.

The reader should recall the definition of $\mathscr{D}_{L^2}{}^1(\Omega)$ which we have given as definition 2.10. $\mathscr{D}'_{L^2}{}^1(\Omega)$ is the dual space of $\mathscr{D}_{L^2}{}^1(\Omega)$ which we have defined in section 4, chapter 2, p. 82. Our Dirichlet problem (P.1) given here can be regarded as a Dirichlet problem in a wider sense compared with the Dirichlet problems which we shall treat in the later sections. The reader should note that in (P.1) we did not impose any condition on the 'shape' of Ω, i.e. the boundedness and/or the smoothness of the boundary of Ω.

Now, if there exists a solution $u \in \mathscr{D}_{L^2}{}^1(\Omega)$ of (3.1), then for an arbitrary $\varphi \in \mathscr{D}(\Omega)$,

$$\langle -\Delta u + c(x)u + \lambda u, \overline{\varphi(x)} \rangle = \langle f, \overline{\varphi(x)} \rangle.$$

† See section 13.

Therefore, by the definition of the derivative of distribution, we have

$$\sum_{i=1}^{n} \left\langle \frac{\partial u}{\partial x_i}, \frac{\partial \bar{\varphi}}{\partial x_i} \right\rangle + \langle c(x)u + \lambda u, \bar{\varphi} \rangle = \langle f, \bar{\varphi} \rangle \quad (\varphi \in \mathscr{D}(\Omega)). \tag{3.11}$$

Furthermore, this relation remains true for an arbitrary $\varphi \in \mathscr{D}_{L^2}{}^1(\Omega)$. In fact, from (3.11) we observe that $u(x), \partial u/\partial x_i \in L^2(\Omega)$.

On the other hand, by the definition of $\mathscr{D}_{L^2}{}^1(\Omega)$, for an arbitrary $\varphi \in \mathscr{D}_{L^2}{}^1(\Omega)$, there is a sequence $\varphi_j(x) \to \varphi(x)$ in $\mathscr{D}_{L^2}{}^1(\Omega)$, $\varphi_j \in \mathscr{D}(\Omega)$, i.e. $\varphi_j(x) \to \varphi(x)$, $\partial \varphi_j/\partial x_i \to \partial \varphi/\partial x_i$ $(j \to \infty)$ is valid in $L^2(\Omega)$. Now, we see that the relation (3.11) which is true for each $\varphi_j(x)$ is also valid for $\varphi \in \mathscr{D}_{L^2}{}^1(\Omega)$ by the limiting process.

Conversely, if (3.11) is valid for an arbitrary $\varphi \in \mathscr{D}_{L^2}{}^1(\Omega)$, then it is obviously valid for $\varphi \in \mathscr{D}(\Omega)$. Reversing the argument, we see that $u \in \mathscr{D}_{L^2}{}^1(\Omega)$ is in fact the solution of (3.1).

Therefore (P.1) is equivalent to:

Given $f \in \mathscr{D}'_{L^2}{}^1(\Omega)$, we wish to obtain $u \in \mathscr{D}_{L^2}{}^1(\Omega)$ which satisfies (3.11) for any $\varphi \in \mathscr{D}_{L^2}{}^1(\Omega)$. (P.1)′

For the time being we postpone the study of (P.1)′ for an arbitrary λ, instead we show that there exists a unique solution for certain λ. In this case we shall see that our problem is reduced to nothing but theorem 2.10 (Riesz theorem) as follows.

First, we consider (3.11) for $\lambda = t$ where $t > -c_0$ and $c_0 = \inf_{x \in \Omega} c(x)$ by (3.10). The left hand side of (3.11) is obviously a positive definite Hermitian form on $\mathscr{D}_{L^2}{}^1(\Omega)$ (see definition 2.11). Also, for an arbitrary $\varphi \in \mathscr{D}_{L^2}{}^1(\Omega)$,

$$m_t \|\varphi\|^2_{1, L^2(\Omega)} \leq \sum \left\langle \frac{\partial \varphi}{\partial x_i}, \frac{\partial \bar{\varphi}}{\partial x_i} \right\rangle + \langle (c(x)+t)\varphi, \bar{\varphi} \rangle \leq M_t \cdot \|\varphi\|^2_{1, L^2(\Omega)}$$

and $m_t, M_t > 0$, i.e. the left hand side of (3.11) is a positive definite Hermitian form which determines an equivalent norm to the norm which we can define on $\mathscr{D}_{L^2}{}^1(\Omega)$.

Let the space equipped with this positive definite Hermitian form be $\mathscr{H}$. Write the left hand side of (3.11) as $(u, \varphi)_{\mathscr{H}}$. Then, we see that $\mathscr{H}$ is a Hilbert space. Furthermore, the space itself and the topology of $\mathscr{H}$ coincide with $\mathscr{D}_{L^2}{}^1(\Omega)$. For this fact the reader should refer to the note of definition 2.11. Now, we see that the right hand side of (3.11) is a continuous antilinear form defined on $\mathscr{H}$. In fact, $\varphi(x)$ belongs to $\mathscr{H}$, and $\varphi(x) \to \langle f, \bar{\varphi} \rangle$

is a continuous form because of the fact that $\mathscr{H}$ and $\mathscr{D}_{L^2}{}^1(\Omega)$ are identical in their spaces and topologies. By Riesz theorem, there exists a unique $u \in \mathscr{H}$, and for $\varphi \in \mathscr{H}$, we have $\langle f, \bar{\varphi} \rangle = (u, \varphi)_{\mathscr{H}}$ i.e. (3.11) is valid.

Again, by Riesz theorem, we see that $\|u\|_{\mathscr{H}} = \|f\|_{\mathscr{H}'}$, where $\mathscr{H}'$ is the dual space of $\mathscr{H}$. From this, there exists a certain k (a positive constant) such that

$$\|u\|_{1, L^2(\Omega)} \leqslant k \|f\|'_{1, L^2(\Omega)} . \tag{3.12}$$

We write the linear mapping: $f \to u$ as $u = G_t \circ f$. We shall call G_t the *Green's operator* of the operator L_t for the Dirichlet problem.

Because of the equivalence of the problems (P.1) and (P.1)′, we have

THEOREM 3.1. Let us assume that (3.10) is true. For $L_t[u] = f$ the Dirichlet problem (P.1) can be uniquely solved if $t > -\inf c(x)$. Therefore, we can establish a correspondence: $G_t : f \to u(x)$ where $u \in \mathscr{D}_{L^2}{}^1(\Omega)$, i.e. G_t is a continuous linear operator from $\mathscr{D}'_{L^2}{}^1(\Omega)$ onto $\mathscr{D}_{L^2}{}^1(\Omega)$.

Note. G_t is a map onto the whole of $\mathscr{D}_{L^2}{}^1(\Omega)$ because L_t is a mapping from $\mathscr{D}_{L^2}{}^1(\Omega)$ to $\mathscr{D}'_{L^2}{}^1(\Omega)$.

The properties of the Green's operator G_t

From the foregoing discussion, we see that

$$\left.\begin{array}{lll} (1) & L_t \circ G_t = I & \text{in } \mathscr{D}'_{L^2}{}^1(\Omega), \\ (2) & G_t \circ L_t = I & \text{in } \mathscr{D}_{L^2}{}^1(\Omega). \end{array}\right\} \tag{3.13}$$

(1) is clear from the definition of G_t. (2) can be shown as follows. First we note that L_t is a one-to-one mapping from $\mathscr{D}_{L^2}{}^1(\Omega)$ to $\mathscr{D}'_{L^2}{}^1(\Omega)$. By (1) we have $L_t \circ G_t \circ L_t u = L_t u$ for $u \in \mathscr{D}_{L^2}{}^1(\Omega)$, therefore, (2) is true.

So far, we have regarded the Green's operator G_t as having $\mathscr{D}'_{L^2}{}^1(\Omega)$ as its domain and $\mathscr{D}_{L^2}{}^1(\Omega)$ as its codomain.

Now, we confine the domain of G_t to $L^2(\Omega)$, and the codomain of G_t to $L^2(\Omega)$ $(\supset \mathscr{D}_{L^2}{}^1(\Omega))$. Then, we have

THEOREM 3.2. The Green's operator G_t is a bounded, positive Hermitian operator defined on $L^2(\Omega)$, and if Ω is bounded it is a completely continuous operator.

Proof. The boundedness is clear. In fact, G_t is a continuous operator from $\mathscr{D}'_{L^2}{}^1(\Omega)$ to $\mathscr{D}_{L^2}{}^1(\Omega)$. Now, we have made the topology of the domain of G_t stronger, and the topology of the codomain of G_t weaker. It is therefore also continuous. In fact, in terms of inequality, from (3.12), if $f \in L^2(\Omega)$, then

$$\|G_t f\|_{L^2(\Omega)} \leqslant \|G_t f\|_{1, L^2(\Omega)} \leqslant k\|f\|'_{1, L^2(\Omega)} \leqslant k\|f\|_{L^2(\Omega)}.$$

The last inequality follows from $\|u\|_{L^2(\Omega)} \geqslant \|u\|'_{1, L^2(\Omega)}$ in the dual space of the space satisfying $\|u\|_{L^2(\Omega)} \leqslant \|u\|_{1, L^2(\Omega)}$.

Hermitian properties

First, if f, $v \in \mathscr{H}$, then

$$(G_t f, v)_{\mathscr{H}} = (f, G_t v)_{\mathscr{H}}.$$

In fact, from the definition of G_t, the left hand side is equal to $(f, v)_{L^2}$. On the other hand, the right hand side becomes

$$(f, G_t v)_{\mathscr{H}} = \overline{(G_t v, f)_{\mathscr{H}}} = \overline{(v, f)_{L^2}} = (f, v)_{L^2}.$$

From this fact, if f, $v \in \mathscr{H}$, we have

$$(G_t f, v)_{L^2} = (f, G_t v)_{L^2}.$$

In fact, the left hand side is equal to $(G_t^2 f, v)_{\mathscr{H}}$, so that, from the foregoing result, $(f, G_t^2 v)_{\mathscr{H}} = \overline{(G_t^2 v, f)_{\mathscr{H}}} = \overline{(G_t v, f)_{L^2}} = (f, G_t v)_{L^2}$ is equal to the right hand side. On the other hand the relation can be established for arbitrary f, $v \in L^2(\Omega)$ because $\mathscr{H} = \mathscr{D}_{L^2}{}^1(\Omega)$ is dense in $L^2(\Omega)$ and G_t is a continuous operator.

Positive definite

From $f \in \mathscr{D}_{L^2}{}^1(\Omega)$, we see that

$$(G_t f, f)_{L^2} = (G_t^2 f, f)_{\mathscr{H}} = (G_t f, G_t f)_{\mathscr{H}}$$

if $f \in \mathscr{H}$. This is also true for $f \in L^2(\Omega)$ because we can use a limiting process. Therefore, $(G_t f, f) \geqslant 0$ and the equality holds if and only if $G_t f = 0$, i.e. $f = 0$.

Complete continuity

In general, we say that a linear operator T is *completely continuous* in a Banach space E if the image of a bounded set B by T is 'relatively compact', i.e. the closure $\overline{T(B)}$ of the image $T(B)$ is a compact set in E. In other words, if $\{f_j\}$ is a bounded sequence of E, then an appropriate sub-sequence $\{g_{j_p}\}$ taken from $g_j = T(f_j)$ becomes a Cauchy sequence of E. For the direct proof of the complete continuity of G_t, we use

THEOREM 3.3 *(Rellich's theorem of choice)*. Let Ω be a bounded open set of R^n, and let $\{f_j\}, f_j \in \mathscr{D}_{L^2}{}^1(\Omega)$, be a bounded sequence, i.e. $\|f_j\|_{1, L^2(\Omega)} \leqslant M < +\infty$. Then, we can choose an appropriate subsequence $\{f_{j_p}(x)\}_{p=1,2,\ldots}$ from $\{f_j\}$ such that it forms a Cauchy sequence of $L^2(\Omega)$.

We can regard this theorem as a L^2-version of the Ascoli–Arzela theorem which deals with a sequence of continuous functions [in a compact metric space].† In fact, from the uniform boundedness of $\partial f_j / \partial x_i$ the equi-continuity of $\{f_j\}$ follows. The proof of the theorem is not complicated but we postpone it until the next section; for the time being we assume that the statement is true.

Now, for $f \in L^2(\Omega)$, we see that

$$\|G_t \cdot f\|_{1, L^2(\Omega)} \leqslant \gamma \|f(x)\|_{L^2(\Omega)},$$

where γ is a positive constant (from (3.12)), therefore, the unit sphere: $\|f\|_{L^2(\Omega)} \leqslant 1$ is mapped by G_t into a bounded set of $\mathscr{D}_{L^2}{}^1(\Omega)$. If Ω is bounded, then the image of G_t is a relatively compact set of $L^2(\Omega)$. Hence, in this case G_t is completely continuous.

Q.E.D. (Theorem 3.2)

Relation between the generalized Dirichlet problem and the Green's operator G_t

Let us go back to the equation

$$L_\lambda[u] \equiv -\Delta u + c(x)u + \lambda u = f(x). \tag{3.1}$$

The problem is that:

$$\left.\begin{array}{l}\text{For } f \in L^2(\Omega), \text{ we wish to obtain the solution } u \in \mathscr{D}_{L^2}{}^1(\Omega) \\ \text{of (3.1).}\end{array}\right\} \tag{P.2}$$

If $u(x)$ is the solution of (P.2), then u satisfies

$$L_\lambda[u] = (-\Delta + c(x) + t)u + (\lambda - t)u = f(x)$$

† Translator's supplement.

where t is fixed as in theorem 3.1. We write the Green's operator of t as G_t. Then, by operating G_t on the equation, from (2) of (3.13) we have

$$u+(\lambda-t)G_t u=G_t f. \tag{3.14}$$

Conversely, for $f\in L^2(\Omega)$, if $u\in L^2(\Omega)$ exists and satisfies (3.14) (note that we do not assume $u\in\mathscr{D}_{L^2}{}^1(\Omega)$), then $u\in\mathscr{D}_{L^2}{}^1(\Omega)$, $G_t u$, $G_t f\in\mathscr{D}_{L^2}{}^1(\Omega)$.

We operate with L_t on (3.14), and from (1) of (3.13) we see

$$L_t[u]+(\lambda-t)u=f(x),$$

so that $L_\lambda[u]=f(x)$.

Summing up we have:

Proposition 3.1. Problem (P.2) is equivalent for $f\in L^2(\Omega)$ to problem (P.3): finding a solution $u\in L^2(\Omega)$ satisfying (3.14). Transforming our problem in this way, we now see that we can derive approximations for the solubility of (3.14). More precisely, we can establish some facts by the use of functional analysis. For example:

(1) If λ satisfies $\operatorname{Im}\lambda\neq 0$, then (3.14) has a unique solution.

(2) If Ω is bounded, G_t is an Hermitian operator which is completely continuous, so that in this case the Hilbert–Schmidt theory is available.

For the convenience of readers who are not familiar with functional analysis, we give some explanation. Note that the following arguments also apply to Banach spaces or more generally to topological vector spaces, but here we confine ourselves to Hilbert spaces.

Concerning $Tu=f$

Let us consider

$$Tu=f, \tag{3.15}$$

an equation in a Hilbert space $\mathscr{H}$, where T is a bounded linear operator. For an arbitrary $f\in\mathscr{H}$ (the right hand side), we consider a problem: seeking conditions for a solution $u\in\mathscr{H}$ which can be uniquely determined. We consider first the necessary conditions.

If (3.15) has a unique solution, then T is bi-continuous because it is a one-to-one mapping from $\mathscr{H}$ onto $\mathscr{H}$ (see theorem 2.5), i.e. there exists a certain constant $c>0$, with

$$\|Tu\|\geqslant c\|u\| \quad (u\in\mathscr{H}). \tag{3.16}$$

Next, for the dual operator T^* of T,

$$T^*:(Tf, g)=(f, T^*g) \qquad (f, g\in\mathscr{H}),^{\dagger}$$

the condition that the only solution of

$$T^*u=0 \tag{3.17}$$

is $u=0$, is a necessary condition. If this is not the case, then there exists a certain $h\neq 0$ and $T^*h=0$. Therefore

$$(f, T^*h)=0 \quad (f\in\mathscr{H}), \quad \text{i.e.} \quad (Tf, h)=0 \quad (f\in\mathscr{H}).$$

This shows that the image set $T\mathscr{H}$ is contained in the ortho-complement of h, so that (3.15) has no solutions for an arbitrary f.

In fact, we can show that the converse of this statement (3.17) is also true, i.e.

LEMMA 3.1. Let us assume (3.16) and (3.17). In this case, the equation (3.15) has a unique u for an arbitrary f, and the correspondence $T^{-1}:f\to u$ is continuous.

Proof. From (3.16) the image $T\mathscr{H}$ is a closed subspace of $\mathscr{H}$. In fact, if $f_j\in T\mathscr{H}\to f_0(j\to\infty)$, then as $f_j=Tu_j$ and $\{u_j\}$ is a Cauchy sequence by (3.16). Let $u_j\to u_0$, then, $f_j=Tu_j\to Tu_0=f_0$. Therefore, $f_0\in T\mathscr{H}$. Next, we see that $T\mathscr{H}$ is dense in $\mathscr{H}$. If not, there exists a certain $h\neq 0$, with

$$(Tf, h)=0 \quad (f\in\mathscr{H}), \quad \text{i.e.} \quad (f, T^*h)=0 \quad (f\in\mathscr{H}).$$

From this $T^*h=0$. This contradicts (3.17), hence, we conclude that $T\mathscr{H}=\mathscr{H}$.

Q.E.D.

From this lemma, we see:

PROPOSITION 3.2. In a Hilbert space $\mathscr{H}$, consider an operator $\lambda I-H$. Let

† T is bounded so that $f\to(Tf, g)$ is a continuous linear form on $\mathscr{H}$. Therefore, by Riesz theorem there exists a unique $g^*\in\mathscr{H}$ for every g satisfying $(Tf, g)=(f, g^*)$. Let $g^*=T^*g$, then T^* is also bounded, and $(T^*)^*=T$.

H be a bounded Hermitian operator, and let

$$m = \inf_{\|f\|=1} (Hf, f), \qquad M = \sup_{\|f\|=1} (Hf, f).$$

Then, for $\lambda \notin [m, M]$, the inverse operator $(\lambda I - H)^{-1}$ exists.

Note. We call the set of λ such that $(\lambda I - H)^{-1}$ is a bounded operator defined on the whole $\mathscr{H}$ the *resolvent set* of H, and the complement of the resolvent set of H the *spectrum* of H. The proposition shows that the spectrum of H lies in the interval $[m, M]$ on the real axis.

Proof. Let $\lambda \notin [m, M]$. We see that

$$((\lambda I - H)f, f) = \lambda \|f\|^2 - (Hf, f).$$

We note that (Hf, f) is always real. So that if we consider the real part and the imaginary part of $\lambda = u + i\sigma$, then the right hand side of this equation is not smaller than $\operatorname{dis}(\lambda, [m, M])\|f\|^2$, where dis means a distance in the λ-plane.

Therefore, by the Schwartz inequality, we have

$$\|(\lambda I - H)f\| \cdot \|f\| \geqslant \operatorname{dis}(\lambda, [m, M])\|f\|^2,$$

i.e.

$$\|(\lambda I - H)f\| \geqslant \operatorname{dis}(\lambda, [m, M])\|f\| \qquad (f \in \mathscr{H}). \tag{3.18}$$

On the other hand, from $(\lambda I - H)^* = \bar{\lambda} I - H^* = \bar{\lambda} I - H$, we have, (3.16) and (3.17) for $T = \lambda I - H$. By the previous lemma we assert that the proposition is true.

Q.E.D.

Note that by (3.18) we also have

$$\|(\lambda I - H)^{-1}\| \leqslant \frac{1}{\operatorname{dis}(\lambda, [m, M])}. \tag{3.19}$$

Note. The Hermitian form (Hf, f) which we considered in the previous proposition is in fact an extension of the quadratic form $\sum a_{ij}\xi_i\bar{\xi}_j$ in a finite-dimensional space. For this form we can prove the following fact:

$$\sup_{\|f\|=1} |(Hf, f)| = \|H\|. \tag{3.20}$$

To see this, we denote the left hand side of (3.20) by N_H. Then, from

$|(Hf,f)|\leqslant\|Hf\|\ \|f\leqslant\|H\|\ \|f\|^2$ obviously $N_H\leqslant\|H\|$. Also, $N_H\geqslant\|H\|$ for the following reason. Let λ be a real number. Then, we have

$$(H(\lambda f+g),\lambda f+g)\leqslant N_H\|\lambda f+g\|^2$$
$$-(H(\lambda f-g),\lambda f-g)\leqslant N_H\|\lambda f-g\|^2.$$

So that, by adding these inequalities:

$$\lambda\{(Hf,g)+(Hg,f)\}\leqslant N_H\{\lambda^2\|f\|^2+\|g\|^2\}$$

Then, we put $g=Hf$. We have $2\lambda\|Hf\|^2\leqslant N_H\{\lambda^2\|f\|^2+\|Hf\|^2\}$, i.e. for $\lambda>0$,

$$\|Hf\|^2\leqslant N_H\tfrac{1}{2}\left\{\lambda\|f\|^2+\frac{1}{\lambda}\|Hf\|^2\right\}.$$

λ is an arbitrary real number satisfying the condition, therefore, we can assume that

$$\lambda=\frac{\|Hf\|}{\|f\|}\quad(f\neq 0).$$

From this $\|Hf\|^2\leqslant N_H\|Hf\|\cdot\|f\|$, i.e.

$$\|Hf\|\leqslant N_H\|f\|.$$

This is true for $f=0$, so that for an arbitrary $f\in\mathscr{H}$ the last inequality is true. Hence, $\|H\|\leqslant N_H$.

Q.E.D.

Let us then apply our proposition to the problem (P.3) for the equation (3.14). First, if λ is taken as $\operatorname{Im}\lambda\neq 0$, then (P.3) has a unique solution. Next, we shall see that (P.3) has a solution under the condition that λ is real and $\lambda>-c_0$ where $c_0=\inf c(x)$. (Recall how we constructed G_t.)

Because (P.2) and (P.3) are equivalent, we have:

THEOREM 3.3.1. Let Ω be an arbitrary open set of R^n.

The problem (P.2) for the equation (3.1) has a unique solution $u\in\mathscr{D}_{L^2}{}^1(\Omega)$ for an arbitrary $f\in L^2(\Omega)$ if $\lambda\notin(-\infty,-c_0]$ where $c_0=\inf c(x)$. Also, in this case we have

$$\|u(x)\|_{L^2(\Omega)}\leqslant\frac{1}{\operatorname{dis}(\lambda,(-\infty,-c_0])}\|f(x)\|_{L^2(\Omega)}.\tag{3.21}$$

Proof. First we note that it is enough to prove (3.21); the rest of the theorem follows immediately from the remark we made just before the theorem. (3.21) can be proved from (3.11).

In fact, if $\lambda \notin (-\infty, -c_0]$, then there exists $u \in \mathscr{D}_{L^2}{}^1(\Omega)$. Let $\varphi = u$ in (3.11), then,

$$(f, u)_{L^2} = \sum_{i=1}^{n} \left\| \frac{\partial u}{\partial x_i} \right\|^2 + ((c(x)+\lambda)u, u)_{L^2}.$$

Therefore, we see that

$$\|f\| \; \|u\| \geqslant \left| ((c(x)+\lambda)u, u) + \sum_{i=1}^{n} \left\| \frac{\partial u}{\partial x_i} \right\|^2 \right|.$$

By considering the real part and imaginary part of the inner product of the right hand side of this inequality, we see that the absolute value of the right hand side is larger than $\operatorname{dis}(\lambda, (-\infty, -c_0])\|u\|^2$, so that,

$$\|f\| \geqslant \operatorname{dis}(\lambda, (-\infty, -c_0])\|u\| .$$

Q.E.D.

Note. From theorem 3.3.1 we see immediately that the spectrum of $\Delta - c(x)$ for the Dirichlet problem lies in $(-\infty, -c_0]$. Now the spectrum of the operator $(-\Delta + c(x))$, lies in $[c_0, \infty)$. We note that in general this seemingly rough estimation of (3.21) cannot improve if Ω is not bounded, i.e. we cannot replace c_0 by a larger value than c_0.

EXAMPLE. Let Ω be the complement of a bounded closed set of R^n, i.e. $R^n \backslash \Omega$ is compact. Then, the left most point of the spectrum of $-\Delta$ for the Dirichlet problem is 0.

If not, there exists a positive constant c and for $u \in \mathscr{D}(\Omega)$ we see that $\|\Delta u\|_{L^2(\Omega)} \geqslant c\|u(x)\|_{L^2(\Omega)}$. If the support of u does not contain the original point, and if we put $u_\varepsilon(x) = (\varepsilon)^{\frac{1}{2}n} u(\varepsilon x)$, then $u_\varepsilon \in \mathscr{D}(\Omega)$, $\|u_\varepsilon(x)\|_{L^2} = \|u(x)\|_{L^2}$, and $\|\Delta u_\varepsilon(x)\|_{L^2} = \varepsilon^2 \|\Delta u(x)\|_{L^2}$ for a sufficiently small ε. So that this is a contradiction.

Next, we shall give a simple account of the Hilbert–Schmidt theory which is useful when G_t is a completely continuous (compact) operator.

The theory of Hilbert–Schmidt

We start from the classical treatment of this theory. We consider the integral equation

$$u(x)-\lambda\int_a^b K(x,s)u(s)\,ds=f(x)$$

where we assume

$$K(x,s)=\overline{K(s,x)},\quad \int\int |K(x,s)|^2\,dx\,ds<+\infty\,.$$

We look at the equation in $L^2(a,b)$. In our case, we note that

$$(I-\lambda H)u=f \tag{3.22}$$

in a Hilbert space $\mathscr{H}$.

Let H be a completely continuous Hermitian operator in $\mathscr{H}$. Let us call μ an *eigenvalue* of H if and only if the homogeneous equation

$$(\mu I-H)\varphi=0\quad\text{or}\quad H\varphi=\mu\varphi \tag{3.23}$$

which is obtained from (3.22) by putting $\mu=1/\lambda$ has a solution $\varphi\neq 0$. Also, we call φ an *eigen-element* corresponding to the eigenvalue μ.

Some authors call φ an eigen-element (or *eigen-solution*) corresponding to the eigenvalue λ when $(I-\lambda H)\varphi=0$ $(\varphi\neq 0)$ for (3.22). To avoid confusion, in this case, we call φ *an eigen-element for the singular value* λ, and if $(I-\lambda H)\varphi=0$ has a unique solution $\varphi=0$, then we call λ a *regular value*.

We prove the following relatively easy proposition.

PROPOSITION 3.3. Let H be a completely continuous Hermitian operator in a Hilbert space. Then the following statements are true.

(1) Eigenvalues are all real. The set of eigen-elements corresponding to an arbitrary eigenvalue $\mu=\mu_0(\neq 0)$ is finite-dimensional (the eigenspace is finite-dimensional). Eigen-elements are mutually orthogonal if they correspond to different eigenvalues.

(2) The set of eigenvalues has at most one accumulating point $\mu=0$.

Proof. From $Hf=\mu f$ $(f\neq 0)$, we have $(Hf,f)=\mu(f,f)$.

From this we show that μ is real. Let $H\varphi_1=\mu_1\varphi_1$, $H\varphi_2=\mu_2\varphi_2$. Then, we have $(H\varphi_1,\varphi_2)=\mu_1(\varphi_1,\varphi_2)$. On the other hand,

$$(H\varphi_1,\varphi_2)=(\varphi_1,H\varphi_2)=(\varphi_1,\mu_2\varphi_2)=\mu_2(\varphi_1,\varphi_2)\,.$$

Therefore, $(\mu_1-\mu_2)(\varphi_1,\varphi_2)=0$, and if $\mu_1\neq\mu_2$, then $(\varphi_1,\varphi_2)=0$.

Next, we show that we have contradiction from

$$H\varphi_j=\mu_j\varphi_j,\ \|\varphi_j\|=1 \quad (\mu_j\to\mu\neq 0).$$

To see this, first we note that $\mu_j\varphi_j=\mu\varphi_j+(\mu_j-\mu)\,\varphi_j$ by decomposition. Then, from $(\varphi_j, \varphi_{j'})=0$ $(j\neq j')$ we see that any subsequence of $\{\mu_j\varphi_j\}$ can not be a convergent sequence, but an appropriate subsequence of $\{H\varphi_j\}$ must be a convergent sequence because H is completely continuous. Finally, from the above argument it is clear that the eigenspace corresponding to a non-zero eigenvalue is finite-dimensional.

Q.E.D.

We now prove the key lemma of the Hilbert–Schmidt theory.

LEMMA 3.2 *(Hilbert)*. If H is a Hermitian operator which is non-zero and completely continuous, then H has an eigenvalue which is not zero.

Proof. Let

$$\|f_j\|=1, \quad |(Hf_j, f_j)|\to\mu=\sup_{\|f\|=1}|(Hf, f)|.$$

From (3.20), we have $\mu=\|H\|\neq 0$. Choosing a subsequence of the sequence if necessary, let

$$(Hf_j, f_j)\to\mu.$$

Then, $$0\leqslant\|Hf_j-\mu f_j\|^2=\|Hf_j\|^2-2\mu(Hf_j, f_j)+\mu^2\to 0.$$
In fact, because $\|Hf_j\|\leqslant\|H\|=\mu$, we see that $Hf_j-\mu f_j\to 0$. We pick an appropriate convergent subsequence Hf_{j_p} from Hf_j. So that $\{f_{j_p}\}$ is itself a convergent sequence: $f_{j_p}\to f_0$ and $\|f_0\|=1$.

Of course we have $(Hf_{j_p}-\mu f_{j_p})\to(Hf_0-\mu f_0)=0$. Now, we see that μ is a non-zero eigenvalue of H. It is apparent from the foregoing, that when $(Hf_j, f_j)\to-\mu$, $-\mu$ is an eigenvalue of H.

Q.E.D.

It is clear that the set of eigenvalues is a bounded set (see proposition 3.2). Let us pick an orthogonal basis in the eigenspace of every eigenvalue μ. Let us also assume that $|\mu_1|\geqslant|\mu_2|\geqslant\cdots\to 0$ for the sequence of eigenvalues $\{\mu_j\}$ $(\mu_j\neq 0)$, if the number of eigenvalues is infinite. Now we construct an orthonormal system in the following way.

We agree that if the dimension of the eigenspace corresponding to $\mu=\mu_1$ is p, then we write μ_1 p times, i.e. the values of $\mu_1, \mu_2, \ldots, \mu_p$ are all equal to the value of μ_1. With this convention, we establish a correspondence between μ_i and an eigen-element $\varphi_i(\|\varphi_i\|=1)$. If the number of repetitions of μ_i is p, then we pick an orthogonal basis in the eigenspace of each eigenvalue. Thus we obtain an orthonormal system $\{\varphi_i\}$. We see that $\{\varphi_i\}$ has the following properties:

THEOREM 3.4 *(The expansion theorem of Hilbert–Schmidt)*. For an arbitrary $f\in\mathscr{H}$, we have

$$Hf=\sum_{i=1}^{\infty}(Hf,\varphi_i)\varphi_i=\sum_{i=1}^{\infty}\mu_i(f,\varphi_i)\varphi_i. \tag{3.24}$$

In particular, if $Hf=0$ has no solution such that $f\neq 0$, then

$$f=\sum_{i=1}^{\infty}(f,\varphi_i)\varphi_i \quad (f\in\mathscr{H}), \tag{3.25}$$

i.e. $\{\varphi_i\}$ is a complete orthonormal system in $\mathscr{H}$.

Proof. It is easy to see that $S=\sum_{i=1}^{\infty}\mu_i(\cdot,\varphi_i)\,\varphi_i$ is a Hermitian operator which is completely continuous. In fact,

$$\left\|\sum_{i=m}^{\infty}\mu_i(f,\varphi_i)\varphi_i\right\|^2=\sum_{i=m}^{\infty}\mu_i^2|(f,\varphi_i)|^2\leqslant\mu_m^2\|f\|^2,$$

by Bessel's inequality. The value of the term on the right of the inequality can be made uniformly small if f runs through a bounded set and m takes a sufficiently large value.

Now, $R=H-S$ is a completely continuous Hermitian operator, and it has zero as its only eigenvalue. In fact, all φ_i $(i=1,2,\ldots)$ are eigen-elements corresponding to the eigenvalue zero of R, so that if R has an eigenvalue μ other than zero, then the corresponding eigen-element $\psi: R\psi=\mu\psi$ is orthogonal to $\{\varphi_i\}_{i=1,2,\ldots}$ (see (1) of proposition 3.3).

Therefore, $R\psi=H\psi$. Hence, μ must be identical with one of $\mu_i(i=1,2,\ldots)$. Since ψ is orthogonal to $\{\varphi_i\}_{i=1,2,\ldots}$ this is a contradiction. From this fact and the previous lemma we have $R=0$. Hence, $H=S$, i.e. (3.24) is true.

We re-write (3.24) as

$$H\left[f-\sum_{i=1}^{\infty}(f,\varphi_i)\varphi_i\right]=0,$$

so that if $H\cdot f=0$ means $f=0$, then (3.25) is true.

Q.E.D.

Note. In other words the result is equivalent to the following.

(1) Let an orthonormal system which consists of eigen-elements corresponding to non-zero eigenvalues be $\{\varphi_i\}$. Then, $\{\varphi_i\}$ forms a complete orthonormal system in $\mathscr{H}_1=\overline{H\mathscr{H}}$ (the closure of the image by H in $\mathscr{H}$) which is a closed subspace of $\mathscr{H}$.

(2) The ortho-complementary space of $\mathscr{H}_1=\overline{H\mathscr{H}}$ is formed by the eigen-elements of H corresponding to the zero eigenvalue, i.e. the condition $(\psi,\varphi_i)=0$ $(i=1, 2, \ldots)$ and the condition $H\psi=0$ are equivalent. (In general, this eigenspace is not finite-dimensional.)

(3) In general, for $f\in\mathscr{H}$ there exists an orthogonal decomposition

$$f=\sum_{i=1}^{\infty}(f,\varphi_i)\varphi_i+\psi, \tag{3.26}$$

where $(\psi,\varphi_i)=0$ $(i=1, 2, \ldots)$, i.e. $H\psi=0$.

Now, we observe the first equation

$$(I-\lambda H)u=f. \tag{3.22}$$

By the previous theorem we have:

THEOREM 3.5. Let H be a completely continuous Hermitian operator in a Hilbert space $\mathscr{H}$. If $\lambda\notin\{1/\mu_i\}_{i=1,2,\ldots}$, i.e. λ is not a singular value of H, then (3.22) has a unique solution

$$u=f+\lambda\sum_{i=1}^{\infty}\frac{\mu_i}{1-\lambda\mu_i}(f,\varphi_i)\varphi_i. \tag{3.27}$$

In particular, if $\mu=0$ is not an eigenvalue of H, then

$$u=\sum_{i=1}^{\infty}\frac{1}{1-\lambda\mu_i}(f,\varphi_i)\varphi_i. \tag{3.28}$$

The necessary and sufficient condition that when $\lambda=1/\mu_k$ (3.22) has at least one solution is that f should be orthogonal with an arbitrary eigen-element of H which corresponds to μ_k.

Proof. Let us consider the case that zero is not an eigenvalue of H, i.e., $\{\varphi_i\}$ is a complete orthonormal system of $\mathscr{H}$.

If there exists such a u, then by equating the Fourier coefficients of both sides of φ_k we have

$$((I-\lambda H)u, \varphi_k)=(f, \varphi_k),$$

i.e.

$$(1-\lambda\mu_k)(u, \varphi_k)=(f, \varphi_k). \tag{3.29}$$

Then, we put $\lambda\notin\{1/\mu_i\}_{i=1,2,\dots}$, and we see that

$$(u, \varphi_k)=\frac{1}{1-\lambda\mu_k}(f, \varphi_k),$$

so that

$$u=\sum_{i=1}^{\infty}\frac{1}{1-\lambda\mu_i}(f, \varphi_i)\varphi_i.$$

Conversely, the fact that this expression is a solution follows clearly from the relation

$$\sum_{i=1}^{\infty}\left|\frac{1}{(1-\lambda\mu_i)}(f, \varphi_i)\right|^2\leqslant\max_i\left|\frac{1}{1-\lambda\mu_i}\right|^2\cdot\|f\|^2<+\infty \quad \text{(Bessel's inequality)},$$

and the completeness of $\{\varphi_i\}$. Next, let 0 be an eigenvalue of H. We put

$$f-\sum_{i=1}^{\infty}(f, \varphi_i)\varphi_i=\psi \tag{3.30}$$

(see (3.26)). We consider the equivalent relation to (3.22)

$$(I-\lambda H)(u-\psi)=\sum_{i=1}^{\infty}(f, \varphi_i)\varphi_i \tag{3.31}$$

since $H\psi=0$. The right hand side is an element of $\mathscr{H}_1$.† Clearly, $H(u-\psi)\in\mathscr{H}_1$, so that $u-\psi\in\mathscr{H}_1$.

† $\mathscr{H}_1=\overline{H\mathscr{H}}$ (see the previous note). In other words, $\mathscr{H}_1$ is a closed subspace determined by $\{\varphi_i\}$.

From this it follows that when we look for a solution of equation (3.31) for $v=u-\psi$, it is immaterial whether we regard (3.31) as an equation in $\mathscr{H}$, or as an equation in $\mathscr{H}_1$. If we view (3.31) as an equation in $\mathscr{H}_1$, then $\{\varphi_i\}$ is a complete system in $\mathscr{H}_1$. Therefore, this is exactly the same case as before. Hence, we have a unique solution

$$u-\psi=\sum_{i=1}^{\infty}\frac{1}{1-\lambda\mu_i}(f,\varphi_i)\varphi_i.$$

Substituting for ψ from (3.30), we have

$$u=f+\sum_{i=1}^{\infty}\left(\frac{1}{1-\lambda\mu_i}-1\right)(f,\varphi_i)\varphi_i$$

and obtain (3.27).

Next, let $\lambda=1/\mu_k$ $(k=s, s+1, \ldots, s+p-1)$, i.e. let the eigenspace of H corresponding μ_k be p-dimensional. If there exists $u\in\mathscr{H}$, then from (3.29) we see that

$$(f,\varphi_k)=0 \quad (k=s, s+1, \ldots, s+p-1). \tag{3.32}$$

Conversely, if (3.32) is satisfied, we can uniquely determine (u, φ_k) in the case $k<s$ or $k\geqslant s+p$, and (u, φ_k) is undetermined if $k=s, s+1, \ldots, s+p-1$.

From this we see that (3.32) has at least one solution with a degree of freedom, i.e. the solution is in the form $\alpha_s\varphi_s+\alpha_{s+1}\varphi_{s+1}+\cdots+\alpha_{s+p-1}\varphi_{s+p-1}$ where the α_i are arbitrary.

Q.E.D.

Note. From the theorem it is clear that $(I-\lambda H)\,u=f$ has a unique solution for an arbitrary f providing that $(I-\lambda H)\,u=0$ only has the solution $u=0$. The fact that the solubility follows from the uniqueness of the solution is fascinating. There were already conjectures in the nineteenth century, but the Fredholm theory of integral equations actually revealed the precise nature of the problem.

Incidentally, the converse of the above statement is also true, i.e. if $(I-\lambda H)\,u=f$ has at least one solution u for f (solubility), then it is a unique solution, (uniqueness). These properties are called the 'Fredholm alternative'. It was proved by Riesz that they are also true if H is completely continuous in case H is not Hermitian (Riesz–Schauder theorem).†

We shall see what happens when H is not completely continuous. The result shows that these properties are no longer true in such cases.

† See section 10.

EXAMPLE. Let $\mathscr{H}=L^2(0,1)$, and let $Hf=xf(x)$, H being a bounded Hermitian operator. Since $\sup_{\|f\|=1}(Hf,f)=\sup_{\|f\|=1}(xf(x),f(x))_{L^2}=1$, $\inf(Hf,f)=0$, so that by proposition 3.2, the spectrum of H lies in $[0,1]$.

Now, let $\mu\in[0,1]$. Then, from $(\mu-x)f(x)=0$, it follows that $f(x)=0$. For an arbitrary ε, there exists $f_\varepsilon\in L^2(0,1)$ with

$$\|(\mu-x)f_\varepsilon(x)\|_{L^2}\leqslant\varepsilon\|f_\varepsilon(x)\|_{L^2}.$$

The inverse of $\mu I-H$† is not continuous, hence, for $\mu\in[0,1]$ and $f\in L^2(0,1)$ the equation $(\mu I-H)u=f$ has no solution (see the inequality (3.16)).

Let us go back to the problem (P. 2), i.e. for $f\in L^2(\Omega)$ we seek a solution $u\in\mathscr{D}_{L^2}{}^1(\Omega)$ for

$$L_\lambda[u]=-\Delta u+c(x)u+\lambda u=f(x). \tag{3.1}$$

According to proposition 3.1 this problem is equivalent to the problem of solving

$$u+(\lambda-t)G_t u=G_t f \tag{3.14}$$

in $L^2(\Omega)$. Incidentally, when $L_\lambda[u]=0$ has a solution $u\in\mathscr{D}_{L^2}{}^1(\Omega)$ $(u(x)\not\equiv 0)$, i.e.

$$Au=\lambda u,\quad A=\Delta-c(x), \tag{3.33}$$

has a solution $u\in\mathscr{D}_{L^2}{}^1(\Omega)$ $(u(x)\not\equiv 0)$, then we call λ an *eigenvalue* for the Dirichlet problem of the operator A, and we call $u(x)$ an *eigenfunction*, where the domain of A is

$$\mathscr{D}(A)=\{u;\,u\in\mathscr{D}_{L^2}{}^1(\Omega),\,Au\in L^2(\Omega)\}.$$

We note that Δ which defines Au is a derivative in the sense of distribution.

From proposition 3.1, we see that (3.33) is equivalent to saying that

$$u+(\lambda-t)G_t u=0 \tag{3.33$'$}$$

has a solution $u\,(\neq 0)$ which belongs to $L^2(\Omega)$. Therefore, to say that $(\lambda,u(x))$ is an eigenvalue and eigenfunction for the Dirichlet problem is equivalent to saying that $(-(\lambda-t),u(x))$ is a singular value and eigenfunction for G_t.

Let us assume that Ω is bounded. In this case, G_t is a completely continuous positive Hermitian operator, and from $G_t f=0$, it follows that $f=0$. So that

† Translator's supplement.

the system of the eigenfunctions corresponding to the non-zero eigenvalues is complete (see theorem 3.4), and $\mu_1 \geqslant \mu_2 \geqslant \mu_3 \geqslant \cdots \to 0$ $(\mu_i > 0)$.†

Therefore, if we define the eigenvalue for (3.33) by the relation

$$\frac{1}{\lambda_i - t} = \mu_i \quad (i=1, 2, \ldots), \tag{3.34}$$

then $t > \lambda_1 \geqslant \lambda_2 \geqslant \lambda_3 \geqslant \cdots \to -\infty$. Also, if we write the eigenfunction of G_t corresponding to λ_i as $u_i(x)$, then $(\lambda_i, u_i(x))$ are the eigen-functions for the eigenvalues of (3.33). We see that $\{u_i(x)\}_{i=1,2,\ldots}$ is a complete orthonormal system in $L^2(\Omega)$.

THEOREM 3.6. Let Ω be bounded. In this case the following statements are true.

(1) The sequence of the eigenvalues satisfying (3.33) is

$$-\inf c(x) = -c_0 > \lambda_1 \geqslant \lambda_2 \geqslant \lambda_3 \geqslant \cdots \to -\infty,$$

i.e. the eigenspace corresponding to each eigenvalue is finite-dimensional, and has an accumulating point at $-\infty$. In the above expression we write λ_k the same number of times as the dimension of the corresponding eigenspace.

Now, if we take eigenfunctions corresponding to different eigenvalues as being mutually orthogonal, and take an orthonormal basis for each eigenspace to establish the correspondence between λ_i and $u_i(x)$, then we see that $\{u_i(x)\}_{i=1,2,\ldots}$ is a complete orthonormal system in $L^2(\Omega)$.

(2) When $\lambda \notin \{\lambda_i\}_{i=1,2,\ldots}$, the Dirichlet problem for (3.1): $(\lambda I - A)u = f$ has a unique solution $u \in \mathscr{D}_{L^2}{}^1(\Omega)$ for an arbitrary $f \notin L^2(\Omega)$, and

$$u(x) = \sum_{i=1}^{\infty} \frac{1}{\lambda - \lambda_i} (f, u_i(x)) u_i(x) \tag{3.35}$$

where the equality is meant in $L^2(\Omega)$.‡

(3) If $\lambda = \lambda_k$, then λ_k has a p-dimensional eigenspace, i.e. if $\lambda_s = \lambda_{s+1} = \cdots = \lambda_{s+p-1} (= \lambda_k)$ then the necessary and sufficient condition that for

† When H is a completely continuous positive Hermitian operator, i.e. $(Hf, f) \geqslant 0$ where equality holds if and only if $f = 0$, then from $(Hf, f) = \sum_{i=1}^{\infty} \mu_i |(f, \varphi_i)|^2$, it follows that $\mu_i > 0$, and $\mu_i \to 0$ since $L^2(\Omega)$ is infinite-dimensional.

‡ In fact, the equality is also meant in $\mathscr{D}_{L^2}{}^1(\Omega)$ (see note 3 on p. 170).

$f \in L^2(\Omega)$ there is a solution of the Dirichlet problem is that

$$(f(x), u_i(x))=0 \quad (i=s, s+1, \ldots, s+p-1). \tag{3.36}$$

In this case (3.35) is obtained by replacing the parts $i=s, s+1, \ldots, s+p-1$ with

$$\alpha_1 u_s(x)+\alpha_2 u_{s+1}(x)+\cdots+\alpha_p u_{s+p-1}(x)$$

where $\alpha_1, \ldots, \alpha_p$ are arbitrary constants.

Proof. (1) is almost clear from our previous explanation. After the proof we shall examine the possibility of estimating the first eigenvalue by an inequality $-c_0>\lambda_1$ (see the note after the proof).

(2) We note that $\lambda \notin \{\lambda_i\}$ is equivalent to saying that $\mu \notin \{\mu_i\}_{i=1,2,\ldots}$ except when $\lambda=t$ by the correspondence $\mu=-1/(\lambda-t)$. Now, we assume $\lambda \neq t$. Then by the previous theorem and (3.14), we have a unique solution

$$u(x)=\sum_{i=1}^{\infty} \frac{1}{1+(\lambda-t)\mu_i}(G_t f, u_i)u_i(x).$$

Note that this is also true for $\lambda=t$.

On the other hand, we have

$$G_t u_i=-\frac{1}{\lambda_i-t} u_i \tag{3.37}$$

and $(G_t f, u_i)=(f, G_t u_i)$.

Now, from these, we have

$$\sum_{i=1}^{\infty} \frac{-1}{\left(1-\dfrac{\lambda-t}{\lambda_i-t}\right)(\lambda_i-t)}(f, u_i)u_i(x)=\sum_{i=1}^{\infty} \frac{1}{\lambda-\lambda_i}(f, u_i)u_i(x).$$

(3) From the previous theorem, we see that the necessary and sufficient condition for (3.14) to have a solution is

$$(G_t f, u_k)=0 \quad (k=s, s+1, \ldots, s+p-1)$$

which is equivalent to $(f, G_t u_k)=0$, and from (3.37), this is equivalent to $(f, u_k)=0$. The rest of (3) is clear if we observe the result of the previous theorem and (2).

Q.E.D.

Note 1. (3.35) is nothing but a representation of the eigenfunction of G_λ, a Green's operator for a Dirichlet problem. For example, if $\lambda=0$ is not an eigenvalue (i.e. if $Au=0$, $u\in\mathscr{D}_{L^2}{}^1(\Omega)$, then $u=0$) then we can write the Green's operator G_0 of A as

$$G_0 f=\sum_{i=1}^{\infty}\frac{(f,u_i)}{\lambda_i}u_i(x)=\int_\Omega\left[\sum_{i=1}^{\infty}\frac{u_i(x)u_i(y)}{\lambda_i}\right]f(y)\,\mathrm{d}y\equiv\int_\Omega G(x,y)f(y)\,\mathrm{d}y$$

since $u_i(x)$ can be taken as a real-valued function. But, of course, this is only a notational observation since we have not yet determined whether $G(x, y)$, the Green's kernel, has actually qualified as a function of (x, y).

Note 2. In the proof of the part (1) of theorem 3.6 we said that the first eigenvalue can be estimated by λ_1 $(<-c_0)$. We shall establish this by using:

LEMMA 3.3 *(Poincaré's inequality).* Let Ω be bounded, and let d=diameter (Ω), i.e. $d=\sup_{(x,y)\in\Omega}\operatorname{dis}(x,y)$. Then, for an arbitrary $u\in\mathscr{D}_{L^2}{}^1(\Omega)$ we have

$$\|u(x)\|_{L^2(\Omega)}\leqslant\frac{d}{\sqrt{2}}\left\|\frac{\partial u}{\partial x_i}(x)\right\|_{L^2(\Omega)}\qquad(i=1,2,\dots,n).\tag{3.38}$$

Proof. Let $u\in\mathscr{D}(\Omega)$. Let us verify (3.38) for the case when $i=1$. We have

$$u(x_1,x_2,\dots,x_n)=\int_{-\infty}^{x_1}\frac{\partial u}{\partial t}(t,x_2,\dots,x_n)\,\mathrm{d}t.$$

From the Schwartz inequality, we have

$$|u(x_1,x_2,\dots,x_n)|^2\leqslant\int_{x_1'}^{x_1}\left|\frac{\partial u}{\partial t}(t,x_2,\dots,x_n)\right|^2\mathrm{d}t\cdot(x_1-x_1')$$
$$\leqslant\int_{-\infty}^{+\infty}\left|\frac{\partial u}{\partial x_1}\right|^2\mathrm{d}x_1\cdot(x_1-x_1'),$$

where x_1' is the lower bound of the coordinate of the support in t, when we regard u as a function of t only.

Therefore, if we integrate the terms which appear on both sides of the inequality with respect to x_1,

$$\int_{-\infty}^{+\infty}|u(x_1,\dots,x_n)|^2\,\mathrm{d}x_1\leqslant\frac{d^2}{2}\int_{-\infty}^{+\infty}\left|\frac{\partial u}{\partial x_1}\right|^2\mathrm{d}x_1$$

on the support of u. We integrate this with respect to $(x_2, \ldots, x_n)$. The result is obviously (3.38). If $u \in \mathscr{D}_{L^2}{}^1(\Omega)$, we use the limiting process $\varphi_j \to u(x)$ in $\mathscr{D}_{L^2}{}^1(\Omega)$ where $\varphi_j \in \mathscr{D}(\Omega)$.

Q.E.D.

Now, if $u \in \mathscr{D}_{L^2}{}^1(\Omega)$ is a solution of $(\lambda I - A)u$, then

$$\sum_{i=1}^{n} \left(\frac{\partial u}{\partial x_i}, \frac{\partial u}{\partial x_i} \right) + ((\lambda + c(x))u, u)_{L^2} = 0,$$

According to Poincaré's inequality, if λ is real we obtain that the left hand side

$$\geqslant \left(\frac{2n}{d^2} + \lambda + \inf c(x) \right) \|u\|_{L^2}{}^2 .$$

From this we have

$$\lambda_1 \leqslant -\inf c(x) - \frac{2n}{d^2}. \tag{3.39}$$

In particular, if Ω is bounded and $A = \Delta$, then we see that $\lambda_1 \leqslant -2n/d^2$, so that $\lambda = 0$ is not an eigenvalue.

Therefore, the Dirichlet problem for $\Delta u = f$ has a unique solution $u \in \mathscr{D}_{L^2}{}^1(\Omega)$ for an arbitrary $f \in L^2(\Omega)$. Also, in this case, irrespective of the sign of $c(x)$, for a sufficiently small d, the Dirichlet problem can be solved since $\lambda_1 < 0$ by (3.39).

Note 3. It looks strange that in the representation of the solution (3.35) $\{u_i\}$ is a complete orthonormal system of $L^2(\Omega)$ while $u \in \mathscr{D}_{L^2}{}^1(\Omega)$. This situation can be clarified by the following argument. First, note that $\{u_i\}$ is a complete orthogonal system in the Hilbert space $\mathscr{H}$ which we introduced at the beginning of this section.

In fact, from

$$\begin{aligned}(u_i, u_k)_{\mathscr{H}} &= \sum_{j=1}^{n} \left(\frac{\partial u_i}{\partial x_j}, \frac{\partial u_k}{\partial x_j} \right) + ((c(x) + t)u_i(x), u_k(x)) \\ &= ((-\Delta + c(x) + t)u_i, u_k) = (t - \lambda_i)\,(u_i(x), u_k(x)),\end{aligned}$$

we have $(u_i, u_k)_{\mathscr{H}} = (t - \lambda_i)\, \delta_i{}^k$, where $\delta_i{}^k$ is the Kronecker delta. Since $\mathscr{H}$ is a subspace of $L^2(\Omega)$, $\{u_i\}$ is a complete system of $\mathscr{H}$ which is identical with $\mathscr{D}_{L^2}{}^1(\Omega)$ in its space as a set and its topology. Therefore, $\{u_i(x)/\sqrt{(t - \lambda_i)}\}$ is a complete orthonormal system of $\mathscr{H}$. From this, we see that:

(i) $\{u_i\}$ form a complete base of $\mathscr{D}_{L^2}{}^1(\Omega)$.

(ii) The necessary and sufficient condition of $u(x)=\sum_{i=1}^{\infty} a_i u_i(x)$ being an element of $\mathscr{D}_{L^2}{}^1(\Omega)$ is

$$\sum_{i=1}^{\infty} |\lambda_i|\,|a_i|^2 < +\infty. \tag{3.40}$$

In fact, if $u \in \mathscr{D}_{L^2}{}^1(\Omega)$, we have

$$\|u\|_{\mathscr{H}^2} = \sum_{i=1}^{\infty} (t-\lambda_i)\cdot|a_i|^2,$$

so that, from $\lambda_i \to -\infty (i\to\infty)$, $t-\lambda_i \sim -\lambda_i$. Hence, the right hand side of (3.35) represents an element of $\mathscr{D}_{L^2}{}^1(\Omega)$.

Next, the necessary and sufficient condition for $f \in L^2(\Omega)$ being an element of $\mathscr{D}(A)$ is $\sum \lambda_i^2 |(f, u_i)|^2 < +\infty$. In this case we can establish the following property,

$$Af = \sum_{i=1}^{\infty} (f, u_i) A u_i(x) = \sum_{i=1}^{\infty} \lambda_i (f, u_i) u_i(x). \tag{3.41}$$

To show this, let f satisfy the above condition, A is a closed operator.† Then let

$$f_j = \sum_{i=1}^{j} (f, u_i) u_i(x).$$

In this case we have

$$f_j \to f, \quad Af_j \to \sum_{i=1}^{\infty} \lambda_i (f, u_i) u_i(x) \quad \text{as} \quad j\to\infty.$$

So that $f \in \mathscr{D}(A)$, hence (3.41) is true.

Next, we assume $f \in \mathscr{D}(A)$. Then, we have $\mathscr{D}(A) = \mathscr{D}(A_t)$. Therefore, $f = G_t \cdot g\,(g \in L^2(\Omega))$. Let us write $g = \sum c_i u_i(x)$, now, we observe

$$f = \sum_{i=1}^{\infty} c_i G_i u_i = \sum_{i=1}^{\infty} \frac{c_i}{t-\lambda_i} u_i(x), \quad \sum \lambda_i^2 \left|\frac{c_i}{t-\lambda_i}\right|^2 < +\infty.$$

† We see that $A_t \equiv A + tI$ is a closed operator since it is the inverse operator of the bounded operator G_t in $L^2(\Omega)$. Hence, A itself is a closed operator.

From this, we see that the Fourier coefficients do in fact satisfy the condition.

Hence, $f \in L^2(\Omega)$. Also the necessary and sufficient condition that $A^j f \in L^2(\Omega)$ $(j=1, 2, \ldots, p)$, i.e. $f \in \mathscr{D}(A^p)$, is

$$\sum_{i=1}^{\infty} (1+|\lambda_i|)^{2p} |(f, u_i)|^2 < +\infty .$$

In this case we establish

$$A^p f = \sum_{i=1}^{\infty} \lambda_i^p (f, u_i) u_i(x) .^\dagger$$

3 *The theorem of Rellich*

Let us prove theorem 3.3, which we left unproved in the previous section. We wish to establish the property that if $f_j \in \mathscr{D}_{L^2}{}^1(\Omega)$, where Ω is bounded and $\| f_j \|_{1, L^2} \leqslant M$, then we can pick a subsequence from $\{f_i\}$ and make it a Cauchy sequence of $L^2(\Omega)$. Various proofs of this are known. Here, we use a method which is based on the idea of a Fourier transform.

Proof. In general, if $f \in L^2(\Omega)$, let us write $\tilde{f}$, an extension of f to the whole space, which takes the value 0 outside Ω. That is

$$\tilde{f}(x) = \begin{cases} f(x) & (x \in \Omega) \\ 0 & (x \in R^n \setminus \Omega) . \end{cases}$$

In this case, if $f \in \mathscr{D}_{L^2}{}^1(\Omega)$, we see that

$$\frac{\partial}{\partial x_i} \widetilde{f(x)} = \widetilde{\frac{\partial}{\partial x_i} f(x)} \quad (i=1, 2, \ldots, n) .$$

In fact, by our hypothesis, we can pick a sequence $\varphi_j \in \mathscr{D}(\Omega) \to f$ in $\mathscr{D}_{L^2}{}^1(\Omega)$,

† From (3.40) and (3.41), for $f \in \mathscr{D}_{L^2}{}^1(\Omega)$, we can easily see that

$$-(Af, f) = \sum_{j=1}^{n} \left(\frac{\partial f}{\partial x_j}, \frac{\partial f}{\partial x_j} \right) + (c(x) f, f) = - \sum_{i=1}^{\infty} \lambda_i |(f, u_i)|^2 .$$

(See (8.87).)

and for $\psi \in \mathscr{D}(R^n)$,

$$\left\langle \widetilde{\frac{\partial f}{\partial x_i}}, \psi(x) \right\rangle = \int_\Omega \frac{\partial f}{\partial x_i} \psi(x)\, dx = \lim_{j\to\infty} \int_\Omega \frac{\partial \varphi_j}{\partial x_j}(x)\psi(x)\, dx$$

$$= -\lim_{j\to\infty} \int_{R^n} \varphi_j(x) \frac{\partial \psi}{\partial x_i}(x)\, dx$$

$$= -\int_{R^n} \widetilde{f(x)} \frac{\partial \psi}{\partial x_i}(x)\, dx = \left\langle \frac{\partial}{\partial x_i} \tilde{f}, \psi \right\rangle.$$

Therefore, $\tilde{f} \in \mathscr{D}_{L^2}{}^1(R^n)$ and

$$\|\tilde{f}(x)\|_{1, L^2(R^n)} = \|f(x)\|_{1, L^2(\Omega)}.$$

Now, let us write the Fourier image of $\tilde{f}_j$ as $\varphi_j(\xi)$ ($\varphi_j \in L^2$), i.e.

$$\varphi_j(\xi) = \int_\Omega f_j(x)\, e^{-2\pi i x\xi}\, dx.$$

If $x \in \Omega$, then we have $\tilde{f}_j = f_j$. By the inversion formula, we have

$$f_j(x) = \int e^{2\pi i x\xi} \varphi_j(\xi)\, d\xi = \int_{|\xi| \leqslant R} e^{2\pi i x\xi} \varphi_j(\xi)\, d\xi + \int_{|\xi| > R} e^{2\pi i x\xi} \varphi_j(\xi)\, d\xi$$

$$\equiv f_j^{(1)}(x) + f_j^{(2)}(x)$$

where we regard the term which corresponds to $f_j^{(2)}(x)$ as a symbolical representation, i.e. the inverse Fourier image of an L^2-function which is obtained from $\varphi_j(\xi)$ by putting its value 0 when $|\xi| \leqslant R$. Therefore, by Plancherel's theorem,

$$\|f_j^{(2)}(x)\|_{L^2}{}^2 = \int_{|\xi|>R} |\varphi_j(\xi)|^2\, d\xi = \int_{|\xi|>R} \frac{1}{1+4\pi^2|\xi|^2} (1+4\pi^2|\xi|^2)|\varphi_j(\xi)|^2\, d\xi$$

$$\leqslant \frac{1}{(2\pi R)^2} \int_{R^n} (1+4\pi^2|\xi|^2)|\varphi_j(\xi)|^2\, d\xi,$$

and the last integral is equal to $\|\tilde{f}_j(x)\|^2{}_{1, L^2(R^n)}$. Hence, by our hypothesis,

$$\|f_j^{(2)}(x)\|_{L^2} \leqslant \frac{M}{2\pi R}$$

where M is a constant.

From this estimation, we can see that the value of the right hand side can be made $<\varepsilon$ for a given $\varepsilon(>0)$ if we take R sufficiently large. Let us fix such an R, then we consider $\{f_j^{(1)}\}$, these are equi-continuous and uniformly bounded functions. In fact, the boundedness follows from

$$\sup|f_j^{(1)}(x)| \leqslant \int_{|\xi|\leqslant R} |\varphi_j(\xi)|\,\mathrm{d}\xi,$$
$$\leqslant \|\varphi_j(\xi)\|_{L^2}\cdot\sqrt{\text{volume}(S_R)} \leqslant M\cdot\sqrt{\text{volume}(S_R)},$$

by the Schwartz inequality. The equi-continuity follows from

$$|f_j^{(1)}(x)-f_j^{(1)}(x')| \leqslant \int_{|\xi|\leqslant R} |\exp(2\pi i x\xi)\,\{1-\exp[2\pi i(x'-x)\xi]\}\varphi_j(\xi)|\,\mathrm{d}\xi$$
$$\leqslant \max_{|\xi|\leqslant R} |1-\exp[2\pi i(x'-x)\xi]|\cdot \int_{|\xi|\leqslant R} |\varphi_j(\xi)|\,\mathrm{d}\xi$$

and the previous inequality.

Therefore, $\bar{\Omega}$ is a bounded closed set, and, by the Ascoli–Arzela theorem, there is a subsequence $\{f_{j_p}{}^{(1)}\}$ of $\{f_j{}^{(1)}\}$ which converges uniformly in $\bar{\Omega}$, i.e., in Ω. Therefore, of course, $\{f_{j_p}\}$ is a Cauchy sequence of $L^2(\Omega)$. From this fact and $\|f_{j_p}{}^{(2)}(x)\|_{L^2(\Omega)}<\varepsilon$, we see that $f_{j_p}(x)\equiv f_{j_p}{}^{(2)}(x)+f_{j_p}{}^{(2)}(x)$ is a Cauchy sequence excluding 2ε, i.e.

$$\overline{\lim_{p,\,q\to\infty}}\ \|f_{j_p}(x)-f_{j_q}(x)\|_{L^2(\Omega)}\leqslant 2\varepsilon,$$

where we take ε progressively as a sequence $\{\varepsilon_i\}$; $\varepsilon_1>\varepsilon_2>\varepsilon_3>\cdots\to 0$. We first take a subsequence $\{f_{1_j}\}_{j=1,\,2,\ldots}$ of $\{f_j\}$ corresponding to ε_1; next, we take a subsequence $\{f_{2_j}\}_{j=1,\,2,\ldots}$, of $\{f_{1_j}\}_{j=1,\,2,\ldots}$ and so on. Then, we choose diagonal elements $\{f_{jj}\}_{j=1,\,2,\ldots}$. This is obviously a convergent sequence of $L^2(\Omega)$.

Q.E.D.

From this theorem, we have:

COROLLARY OF THEOREM 3.3. Let Ω be a bounded open set. If $\{f_j\}$ is an arbitrary bounded sequence of $\mathscr{D}_{L^2}{}^m(\Omega)$, then we have a Cauchy sequence of $\mathscr{D}_{L^2}{}^{m-1}(\Omega)$, $\{f_{j_p}\}$ which is appropriately selected from $\{f_j\}$.

Proof. Because the sequence $\{\mathrm{D}^\alpha f_j(x)\}_{j=1,2,\ldots}(|\alpha|\leqslant m-1)$ is bounded in $\mathscr{D}_{L^2}{}^1(\Omega)$, by theorem 3.3, there is a Cauchy subsequence $\{\mathrm{D}^\alpha f_{j_p}(x)\}_{p=1,\,2,\ldots}$

($|\alpha| \leqslant m-1$) of $L^2(\Omega)$. We note that this subsequence can be taken commonly for all $\alpha: |\alpha| \leqslant m-1$, i.e. $\{f_{j_p}\}$ is a Cauchy sequence of $\mathscr{D}_{L^2}{}^{m-1}(\Omega)$. Of course, from the completeness of $\mathscr{D}_{L^2}{}^{m-1}(\Omega)$ we see that $f_{j_p}(x) \to f_0(x) \in \mathscr{D}_{L^2}{}^{m-1}(\Omega)$.

In Rellich's choice theorem, we assumed that $f_j \in \mathscr{D}_{L^2}{}^1(\Omega)$. If we now assume that $f_j \in \mathscr{E}_{L^2}{}^1(\Omega)$, the situation is different. For example, the proof previously given is no longer valid. If Ω is bounded, however, we can impose some restrictions on the 'shape' of the boundary of Ω in order to make theorem 3.3 remain true. This is one of the important results obtained by Sobolev. We shall discuss his theory in section 6. We have to consider the case where Ω is not bounded. In fact if $\Omega = R^n$, theorem 3.3 (Rellich), p. 154, is not true.

What then does happen if Ω is not bounded? In fact, if $f \in \mathscr{D}$, we write $f_j(x) = f(x - jx_0)$ where x_0 is arbitrary but $|x_0| \neq 0$. Then, $\{f_j\}$ has no subsequence which is convergent in L^2.

THEOREM 3.7. Let Ω be an arbitrary open set (not necessarily bounded), and let $\{f_j\}_{j=1,2,\ldots}$ be a bounded sequence of $\mathscr{E}_{L^2}{}^1(\Omega)$. Then, we can pick an appropriate Cauchy subsequence $\{f_{j_p}\}_{p=1,2,\ldots}$ in $L^2{}_{\text{loc}}(\Omega)$, i.e. there exists $f_0 \in L^2{}_{\text{loc}}(\Omega)$ such that

$$\int_K |f_{j_p}(x) - f_0(x)|^2 \, dx \to 0 \quad (p \to \infty)$$

on an arbitrary compact set K of Ω.

Proof. We consider a sequence of compact sets $K_1 \Subset K_2 \Subset K_3 \Subset \cdots \to \Omega$, where $\Subset$ means that K_i consists of interior points of K_{i+1} and each point of Ω is contained in a certain K_n. Then, in this case, we can easily see that there exists a certain K_n for an arbitrary compact set K, and $K \subset K_n$. Now, we take $\alpha_n \in \mathscr{D}(\Omega)$ for such K_n and make $\alpha_n(x) \equiv 1$ on K_n.

First, we consider $\{\alpha_1(x) f_j(x)\}_{j=1,2,\ldots}$. This is a bounded sequence of $\mathscr{D}_{L^2}{}^1(\Omega_1)$ where Ω_1 is a bounded open set such that $\Omega_1 \subset \Omega$. By Rellich's choice theorem, we see that there exists a Cauchy subsequence $\{\alpha_1(x) f_{1p}(x)\}_{p=1,2,\ldots}$ in $L^2(G_1)$, and, therefore, in $L^2(\Omega)$. Next, we pick a convergent subsequence from $\{\alpha_2(x) f_{1p}(x)\}_{p=1,2,\ldots}$ and so on.

Let us write the diagonal elements of the sequence of subsequences $\{f_m(x)\}$ which has been obtained in this way. Then $\{f_m(x)\}_{m=1,2\ldots}$ is a convergent sequence in the sense of L^2 on an arbitrary K_n.

Q.E.D.

We can generalize this theorem, as we did theorem 3.3, to obtain the corollary of theorem 3.3 (see section 3).

DEFINITION 3.1. We define the space $\mathscr{E}_{L^2}{}^m{}_{(\mathrm{loc})}(\Omega)$ by forming the set of all f which satisfies $\alpha(x)f\in\mathscr{D}_{L^2}{}^m$ for an arbitrary $\alpha\in\mathscr{D}(\Omega)$. The topology of this space is given by the following: $f_j(x)\to 0$ in $\mathscr{E}_{L^2}{}^m{}_{(\mathrm{loc})}(\Omega)$ if and only if for an arbitrary $\alpha\in\mathscr{D}(\Omega)$, $\alpha f_j\to 0$ in $\mathscr{D}_{L^2}{}^m$. Thus we have a Fréchet space having an inclusion property $\mathscr{E}_{L^2}{}^m(\Omega)\subset\mathscr{E}_{L^2}{}^m{}_{(\mathrm{loc})}(\Omega)$.

COROLLARY OF THEOREM 3.7. Let Ω be an arbitrary open (not necessarily bounded) set. If $\{f_j\}$ is a bounded sequence of $\mathscr{E}_{L^2}{}^m{}_{(\mathrm{loc})}(\Omega)$, we can choose an appropriate subsequence $\{f_{j_p}\}$ from $\{f_j\}$ which is a Cauchy sequence of $\mathscr{E}_{L^2}{}^{m-1}{}_{(\mathrm{loc})}(\Omega)$.

Note. In this case the boundedness of $\{f_j\}$ means that for an arbitrary $\alpha\in\mathscr{D}(\Omega)$, $\|\alpha f_j\|_{m,L^2}\leqslant M(\alpha)<+\infty$.

4 *Trace on boundaries (boundary values in the wider sense)*

We have not imposed any restriction on the property of the open set Ω in the previous argument. In this section we shall impose the condition that Ω is a domain surrounded by a finite number of smooth hypersurfaces S_i which are mutually non-intersecting. We shall now explain what we mean by '*smooth*' for a hypersurface S.

By a hypersurface S is C^m-class $(m\geqslant 1)$ we mean that for an arbitrary $x_0\in S$, if we take V, a neighbourhood of x_0, sufficiently small (for instance V can be an open sphere with centre x_0), then there exists a function $\varphi\in C^m$, and S can be expressed by the equation

$$\varphi(x)=0 \quad \text{where} \quad \sum_{i=1}^{n}\left|\frac{\partial\varphi}{\partial x_i}(x)\right|\neq 0,\ x\in V$$

in V. We note that $\varphi(x)$ is not uniquely determined. In this case, we put $x_n'=\varphi(x)$ and regard $(x_1',\ldots,x_{n-1}')$ as the value of an appropriate function Φ of C^m-class (the function can be made linear). Then, the Jacobian of the transformation $x'=\Phi(x)$ is non-zero in a certain neighbourhood $V'(\subset V)$ of x_0. Therefore, if necessary, we can take V sufficiently small, then, by the transformation $x'=\Phi(x)$, V is mapped into an open neighbourhood V' of $x_0'(=\Phi(x_0))$ where the mapping is bijective, and the inverse mapping Ψ is also a C^m-class function.

We say that a hypersurface S is *closed* if and only if S is compact, i.e.

for an arbitrary open covering of S, we can choose a finite subcovering which itself is an open covering. Now, given a closed hypersurface S, we consider an open covering $\{V_i\}_{i=1,\dots,p}$ satisfying the finiteness condition which we have just stated. In this case, we can choose $\alpha_i \in \mathscr{D}(V_i)$ $(0 \leqslant \alpha_i(x) \leqslant 1)$ such that $\sum_{i=1}^{p} \alpha_i(x) = 1$ is true not only on S but also in a neighbourhood of S

Let Ω be the interior domain of S (or the exterior domain of S). Let $k \leqslant m$, and let $f \in \mathscr{E}_{L^2}{}^k(\Omega)$. Then, we have a decomposition

$$f(x) = \sum_{i=1}^{p} \alpha_i(x) f(x) + \left(1 - \sum_{i=1}^{p} \alpha_i(x)\right) f(x). \tag{3.42}$$

The first term on the right hand side, $\alpha_i(x)f(x)$, is a function having its support in the interior of V_i and $\alpha_i(x)f \in \mathscr{E}_{L^2}{}^k(V_i \cap \Omega)$. The second term on the right hand side belongs to $\mathscr{E}_{L^2}{}^k(\Omega)$, but its support does not intersect the boundary S of Ω. Therefore, we can ignore this when we are interested in the behaviour of f in a neighbourhood of the boundary of Ω.

Let us consider $\alpha_i(x)f(x)$ in a transformed form $g_i(x') \equiv (\alpha_i f)(\Psi_i(x'))$ by the local transformation $x' = \Phi_i(x)$ $(x \in V_i)$. We note that by this transformation the boundary $S \cap V_i$ is mapped into a part of a hyperplane of (x')-space: $\{x_n' = 0\} \cap V_i'$.

LEMMA 3.4. Let Ω be a bounded open set of (x)-space. Let $\Phi : x \to x'$ be a one-to-one mapping from Ω onto an open set Ω' of (x')-space including its boundary, and let Φ be a C^m-class $(m \geqslant 1)$ function, having as its inverse $\Psi : x' \to x$, i.e. Φ and Ψ can be extended to appropriate new open sets $\tilde{\Omega}(\supset \bar{\Omega})$ and $\tilde{\Omega}'(\supset \bar{\Omega}')$.† In this case, the function $g(x') \equiv f(\Psi(x'))$ which is transformed from an arbitrary function $f \in \mathscr{E}_{L^2}{}^k(\Omega)$ $(k \leqslant m)$ belongs to $\mathscr{E}_{L^2}{}^k(\Omega')$, and this linear correspondence $f(x) \to g(x')$ is continuous.

If we exchange x and x', the inverse of this statement is true, i.e. for an arbitrary $g \in \mathscr{E}_{L^2}{}^k(\Omega')$, $f \equiv g \circ \Phi \in \mathscr{E}_{L^2}{}^k(\Omega)$ and this correspondence $f \to g$ is continuous.

Proof. We sketch the outline of the proof. We assume that, $f \in \mathscr{E}_{L^2}{}^1(\Omega)$,

† Translator's note: the original notation has been retained.

and $g(x')=f(\psi_1(x'), \psi_2(x'), \ldots, \psi_n(x'))$, then we have

$$\frac{\partial}{\partial x_i'} g(x')=\frac{\partial \psi_1}{\partial x_i}\cdot\frac{\partial f}{\partial x_1}+\frac{\partial \psi_2}{\partial x_i'}\cdot\frac{\partial f}{\partial x_2}+\cdots+\frac{\partial \psi_n}{\partial x_i'}\cdot\frac{\partial f}{\partial x_n}. \tag{3.43}$$

We note that (3.43) is true for

$$f_\varepsilon(x)=\rho_\varepsilon * f(x)=\int_\Omega \rho_\varepsilon(x-y)f(y)\,\mathrm{d}y \in \mathscr{E}^\infty,$$

and if $\varepsilon\to+0$, we have

$$f_\varepsilon(x)\to f(x) \quad (\partial f_\varepsilon/\partial x_j \to \partial f/\partial x_j)$$

on an arbitrary compact set of Ω in the sense of L^2, i.e. $f_\varepsilon(x)\to f(x)$ in $\mathscr{E}_{L^2}{}^1{}_{\mathrm{loc}}(\Omega)$ (see proposition 1.3). From this fact we see that $g_\varepsilon(x')\to g(x')$ in $L^2{}_{\mathrm{loc}}(\Omega')$ and

$$\frac{\partial f_\varepsilon}{\partial x_j}(\psi_1(x'),\ldots,\psi_n(x'))\to\frac{\partial f}{\partial x_j}(\psi_1(x'),\ldots,\psi_n(x')) \quad \text{in} \quad L^2{}_{\mathrm{loc}} \quad (\Omega').$$

Therefore, if $\varepsilon\to 0$, we see that (3.43) is true for an element of $\mathscr{D}'(\Omega)$.

Q.E.D.

Trace of $\mathscr{E}_{L^2}{}^1(R_+{}^n)$ on the hyperplane $x_n=0$

Under this sub-heading, we made an observation of a rather different nature from what has gone immediately before, so although we shall use the same notation which we used previously for different entities, there should be no confusion. For example we write x instead of x';

$$R_+{}^n=\{x;\, x_n>0\}.$$

Also, we use a simplification convention; $(x_1,\ldots,x_{n-1})=x'$. Now, we establish the following:

LEMMA 3.5. Let $f\in\mathscr{E}_{L^2}{}^1(R_+{}^n)$.

(1) For $x'\in R^{n-1}$ there exists a finite limit of

$$\lim_{x_n\to+0} f(x_1,\ldots,x_{n-1},x_n)=f(x',+0)=\varphi(x')$$

almost everywhere.

(2) $\varphi = L^2(R^{n-1})$ and $\|\varphi(x')\|_{L^2(R^{n-1})} \leqslant C\|f(x)\|_{1,L^2(R_+{}^n)}$ where C is a positive integer which is independent of $f(x)$.

Proof. Let $\alpha(x_n)$ be a C^∞-class function which is equal to 1 in the neighbourhood of $x_n=0$ and equal to 0 for $x_n \geqslant 1$. Let us then fix α. It is enough for us to prove the theorem for $\alpha(x_n)f(x)$ instead of $f(x)$. Therefore, we can assume that $f \in \mathscr{E}_{L^2}{}^1(R_+{}^n)$ and the support of f lies within $x_n \leqslant 1$.

Now, from

$$\iint \left|\frac{\partial f}{\partial x_n}\right|^2 dx'\, dx_n < +\infty$$

and Fubini's theorem, we have that

$$\int_0^1 \left|\frac{\partial f}{\partial x_n}(x', x_n)\right|^2 dx_n < +\infty$$

is true for x' almost everywhere. Hence, from the Schwartz inequality we see that

$$\int_0^1 \left|\frac{\partial f}{\partial x_n}(x', x_n)\right| dx_n < +\infty .$$

On the other hand, f and $\partial f/\partial x_n \in L^1{}_{\text{loc}}(R_+{}^n)$, so that by Nikodym's theorem (theorem 2.7), we can establish that for x' almost everywhere, $f(x', x_n)$ is a function of x_n which is absolutely continuous† in $(0, 1]$.

In this case, the ordinary meaning of $\partial f/\partial x_n$ and the meaning of the same function in the sense of distribution coincide. Therefore, the following equality holds:

$$f(x', \varepsilon) = -\int_\varepsilon^1 \frac{\partial f}{\partial x_n}(x', x_n)\, dx_n \quad (\varepsilon > 0).$$

As $\varepsilon \to +0$,

$$\varphi(x') = f(x', +0) = -\int_0^1 \frac{\partial_J}{\partial x_n}(x', x_n)\, dx_n$$

has a finite limit. We apply the Schwartz inequality to this, then

$$|\varphi(x')|^2 \leqslant \int_0^1 \left|\frac{\partial f}{\partial x_n}(x', x_n)\right|^2 dx_n .$$

† I.e. for an arbitrary $\varepsilon(>0)$, it is absolutely continuous in $[\varepsilon, 1]$ almost everywhere.

If we integrate this with respect to x' over R^{n-1}, then we have

$$\|\varphi(x')\|_{L^2} \leqslant \left\|\frac{\partial f}{\partial x_n}\right\|_{L^2(R_+{}^n)}.$$

Q.E.D.

Let us examine the results which can be derived from lemma 3.5 in the case when the boundary of Ω consists of a closed hypersurface S. For simplicity, we consider a case in which $f \in \mathscr{E}_{L^2}{}^1(\Omega)$ approaches the hypersurface S along the direction of a normal of S.

As is well known, if S is of the C^2-class, then for a sufficiently small neighbourhood V of S, there exists a unique normal of S which contains an arbitrary point of V. If necessary, we can make the neighbourhood even smaller in order to obtain $\{V_i, \alpha_i(x)\}$, and a one-to-one and invertable mapping $\Phi: x \to x'(x \in V_i)$ which maps $V_i \cap \Omega$ onto $V_i' \cap R_+{}^n$ for every V_i, and normal congruence of S in V_i onto x' (const.).

From the previous two lemmas we now have:

THEOREM 3.8. Let Ω be a domain which has a closed C^2-class hypersurface as its boundary. In this case, for an arbitrary $f \in \mathscr{E}_{L^2}{}^1(\Omega)$, we define its boundary value as the limit of $f(x)$ along a normal of S. Then, there exists the finite boundary value of $f(x)$ on S almost everywhere. If we write this as $\varphi(\xi)$, then $\varphi \in L^2(S)$ satisfying

$$\|\varphi(\xi)\|_{L^2(S)} \leqslant C\|f(x)\|_{1, L^2(\Omega)}, \tag{3.44}$$

where C is a positive constant.

Note that we choose the normal congruence in order to define the boundary value, but this method is just conventional. In fact, in the case when we have a set of transversal curves for S at a neighbourhood of the surface S (we shall explain the meaning of this later), the limit of $f(x)$ along these curves does exist in S almost everywhere. Let us examine this statement more closely.

By saying that a C^1-class curve passing through a point x_0 of S is *transversal* with respect to S, we mean that the tangential line of the curve at x_0 does not lie in the tangential plane of S at x_0, i.e. the curve cuts across S. Now, let V be a certain neighbourhood of x_0, with x_0 in S where S is defined by $\varphi(x) = 0$.

In V, let us consider a system of curves $\psi_1(x)=C_1$, $\psi(_2x)=C_2,\ldots$, $\psi_{n-1}(x)=C_{n-1}$, where $(C_1,\ldots,C_{n-1})$ are given by real parameters and ψ_i, $i=1,\ldots,n-1$, are all C^1-class functions. The condition that the system of these curves is *transversal* for S is equivalent to the claim that following determinant is non-zero

$$\begin{vmatrix} \frac{\partial\varphi}{\partial x_1} & \cdots & \frac{\partial\varphi}{\partial x_n} \\ \frac{\partial\psi_1}{\partial x_1} & \cdots & \frac{\partial\psi_1}{\partial x_n} \\ \frac{\partial\psi_{n-1}}{\partial x_1} & \cdots & \frac{\partial\psi_{n-1}}{\partial x_n} \end{vmatrix} \neq 0.$$

So that the transformation $x'=\Phi(x)$ defined by $x_n'=\varphi(x)$, $x_i'=\psi_i(x)$ $(i=1,2,\ldots,n-1)$ is non-singular, and the curves are mapped into straight lines which are parallel to the x_n'-axis. We therefore call a system of curves which satisfies this condition *a regular transversal system of curves of S.*

The previous theorem is also valid for this system of curves. Nevertheless, in this case, we are not yet sure that the boundary value $\varphi(\xi)$ of $f(x)$ which is taken along the normal congruence is identical as an element of $L^2(S)$, with the boundary value of $f(x)$ which is taken along the system of curves, i.e. we have to prove that $\varphi(\xi)$ is actually qualified as 'the boundary value of $f(x)$' as stated in theorem 3.8. Let us show this first. What we need is the following:

THEOREM 3.9 (Trace operator). Let Ω be a domain (outer or inner) having the boundary S which is a closed hypersurface of the C^2-class. Then, there exists a unique linear operator γ with the following properties.

(1) If $f(x)$ is once continuously differentiable in $\bar{\Omega}$,† then $(\gamma f)(\xi)$ is identical with the boundary value of $f(x)$ on S.

(2) $f\in\mathscr{E}_{L^2}{}^1(\Omega)\to\gamma f\in L^2(S)$ is continuous.

In this case we call γ a *trace operator* on S.

Proof. It is enough to show the uniqueness of γ because the existence of γ has been shown already in theorem 3.8.

† In other words, $f\in C^1(\overline{\Omega})$ in the sense of definition 3.2.

Proof of the uniqueness of a trace operator. Let us explain the ideas which we are going to use when we deal with a classical boundary value problem. Let Ω be a domain having a boundary S of C^m-class.

DEFINITION 3.2 ($C^m(\bar{\Omega})$). By $f \in C^m(\bar{\Omega})$ we mean that $f \in \mathscr{E}^m(\Omega)$ and there exists an extension of f: $\tilde{f} \in \mathscr{E}^m(\tilde{\Omega})$ beyond S where $\tilde{\Omega}$ is an open set which contains $\bar{\Omega}$.
Alternatively we can define the same $C^m(\bar{\Omega})$ as:

DEFINITION 3.3 ($C^m(\bar{\Omega})$). By $f \in C^m(\bar{\Omega})$ we mean that $D^\alpha f(x)$ ($|\alpha| \leqslant m$) are continuous functions which are respectively defined on Ω including its boundary. These two definitions coincide when the boundary S is smooth, i.e.

THEOREM 3.10. Let Ω be a domain having as boundary a closed hypersurface S of the C^m-class. If $D^\alpha f(x)$ ($|\alpha| \leqslant m$) is continuous on Ω including its boundary, then f can be extended beyond S as a function of the C^m-class.

Proof. We consider the partition of the unity $\{V_i, \alpha_i(x)\}$ which we have observed at the beginning of section 4. Let us pick one of them, $\alpha_i(x)f(x)$. We transform this by an invertible transformation $x' = \Phi(x)$. What we have to prove is the following:

LEMMA 3.6. Let $f \in \mathscr{E}^m(R_+{}^n)$. If every derivative $D^\alpha f(x)$ ($|\alpha| \leqslant m$) is continuous on $R_+{}^n$ including the hyperplane $x_n = 0$, then there exists an extension $\tilde{f} \in \mathscr{E}^m(R^n)$ of f.

Proof. Let us write $x' = (x_1, \ldots, x_{n-1})$. We extend $f(x', x_n)$ to the lower half-space $x_n < 0$ by putting

$$f(x', x_n) = \sum_{\nu=1}^{m+1} a_\nu f(x', -\nu x_n) \tag{3.45}$$

where $(a_1, a_2, \ldots, a_{m+1})$ is the solution of

$$1 = \sum_{\nu=1}^{m+1} (-\nu)^j a_\nu \quad (j = 0, 1, 2, \ldots, m). \tag{3.46}$$

The coefficients of $\{a_\nu\}$ form a Vandermonde's determinant, so that $\{a_\nu\}$ can be uniquely determined. From this fact we have

$$\lim_{x_n\to -0}\left(\frac{\partial}{\partial x_n}\right)^j f(x', x_n) = \lim_{x_n\to +0}\left(\frac{\partial}{\partial x_n}\right)^j f(x', x_n) \quad (j=0, 1, 2, \ldots, m).$$

Therefore, if we put $f(x', 0)=f(x', +0)\ (=f(x', -0))$,

$$\left(\frac{\partial}{\partial x_n}\right)^j f(x', x_n) \quad (j=0, 1, \ldots, m)$$

are all continuous functions defined on R^n. Furthermore, from our hypothesis, we have

$$\lim_{x_n\to -0}\left(\frac{\partial}{\partial x'}\right)^\alpha\left(\frac{\partial}{\partial x_n}\right)^j f(x', x_n) = \lim_{x_n\to +0}\left(\frac{\partial}{\partial x'}\right)^\alpha\left(\frac{\partial}{\partial x_n}\right)^j f(x', x_n) \quad (|\alpha|+j\leqslant m),$$

where the convergences are uniformly compact in $R_{x}'^{\,n-1}$, i.e. if we consider

$$x_n \to \left(\frac{\partial}{\partial x_n}\right)^j f(x', x_n) \in \mathscr{E}^{m-j}(R_{x'}{}^{n-1}),$$

then this is a convergent sequence elementwise of the function space $\mathscr{E}^{m-j}(R_{x'}{}^{n-1})$ as $x_n\to 0$.

On the other hand, we have

$$\left(\frac{\partial}{\partial x_n}\right)^j f(x', x_n) \to \left(\frac{\partial}{\partial x_n}\right)^j f(x', 0) \quad \text{in} \quad \mathscr{E}^{m-j}(R_{x'}{}^{n-1}) \quad \text{as} \quad x_n\to 0.$$

This shows the continuity of

$$\left(\frac{\partial}{\partial x'}\right)^\alpha\left(\frac{\partial}{\partial x_n}\right)^j f(x', x_n) \quad (|\alpha|+j\leqslant m).$$

Q.E.D.

Let us go back to the space L^2. We need some preparation before proceeding. Let $f\in\mathscr{E}_{L^2}{}^m(R_+{}^n)$. In general, if $g(x)$ is defined on $R_+{}^n$, we put

$$\tilde{g}(x) = \begin{cases} g(x) & (x>0), \\ 0 & (x<0). \end{cases}$$

$\tilde{f}$ is regarded as a distribution defined on the whole space R^n which is an extension of $f \in \mathscr{E}_{L^2}{}^1(R_+{}^n)$ covering also the lower half-space by taking the value zero there. Note that the difference of $\tilde{f}$ and f is a distribution. One of the reasons is that in general

$$\frac{\partial}{\partial x_n}\tilde{f} \neq \left(\frac{\partial f}{\partial x_n}\right)^{\sim}.$$

In fact, for $\varphi \in \mathscr{D}(R^n)$,

$$\begin{aligned}\left\langle \frac{\partial}{\partial x_n}\tilde{f}(x), \varphi(x)\right\rangle &= -\int_{x_n>0} f(x)\frac{\partial \varphi}{\partial x_n}(x)\,dx\\ &= -\int dx' \int_{x_n>0} f(x', x_n)\frac{\partial \varphi}{\partial x_n}(x', x_n)\,dx_n.\end{aligned}$$

Now, we use lemma 3.5 to perform an integration by parts. We have:

$$\int f(x', +0)\,\varphi(x', 0)\,dx' + \int_{x_n>0}\frac{\partial f}{\partial x_n}\varphi(x)\,dx.$$

Therefore

$$\frac{\partial}{\partial x_n}\tilde{f}(x) = f(x', +0)\otimes\delta_{x_n} + \left(\frac{\partial f}{\partial x_n}\right)^{\sim}. \tag{3.47}$$

If $f \in \mathscr{E}_{L^2}{}^2(R_+{}^n)$, then $\partial f/\partial x_n \in \mathscr{E}_{L^2}{}^1(R_+{}^n)$, so that from this fact we have

$$\frac{\partial^2}{\partial x_n{}^2}\tilde{f}(x) = f(x', +0)\otimes\delta_{x_n}{}' + \frac{\partial f}{\partial x_n}(x', +0)\otimes\delta_{x_n} + \left(\frac{\partial^2 f}{\partial x_n{}^2}\right)^{\sim}.$$

We see that for $f \in \mathscr{E}_{L^2}{}^m(R_+{}^n)$ $(p \leqslant m)$,

$$\begin{aligned}\left(\frac{\partial}{\partial x_n}\right)^p \tilde{f}(x) = &\left(\left(\frac{\partial}{\partial x_n}\right)^p f(x)\right)^{\sim} + f(x', +0)\otimes\delta_{x_n}{}^{(p-1)}\\ &+ \frac{\partial}{\partial x_n} f(x', +0)\otimes\delta_{x_n}{}^{(p-2)} + \cdots\\ &\cdots + \left(\frac{\partial}{\partial x_n}\right)^{p-1} f(x', +0)\otimes\delta_{x_n}.\end{aligned} \tag{3.48}$$

On the other hand we have† obviously

$$\left(\frac{\partial}{\partial x'}\right)^{\alpha}\tilde{f}(x) = \left(\left(\frac{\partial}{\partial x'}\right)^{\alpha} f(x)\right)^{\sim} \quad (|\alpha| \leqslant m). \tag{3.49}$$

† From Nikodym's theorem (theorem 2.7) we have

$$\left(\frac{\partial f}{\partial x_j}\right)^{\sim} = \frac{\partial}{\partial x_j}\tilde{f} \quad (j = 1, 2, \ldots, n-1)$$

if f, $\partial f/\partial x_j \in L^2(R_+{}^n)$.

From this, for $p+|\alpha|\leqslant m$

$$\begin{aligned}D_{x_n}{}^p D_{x'}{}^\alpha \tilde{f}(x) &= D_{x_n}\,(D_{x'}{}^\alpha f(x))^\sim \\ &= (D_{x_n}{}^p D_{x'}{}^\alpha f(x))^\sim + \sum_{j=0}^{p-1} D_{x_n}{}^j\, D_{x'}{}^\alpha f(x', +0)\otimes\delta_{x_n}{}^{(p-1-j)} \\ &\qquad \text{(by (3.48))}. \qquad (3.50)\end{aligned}$$

The last result must be understood in the sense of $D_{x_n}{}^j D_{x'}{}^\alpha f(x', x_n) \in L^2(R^{n-1})$ as $x_n \to +0$ fixing x'. If we write $D_{x_n}{}^j f(x', +0) = \varphi_j(x')$ (a trace of j-degree), then this is equal to $D_{x'}{}^\alpha \varphi_j(x')$, i.e. the last term T on the right hand side of (3.50) satisfies

$$\begin{aligned}\langle T, \psi(x)\rangle &= \sum_{j=0}^{p-1} (-1)^{p-1-j} \int_{R^{n-1}} D_{x'}{}^\alpha \varphi_j(x')\, D_{x_n}{}^{p-1-j}\psi(x', 0)\, dx'. \\ &= \sum_{j=0}^{p-1} (-1)^{p+|\alpha|-1-j} \int_{R^{n-1}} \varphi_j(x')\, D_{x_n}{}^{p-1-j}\, D_{x'}{}^\alpha \psi(x', 0)\, dx'.\end{aligned}$$

From this we have

Proposition 3.4. $f \in \mathscr{E}_{L^2}{}^m(R_+{}^n)$ has an extension $F \in \mathscr{E}_{L^2}{}^m(R^n)$ to the whole space R^n.

Proof. The proof is quite similar to lemma 3.6 except that we have to replace $m+1$ by m, i.e. if $x_n<0$, then

$$g(x) = \sum_{\nu=1}^{m} a_\nu f(x', -\nu x_n), \qquad (3.51)$$

where

$$\sum_{\nu=1}^{m} (-\nu)^j a_\nu = 1 \quad (j=0, 1, 2, \ldots, m-1).$$

If we define $g(x)$ in this way, $g \in \mathscr{E}_{L^2}{}^m(R_-{}^n)$, and its trace into $x_n=0$ coincides with that of f to the degree $(m-1)$, i.e.

$$\lim_{x_n\to+0}\left(\frac{\partial}{\partial x_n}\right)^j f(x', x_n) = \lim_{x_n\to-0}\left(\frac{\partial}{\partial x_n}\right)^j g(x) \quad (j=0, 1, 2, \ldots, m-1),$$

for x' almost everywhere.

So that if

$$\tilde{g}(x)=\begin{cases}0 & (x_n>0),\\ g(x) & (x_n<0),\end{cases}$$

then from (3.50), for $|\alpha|\leqslant m$

$$D^\gamma(\tilde{f}(x)+\tilde{g}(x))=D^\gamma\tilde{f}(x)+D^\gamma\tilde{g}(x)=(D^\gamma f(x))^\sim+(D^\gamma g(x))^\sim, \tag{3.52}$$

because the terms containing T (see (3.50)) cancel each other. In fact, for $\tilde{g}(x)$, we have

$$\frac{\partial}{\partial x_n}\tilde{g}(x)=-g(x', -0)\otimes\delta_{x_n}+\left(\frac{\partial g}{\partial x_n}\right)^\sim.$$

If we compare this with (3.47), we observe that the term of the distribution which lies on $x_n=0$ takes opposite signs in these two cases. The same situation occurs in the case of the derivatives of distribution of higher degrees. From these facts, if we define $F(x)=\tilde{f}(x)+\tilde{g}(x)$, we see that $F(x)\equiv f(x)$ in $R_+{}^n$, and from (3.52) we have that $F\in\mathscr{E}_{L^2}{}^m(R^n)$.

Q.E.D.

Note. Let us pick a point x_0 on the hyperplane $x_n=0$, and let V be a neighbourhood of x_0. If the support of $f\in\mathscr{E}_{L^2}{}^m(R_+{}^n)$ lies in the interior of V, then the support of F also lies in the interior of V. In fact, we can define the product of the old $F(x)$ with an appropriate $\alpha(x)$ ($\alpha\in\mathscr{D}(V)$) as the new $F(x)$.

We have the following theorem in relation to theorem 3.10.

THEOREM 3.11. Let Ω be a domain having as its boundary a closed hypersurface of the C^m-class. For $f\in\mathscr{E}_{L^2}{}^m(\Omega)$ we have an extension $F\in\mathscr{E}_{L^2}{}^m(R^n)$.

Proof. Let us consider the partition of unity $\{V_i, \alpha_i(x)\}$ which we observed at the beginning of this section. We apply proposition 3.4 to a transformed function $g_i\in\mathscr{E}_{L^2}{}^m(R_+{}^n)$ for each $\alpha_i(x)f(x)$ taking account of the 'note' which we have just given. Also we consider the extension of $g_i'(x)$ in the original space. Let us call this $\tilde{f}_i(x)$. Then, corresponding to (3.42) we define

$$F(x)=\sum_{i=1}^{p}\tilde{f}_i(x)+(1-\sum\alpha_i(x))\,f(x).$$

By lemma 3.4, we see that $F(x)$ satisfies our requirements.

Q.E.D.

Let $f \in \mathscr{E}_{L^2}{}^m(\Omega)$. If $F(x)$ does exist, as theorem 3.11 claims, we put

$$f_\varepsilon(x) = \rho_\varepsilon * F = \int \rho_\varepsilon(x-y)\, F(y)\, \mathrm{d}y .$$

We see $f_\varepsilon \in \mathscr{E}^\infty$, and $f_\varepsilon(x) \to F(x)$ in $\mathscr{E}_{L^2}{}^m(R^n)$ as $\varepsilon \to 0$ (theorem 2.8). Therefore, immediately,

$$f_\varepsilon(x) \to f(x) \quad \text{in} \quad \mathscr{E}_{L^2}{}^m(\Omega). \tag{3.53}$$

Summing up we have:

COROLLARY.

(1) Let Ω satisfy the condition of theorem 3.11. Then $C^\infty(\bar{\Omega}) \cap \mathscr{E}_{L^2}{}^m(\Omega)$ is dense in $\mathscr{E}_{L^2}{}^m(\Omega)$, where $f \in C^\infty(\bar{\Omega})$ has the same meaning as in definition 3.2, i.e. it is a restriction to Ω of a C^∞-function which is defined over the whole space.

(2) The *trace operator* γ (which was defined in theorem 3.9) is unique.

Proof. (1) is already proved. So that we must show (2). Let us put $m=1$ in (1). In this case we see that $C^1(\bar{\Omega})$ is dense in $\mathscr{E}_{L^2}{}^1(\Omega)$. Recall that we claim γ is a continuous linear operator from $\mathscr{E}_{L^2}{}^1(\Omega)$ to $L^2(S)$.

Therefore, γ can be determined as a unique extension of a continuous linear operator which is defined on a dense subspace of the whole space.

Q.E.D.

5 *Characterization of* $\mathscr{D}_{L^2}{}^1(\Omega)$

We have demonstrated the existence (theorem 3.8) and uniqueness of the trace operator γ in the previous section. We now give the characterization of $\mathscr{D}_{L^2}{}^1(\Omega)$ by using this boundary value in a wider sense. We note that $f \in \mathscr{D}_{L^2}{}^1(\Omega)$ means, roughly speaking, that f is a function which takes the value 0 on the boundary of Ω.

We state this in the following precise form:

THEOREM 3.12. Let Ω be an inner or outer domain having as boundary a closed hypersurface S of the C^2-class. The necessary and sufficient condition for $f \in \mathscr{D}_{L^2}{}^1(\Omega)$, given $f \in \mathscr{E}_{L^2}{}^1(\Omega)$, is either of the following conditions:

(1) $\qquad \gamma f = 0,$

(2) $\qquad \partial \tilde{f}/\partial x_i = (\partial f/\partial x_i)^\sim \quad (i=1, 2, \ldots, n),$

where $\tilde{f}$ is defined as an extension of f which takes the value 0 in the complement of Ω.

Proof. We can easily see that $f \in \mathscr{D}_{L^2}{}^1(\Omega)$ satisfies (1) and (2). In fact, $f_j(x) \in \mathscr{D}(\Omega) \to f(x)$ in $\mathscr{E}_{L^2}{}^1(\Omega)$, and $\gamma f_j = 0$ $(j=1,2,\ldots)$. Therefore, $\gamma f = 0$ because γ is a continuous operator. Also, f satisfies (2) (we showed this at the beginning of section 3).

The converse is a little more complicated. First, we show that (1) and (2) are equivalent. We proceed as follows. We note that for $f \in \mathscr{E}_{L^2}{}^1(\Omega)$ and $\varphi \in \mathscr{D}(R^n)$, the Green's formula

$$\int_\Omega \frac{\partial f}{\partial x_i}\,\varphi\,\mathrm{d}x = -\int_S \gamma f \cdot \varphi \cos\gamma_i\,\mathrm{d}S - \int_\Omega f \frac{\partial \varphi}{\partial x_i}\,\mathrm{d}x \tag{3.54}$$

is valid, where γ_i is the angle between the inner normal of S and the x_i-axis.

To see this, we first observe that it is true when $f \in \mathscr{D}(R^n)$. On the other hand, by the corollary of theorem 3.11, there exists a series

$$f_p(x) \in \mathscr{D}(R^n) \to f(x) \quad \text{in} \quad \mathscr{E}_{L^2}{}^1(\Omega).$$

(This may not be immediately obvious but it does follow from the proof given there.) Therefore, if we replace f by f_p, and $p \to \infty$, we have (3.54) by $\gamma f_p \to \gamma f$ in $L^2(S)$; we see that (3.54) is in the form of

$$\left\langle \left(\frac{\partial f}{\partial x_i}\right)^{\sim}, \varphi \right\rangle = \left\langle \frac{\partial}{\partial x_i}\tilde{f}, \varphi \right\rangle - \int_S \gamma f \cdot \varphi \cos\gamma_i\,\mathrm{d}S. \tag{3.55}$$

Hence, if (2) is true

$$\int_S \gamma f \cdot \varphi \cos\gamma_i\,\mathrm{d}S = 0 \quad (i=1,2,\ldots,n)$$

is also true for an arbitrary $\varphi \in \mathscr{D}(R^n)$. Hence, $\gamma f = 0$ in $L^2(S)$.

Conversely, by (3.55) it is clear that from $\gamma f = 0$ (2) follows. Now, the fact that if $\gamma f = 0$ then $f \in \mathscr{D}_{L^2}{}^1(\Omega)$ still remains to be proved. To do this, we pick the partition of unity $\{V_i, \alpha_i(x)\}$ in the neighbourhood of S. For an arbitrary $\varepsilon'(>0)$, there exists $\psi_i \in \mathscr{D}(\Omega)$ and

$$\|\alpha_i f - \varphi_i\|_{1, L^2(\Omega)} < \varepsilon' \quad (i=1,2,\ldots,p). \tag{3.56}$$

In fact, the second term on the right hand side of (3.42) (the resolution of $f(x)$), after the operation of a mollifier, satisfies a relation similar to (3.56).

Roughly speaking, $\psi_i(x)$ can be obtained from $\alpha_i(x)f(x)$ by a transformation in the direction of the inner normal at x and the subsequent operation of a mollifier in order to obtain a C^∞-function. We shall explain this in detail in what follows.

If $\gamma f=0$, then $\gamma(\alpha_i f)=0$ so that the image $g_i(x')$ of Φ_i by the transformation formula $x'=\Phi_i(x)$ of $\alpha_i f$ satisfies $\gamma g_i=0$, $g_i\in\mathscr{E}_{L^2}{}^1(R_+{}^n)$.
In this case we have the following:

LEMMA 3.7. Let $\psi\in\mathscr{E}_{L^2}{}^1(R_+{}^n)$ and let $\gamma\psi=0$. Then $\tilde{\psi}(x', x_n-\varepsilon)\in\mathscr{E}_{L^2}{}^1(R_+{}^n)$ and $\tilde{\psi}(x', x_n-\varepsilon)\to\psi(x', x_n)$ in $\mathscr{E}_{L^2}{}^1(R_+{}^n)$ as $\varepsilon\to+0$.

The proof of this lemma is omitted because it is easy.

We consider

$$\rho_{\frac{1}{2}\varepsilon} \underset{(x')}{*} g_i(x_1', \dots, x_{n-1}', x_n'-\varepsilon)$$

by using a mollifier. By this lemma, we see that its support lies in $x_n'\geqslant\frac{1}{2}\varepsilon$, and it tends to $g_i(x')$ as $\varepsilon\to+0$ by the topology of $\mathscr{E}_{L^2}{}^1$. Therefore, if we put $\psi_i(x)$ as the image by the inverse transformation of $\Phi_i(x)$ of this function, we see that $\psi_i(x)$ satisfies (3.56) when we make ε sufficiently small. Hence, from $\gamma f=0$, it follows that $f\in\mathscr{D}_{L^2}{}^1(\Omega)$.

Q.E.D.

6 *Properties of* $\mathscr{E}_{L^2}{}^m(\Omega)$

We have proved the uniqueness of a trace operator in section 4. We shall explain the method which we have used there and state some results which we can establish from our observations. In this section, we follow the convention that Ω is an inner or outer domain having as its boundary S a closed hypersurface of the C^m-class. First, if we examine the proof of proposition 3.4, we see that theorem 3.11 can be re-written as follows:

THEOREM 3.13. There exists a certain linear extension of a continuous linear operator $\Phi: F=\Phi\cdot f\in\mathscr{E}_{L^2}{}^m(R^n)$ where $f\in\mathscr{E}_{L^2}{}^m(\Omega)$, i.e. for a constant C

$$\|F(x)\|_{m, L^2}\leqslant C\|f(x)\|_{m, L^2(\Omega)}, \tag{3.57}$$

where C is dependent on Ω and Φ, but independent of f. In particular, if Ω is a bounded domain, we can assume that the support of $F(x)$ lies in an arbitrary bounded open set Ω_1 whose interior contains $\bar{\Omega}$, i.e.

$$F \in \mathscr{D}_{L^2}{}^m(\Omega_1).$$

Note. Φ is *not* uniquely determined; there are infinitely many Φ.

From this theorem with Rellich's theorem, we have:

THEOREM 3.14. Let Ω be an inner domain which is surrounded by a closed hypersurface S of the C^m-class as its boundary. Alternatively, we can take a bounded domain which is surrounded by a finite set of mutually non-intersecting closed hypersurfaces S_i of the C^m-class. In this case, for an arbitrary bounded sequence $f_j \in \mathscr{E}_{L^2}{}^m(\Omega)$, there exists a convergent sub-sequence f_{j_p} of f_j of $\mathscr{E}_{L^2}{}^{m-1}(\Omega)$.

Proof. Ω is bounded according to our assumption. We define Φ by choosing Ω_1 as in the previous theorem 3.13. We see that $F_j = \Phi \cdot f \in \mathscr{D}_{L^2}{}^m(\Omega_1)$.

From (3.57), $\{F_j\}$ is a bounded sequence of $\mathscr{D}_{L^2}{}^m(\Omega_1)$. Therefore, by Rellich's theorem (corollary of theorem 3.3), if we choose an appropriate subsequence $\{F_{j_p}\}$ from $\{F_j\}$, it forms a Cauchy sequence of $\mathscr{E}_{L^2}{}^{m-1}(\Omega)$.

Q.E.D.

THEOREM 3.15. Let Ω be a domain (bounded or not bounded) having a finite set of mutually non-intersecting closed hypersurfaces S_i of the C^m-class as its boundary.

In this case, from $f \in \mathscr{E}_{L^2}{}^m(\Omega)$ where $(m = [\frac{1}{2}n] + 1 + p, p = 0, 1, 2, \ldots)$, it follows that $f(x) \in \mathscr{B}^p(\bar{\Omega})$ and

$$|f(x)|_p \leqslant c \|f(x)\|_{m, L^2(\Omega)}, \tag{3.58}$$

where c is a positive constant. $\mathscr{B}^p(\bar{\Omega})$ is defined as the set of the functions which are the restrictions of all functions belonging to $\mathscr{B}^p(R^n)$ to Ω.

Proof. Let us consider Φ satisfying the condition of theorem 3.13. Then, we apply Sobolev's lemma (theorem 2.8 (2.25)) to $F(x) = \Phi \cdot f(x)$.

Q.E.D.

7 *Improving the estimation of γf*

We have demonstrated an estimation formula (3.44) for γf. In fact, we can make it sharper if we consider the estimation introducing the following function space:

DEFINITION 3.4 *(Function space $\mathscr{D}_{L^2}{}^s, -\infty < s < +\infty$).* By $f \in \mathscr{D}_{L^2}{}^s$ we mean that $f \in \mathscr{S}'$ and $(1+|\xi|)^s \hat{f}(\xi) \in L^2$. We introduce as a norm for this space

$$\|f(x)\|_s = \|(1+|\xi|)^s \hat{f}(\xi)\|_{L^2}, \tag{3.59}$$

i.e. the inner products of the space are defined as

$$(f, g)_s = ((1+|\xi|)^s \hat{f}(\xi), (1+|\xi|)^s \hat{g}(\xi))_{L^2}.$$

$\mathscr{D}_{L^2}{}^s$ is a Hilbert space, and $\mathscr{D}_{L^2}{}^0 = L^2$. If $s = 0, 1, 2, \ldots$, then these spaces coincide with the previously defined spaces.

PROPOSITION 3.5 *(Properties of $\mathscr{D}_{L^2}{}^s$).*

(1) If $s < s'$, then $\mathscr{D}_{L^2}{}^{s'} \subset \mathscr{D}_{L^2}{}^s$ where $\mathscr{D}_{L^2}{}^{s'}$ is dense in $\mathscr{D}_{L^2}{}^s$.†

(2) The dual space of $\mathscr{D}_{L^2}{}^s$ is $(\mathscr{D}_{L^2})^{-s}$.

(3) If $0 < s < s'$, then for an arbitrary ε (<0), there exists $C(\varepsilon; s, s')$, and for $f \in \mathscr{D}_{L^2}{}^{s'}$

$$\|f(x)\|_s \leqslant \varepsilon \|f(x)\|_{s'} + C(\varepsilon; s, s')\|f(x)\|_0. \tag{3.60}$$

We note that $\|\cdot\|_0 = \|\cdot\|_{L^2}$.

Proof.

(1) The beginning half of the theorem is clear from (3.59), and the rest of the theorem is clear from (3) of theorem 2.8.

(2) Let $T \in \mathscr{D}_{L^2}{}^{s'}$. By Riesz theorem, there exists a unique $\psi(\xi)$, with $(1+|\xi|)^s \psi(\xi) \in L^2$. Also, for $f \in \mathscr{D}_{L^2}{}^s$,

$$\langle T, f \rangle = ((1+|\xi|)^s \hat{f}(\xi), (1+|\xi|)^s \psi(\xi))_{L^2},$$

and $\|(1+|\xi|)^s \psi(\xi)\|_{L^2}$ is equal to the norm of T in $\mathscr{D}_{L^2}{}^{s'}$. Now since $\mathscr{S}$ is dense in $\mathscr{D}_{L^2}{}^s$ we may assume that $f \in \mathscr{S}$. Then, from the definition of Fourier transform, we have

$$\langle T, f \rangle = \langle (1+|\xi|)^{2s} \overline{\psi(\xi)}, \hat{f}(\xi) \rangle,$$

so that

$$\mathscr{F}[T] = (1+|\xi|)^{2s} \overline{\psi(\xi)}.$$

The term on the extreme right is equal to the norm of T in $\mathscr{D}_{L^2}{}^{s'}$ as we have observed before. The converse of (2) is obvious.

† Translator's note: the original says $\mathscr{D}$ *is dense in* $\mathscr{D}_{L^2}{}^s$.

Hence

$$\mathscr{F}[T]=(1+|\xi|)^{2s}\,\overline{\psi(-\xi)},$$

therefore

$$\|\mathscr{F}[T]\|_{-s}=\|(1+|\xi|)^{s}\,\overline{\psi(-\xi)}\|_{L^2}=\|(1+|\xi|)^{s}\,\psi(\xi)\|_{L^2}.$$

(3) Let C be sufficiently large for a given $\varepsilon(>0)$. Then we have

$$(1+|\xi|)^{2s}\leqslant\varepsilon^2(1+|\xi|)^{2s'}+C^2 \qquad (\xi\in R^n).$$

Integrating both sides of this inequality after multiplying by $|\hat{f}(\xi)|^2$, we get:

$$\|f\|_s^2\leqslant\varepsilon^2\|f\|_{s'}^2+C^2\|f\|_0^2\leqslant(\varepsilon\|f\|_{s'}+C\|f\|_0)^2.$$

Q.E.D.

We shall now deal with our original problem. Let us assume the boundary of Ω consists of a closed hypersurface S of the C^2-class. We consider $\alpha_i f$ which we obtained by the partition of unity $\{V_i, \alpha_i(x)\}$ in a neighbourhood of S. (See the beginning of section 4.) In this case, with the transformation $x'=\Phi(x)$, we can regard Ω as $R_+{}^n$; we shall change our notation as follows:

$$\left.\begin{array}{l} x=(x_1,\dots,x_{n-1}) \\ y=x_n \end{array}\right\}$$

and $f(x,y)\in\mathscr{E}_{L^2}{}^1(R_+{}^2)$. Furthermore, according to (3.51) – in this case $g(x,y)=f(x,-y)$ – if we extend $f(x,y)$ to the whole space (let us call this the extension $F(x,y)$), we have $F\in\mathscr{E}_{L^2}{}^1(R^n)$ and $\|F(x,y)\|_1=2^{\frac{1}{2}}\|f(x,y)\|_1$ by proposition 3.4.

Let us assume $F\in\mathscr{D}$, $\mathscr{D}$ being dense in $\mathscr{D}_{L^2}{}^1(=\mathscr{E}_{L^2}{}^1(R^n))$, and let the dual space of (x,y) be $(\xi,\eta)=(\xi_1,\dots,\xi_{n-1},\eta)$. Now we observe the following Fourier transform; we write

$$\iint \mathrm{e}^{-2\pi i x\xi}\,\mathrm{e}^{-2\pi i y\eta}F(x,y)\,\mathrm{d}x\,\mathrm{d}y=\int_{-\infty}^{+\infty}\mathrm{e}^{-2\pi i y\eta}\hat{F}(\xi,y)\,\mathrm{d}y=\hat{F}(\xi,\eta).$$

The trace of $F(x,y)$ on hyperplane $y=0$ is $F(x,0)$. Therefore, we have

$$\varphi(x)=F(x,0)$$

$$\hat{F}(\xi,y)=\int_{-\infty}^{+\infty}\mathrm{e}^{2\pi i y\eta}\hat{F}(\xi,\eta)\,\mathrm{d}\eta,$$

so that

$$\Phi(\xi) = \hat{F}(\xi, 0) = \int_{-\infty}^{+\infty} \hat{F}(\xi, \eta)\, \mathrm{d}\eta,$$

$$|\Phi(\xi)| \leqslant \int_{-\infty}^{+\infty} |\hat{F}(\xi, \eta)|\, \mathrm{d}\eta.$$

We shall consider the following decomposition of the right hand side of this inequality:

$$\int_{-\infty}^{+\infty} \{(1+|\xi|+|\eta|)^{\frac{1}{2}\delta}\, |\hat{F}(\xi, \eta)|^{\frac{1}{2}}\} \cdot \{(1+|\xi|+|\eta|)^{\frac{1}{2}}\, |\hat{F}(\xi, \eta)|^{\frac{1}{2}}\} \times (1+|\xi|+|\eta|)^{-\frac{1}{2}-\frac{1}{2}\delta}\, \mathrm{d}\eta$$

where $0<\delta\leqslant 1$.

Now, we can apply Hölder's inequality; if $\sum 1/p_i = 1$ $(p_i>0)$, then for $f_i(x)\geqslant 0$, we have

$$\int f_1(x) \ldots f_n(x)\, \mathrm{d}x \leqslant \left(\int f_1(x)^{p_1}\, \mathrm{d}x\right)^{1/p_1} \cdots \left(\int f_n(x)^{p_n}\, \mathrm{d}x\right)^{1/p_n}. \tag{3.61}$$

Therefore, the above integral can be estimated by

$$|\Phi(\xi)| \leqslant \left(\int (1+|\xi|+|\eta|)^{2\delta}\, |\hat{F}(\xi, \eta)|^2\, \mathrm{d}\eta\right)^{\frac{1}{4}} \left(\int (1+|\xi|+|\eta|)^{2}\, |\hat{F}(\xi, \eta)|^2\, \mathrm{d}\eta\right)^{\frac{1}{4}} \times \left(\int (1+|\xi|+|\eta|)^{-1-\delta}\, \mathrm{d}\eta\right)^{\frac{1}{2}}$$

by putting $p_1=p_2=4$, and $p_3=2$. On the other hand

$$\int_{-\infty}^{+\infty} \frac{\mathrm{d}\eta}{(1+|\xi|+|\eta|)^{1+\delta}} = \frac{2}{\delta(1+|\xi|)^{\delta}}.$$

By writing the first and the second factors $\Phi_1(\xi)^{\frac{1}{4}}$, $\Phi_2(\xi)^{\frac{1}{4}}$ respectively, we have

$$\int (1+|\xi|)^{\delta}\, |\Phi(\xi)|^2\, \mathrm{d}\xi \leqslant \frac{2}{\delta} \int \Phi_1(\xi)^{\frac{1}{2}}\, \Phi_2(\xi)^{\frac{1}{2}}\, \mathrm{d}\xi$$

$$\leqslant \frac{2}{\delta} \left(\int \Phi_1(\xi)\, \mathrm{d}\xi\right)^{\frac{1}{2}} \left(\int \Phi_2(\xi)\, \mathrm{d}\xi\right)^{\frac{1}{2}},$$

i.e.

$$\int (1+|\xi|)^{\delta}\, |\Phi(\xi)|^2\, \mathrm{d}\xi \leqslant \frac{2}{\delta} \left(\iint (1+|\xi|+|\eta|)^{2\delta}\, |\hat{F}(\xi, \eta)|^2\, \mathrm{d}\xi\, \mathrm{d}\eta\right)^{\frac{1}{2}} \times \left(\iint (1+|\xi|+|\eta|)^{2}\, |\hat{F}(\xi. \eta|)^2\, \mathrm{d}\xi\, \mathrm{d}\eta\right)^{\frac{1}{2}}.$$

We now represent this by the norm of the function space which we introduced at the beginning of this section. We have

$$\|\varphi(x)\|_{\frac{1}{2}\delta}{}^2 \leqslant \frac{c(n)}{\delta} \|F(x, y)\|_\delta \, \|F(x, y)\|_1 \quad (0<\delta\leqslant 1), \tag{3.62}$$

because $|\xi|+|\eta|$ is equivalent to $\sqrt{(|\xi|^2+\eta^2)}$, where $c(n)$ is a constant depending only on the dimension n of the space. We have dealt with our problem when $F\in\mathscr{D}$, but we can prove the same fact when $F\in\mathscr{D}_{L^2}{}^1$ by a limiting process.

Now, if we put $\delta=\frac{1}{2}$ in (3.62) and apply the inequality (3.60) to $\|F(x, y)\|_{\frac{1}{2}}$, we see that

$$\|\varphi(x)\|_{L^2}{}^2 \leqslant \|\varphi(x)\|_{\frac{1}{4}}{}^2 \leqslant 2c(n)\,[\varepsilon\|F\|_1 + c(\varepsilon)\|F\|_0]\|F\|_1.$$

Summing up, we have:

PROPOSITION 3.6. Let $f\in\mathscr{E}_{L^2}{}^1(R_+{}^n)$. Then, the following are true.

(1) $\|\varphi(x)\|_{\frac{1}{4}} \leqslant c\|f(x, y)\|_1$, where c is a constant.

(2) For an arbitrary $\varepsilon\,(>0)$, there exists $c(\varepsilon, n)$ satisfying

$$\|\varphi(x)\|_{L^2} \leqslant \varepsilon\|f(x, y)\|_1 + c(\varepsilon, n)\|f(x, y)\|_{L^2}.$$

Finally, we have:

THEOREM 3.16 Let Ω be a domain having a closed hypersurface S of the C^2-class as its boundary. For $f\in\mathscr{E}_{L^2}{}^1(\Omega)$, we have

$$\|\gamma f\|_{L^2(S)} \leqslant \varepsilon\|f(x)\|_{1,\,L^2(\Omega)} + C(\varepsilon, \Omega)\|f(x)\|_{L^2(\Omega)},$$

where $\varepsilon(>0)$ can be arbitrarily small (consequently $C(\varepsilon, \Omega)$ is large).

Note. In this theorem we assume S is closed. One may wonder what happens if we remove this assumption. In general, it is difficult to obtain the same result as above because there exist an infinite number of partitions of a unit $\{V_i, \alpha_i(x)\}$ in a neighbourhood of S, and also because without a kind of uniformity concerning the sequence $\{\alpha_i(x)\}$, we can hardly obtain the same result as above when we interpret the result for $\mathscr{E}_{L^2}{}^1(R_+{}^n)$ as the result for a neighbourhood of a surface.

Let us consider a particular case of S. For example, if Ω is a cylindrical domain, the theorem is true. In this case, to be precise, the boundary S is $S=\Gamma\times(-\infty, +\infty)$, and Γ is a closed hypersurface of the C^2-class in R^{n-1}.

8 *Boundary value problems for elliptic differential equations of second order*

We have discussed the Dirichlet problem for $A=-\Delta+c(x)$. The reader may have an impression that the problem itself is too specific. In this section we shall give an account of the problem

$$A+\lambda I\equiv-\Delta+\sum_{i=1}^{n} a_i(x)\frac{\partial}{\partial x_i}+c(x)+\lambda, \tag{3.63}$$

a slightly generalized form of the equation (3.1), where we assume

$$a_i\in\mathscr{B}^1(\Omega), \tag{3.64}$$

$c(x)$ is a bounded function, and a_i and c are not necessarily real-valued functions, i.e. they may be complex-valued functions.

Next, we consider the generalized boundary value problem of the third kind including the Neumann problem

$$\frac{du}{dn}+\sigma(x)\,u=0 \quad (x\in S), \tag{3.65}$$

where $\sigma(x)$ is a bounded, measurable, complex-valued function defined on S, and du/dn is the derivative in the direction of the inner normal of u.

Is it possible that this problem can be treated by the same method which we have used for the Dirichlet problem? First, we compare (3.63) with (3.1). For A in (3.63), we define A^* by

$$A^*v=-\Delta v-\sum_{i=1}^{n}\frac{\partial}{\partial x_i}(a_i(x)\,v)+c(x)\,v.$$

In this case we call A^* the *formally conjugate operator of* A. If $A=A^*$, we say that A is *formally self-adjoint*. Obviously, we have

$$(Au, v)=(u, A^*v) \quad (u, v\in\mathscr{D}(\Omega)).$$

We note that if the operator $A=-\Delta+c(x)$, where c is a real-valued function, then $A=A^*$, but if A is in the form of (3.63), then, in general, it is not formally self-adjoint. So that we see that the observations we made in sections 2 and 3 are no longer available. We come back to this point in the next section. On the other hand, we see that in (3.63), if $A=A^*$, then the arguments and results which we have described in sections 2 and 3 are immediately applicable.

EXAMPLE 1. Let

$$A=\sum_{j=1}^{n}\left(\frac{1}{\mathrm{i}}\frac{\partial}{\partial x_j}-b_j(x)\right)^2+c(x)\quad(\mathrm{i}=\sqrt{-1})\tag{3.66}$$

where $b_j\in\mathscr{B}^1(\Omega)$, c is bounded, and both functions are real-valued. It is obvious that $A=A^*$. Furthermore, for $u\in\mathscr{D}_{L^2}{}^1(\Omega)$, we can establish

$$\langle Au,\bar{u}\rangle=\sum_{j=1}^{n}\left\|\frac{\partial u}{\partial x_j}-ib_ju\right\|_{L^2}^2+(c(x)\,u,u)\geqslant\tfrac{1}{2}\sum_{j=1}^{n}\left\|\frac{\partial u}{\partial x_j}\right\|^2-\gamma\|u\|^2$$

where γ is an appropriate positive constant.

From this, we see that we can treat the Dirichlet problem for (3.66) as in the case $A=-\Delta+c(x)$. Therefore, the results which we have established in sections 2 and 3 are available for this equation.

EXAMPLE 2.

$$Au=-\sum_{i,j=1}^{n}\frac{\partial}{\partial x_i}\left(a_{ij}(x)\frac{\partial}{\partial x_j}u\right)+c(x)\,u\,,$$

where $a_{ij}\in\mathscr{B}^1(\Omega)\,(a_{ij}(x)\equiv a_{ji}(x))$, and a_{ij} and c are real-valued functions. Furthermore, we assume *uniform ellipticity*

$$\sum a_{ij}(x)\,\xi_i\xi_j\geqslant\delta|\xi|^2\quad(x\in\Omega,\,\xi\in R^n)\tag{3.67}$$

where δ is a positive constant. In this case A is obviously formally self-adjoint, and for $u\in\mathscr{D}_{L^2}{}^1(\Omega)$, we can write

$$\langle Au,\bar{u}\rangle=\sum_{i,j}\left(a_{ij}\frac{\partial u}{\partial x_i},\frac{\partial u}{\partial x_j}\right)+(c(x)u,u)\,.$$

If we write $u(x)=u_1(x)+\mathrm{i}u_2(x)$ where $u_1(x)$ and $u_2(x)$ are respectively the real and imaginary parts of $u(x)$, then under our assumption (3.67), we have

$$\sum_{i,j}\left(a_{ij}\frac{\partial u}{\partial x_i}, \frac{\partial u}{\partial x_j}\right) \geqslant \delta \sum_{i=1}^{n}\left\|\frac{\partial u}{\partial x_i}\right\|^2_{L^2(\Omega)}.$$

We see that this is a similar case to the Dirichlet problem for $A = -\Delta + c(x)$.

Next, we consider the equation (3.1) corresponding to the boundary condition (3.65). Let us explain the significance of (3.65). By $\mathrm{d}u/\mathrm{d}n$ we understand

$$\left.\frac{\mathrm{d}u}{\mathrm{d}n}\right|_S = \sum_{i=1}^{n} \cos\alpha_i \cdot \gamma\left(\frac{\partial u}{\partial x_i}\right), \tag{3.68}$$

where α_i is the angle between the inner normal of S and the x_i-axis, and γ is the trace operator which we considered in section 4.

The above equation (3.68) will be meaningful if $u \in \mathscr{E}_{L^2}{}^2(\Omega)$. Let Ω be an inner or exterior domain of a C^2-class closed hypersurface S. Also, for simplicity, we consider that σ in (3.65) is a bounded real-valued function defined on S. Now, on the right hand side of (3.1), we assume $f \in L^2(\Omega)$, and pick the solution of (3.1) $u \in \mathscr{E}_{L^2}{}^2(\Omega)$ satisfying the boundary condition (3.65). In this case, for an arbitrary $\varphi \in \mathscr{D}(R^n)$, we have

$$\langle -\Delta u + c(x)u + \lambda u, \overline{\varphi(x)}\rangle = \langle f(x), \overline{\varphi(x)}\rangle.$$

We apply Green's theorem to this to obtain the following result. For arbitrary $u \in \mathscr{E}_{L^2}{}^2(\Omega)$ and $\varphi \in \mathscr{E}_{L^2}{}^1(\Omega)$, we have

$$\langle -\Delta u, \bar{\varphi}\rangle = \int_S \gamma\left(\frac{\mathrm{d}u}{\mathrm{d}n}\right)\gamma\bar{\varphi}\,\mathrm{d}S + \sum_{i=1}^{n}\left\langle\frac{\partial u}{\partial x_i}, \frac{\partial \bar{\varphi}}{\partial x_i}\right\rangle \tag{3.69}$$

where $\gamma(\mathrm{d}u/\mathrm{d}n)$ is understood to be in the sense of (3.68).

In fact, let $\varphi \in \mathscr{D}$. Then, if $u \in \mathscr{D}$, (3.69) holds. If u satisfies our assumption, then, from the corollary of theorem 3.11, there exists a sequence $u_j \in \mathscr{D}$ satisfying $u_j(x) \to u(x)$ in $\mathscr{E}_{L^2}{}^2(\Omega)$. In this case, from the continuity of γ, we have

$$\gamma\left(\frac{\mathrm{d}u_j}{\mathrm{d}n}\right) \to \gamma\left(\frac{\mathrm{d}u}{\mathrm{d}n}\right) \quad \text{in} \quad L^2(S).$$

Therefore, by a limiting process, we can establish (3.69). Next, if $\varphi \in \mathscr{E}_{L^2}{}^1(\Omega)$, then, by a similar limiting process, which we applied in the previous case, we establish (3.69).

Using the boundary condition (3.65), we see that, for an arbitrary

$\varphi \in \mathscr{E}_{L^2}{}^1(\Omega)$, $u(x)$ satisfies

$$\sum_{i=1}^{n}\left(\frac{\partial u}{\partial x_i}, \frac{\partial \varphi}{\partial x_i}\right)+((c(x)+\lambda)\, u, \varphi)-\int_S \sigma(x)\, u \cdot \bar{\varphi}\, \mathrm{d}S=\langle f, \bar{\varphi}\rangle, \qquad (3.70)$$

where $u \cdot \bar{\varphi}$ under the integral sign is a trace on the boundary S.

Conversely, if we assume that $u(x)$ satisfies (3.70), then if $\varphi \in \mathscr{D}(\Omega)$, the integral in (3.70) disappears and

$$\langle -\Delta u+(c(x)+\lambda)\, u, \bar{\varphi}\rangle=\langle f, \bar{\varphi}\rangle \quad (\varphi \in \mathscr{D}(\Omega)),$$

i.e.

$$-\Delta u+c(x)\, u+\lambda u=f.$$

This is a solution of (3.1).

To prove that $u(x)$ satisfies (3.65), we have to show that $u \in \mathscr{E}_{L^2}{}^2(\Omega)$. Let us assume that this is true. In this case, using the equation (3.69) we transform (3.70). Then, by the result we obtained above, we have

$$\int_S \left(\frac{\mathrm{d}u}{\mathrm{d}n}+\sigma u\right) \bar{\varphi}\, \mathrm{d}S=0$$

for an arbitrary $\varphi \in \mathscr{E}_{L^2}{}^1(\Omega)$. If φ runs through $\mathscr{E}_{L^2}{}^1(\Omega)$, $\gamma\bar{\varphi}$ forms a dense set of $L^2(S)$, so that $(\mathrm{d}u/\mathrm{d}n)+\sigma u=0$ is true almost everywhere in S. To find a solution for (3.70) in $\mathscr{E}_{L^2}{}^1(\Omega)$ is a problem similar to the Dirichlet one. First, we note that the right hand side of (3.70) is a Hermitian form defined on $\mathscr{E}_{L^2}{}^1(\Omega)$. There exists t_0 and if $\lambda>t_0$ (λ is a real number), it is not only a positive-definite Hermitian form but also it defines a norm which is equivalent to that of $\mathscr{E}_{L^2}{}^1(\Omega)$. We show this as follows:

If $\sigma(x) \leqslant 0$ it is obvious. If the sign $\sigma(x)$ is not definite, using theorem 3.16, we have

$$-\int_S \sigma(x)|u|^2\, \mathrm{d}S \geqslant -\sup \sigma(x) \int_S |u|^2\, \mathrm{d}S \geqslant -\sup \sigma(\varepsilon\|u\|_1{}^2+C(\varepsilon, \Omega)\|u\|_{L^2}{}^2)$$

as an estimation of its lower bound, and

$$\sum_{i=1}^{n}\left\|\frac{\partial u}{\partial x_i}\right\|^2+((c(x)+\lambda)\, u, u)-\int_S \sigma(x)|u|^2\, \mathrm{d}S$$

$$\geqslant(1-\varepsilon \sup \sigma) \sum_{i=1}^{n}\left\|\frac{\partial u}{\partial x_i}\right\|^2+[\lambda-c_0-c(\varepsilon)]\|u\|^2.$$

Suppose ε satisfies $\varepsilon\cdot\sup\sigma(x)<\frac{1}{2}$. For such an ε let us take λ which satisfies $\lambda-c_0-c(\varepsilon)>0$. We see that the left hand side is equivalent to the square of the norm of $\mathscr{E}_{L^2}{}^1(\Omega)$. Therefore, if we take $\lambda=t$ sufficiently large, we can affirm the existence of a Green's operator G_t just as we did in the case of the Dirichlet problem. In particular, if Ω is an inner domain, G_t becomes a completely continuous Hermitian operator on $L^2(\Omega)$. This can be proved from theorem 3.14, an extension of Rellich's theorem. We shall return to this point later.

Note. In the above argument, we assumed that S is a closed surface. This is not a necessary condition. For example, the above argument is true for the cylindrical domain which we mentioned in the previous section. It is also true for Ω which is a half space.

9 *Dirichlet problems for the general second-order elliptic operator*

In this section we treat a type of Dirichlet problem. Our idea is as follows: given a form $A[u, v]$ where A is linear with respect to u, and anti-linear with respect to v, we decompose this into two Hermitian forms and use the results which we developed in sections 2 and 3.

First, we consider the Dirichlet problem for the operator

$$A=-\sum_{i,j} a_{ij}(x)\frac{\partial^2}{\partial x_i\partial x_j}+\sum_i a_i(x)\frac{\partial}{\partial x_i}+c(x), \tag{3.71}$$

where a_{ij} is a real-valued function, such that $a_{ij}\equiv a_{ij}$, $a_{ij}\in\mathscr{B}^2(\Omega)$, $a_i\in\mathscr{B}^1(\Omega)$, and c is a bounded function. Furthermore, we assume uniform ellipticity (3.67). For (3.71), if we define A^* as

$$A^*v=-\sum_{i,j}\frac{\partial^2}{\partial x_i\partial x_j}(a_{ij}(x)\,v)-\sum\frac{\partial}{\partial x_i}(\overline{a_i(x)}\,v)+\overline{c(x)}\,v, \tag{3.72}$$

we have

$$(Au, v)=(u, A^*v)\quad (u, v\in\mathscr{D}(\Omega)),$$

where A is not necessarily 'formally' self-adjoint. So that we decompose

$$A=\tfrac{1}{2}(A+A^*)+\mathrm{i}\frac{1}{2\mathrm{i}}(A-A^*)\equiv A_1+\mathrm{i}A_2 \tag{3.73}$$

where A_1, A_2 are formally self-adjoint.†

† Obviously, $(A^*)^*=A$. This can be seen by a direct calculation from A^*, but another way to look at this is as follows: Let $P(x, \partial/\partial x)$ be a differential operator with its coefficients continuous functions. If, for an arbitrary $u\in\mathscr{D}(\Omega)$, $P(x, \partial/\partial x)u(x)\equiv0$, then, $P=0$, i.e. if $P=\Sigma\, a_\nu(x)\,(\partial/\partial x)^\nu$, it must be that $a_\nu(x)\equiv0$.

Roughly speaking, this decomposition can be regarded as similar to the decomposition of a complex number into its real part and imaginary part. In fact, $(Au, u) = (A_1u, u) + \mathrm{i}(A_2u, u)$ shows this is true. Now, we have

$$\langle (A + A^*)\, u, \bar{u}\rangle \equiv 2(A_1u, u) \geqslant \delta \sum_{i=1}^{n} \left\| \frac{\partial u}{\partial x_i} \right\|^2 - C\|u\|^2,$$

for an arbitrary $u \in \mathscr{D}(\Omega)$, where C is a constant. Now we prove this inequality as follows: we can write

$$A_1 u = -\sum_{i,j} \frac{\partial}{\partial x_j}\left(a_{ij}(x) \frac{\partial}{\partial x_j} u \right) + \sum b_i(x) \frac{\partial}{\partial x_i} u + b(x)\, u\,.$$

The first term on the right hand side has already been discussed in example 2, section 8. On the other hand,

$$\left| \left(b_i(x) \frac{\partial u}{\partial x_i}, u \right) \right| \leqslant \sup |b(x)| \cdot \left\| \frac{\partial u}{\partial x_i} \right\| \|u\| \leqslant \sup |b(x)| \left(\varepsilon \left\| \frac{\partial u}{\partial x_i} \right\|^2 + \frac{1}{2\varepsilon} \|u\|^2 \right),$$

where $\varepsilon > 0$ is arbitrary. Therefore, we have proved the above inequality.

Let t be a real number, and let $u \in \mathscr{D}_{L^2}{}^1(\Omega)$, $f \in L^2(\Omega)$, with

$$(A + tI)\, u = f\,.$$

By (3.73) we have

$$\langle (A_1 + tI)\, u, \bar{\varphi}\rangle + \mathrm{i}\langle A_2 u, \bar{\varphi}\rangle = \langle f, \bar{\varphi}\rangle\,. \tag{3.74}$$

Then, by a limiting process, $\langle (A_1 + tI)u, \bar{\varphi}\rangle$ is a positive definite Hermitian form defined on $\mathscr{D}_{L^2}{}^1(\Omega)$ (if we take t sufficiently large), and this defines a norm which is equivalent to $D_{L^2}{}^1(\Omega)$.

Let us call a Hilbert space which has this Hermitian form as inner product $\mathscr{H}$.† Let us then write $\langle (A_1 + tI)u, \bar{v}\rangle = (u, v)_{\mathscr{H}}$. Note that for the Hermitian form $\langle A_2 u, \bar{\varphi}\rangle$,

$$|\langle A_2 u, \bar{u}\rangle| \leqslant C\|u\|_1{}^2 \leqslant C\|u\|_{\mathscr{H}}{}^2 \tag{3.75}$$

where $u \in \mathscr{D}_{L^2}{}^1(\Omega)$. First, we prove:

THEOREM 3.17. Let $B[u, v]$ be a bounded Hermitian form defined on $\mathscr{H}$, i.e. for a positive number M, $-M\|u\|^2 \leqslant B[u, u] \leqslant M\|u\|^2$. In this case, there

† Of course, $\mathscr{H} = \mathscr{D}_{L^2}{}^1(\Omega)$ as a space and its topology.

exists a unique H, a bounded Hermitian operator, and $B[u, v]=(Hu, v)$ is valid for arbitrary $u, v \in \mathscr{H}$. Also $\|H\| = \sup_{\|u\|=1} |B[u, u]|$.

Proof. By the conditions we imposed, we see that there exists C with $|B[u, v]| \leqslant C\|u\|\,\|v\|$. In fact if we take t sufficiently large, $B[u, v]+t[u, v]$ becomes a positive definite Hermitian form which defines a norm equivalent to that of $\mathscr{H}$. Therefore, from $\frac{1}{2}(|a+b|^2) \leqslant |a|^2+|b|^2$ we have

$$\tfrac{1}{2}|B[u, v]|^2 \leqslant |B[u, v]+t(u, v)|^2 + t^2|(u, v)|^2,$$

where the first term of the right hand side can be estimated by $(B[u, u]+t\|u\|^2)(B[v, v]+t\|v\|^2)$ because of the validity of the Schwartz inequality for a positive definite Hermitian form.

Hence, it is estimated by $C'\|u\|^2\|v\|^2$, so that we have $|B[u, v]| \leqslant C\|u\|\,\|v\|$. For this reason, for a fixed u, if we consider $v \to B[u, v]$, we see that this is a continuous antilinear form, so that by Riesz theorem, we can write $B[u, v]=(Hu, v)$ and there is a $g=H\cdot u$ which is uniquely determined.

It is obvious that H is linear, also it is bounded because $\|Hu\| = \sup_{\|v\|=1} \|B[u, v]\| \leqslant C\|u\|$. That H is Hermitian can be deduced from the fact that $(Hu, u)=B[u, u]$ always takes a real value.† Taking (3.20) into account, we have

$$\|H\| = \sup_{\|u\|=1} |(Hu, u)| = \sup_{\|u\|=1} |B[u, u]|.$$

Q.E.D.

Note. The meaning of the theorem is that the endowment of a bounded Hermitian form and that of a bounded Hermitian operator have essentially the same effect. We call H an *operator generated by* $B[u, v]$. Now, by (3.75) and this theorem, we can see the existence of a bounded Hermitian operator H of $\mathscr{H}$, and

$$\langle A_2 u, \bar{\varphi}\rangle = (Hu, \varphi)_{\mathscr{H}}.$$

Also we see that

$$\langle f, \bar{\varphi}\rangle = (Cf, \varphi)_{\mathscr{H}},$$

where C is a continuous operator from $\mathscr{D}_{L^2}'^1(\Omega)$ to $\mathscr{D}_{L^2}^1(\Omega)=\mathscr{H}$. Therefore, we can write

$$((I+\mathrm{i}H)\,u, \varphi)_{\mathscr{H}} = (Cf, \varphi)_{\mathscr{H}} \quad (\varphi \in \mathscr{H})$$

† In fact, if we consider $4(Hu, v) = (H(u+v), u+v) - (H(u-v), u-v) + \mathrm{i}(H(u+\mathrm{i}v), u+\mathrm{i}v) - \mathrm{i}(H(u-\mathrm{i}v), u-\mathrm{i}v)$ it follows that $(Hv, u) = \overline{(Hu, v)}$.

instead of (3.74). Then, we have

$$(I+\mathrm{i}H)\,u=Cf, \tag{3.76}$$

a linear equation in $\mathscr{H}$.

By proposition 3.2, we know that $(I+\mathrm{i}H)^{-1}$ exists, so that

$$u=G_t f=(I+\mathrm{i}H)^{-1}Cf \tag{3.77}$$

is a solution. It is obvious that for G_t the fundamental properties: $L_t\circ G_t=I$ in $\mathscr{D}'_{L^2}{}^1(\Omega)$, $G_t\circ L_t=I$ in $\mathscr{D}_{L^2}{}^1(\Omega)$ can be established. Therefore, proposition 3.1 is also true in this case. In fact, if A satisfies (3.71), for

$$(\lambda I+A)\,u=f(x) \quad (f\in L^2(\Omega)), \tag{3.78}$$

obtaining $u\in\mathscr{D}_{L^2}{}^1(\Omega)$ and obtaining a solution for

$$u+(\lambda-t)\,G_t u=G_t f \tag{3.79}$$

in $L^2(\Omega)$ are equivalent.†

Referring to (3.79), we see that G_t is a bounded operator in $L^2(\Omega)$, and in particular, G_t is also completely continuous if Ω is bounded. But note that in this case G_t is *not* a Hermitian. First we consider the case in which Ω is bounded.

According to the Riesz–Schauder theory (see section 10), a condition that (3.79) should have a solution u for $G_t f$ (arbitrary) is that $u+(\lambda-t)\,G_t u=0$ has no solution except $u=0$.‡ Next, the necessary and sufficient condition for (3.79) to have at least one solution for a given f is that every solution v of the conjugate equation of (3.79),

$$v+(\bar{\lambda}-t)\,G_t^* v=0 \tag{3.80}$$

is orthogonal to $G_t f$, where G_t^* is the conjugate operator of G_t when G_t is regarded as a bounded operator of $L^2(\Omega)$.

Now, we have

$$(G_t f, v)=(f, G_t^* v)=-(\lambda-t)^{-1}\,(f, v),$$

† The domain of A is the whole set of u such that $u\in\mathscr{D}_{L^2}{}^1(\Omega)$ and $Au\in L^2(\Omega)$.

‡ The right hand side of (3.79) has a form $G_t f$. Therefore, the space formed by the right hand side term is a real subspace of $L^2(\Omega)$. The condition above, that $-(\lambda-t)$ is not a singular value of G_t is a sufficient condition for the existence of solutions of (3.79) for an arbitrary $f\in L^2(\Omega)$, but we cannot yet conclude that this is a necessary condition.

so that the above condition is equivalent to $(f, v)=0$. Also, in this case the statement claiming that $-(\lambda-t)$ is a singular value of G_t and the statement claiming that $-(\bar{\lambda}-t)$ is singular value of G_t^* are equivalent. Therefore, the necessary and sufficient condition for (3.79) to have a solution for an arbitrary $f \in L^2(\Omega)$ is that $-(\lambda-t)$ is not a singular value of G_t in (3.79). This is equivalent to saying that λ is not an eigenvalue of the operator $-A$ of the Dirichlet problem.

On the other hand, the G_t^* which appears in (3.80) is in fact a Green's operator for the Dirichlet problem of a formal conjugate operator (A^*+tI). Therefore, for

$$(\bar{\lambda}I+A^*)\, v(x)=g(x) \tag{3.81}$$

the fact that $\lambda=\bar{\lambda}_0$ is an eigenvalue for the Dirichlet problem of $-A^*$ and the fact $\lambda=\lambda_0$ is an eigenvalue for the Dirichlet problem of $-A$ in (3.78) are equivalent, and the dimensions of the two eigenspaces (finite) corresponding to these eigenvalues are equal. Note that in order to obtain the following result we have used the Riesz–Schauder theorem which we shall prove in section 10.

Summing up, our argument:

THEOREM 3.18. Let Ω be bounded. The following statements are true for the Dirichlet problems for (3.78) and (3.81).

(1) The eigenvalues of $-A$ in (3.78) consist of $\{\lambda_\nu\}_{\nu=1,2,\ldots}$; $|\lambda_1|\leqslant|\lambda_2| \leqslant|\lambda_3|\leqslant\cdots\to+\infty$ where $+\infty$ is considered to be at *most* an accumulation point. In this case the eigenvalues of the operator $-A^*$ are given as $\{\bar{\lambda}_\nu\}_{\nu=1,2,\ldots}$.

(2) The dimension of the eigenspace which corresponds to λ_ν is finite, and is the dimension of the eigenspace of A^* corresponding to $\bar{\lambda}_\nu$.

(3) If $\lambda\notin\{\lambda_\nu\}$, (3.78) has a unique solution $u\in\mathscr{D}_{L^2}{}^1(\Omega)$ for an arbitrary $f\in L^2(\Omega)$, the correspondence: $f(x)\to u(x)$ thus obtained is a bounded operator in $L^2(\Omega)$. A similar argument can be applied to (3.81).

(4) If $\lambda=\lambda_\nu$, the necessary and sufficient condition for (3.78) to have at least one solution for a given $f\in L^2(\Omega)$ is that f is orthogonal to the eigenspace of $-A^*$ which corresponds to $\lambda=\bar{\lambda}_\nu$ in (3.81).

REMARK 1 *(About G_t^*)*. G_t^* is defined as the conjugate operator: $(G_t u, v) =(u, G_t^* v)$ in $L^2(\Omega)$ where $u, v\in L^2(\Omega)$. The following statements are true.

(1)[†] $G_t{}^*$ has a unique extension as a continuous operator from $\mathscr{D}'_{L^2}{}^1(\Omega)$ to $\mathscr{D}_{L^2}{}^1(\Omega)$, and for arbitrary $u, v \in \mathscr{D}'_{L^2}{}^1(\Omega)$, $\langle G_t u, \bar{v}\rangle = \langle u, G_t{}^*\bar{v}\rangle$.

(2) Let $A + tI = L_t$. Then,

$$L_t{}^* \circ G_t{}^* = I \quad \text{in} \quad \mathscr{D}'_{L^2}{}^1(\Omega), \qquad G_t{}^* \circ L_t{}^* = I \quad \text{in} \quad \mathscr{D}_{L^2}{}^1(\Omega),$$

i.e. $G_t{}^*$ is a Green's operator for the Dirichlet problem of $L_t{}^*$.

Proof.

(1) Let $u, v \in L^2(\Omega)$. Then,

$$|\langle G_t u, v\rangle| \leqslant \|G_t u\|_{1, L^2(\Omega)} \|v\|'_{1, L^2(\Omega)} \leqslant C\|u\|'_{1, L^2(\Omega)} \|v\|'_{1, L^2(\Omega)},$$

where $L^2(\Omega)$ is dense in $\mathscr{D}'_{L^2}{}^1(\Omega)$, and the conjugate space of $\mathscr{D}'_{L^2}{}^1(\Omega)$ is $\mathscr{D}_{L^2}{}^1(\Omega)$. We see that $v \to G_t{}^* v$ has a unique extension as a continuous mapping from $\mathscr{D}'_{L^2}{}^1(\Omega)$ to $\mathscr{D}_{L^2}{}^1(\Omega)$. It is obvious that the relation is still preserved by this extension.

(2) Let $u \in \mathscr{D}(\Omega)$ and $v \in \mathscr{D}'_{L^2}{}^1(\Omega)$. Then,

$$\langle u, \bar{v}\rangle = \langle G_t \circ L_t u, \bar{v}\rangle = \langle L_t u, \overline{G_t{}^* v}\rangle = \langle u, \overline{L_t{}^* \circ G_t{}^* v}\rangle .$$

But, $\mathscr{D}(\Omega)$ is dense in $\mathscr{D}_{L^2}{}^1(\Omega)$, so that $v = L_t{}^* \circ G_t{}^* v$ in $\mathscr{D}'_{L^2}{}^1(\Omega)$. The second relation can be proved in a similar way.

Q.E.D.

REMARK 2 *(Concerning a resolvent set)*. When we assume that Ω is bounded, we obtain theorem 3.18 which corresponds to the first, second and third of Fredholm's theorems. On the other hand, what would happen if Ω were not bounded? In the author's view, it seems to be difficult to express the conditions for (3.79) to have solutions in terms of A and A^*. For this reason, we shall only give the range of λ for which a Dirichlet problem has a unique solution, and the estimation of G_λ, a Green's operator for such λ. The following argument is valid whether Ω is bounded or not.

First choose c_0 satisfying

$$\langle (A_1 + c_0 I)\, u, \bar{u}\rangle \geqslant \delta \sum_{i=1}^{n} \left\|\frac{\partial u}{\partial x_i}\right\|^2 + \|u\|^2$$

for $u \in \mathscr{D}_{L^2}{}^1(\Omega)$. We separate the possibilities of the position of λ as follows.

[†] For an arbitrary distribution $T \in \mathscr{D}'(\Omega)$, we define $\bar{T} \in \mathscr{D}'(\Omega)$ as satisfying $\langle \bar{T}, \varphi\rangle = \overline{\langle T, \bar{\varphi}\rangle}$. It is obvious that from $T_j \to T$ in $\mathscr{D}'_{L^2}{}^1(\Omega)$, we have $\bar{T}_j \to \bar{T}$ in $\mathscr{D}'_{L^2}{}^1(\Omega)$.

(1) Let $\lambda = t + i\mu$ where $\operatorname{Re}\lambda \geqslant t_0 (> c_0)$. We decompose

$$\langle (A+\lambda I)\,u,\,\bar{\varphi}\rangle = \langle (A_1 + tI)\,u,\,\bar{\varphi}\rangle + i\langle (A_2+\mu I)\,u,\,\bar{\varphi}\rangle .$$

We see now the existence of a Green's operator G_λ is guaranteed.
Next,

$$|\langle (A_2+\mu I)\,u,\,\bar{u}\rangle| \geqslant \big|\,|\mu|\;\|u\|^2 - |\langle A_2 u,\,\bar{u}\rangle|\,\big|$$

and

$$|\langle A_2 u,\,\bar{u}\rangle| \leqslant c\left(\sum \left\|\frac{\partial u}{\partial x_i}\right\|^2\right)^{\frac{1}{2}} \|u\| + c\|u\|^2 .$$

From this, for a certain $c_1 (>0)$

$$|\langle (A+\lambda I)\,u,\,\bar{u}\rangle| \geqslant \tfrac{1}{2}(|\lambda - c_0| + t - c_0 - c_1)\|u\|^2 .$$

Therefore, by the Schwartz inequality,

$$\|(A+\lambda I)\,u\| \geqslant \tfrac{1}{2}(|\lambda - c_0| + t - c_0 - c_1)\|u\| .$$

Let $t_0 > c_0 + c_1$. In this case the right hand side of the above inequality is larger than $\frac{1}{2}|\lambda - c_0|\;\|u\|$.

(2) If $\operatorname{Re}\lambda \leqslant t_0$. Let $\lambda = t_0 + i\mu$. Considering (3.79), we have the equation

$$u + (\lambda - t_0 - i\mu)\,G_{t_0+i\mu}u = G_{t_0+i\mu}f .$$

Therefore,

$$\|G_{t_0+i_\mu}\| \leqslant 2|t_0 - c_0 + i\mu|^{-1} .$$

Let us take ρ as satisfying $0<\rho<1$, and consider λ satisfying

$$2|\lambda - (t_0 + i\mu)| \leqslant \rho|(t_0 - c_0) + i\mu| .$$

Then, for such λ the equation has a solution which is the sum of a Neumann series. We now consider μ as the variable. We have:

THEOREM 3.19. For appropriate a and $\theta (0<\theta<\frac{1}{2}\pi)$, let us define a fan-shaped domain Σ as follows. Take the point $(a, 0)$ as origin in the λ-plane. Draw two half-lines from the origin which have angles θ and $-\theta$ measured from the negative axis of the real coordinate. We now have Σ. Thus obtained, Σ has the following properties.

(1) The spectra of $-A$ and $-A^*$ for a Dirichlet problem are in Σ.

(2) A Green's operator G_λ is regular for λ as an operator in $L^2(\Omega)$ in the complementary set of Σ where

$$\|G_\lambda\| \leqslant \frac{C}{(|\lambda|+1)}, \tag{3.82}$$

C being an arbitrary constant.

Note. When Ω is bounded we have a more powerful theorem. In this case the alternative theorem of Fredholm is valid, so that we examine the range of λ for which the problem has a unique solution, i.e. the range of λ such that

$$\|(A+\lambda I)\, u\| \geqslant C(\lambda)\|u\| \quad (u \in \mathscr{D}(A))^\dagger$$

where $C(\lambda)$ is an appropriate positive constant which can vary with λ. Let $\lambda = \sigma + i\mu$. If $\sigma \to -\infty$, we consider μ such that the above estimation is valid. Then, obviously:

THEOREM 3.20. If Ω is bounded, the point spectrum $\{\lambda_\nu\}$ of $-A$ for a Dirichlet problem lies in the interior of a parabola: $a(\sigma-c)+\mu^2 = 0 (a>0, \lambda=\sigma+i\mu)$.
The proof is left to the reader.

10 *Fredholm's alternative theorem for a completely continuous operator*

The theorem of Riesz–Schauder is to be found in most advanced texts on functional analysis. We give a proof here for the benefit of readers who may not be familiar with this subject. We can establish the theorem in an abstract Banach space, but we prove it for a Hilbert space. Let us begin with the following.

LEMMA 3.8 *(Solubility ⇒ Uniqueness)*. Let A be a completely continuous operator in a Hilbert space $\mathscr{H}$. If a linear equation in $\mathscr{H}$,

$$(I-A)\, u = f,$$

has at least one solution u for an arbitrary $f \in \mathscr{H}$, then u is uniquely determined for each f, i.e. from $(I-A)u = 0$ it follows that $u = 0$.

† I.e. u, where $u \in \mathscr{D}_{L^2}{}^1(\Omega)$ and $Au \in L^2(\Omega)$.

Proof (*First step*). Let $I-A=T$ and let a set of f be $\mathcal{M}_n$ where f satisfies $T^n f=0$. Obviously,

$$\mathcal{M}_0=\{0\}\subseteq\mathcal{M}_1\subseteq\mathcal{M}_2\subseteq\mathcal{M}_3\subseteq\cdots\subseteq\mathcal{M}_n\subseteq\mathcal{M}_{n+1}\subseteq\cdots,$$

T^n is a continuous operator of $\mathcal{H}$, so that $\mathcal{M}_n$ is a closed subspace. (We consider $T^0=I$.) If A is completely continuous, for a certain n, $\mathcal{M}_n=\mathcal{M}_{n+1}=\mathcal{M}_{n+2}=\cdots$. We prove this as follows: for a certain n, if $\mathcal{M}_n=\mathcal{M}_{n+1}$, $\mathcal{M}_{n+1}=\mathcal{M}_{n+2}=\mathcal{M}_{n+3}=\cdots=0$ because if $\mathcal{M}_{n+1}\subset\mathcal{M}_{n+2}$, there is an f satisfying $T^{n+1}f\neq0$, $T^{n+2}f=0$, so that $T^n(Tf)\neq0$, $T^{n+1}(Tf)=0$. This is a contradiction to the fact that $\mathcal{M}_n=\mathcal{M}_{n+1}$.

Therefore, we deny the statement the sequence contains no equality, i.e. $\mathcal{M}_n\underset{\neq}{\subset}\mathcal{M}_{n+1}(n=0, 1, 2,\ldots)$. In this case, by an induction process, we can construct a sequence $\{\varphi_n\}$ such that $\varphi_n\in\mathcal{M}_n$, φ_n is orthogonal with $\mathcal{M}_{n-1}$, and $\|\varphi_n\|=1$. We put $n>m$, then,

$$A\varphi_n-A\varphi_m=(I-T)\,\varphi_n-(I-T)\,\varphi_m=\varphi_n-(T\varphi_n+\varphi_m-T\varphi_m),$$

so that

$$(T\varphi_n+\varphi_m-T\varphi_m)\in\mathcal{M}_{n-1}.$$

Therefore

$$\begin{aligned}\|A\varphi_n-A\varphi_m\|^2&=\|\varphi_n-(T\varphi_n+\varphi_m-T\varphi_m)\|^2\\&=\|\varphi_n\|^2+\|T\varphi_n+\varphi_m-T\varphi_m\|^2\geqslant\|\varphi_n\|^2=1.\end{aligned}$$

This contradicts the fact that A is completely continuous.

(*Second step*) Let us now prove the lemma by a contradiction. Suppose T is not a bijection. In this case, there is a certain $u_1\neq0$, and $Tu_1=0$. By our hypothesis, $T(\mathcal{H})=\mathcal{H}$, so that there exists u_2 and $u_1=Tu_2$. Next we see that there exists u_3 and $u_2=Tu_3$. Continuing this construction, we finally obtain a sequence $\{u_n\}$ such that $u_n=Tu_{n+1}$.

On the other hand, $u_1\notin\mathcal{M}_0$, $u_1\in\mathcal{M}_1$ so that $u_2\notin\mathcal{M}_1$, $u_2\in\mathcal{M}_2$. In general $u_n\notin\mathcal{M}_{n-1}$, but $u_n\in\mathcal{M}_n$. This is a contradiction to the previous result.

Q.E.D.

LEMMA 3.9 *(Uniqueness ⇒ Solubility)*. Let A be completely continuous. If the linear mapping $T=I-A$ is one-to-one, then T maps $\mathcal{H}$ onto $\mathcal{H}:T(\mathcal{H})=\mathcal{H}$.

Proof. We show the image $T(\mathcal{H})$ is dense. Now we repeat the same argument (see, for example, the proof of lemma 3.1). For a bounded operator T, the statements that $T(\mathscr{H})$ is dense and that T^* is a bijection are equivalent. Therefore, we wish to show that $T^*(=I-A^*)$ is a bijection.

By our hypothesis $T(=(T^*)^*)$ is one-to-one, so that $T^*(\mathscr{H})$ is dense, and A^* is completely continuous.† Therefore, the image $T^*(\mathscr{H})$ is also closed. Hence, $T^*(\mathscr{H})=\mathscr{H}$. From the previous lemma, we see that T^* is one-to-one mapping.

Finally, we demonstrate that the image $T(\mathscr{H})$ is closed if A is completely continuous. To do this, we first consider T as a one-to-one mapping. Assume $Tu_n \to f_0$. In this case, if $\{u_n\}$ contains a bounded subsequence, we write it as u_n, where u_n is assumed to be bounded. Then $\{Au_n\}$ contains a Cauchy sequence $\{Au_{n_p}\}$, so that $Tu_{n_p}=u_{n_p}-Au_{n_p}$. From this we see that $\{u_{n_p}\}$ itself is a convergent sequence. Let $u_{n_p}\to u_0$, then, $Tu_{n_p}\to Tu_0=f_0$. Suppose $\|u_n\|\to+\infty$, and write $v_n=u_n\|u_n\|^{-1}$. We see that $\|v_n\|=1$ $(n=1, 2,\ldots)$ and also $Tv_n\to 0$. So that from the above discussion we conclude that there is a subsequence such that $Tv_0=0$ and $\|v_0\|=1$. This contradicts the fact that T is one-to-one.

That is, if T is one-to-one, then from $Tu_n\to f$ it follows that $\{\|u_n\|\}$ is bounded. If T is not one-to-one, we put $\mathscr{N}=\{u:Tu=0\}$. In this case if we restrict the domain of T to $\mathscr{H}\ominus\mathscr{N}$ (the ortho-complementary space of $\mathscr{N}$), we see that the restriction of T is one-to-one. Also, $T(\mathscr{H}\ominus\mathscr{N})=T(\mathscr{H})$. From the result which we obtained before we see that $T(\mathscr{H}\ominus\mathscr{N})$ is closed. Therefore $T(\mathscr{H})$ is closed.

Q.E.D.

We now have:

COROLLARY. Under the same conditions as the previous lemma, there exist T^{-1}, $(T^*)^{-1}$, and these are bounded operators (see lemma 3.1). More exactly, if A is completely continuous, and if $u=0$ follows from $(I-A)u=0$, then the equations $(I-A)u=f$, $(I-A^*)v=g$ have unique solutions u and v for arbitrary f, g, and the correspondence between them is continuous.

---

† If A is completely continuous, so is A^*. To see this we first note the definition $(Af, g) = (f, A^*g)$. If g_j is a bounded sequence of $\mathscr{H}$, we can choose a subsequence $\{g_{j_p}\}$ which is weakly convergent (theorem 2.1). If f runs through a unit sphere, $\overline{\{Af\}}$ is a compact set. By the definition, we see that $\{A^*g_{j_p}\}$ is uniformly convergent on the unit sphere $\|f\|=1$, i.e. $\{A^*g_{j_p}\}$ is a strongly convergent sequence.

LEMMA 3.10.[†] Let A be completely continuous, and let $T=I-A$. We write $\mathcal{N}$ for the space formed with f satisfying $Tf=0$. Let the ortho-complementary space of the image of $T(\mathscr{H})$ be $\mathcal{M}$. Then, $\dim(\mathcal{N})=\dim(\mathcal{M})<+\infty$.

Note. We call $\mathcal{N}$ the *null space* of an operator T. We can now ascertain the dimension of the *null space* of T and the co-dimension of the image $T(\mathscr{H})$.

Proof. It is clear that $\mathcal{N}$ is finite-dimensional.[‡] Let $\{\varphi_1, \ldots, \varphi_n\}$ be a base of $\mathcal{N}$. Next, let $\{\psi_1, \ldots, \psi_m\}$ be a base of $\mathcal{M}=\mathscr{H}\ominus T(\mathscr{H})$. We will show that $m\leqslant n$. To prove this, we can show that if $m>n$, then a contradiction arises.

Let $m>n$ and consider

$$\tilde{T}=I-\left(A-\sum_{i=1}^{n}(\cdot, \varphi_i)\psi_i\right)=T+\sum_{i=1}^{n}(\cdot, \varphi_i)\psi_i.$$

We can see that $A-\sum(\cdot, \varphi_i)\psi_i$ is a completely continuous operator. Now $\tilde{T}$ is a one-to-one mapping. In fact, from $\tilde{T}f=0$, we have $Tf=0$ and $\sum(f, \varphi_i)\psi_i=0$. Therefore, it follows that $(f, \varphi_i)=0$ $(i=1, 2, \ldots, n)$, because the image $T(\mathscr{H})$ and ψ_i are orthogonal. Hence $f=0$. From the fact that $\tilde{T}$ is one-to-one, and the previous lemma, we see that $\tilde{T}(\mathscr{H})=\mathscr{H}$. On the other hand, this image does not belong to a space spanned by ψ_{n+1}. This is a contradiction, hence, $m\leqslant n$.

We now wish to prove $m\geqslant n$. To see this it is enough to note that $\mathcal{M}$ is the null space for T^*, and $\mathcal{N}$ is the ortho-complementary space of $T^*(\mathscr{H})$.

Q.E.D.

As in the case of a Hermitian operator, let us consider the equations

$$(I-\lambda A)\,u=f, \tag{3.83}$$

$$(I-\lambda A^*)\,v=g, \tag{3.84}$$

by introducing a complex λ, where A is a completely continuous linear operator of a Hilbert space $\mathscr{H}$, and A^* is the conjugate operator of A in $\mathscr{H}$. In this case, a proposition similar to proposition 3.3 can be established, but the orthogonality of the system of eigenfunctions is no longer valid, and the eigenvalues are not necessarily real.

[†] Schauder proved the lemma for a Banach space. We can see our proof is straightforward in the case of a Hilbert space. Riesz essentially worked out the foregoing lemmas including this lemma.

[‡] Cf. the proof of lemma 3.3.

Let $\varphi_1,\ldots, \varphi_n$ be a set of arbitrary non-zero solutions of the homogeneous equation corresponding to different singular values $\lambda_1,\ldots, \lambda_n$. Then, they are mutually linearly independent. This can be proved by induction for n. Summing up:

THEOREM 3.21.

(1) The set of the singular values of A is at most countable and it does not accumulate other than at the point at infinity $\lambda=\infty$. Let $\{\lambda_\nu\}_{\nu=1,2,\ldots}$ be such a sequence. If $\lambda\notin\{\lambda_\nu\}$, the equation (3.83) has a unique solution for an arbitrary f, and $u=(I-\lambda A)^{-1}f$ is a bounded operator.

(2) The singular values of A^* are $\{\bar{\lambda}\}_{\nu=1,2,\ldots}$, and the dimension (finite) of the eigenspace of A corresponding to λ_ν and that of A^* corresponding to $\bar{\lambda}_\nu$ are equal. Also the eigenfunctions for different singular values are mutually independent.

(3) If $\lambda=\lambda_\nu$, the necessary and sufficient condition for the existence of at least one solution u for (3.83) for a given f is that the eigenspace of A^* for$\bar{\lambda}_\nu$ and f are orthogonal with each other.

Proof. This is obvious from lemmas 3.8, 3.9 and 3.10. (The null space of $(I-\bar{\lambda}A^*)$ and the ortho-complementary space of the image $(I-\lambda A)(\mathscr{H})$ with respect to $\mathscr{H}$ coincide as we noted in the proof of lemma 3.10.)

Q.E.D.

11 *Differentiability of a solution*

We have discussed the existence problems of the Dirichlet and Neumann problems, but so far we have not been concerned with the problem of the differentiability of solutions assuming they are obtained somehow; we shall now discuss this problem. To clarify the situation, we shall treat some simple examples.

First, consider the equation

$$-\Delta u+c(x)\,u+\lambda u=f(x) \tag{3.1}$$

which we have already discussed. We knew that if $u(x)$ is a solution for a Dirichlet problem, $u\in\mathscr{D}_{L^2}{}^1(\Omega)$ for $f\in L^2(\Omega)$, $Au\in L^2(\Omega)$, and, being regarded as an element of $L^2(\Omega)$, u satisfies (3.1), where the $-\Delta u$ which appears in Au should be understood as a distribution in Ω. The first question is: is $u(x)$ itself smooth if $f(x)$ is sufficiently smooth? The answer to this question is yes, providing $c(x)$ is also smooth. The next question, which is connected with the first question is: is $u(x)$ smooth including the boundary of Ω under the same conditions? The answer to the second question is also affirmative, i.e. the more we impose the smoothness of the boundary S of Ω, $c(x)$, $f(x)$ in $\bar{\Omega}$, the more we obtain the smoothness of $u(x)$ in $\bar{\Omega}$. The boundary value problem of the third kind has the same properties. In fact, the smoothness of $u(x)$ in Ω (we call this property *interior regularity*) is totally independent on the boundary conditions of the problem. We shall see this shortly. On the other hand, when we discuss the smoothness in $\bar{\Omega}$, we need to examine the smoothness of $\sigma(x)$ in the boundary conditions of

$$\frac{\mathrm{d}u}{\mathrm{d}n}+\sigma(x)\,u=0 \tag{3.65}$$

besides the smoothnesses which we assumed to exist in the previous cases.

In this section we describe only interior regularity of u, and we shall look at regularity of the whole, including the boundary, in the next section. Let us examine the differentiability of a solution $u(x)$ of (3.1). Recall definition 3.1. That is $u\in\mathscr{E}_{L^2}{}^m{}_{(\mathrm{loc})}(\Omega)$ means, for an arbitrary $\alpha\in\mathscr{D}(\Omega)$, $\alpha u\in\mathscr{D}_{L^2}{}^m$. For simplicity we assume $c\in C^\infty$. Then we have:

THEOREM 3.22. Let $u\in L^2{}_{\mathrm{loc}}(\Omega)$ satisfy (3.1) in an open set Ω. If $f\in\mathscr{E}_{L^2}{}^m{}_{(\mathrm{loc})}(\Omega)$ then $u\in\mathscr{E}_{L^2}{}^{m+2}_{(\mathrm{loc})}(\Omega)$, where $m=0, 1, 2, \ldots$.

Note. No boundary condition is imposed on $u(x)$.

Proof (*First step*). We see that if

$$(-\Delta+1)\,u(x)=f(x)$$

and if $u\in\mathscr{S}'$, $f\in\mathscr{D}_{L^2}{}^s$, then $u\in\mathscr{D}_{L^2}{}^{s+2}$ where s is an arbitrary real number (definition 3.4). By a Fourier transformation,

$$(1+4\pi^2|\xi|^2)\,\hat{u}(\xi)=\hat{f}(\xi),$$

i.e.

$$\hat{u}(\xi)=(1+4\pi^2|\xi|^2)^{-1}\hat{f}(\xi).$$

From this we have

$$\|u\|_{s+2}=\|(1+|\xi|)^{s+2}\,\hat{u}(\xi)\|_{L^2}=\|(1+|\xi|)^{s+2}\,(1+4\pi^2|\xi|^2)^{-1}\,\hat{f}(\xi)\|_{L^2}$$
$$\leqslant C\|(1+|\xi|)^s\,\hat{f}(\xi)\|_{L^2}=C\|f\|_s.$$

(*Second step*) Before carrying out a direct proof, we note that if s is an integer (either positive or negative), and $f\in\mathscr{D}_{L^2}{}^s$, then $\partial f/\partial x_i\in\mathscr{D}_{L^2}{}^{s-1}$, and if $c\in\mathscr{B}^\infty$ then† $c(x)f\in\mathscr{D}_{L^2}{}^s$.

Let us multiply both sides of (3.1) $\alpha\in\mathscr{D}(\Omega)$. Then, we have

$$(-\Delta+1)(\alpha u)=\alpha f-2\sum\alpha_i(x)\frac{\partial u}{\partial x_i}-\Delta\alpha(x)\,u-(c(x)+\lambda-1)(\alpha u)$$

where $\alpha_i=\partial\alpha/\partial x_i$. If we put the right hand side term equal to $g(x)$, then by our hypothesis $g\in\mathscr{D}_{L^2}{}^{-1}$. In fact, for $\beta\in\mathscr{D}(\Omega)$,

$$\beta(x)\frac{\partial u}{\partial x_i}=\frac{\partial}{\partial x_i}(\beta u)-\frac{\partial\beta}{\partial x_i}u\in\mathscr{D}_{L^2}{}^{-1}.$$

From the result of the first step, $\alpha u\in\mathscr{D}_{L^2}{}^1$ where α is arbitrary. Therefore, $u\in\mathscr{E}_{L^2}{}^1{}_{(\text{loc})}$, from this $g\in L^2$. Then, again by the result of the first step, $u\in\mathscr{E}_{L^2}{}^2{}_{(\text{loc})}$, and so on.

Q.E.D.

Note. If c is simply k-times continuously differentiable, we replace $m+2$ by $\min(k,m)+2$. Then theorem 3.22 is valid.

COROLLARY. If $f\in\mathscr{E}_{L^2}{}^{[\frac{1}{2}n]+1+p}{}_{(\text{loc})}(\Omega)$, then $u\in\mathscr{E}^{p+2}(\Omega)$ where $p=-1,0,1,2\ldots$. In particular, if $f\in\mathscr{E}^\infty(\Omega)$, then $u\in\mathscr{E}^\infty(\Omega)$.

Proof. This is just another form of Sobolev's theorem 2.8.

Q.E.D.

The case of variable coefficients

The same result can be established if the coefficients are variables. The basic principle of the proof of this fact is also used for elliptic operators of high orders. In order to preserve generality, we consider a general elliptic type equation:

DEFINITION 3.5. If a differential operator defined on an open set Ω

$$L(x,\mathrm{D})=\sum_{|\alpha|\leqslant m}a_\alpha(x)\,\mathrm{D}^\alpha$$

† If $s\geqslant0$, the statement is equivalent to lemma 2.4. If $s<0$, we can consider this case as $s\geqslant0$ by the conjugate relation.

satisfies

$$|\sum_{|\alpha|=m} a_\alpha(x)\,\xi^\alpha| \geqslant C|\xi|^m \quad (\xi \in R^n), \tag{3.85}$$

we say that L is *an elliptic operator*, where C is an appropriate positive constant, which may depend on x as x varies. Note that (3.85) is equivalent to $\sum a_\alpha(x)\xi^\alpha \neq 0 (\xi \neq 0)$, where $a_\alpha(x)$ ($|\alpha|=m$) is not necessarily a real number.

For simplicity, we take all coefficients as belonging to C^∞.† We have:

THEOREM 3.23. Let L be an elliptic operator on Ω, and let $u \in L^2_{\text{loc}}$ be an arbitrary solution for $L(x, D)u(x)=f(x)$. Then, if $f \in \mathscr{E}_{L^2}{}^s{}_{(\text{loc})}$, then $u \in \mathscr{E}_{L^2}{}^{s+m}{}_{(\text{loc})}$, where $s=0, 1, 2, \ldots$ and m is the order of the operator L.

Proof. The proof is a little complicated, so we shall prove the theorem step by step. Without loss of generality, we can assume that Ω contains the origin of the space. Then, we prove the theorem in the neighbourhood of the origin. First, we see:

(A) Given $s \geqslant 0$, if we take δ sufficiently small, then there exists a positive constant $C(s, \delta)$ and for an arbitrary $\varphi \in \mathscr{D}_{L^2}{}^s \cap \mathscr{E}'(B_\delta)$ we have the following inequality:‡

$$\|L[\varphi]\|_{s-m} \geqslant C(s, \delta)\|\varphi\|_s \quad (s=0, 1, 2 \ldots), \tag{3.86}$$

where B_δ is a sphere with the radius δ, centre the origin. We prove this fact later as a lemma.

The inequality means that L is in fact an isomorphism from $\mathscr{D}_{L^2}{}^s$ to $\mathscr{D}_{L^2}{}^{s-m}$ because the reverse inequality also holds if we make the support of φ sufficiently small. Therefore, for a sequence $\varphi_j \in \mathscr{D}_{L^2}{}^s$, if $L[\varphi_j]$ forms a Cauchy sequence in $\mathscr{D}_{L^2}{}^{s-m}$, then $\{\varphi_j\}$ is also a Cauchy sequence in $\mathscr{D}_{L^2}{}^s$.

(B) The relation between a derivative and a difference quotient: let $u \in \mathscr{D}_{L^2}{}^m$ where m is an arbitrary real number. Then,

$$\frac{u(x+h_j)-u(x)}{h} \to \frac{\partial u}{\partial x_j} \quad \text{in} \quad \mathscr{D}_{L^2}{}^{m-1}$$

$h_j=(0, \ldots, 0, h, 0, \ldots, 0)$, where h occurs in the jth place. We prove this as

† This is just a temporary assumption. Later we shall see that $a_\alpha \in C^{|\alpha|+s}$.

‡ $\mathscr{E}'(B_\delta)$ means that the support of φ belongs to B_δ.

follows. First we note that

$$\frac{u(x+h_j)-u(x)}{h}-\frac{\partial u}{\partial x_j}\underset{\mathscr{F}}{\to}\{[\exp(2\pi i\xi_j h)-1]h^{-1}-2\pi i\xi_j\}\,\hat{u}(\xi)\ (\text{use (2.64)}).$$

From our hypothesis, $(1+|\xi|)^m\hat{u}(\xi)\in L^2$. Also there is an M such that

$$|[\exp(2\pi i\xi_j h)-1]\,h^{-1}-2\pi i\xi_j|\leqslant M(1+|\xi|)$$

where the left hand term tends to 0 as $h\to 0$. This follows from Lebesgue's theorem.

Conversely, let $h\to 0$. If

$$\frac{u(x+h_j)-u(x)}{h}\to\psi(x)\ \text{ in }\ \mathscr{D}_{L^2}{}^m,$$

then

$$\psi(x)=\partial u/\partial x_j\in\mathscr{D}_{L^2}{}^m.$$

In fact, the distribution of the left hand side term tends to $\partial u/\partial x_j$ by the topology of $\mathscr{D}'$.

(C) Let $L[u]=f$, and let $u\in\mathscr{D}_{L^2}{}^{s+m-1}$ and $f\in\mathscr{D}_{L^2}{}^s$, then, $u\in\mathscr{D}_{L^2}{}^{s+m}$. Let $s\geqslant -m+1$. Assume that the support of u lies in the interior of B_δ which can be determined by (3.86) after replacing s by $s-1$. This can be proved as follows. From

$$L(x,\mathrm{D})\,u(x)=f(x),\qquad L(x+h_j,\mathrm{D})\,u(x+h_j)=f(x+h_j),$$

we have

$$\frac{L(x+h_j,\mathrm{D})-L(x,\mathrm{D})}{h}\,u(x+h_j)+L(x,\mathrm{D})\left[\frac{u(x+h_j)-u(x)}{h}\right]=\frac{f(x+h_j)-f(x)}{h}.$$

Let us now examine what happens if $h\to 0$. The right hand side term is a convergent sequence in $\mathscr{D}_{L^2}{}^{s-1}$ by (B), and the first term on the left hand side is also convergent in the same topology. In fact, the mapping: $(a(x),f)\to a(x)f$ is continuous from $\mathscr{B}^{|s|}\times\mathscr{D}_{L^2}{}^s$ to $\mathscr{D}_{L^2}{}^s$. This is clear for $s\geqslant 0$. If $s<0$, then from $(\mathscr{D}_{L^2}{}^s)'=\mathscr{D}_{L^2}{}^{-s}$, we have $\langle a(x)f,\varphi\rangle=\langle f,a(x)\varphi\rangle$ by putting $\varphi\in\mathscr{D}_{L^2}{}^{-s}$. Therefore,

$$|\langle a(x)\,f,\varphi\rangle|\leqslant\|f\|_s\|a(x)\,\varphi(x)\|_{-s}\leqslant C|a(x)|_{-s}\|f\|_s\|\varphi\|_{-s}.$$

Considering

$$\mathrm{D}^{\alpha}u(x+h_j)\to\mathrm{D}^{\alpha}u(x) \quad \text{in} \quad \mathscr{D}_{L^2}{}^{s-1} \qquad (|\alpha|\leqslant m),$$

we see that the first term on the left hand side is a convergent sequence in $\mathscr{D}_{L^2}{}^{s-1}$. Hence,

$$L\left[\frac{u(x+h_j)-u(x)}{h}\right]$$

is convergent in $\mathscr{D}_{L^2}{}^{s-1}$. So that by (A)

$$\frac{u(x+h_j)-u(x)}{h}$$

is itself convergent in $\mathscr{D}_{L^2}{}^{s-1+m}$. Finally, by (B)

$$\partial u/\partial x_j\in\mathscr{D}_{L^2}{}^{s-1+m} \qquad (j=1, 2, \ldots, n),$$

so that $u\in\mathscr{D}_{L^2}{}^{s+m}$. (This is obvious from the estimation of $\hat{u}(\xi)$, the Fourier image of $u(x)$.)

(D) The theorem itself follows immediately from (C). Let $\alpha\in\mathscr{D}$. We assume the support of α lies in the interior of B_δ, otherwise α is arbitrary (we put $s=-m+1$).

From $\alpha(x)L(x, \mathrm{D})u(x)=\alpha(x)f(x)$ we see that

$$L(x, \mathrm{D})\,(\alpha u)=\alpha(x)\,f(x)+\sum_{|\nu|\geqslant 1}\frac{\mathrm{D}^{\nu}\alpha(x)}{\nu!}\,L^{(\nu)}(x, \mathrm{D})\,u$$

by using Leibniz formula, where $L^{(\nu)}(x, \mathrm{D})$ is defined as

$$L^{(\nu)}(x, \xi)=\left(\frac{\partial}{\partial\xi}\right)^{\nu}L(x, \xi).$$

It is, therefore, a differential operator of degree $(m-|\nu|)$.

But $|\nu|\geqslant 1$, so that all the differential operators which appear in the second term on the right hand side are of degree less than degree $(m-1)$. By the hypothesis of the theorem, the last term on the right hand side belongs to $\mathscr{D}_{L^2}{}^{-m+1}$. Because $\alpha f\in L^2$, we have $\alpha u\in\mathscr{D}_{L^2}{}^1$ by using the result of (C) (put $s=-m+1$), i.e. $u\in\mathscr{E}_{L^2}{}^1{}_{(\mathrm{loc})}$. If $m\geqslant 2$, then using the result just obtained, we see that the right hand term belongs to $\mathscr{D}_{L^2}{}^{-m+2}$. From this $\alpha u\in\mathscr{D}_{L^2}{}^2$.

Hence, $u\in\mathscr{E}_{L^2}{}^2{}_{(\mathrm{loc})}$. This may be repeated until we obtain $u\in\mathscr{E}_{L^2}{}^m{}_{(\mathrm{loc})}$. If $\alpha f\in\mathscr{D}_{L^2}{}^s(s\geqslant 1)$, we can use the same process.

Q.E.D.

We now prove (3.86) in (A). First, we prove a simple case:

LEMMA 3.11. Let $L_0=\sum_{|\alpha|=m} a_\alpha D^\alpha$ be an elliptic operator (definition 3.5) of the equation having constant coefficients. Given an integer $s(\geqslant -m)$, if δ is sufficiently small, for an arbitrary $\varphi\in\mathscr{D}_{L^2}{}^{s+m}\cap\mathscr{E}'(B_\delta)$ we have an inequality $\|L_0[\varphi]\|_s\geqslant C\|\varphi\|_{s+m}$ where C is a positive constant.

Proof. We have

$$\|L_0[\varphi]\|_s=\|(1+|\xi|)^s|\sum a_\alpha(2\pi\xi)^\alpha|\phi(\xi)\|_{L^2}\geqslant C_0\|(1+|\xi|)^s|\xi|^m\phi(\xi)\|_{L^2}.$$

Therefore, we have to show the right hand side term is larger than $\|\varphi\|_{s+m}$ ignoring the positive constant multiplier. We put

$$(1+|\xi|)^s|\xi|^m=(1+|\xi|)^{s+m}+\sigma(\xi)(1+|\xi|)^{s+m}.$$

We see that $\sigma(\xi)$ is bounded and $O(|\xi|^{-1})$ as $|\xi|\to\infty$. We have

$$\|(1+|\xi|)^s|\xi|^m\phi(\xi)\|\geqslant\|\varphi\|_{s+m}-\|\sigma(\xi)(1+|\xi|)^{s+m}\phi(\xi)\|$$

where the second term on the right hand side is less than $\|\varphi\|_{s+m}$ if we make the support of φ sufficiently small. To see this we proceed as follows.

Let us write $\psi(x)$ for the inverse Fourier image of the function which appears as the second term on the right hand side. We see that

$$\psi(x)=\int_{|\xi|\leqslant R} e^{2\pi ix\xi}\sigma(\xi)\,(1+|\xi|)^{s+m}\phi(\xi)\,d\xi+\int_{|\xi|>R}\dots d\xi\equiv\psi_1(x)+\psi_2(x).$$

By Plancherel's theorem,

$$\|\psi_2(x)\|\leqslant\sup_{|\xi|\geqslant R}|\sigma(\xi)|\;\|\varphi\|_{s+m}.$$

Let us choose R satisfying $\sup_{|\xi|\geqslant R}|\sigma(\xi)|\leqslant\frac{1}{3}$. For $\psi_1(x)$, we have

$$\|\psi_1(x)\|^2\leqslant C(R)\int_{|\xi|\leqslant R}|\phi(\xi)|^2\,d\xi.$$

On the other hand, from $|\phi(\xi)| \leqslant \int |\varphi(x)|\,dx$, using the Schwartz inequality, we have

$$\|\psi_1(x)\|^2 \leqslant m(B_\delta)C'(R)\|\varphi\|^2 \leqslant m(B_\delta)C'(R)\|\varphi\|_{s+m}^{\;2},$$

because $|\phi(\xi)|^2 \leqslant m(B_\delta)\|\varphi\|^2$, where $s+m \geqslant 0$.

Therefore, if we make δ sufficiently small,

$$\|\psi_1+\psi_2\| \leqslant \tfrac{2}{3}\|\varphi\|_{s+m}.$$

Q.E.D.

LEMMA 3.12. The previous lemma is also valid for the case of variable coefficients.

Proof. Let $L_0 = \sum_{|\alpha|=m} a_\alpha(0)D^\alpha$. We decompose L into $L = L_0 + (L - L_0)$. Then, we see that

$$\|L\varphi\|_s \geqslant \|L_0\varphi\|_s - \|(L-L_0)\,\varphi\|_s.$$

From the previous lemma, we have to prove that the second term of the right hand side becomes arbitrarily small compared with $\|\varphi\|_{s+m}$ if we make the support of φ sufficiently small.

We consider

$$\|(L-L_0)\varphi\|_s \leqslant \sum_{|\alpha|=m} \|[a_\alpha(x)-a_\alpha(0)]\,D^\alpha\varphi\|_s + \sum_{|\alpha|\leqslant m-1} \|a_\alpha(x)\,D^\alpha\varphi\|_s.$$

Observe that this is nothing but the previous inequality if $s \geqslant 0$. (We apply Poincaré's inequality (3.38).) If $s<0$, then we use a conjugate relation. For example, let us consider $\|[a_\alpha(x)-a_\alpha(0)]D^\alpha\varphi\|_s$. Let $\psi \in \mathscr{D}$. Then we have

$$\langle [a_\alpha(x)-a_\alpha(0)]\,D^\alpha\varphi, \psi(x)\rangle = (-1)^{|\alpha_2|}\langle D^{\alpha_1}\varphi, D^{\alpha_2}\{[a_\alpha(x)-a_\alpha(0)]\,\psi(x)\}\rangle$$

where $|\alpha_1| = m+s(<m)$, $|\alpha_2| = -s$. We see that in absolute terms this cannot exceed $\varepsilon\|\varphi\|_{m+s}\|\psi\|_{-s}$. ε can be taken to be as small a number as we wish, like δ of B_δ. Because $\mathscr{D}$ is dense in $\mathscr{D}_{L^2}{}^{-s}$, we have $\|[a_\alpha(x)-a_\alpha(0)]D^\alpha\varphi\|_s \leqslant \varepsilon\|\varphi\|_{m+s}$.

Q.E.D.

12 *Differentiability of a solution in the neighbourhood of a boundary*†

Let us separate our study into the case of the Dirichlet problem and that of

† We follow closely the argument of Nirenberg: 'Remarks on strongly elliptic partial differential equations', *Comm. Pure Appl. Math.* 8 (1955).

the boundary value problem of the third kind (generalized Neumann problem). First we treat the case of a Dirichlet problem. We look at only higher order ($\geqslant 3$) cases because the second order case is not much different. We specify the problem for a so-called 'strongly elliptic operator' because of difficulties which occur when we treat a general elliptic operator as we defined in definition 3.5.

DEFINITION 3.6. When a differential operator of $2m$-degree, $A=\sum a_\alpha(x)D^\alpha$ satisfies

$$\operatorname{Re}(-1)^m \sum_{|\alpha|=2m} \alpha_\alpha(x)\,\xi^\alpha > \delta|\xi|^{2m} \quad (\delta>0). \tag{3.87}$$

We call A a *strongly elliptic operator*. Putting this in easier terms, this means that if we separate the coefficient of the highest ordered term as a real part and an imaginary part $a_\alpha(x)=a_\alpha^{(1)}(x)+ia_\alpha^{(2)}(x)$, then

$$(-1)^m \sum_{|\alpha|=2m} a_\alpha^{(1)}(x)\,\xi^\alpha > \delta|\xi|^{2m}. \tag{3.87$'$}$$

In this case, the Dirichlet problem for A means that given $f\in L^2(\Omega)$ we look for $u\in \mathscr{D}_{L^2}{}^m(\Omega)$ satisfying $Au=f$. The problem is essentially the same as the problem we treated before. We shall give the existence theorem later. Here, we examine the differentiability of a solution u in the neighbourhood of a boundary S.

Let x_0 be a point taken on the boundary S of Ω, and let S be a hypersurface of the C^∞-class in the neighbourhood of x_0. Assume that the coefficients of A are of the C^∞-class in $\bar{\Omega}$. In this case we construct a bijection which maps S into the hyperplane and also this mapping and the inverse mapping are both C^∞-class functions. Note that the strong ellipticity of an operator is invariant under a coordinate change. By using this mapping we prove the following:

PROPOSITION 3.7. Let u be a function defined on a semi-ball ΣR: $\{x;|x|<R,\ x_n>0\}$ which is centered on the origin. Let us assume the $u\in \mathscr{E}_{L^2}{}^m(\Sigma_R)$ and, moreover, for an arbitrary $\alpha\in\mathscr{D}(B_R)$, we have $\alpha u\in \mathscr{D}_{L^2}{}^m(\Sigma_R)$ where $Au=f$, $f\in L^2(\Sigma_R)$ and A is a strong elliptic operator defined on Σ_R. In this case, if $\delta(<R)$ is sufficiently small, then $u\in\mathscr{E}_{L^2}{}^{2m}(\Sigma_\delta)$. Furthermore, if $f\in\mathscr{E}_{L^2}{}^s(\Sigma_R)(s=0, 1, 2, \ldots)$, then $u\in\mathscr{E}_{L^2}{}^{2m+s}(\Sigma_{\delta-\varepsilon})$ for $\varepsilon>0$.

Proof. The proof is complicated so we separate it into steps.

(*First step*) We prove the following fact. For sufficiently small δ,

$$|\langle A\varphi, \bar{\varphi}\rangle| \geqslant C\|\varphi\|_m^2, \tag{3.88}$$

where $\varphi \in \mathscr{D}_{L^2}{}^m(\Sigma_\delta)$ (arbitrary) and C is a positive constant. The proof of this fact is as follows. First, let

$$A = \tfrac{1}{2}(A + A^*) + \mathrm{i}\,\frac{1}{2\mathrm{i}}\,(A - A^*) \equiv A_1 + \mathrm{i}A_2 \qquad \text{(see section 9).}$$

We have

$$|\langle A\varphi, \bar{\varphi}\rangle| \geqslant \mathrm{Re}\langle A\varphi, \bar{\varphi}\rangle = \langle A_1\varphi, \bar{\varphi}\rangle.$$

Let

$$A_1(x, \mathrm{D}) = \sum a_\alpha{}^{(1)}(x)\,\mathrm{D}^\alpha \qquad \text{(definition 3.6)}$$

and let

$$A_0 = \sum_{|\alpha| = 2m} a_\alpha{}^{(1)}(0)\,\mathrm{D}^\alpha,$$

so that

$$\langle A_1\varphi, \bar{\varphi}\rangle = \langle A_0(\mathrm{D})\,\varphi, \bar{\varphi}\rangle + \langle (A_1 - A_0)\,\varphi, \bar{\varphi}\rangle.$$

The first term on the right hand side is

$$\langle A_0(\mathrm{D})\,\varphi, \bar{\varphi}\rangle = \langle \sum a_\alpha(0)(2\pi\mathrm{i}\xi)^\alpha \phi(\xi), \overline{\phi(\xi)}\rangle \geqslant \delta\langle |2\pi\xi|^{2m}\,\overline{\phi(\xi)}, \phi(\xi)\rangle$$
$$\geqslant \delta' \sum_{|\alpha| = m} \|\mathrm{D}^\alpha\varphi\|^2 \qquad \text{(by (3.87)')}.$$

If Σ_δ is small enough, then the Poincaré inequality gives

$$\sum_{|\alpha| = m} \|\mathrm{D}^\alpha\varphi\|^2 \geqslant C\|\varphi\|_m^2 \qquad (C > 0).$$

Next we have to deal with $\langle (A_1 - A_0)\varphi, \bar{\varphi}\rangle$. In this case, we can use the same inequality and we see that for an arbitrary $\varepsilon(>0)$, the absolute value of this factor becomes smaller than $\varepsilon\|\varphi\|_{m^2}$. This proves (3.88). From this we have

$$\|A\varphi\|_{-m} \geqslant C\|\varphi\|_m \tag{3.89}$$

where $\|A\varphi\|_{-m}$ means the norm of $A\varphi$ in $\mathscr{D}'_{L^2}{}^m(\Sigma_\delta)$.

Now for $\alpha u \in \mathscr{D}_{L^2}{}^m(\Sigma_\delta)$, we can apply the same argument as theorem 3.23. In this case we consider only h_j for $j = 1, 2, \ldots, n-1$. That is we only look at derivatives which are taken in the parallel direction of hyperplanes.

Therefore, if we assume $f \in L^2(\Sigma_R)$, then

$$\frac{\partial}{\partial x_j}(\alpha u) \in \mathscr{D}_{L^2}{}^m(\Sigma_\delta) \qquad (j=1, 2, \ldots, n-1).$$

On the other hand, if we take the differential of $Au=f$, we have

$$A\left(\frac{\partial u}{\partial x_j}\right)+A_j u=\frac{\partial f}{\partial x_j}, \qquad A_j(x, \mathrm{D})=\frac{\partial A}{\partial x_j}(x, \mathrm{D}).$$

So that, for an arbitrary $\alpha \in \mathscr{D}(B_\delta)$,

$$\frac{\partial}{\partial x_k}\left(\alpha \frac{\partial u}{\partial x_j}\right) \in \mathscr{D}_{L^2}{}^m(\Sigma_\delta) \qquad (j, k=1, 2, \ldots, n-1).$$

Hence, for an arbitrary $\alpha \in \mathscr{D}(B_\delta)$, we have†

$$\left(\frac{\partial}{\partial x'}\right)^\gamma (\alpha u) \in \mathscr{D}_{L^2}{}^m(\Sigma_\delta) \qquad (|\gamma| \leqslant m, \, x'=(x_1, \ldots, x_{n-1})).$$

If we assume $f \in \mathscr{E}_{L^2}{}^s(\Sigma_R)$, then the above argument is valid when $|\gamma| \leqslant m+s$.

(*Second step*) Let us start proving the proposition. First, we assume $f \in L^2(\Sigma_R)$ and take $\alpha \in \mathscr{D}(B_\delta)$. From $\alpha(x)Au=\alpha f$, we have

$$\mathrm{D}_y{}^m[a_0(x)\, \mathrm{D}_y{}^m(\alpha u)+\sum c_{\gamma, j}(x)\, \mathrm{D}_{x'}{}^\gamma\, \mathrm{D}_y{}^j u]+B(x, \mathrm{D})\, u=\alpha f \qquad (3.90)$$

where we put $x_n=y$, and $\partial/\partial x_n=\mathrm{D}_y$. In (3.90):

(1) Σ on the left hand side is taken over $|\gamma|+j \leqslant m$, $0 \leqslant j \leqslant m-1$. The support of $c_{\gamma,j}(x)$ is contained in that of $\alpha(x)$. Furthermore, we have $|\alpha_0(x)|>\sigma>0$ $(x \in \Sigma_\delta)$.

(2) $B(x, \mathrm{D})$ is a differential operator of $2m$-degree. It is of m-degree with respect to D_y. The coefficient of $B(x, \mathrm{D})$ is contained in that of $\alpha(x)$ as a support.

Now, for the time being we assume the following 'interpolation' lemma, which we shall prove in the next section. Let $g(x', y)$ and $\mathrm{D}_{x'}{}^\gamma g(x', y)$ $(|\gamma| \leqslant m+s)$ belong to $L^2(R_+{}^n)$. Then, if $\mathrm{D}_y{}^m g$ and its derivates of s-degree inclusive with respect to x' belong to $L^2(R_+{}^n)$, we have

$$\mathrm{D}_{x'}{}^\nu \mathrm{D}_y{}^j g(x', y) \in L^2(R_+{}^n) \quad (|\nu|+j \leqslant m).$$

We let the function inside the square brackets [ ] on the left hand side of

† This is equivalent to saying that, for an arbitrary $\alpha \in \mathscr{D}(B_\delta)$, $\alpha(x)(\partial/\partial x')u \in \mathscr{D}_{L^2}{}^m(\Sigma_\delta)$ $(|\gamma| \leqslant m)$.

(3.90) be $g(x', y)$. This satisfies the condition of the lemma (from the result obtained in the first step). Now, we consider $D_y g(x', y)$, putting

$$D_y g(x', y) = a_0(x)\, D_y^{m+1}(\alpha u) + \psi_1(x).$$

From the previous result in the first step, $\psi_1 \in L^2$ up to the derivative of $(m-1)$-degree inclusive with respect to x'. We can establish a similar result about $D_y g$. So that, $D_y^{m+1}(\alpha u)$ belongs to $L^2(\Sigma_\delta)$ up to the derivative of $(m-1)$-degree inclusive with respect to x'. Therefore, for an arbitrary $\alpha \in \mathscr{D}(B_\delta)$, we have

$$\alpha D_{x'}^{\gamma}\, D_y^{m+1} u \in L^2(\Sigma_\delta) \qquad (|\gamma| + m + 1 \leqslant 2m\ ).$$

We observe $D_y^2 g = a_0(x) D_y^{m+2}(\alpha u) + \psi_2(x)$ as the case of $D_y g$. Then, we have $\alpha D_{x'}^{\gamma} D_y^{m+2} u \in L^2(\Sigma_\delta)$ $(|\gamma| + m + 2 \leqslant 2m)$.

If we repeat this process, then we see that, for an arbitrary $\alpha \in \mathscr{D}(B_\delta)$,

$$\alpha(x)\, D_x^{\gamma} u \in L^2(\Sigma_\delta) \qquad (|\gamma| \leqslant 2m).$$

This shows $u \in \mathscr{E}_{L^2}^{2m}(\Sigma_{\delta-\varepsilon})$ for an arbitrarily small $\varepsilon(>0)$.

Next, let $f \in \mathscr{E}_{L^2}^{s}(\Sigma_R)$. In this case we modify the 'interpolation' lemma as follows. Let $g(x', y)$ and $D_{x'}^{\gamma} g(x', y) \in L^2(R_+^n)$ $(|\gamma| \leqslant m+s)$. If $D_y^m g \in L^2(R_+^n)$ up to the derivative of s-degree inclusive with respect to x', then $D_{x'}^{\alpha}\, D_y^{j} g \in L^2(R_+^n)$ $(0 \leqslant j \leqslant m,\ |\alpha| + j \leqslant m+s)$. Using this fact, we have for an arbitrary $\alpha \in \mathscr{D}(B_\delta)$,

$$\alpha(x)\, D_{x'}^{\gamma}\, D_y^{j} u(x', y) \in L^2(\Sigma_\delta) \qquad (|\gamma| + j \leqslant 2m+s,\ j \leqslant 2m).$$

Now, we return to the original $Au = f$ checking derivatives along the y-axis. We conclude that the relationship holds for $0 \leqslant j \leqslant 2m+s$. Hence, for an arbitrary $\varepsilon(>0)$, we have $u \in \mathscr{E}_{L^2}^{2m+s}(\Sigma_{\delta-s})$.

Q.E.D.

We summarize the results which we have obtained so far. Let A be a strongly elliptic operator of $2m$-degree whose coefficients are sufficiently smooth on $\bar{\Omega}$. Let $u(x)$ be a solution of the Dirichlet problem:

$$Au = f \qquad (u \in \mathscr{D}_{L^2}^{m}(\Omega)).$$

Then, from the result obtained in the previous section 11 (about interior regularity), and from theorem 3.15, we have:

THEOREM 3.14. Let Ω be an interior domain surrounded by a sufficiently smooth closed hypersurface S. Let $f \in L^2(\Omega)$, then, $u \in \mathscr{E}_{L^2}{}^{2m}(\Omega)$. If, further, we assume $f \in \mathscr{E}_{L^2}{}^{s}(\Omega)$, then $u \in \mathscr{E}_{L^2}{}^{2m+s}(\Omega)$. In particular, if $s = [\frac{1}{2}n]+1$, then u is continuously differentiable $2m$-times including the boundary and is 0 up to the derivative of $(m-1)$-degree inclusive on the boundary.

Note. In this theorem 3.24 we did not mention strictly the smoothness of the coefficients of A includes that of the boundary, and the smoothness of S itself. However, for the smoothness of the coefficients of A it is enough to assume that they are continuously differentiable $(2m+s)$ times, and for the smoothness of S it is enough to assume that S is of the C^{2m+s}-class.†

Next, what about the case of an exterior domain? The last part of the theorem is also true in this case; to prove $u \in \mathscr{E}_{L^2}{}^{2m+s}(\Omega)$, we consider Gårding's inequality (we shall explain this later), and assume uniform ellipticity – this means that, in definition 3.6, δ in (3.87) is independent of x – and also assume the boundedness of the coefficients of A up to a sufficiently high order.

The solution of the boundary value problem of the third kind

In the last half of section 8, we gave the reason why seeking the solution of the boundary value problem of the third kind is equivalent to seeking that of (3.70). There we assumed that the solution of (3.70), $u \in \mathscr{E}_{L^2}{}^2(\Omega)$. We shall now prove this.

In this case, we cannot regard the main part of the equation as $-\Delta$ because we transform a part of the boundary into a hyperplane as we have done in the case of the Dirichlet problem. Taking this into account, and to emphasize the general principles of the following proof we choose a slightly generalized equation

$$\sum_{i=1}^{n}\left(\frac{\partial u}{\partial x_i}, \frac{\partial \varphi}{\partial x_i}\right)+\left(\sum_i a_i(x)\frac{\partial u}{\partial x_i}+c(x)\,u,\ \varphi\right)-\int_S \sigma(x)\,u\bar{\varphi}\,dS=(f, \varphi), \tag{3.91}$$

where $a_i(x)$, $c(x)$, $\sigma(x)$ are smooth but not necessarily real-valued functions.

---

† We can show this under a weaker condition. For example, let S be of C^2-class, and let $c(x)$ be bounded and measurable in Ω where $u(x)$ satisfies $(-\Delta+c(x))u(x)=f(x)$. Then, from $u \in \mathscr{D}_{L^2}{}^1(\Omega)$, $f \in L^2(\Omega)$ it follows that $u \in \mathscr{E}_{L^2}{}^2(\Omega)$. In fact, if we look at step (C) of theorem 3.23, letting the transformed operator $-\Delta$ be $L(x, \mathrm{D})$ and $L(x, \mathrm{D})u = f(x) - c(x)u$, then the right hand side of this equation is L^2, so that $u \in \mathscr{E}_{L^2}{}^2(\Omega)$ if we follow the proof of theorem 3.24.

This form is, of course, closely related to the problem

$$L[u]\equiv -\Delta u+\sum_{i=1}^{n} a_i(x)\frac{\partial u}{\partial x_i}+c(x)\,u=f; \qquad \frac{\mathrm{d}u}{\mathrm{d}n}+\sigma(x)\,u=0 \quad (x\in S). \tag{3.91'}$$

To solve this problem we shall require an existence theorem; we prove this theorem at the beginning.

First we let the left hand side of (3.91) be $B[u, \varphi]$. Then, $B[u, \varphi]$ is linear for $u\in\mathscr{E}_{L^2}{}^1(\Omega)$, and anti-linear for $\varphi\in\mathscr{E}_{L^2}{}^1(\Omega)$. Using the Schwartz inequality, we see

$$\mathrm{Re}(B[u, u]+t\|u\|_{L^2}{}^2)\geqslant C\|u\|_1{}^2 \quad (C>0), \tag{3.92}$$

for a large positive number t (see the last half of section 8). Let us fix t by putting $B[u, \varphi]+t(u, \varphi)_{L^2}=B_t[u, \varphi]$. Now,

$$|B_t[u, \varphi]|\leqslant C\|u\|_1\|\varphi\|_1 \tag{3.93}$$

is obvious. By Riesz theorem, we can determine a bounded linear operator $\tilde{A}$ in $\mathscr{E}_{L^2}{}^1(\Omega)$ such that $B_t[u, \varphi]=(\tilde{A}u, \varphi)_{1, L^2(\Omega)}$. From the result obtained in section 9, we decompose $\tilde{A}$ as the sum of two Hermitian operators:

$$\tilde{A}=\tfrac{1}{2}(A+A^*)+\mathrm{i}\frac{1}{2\mathrm{i}}(A-A^*)\equiv A_1+\mathrm{i}A_2.$$

Because $(A_1u, u)_{1, L^2(\Omega)}=\mathrm{Re}\,B_t[u, u]$, $B_t[u, \varphi]$ is a positive Hermitian form defined on $\mathscr{E}_{L^2}{}^1(\Omega)$, so that the norm defined by this form is equivalent to that of $\mathscr{E}_{L^2}{}^1(\Omega)$. We write $\mathscr{H}$ for $\mathscr{E}_{L^2}{}^1(\Omega)$ equipped with this Hermitian form. In this case, as we said in section 9, there exists a bounded Hermitian operator H in $\mathscr{H}$ (by (3.93)), with

$$(A_2u, \varphi)_{1, L^2(\Omega)}=(Hu, \varphi)_{\mathscr{H}}.$$

Therefore,

$$B_t[u, \varphi]=((I+\mathrm{i}H)\,u, \varphi)_{\mathscr{H}}.$$

Hence, the problem allows the same treatment in principle if we replace $\mathscr{D}_{L^2}{}^1(\Omega)$ by $\mathscr{E}_{L^2}{}^1(\Omega)$.

In this case, we note that

$$\|u\|_{\mathscr{H}}\leqslant\|(I+\mathrm{i}H)\,u\|_{\mathscr{H}}=\sup_{\varphi}|B_t[u, \varphi]|$$

where 'sup' is the upper bound of the value for φ on the unit sphere of $\mathscr{H}$. From this we see that the equation obtained by adding $t(u, \varphi)$ to the left hand side of (3.91) has a Green's operator $G_t f=(I+iH)^{-1}Cf$ which is a continuous operator from $(\mathscr{E}_{L^2}{}^1(\Omega))'$ to $\mathscr{E}_{L^2}{}^1(\Omega)$. In particular, if Ω is bounded, this operator is completely continuous as an operator of $L^2(\Omega)$.

Next, we consider the differentiability of a solution in the neighbourhood of the boundary S. Let us assume S is sufficiently smooth. Let x_0 be a point of S, and let $B_\delta(x_0)=|x-x_0|<\delta$, an open sphere. In this case, if, in (3.92), the support of u is in $B_\delta(x_0)$, then $\operatorname{Re} B[u, u]\geqslant C\|u_1\|^2$, i.e. t can be ignored if δ is sufficiently small. In fact, we see that a generalized Poincaré inequality holds. (See the proof of lemma 3.3. We map S into a hyperplane by the transformation of the neighbourhood of x_0.) From this, if δ is sufficiently small, for $u\in\mathscr{E}_{L^2}{}^1(\Omega)\cap\mathscr{E}'(B_\delta(x_0))$, we have

$$\|u\|_1\leqslant C\sup_{\varphi}|B[u, \varphi]|, \tag{3.94}$$

where 'sup' is taken when φ runs through the unit sphere of $\mathscr{E}_{L^2}{}^1(\Omega)$. Therefore, $\{u_j(x)\}$ is a sequence which satisfies (3.94), and if $B[u_j, \varphi]$ is uniformly convergent on the unit sphere of $\mathscr{E}_{L^2}{}^1(\Omega)$, then $\{u_j(x)\}$ is a convergent sequence of $\mathscr{E}_{L^2}{}^1(\Omega)$, i.e. $B[u, \varphi]$ is in fact an inner product of $\mathscr{E}_{L^2}{}^1(\Omega)$.

On the other hand, if $\alpha\in\mathscr{D}(B_\delta(x_0))$ and α is a real-valued function, then from

$$\left(\frac{\partial}{\partial x_i}u, \frac{\partial}{\partial x_i}(\alpha\varphi)\right)=\left(\frac{\partial}{\partial x_i}(\alpha u), \frac{\partial}{\partial x_i}\varphi\right)+\left(\alpha_i\frac{\partial}{\partial x_i}u, \varphi\right)-\left(\alpha_i u, \frac{\partial}{\partial x_i}\varphi\right)$$

where

$$\alpha_i=\partial\alpha/\partial x_i$$

it follows that

$$\left.\begin{aligned} &B[u, \alpha\varphi]=B[\alpha u, \varphi]+C[u, \varphi]\\ &C[u, \varphi]=\sum_{i=1}^{n}\left\{\left(\alpha_i\frac{\partial}{\partial x_i}u, \varphi\right)-\left(\alpha_i u, \frac{\partial}{\partial x_i}\varphi\right)-(a_i\alpha_i u, \varphi)\right\}\end{aligned}\right\}. \tag{3.95}$$

Now we transform a neighbourhood of x_0 to map S into the hyperplane $x_n=0$, and x_0 into the origin. We assume that the transformation is sufficiently regular, including its inverse. (If necessary, we take a smaller $B_\delta(x_0)$ which is contained in the old $B_\delta(x_0)$.) By this transformation, $B[u, \varphi]$ and $C[u, \varphi]$ are mapped into $\tilde{B}[u, \varphi]$ and $\tilde{C}[u, \varphi]$ respectively. For example,

$$\tilde{B}[u, \varphi]=\sum_{i=1}^{n}(E_i u, E_i\varphi)+(Fu, \varphi)+\int_{x_n=0}\rho(x')\,u(x')\,\overline{\varphi(x')}\,dx',$$

where E_i and F are first-order differential operators, and their coefficients are sufficiently smooth. Obviously, for $\tilde{B}[u, \varphi]$, we see that (3.94) is valid because $\|u\|_1$ is transformed into an equivalent norm. For simplicity, we write B and C instead of $\tilde{B}$ and $\tilde{C}$ respectively.

With this new notation, we have

$$B[\alpha u, \varphi]=(j(x)\,\alpha(x)\,f(x), \varphi(x))-C[u, \varphi] \tag{3.96}$$

from (3.91) using (3.94), where $j(x)$ is a Jacobian. If we use the same notation as at the beginning half of this section, we can write $\alpha\in\mathscr{D}(B_\delta)$ and $\varphi\in\mathscr{E}_{L^2}{}^1(\Sigma_\delta)$. Furthermore, we can assume $\varphi(x)=0$ in the neighbourhood of $|x|=\delta$. $\varphi(x)$ is arbitrary under the same condition, so that, using the same notation as at the beginning half of this section replacing $\varphi(x)$ by $\varphi(x+h_j)$ $(j=1, 2, \ldots, n-1)$, reducing (3.94), and dividing by h, we see that

$$B[\alpha u, \Delta_j\varphi]=(j(x)\,(\alpha f)\,(x), \Delta_j\varphi(x))-C[u, \Delta_j\varphi],$$

where $\Delta_j g(x)=(g(x+h_j)-g(x))/h$.

We now extend the notation to a differential operator, i.e. let $\Delta_j E(x, \mathrm{D})$ be a differential operator, taking the difference quotient of each coefficient of E. It is easy to see that if

$$B[\alpha u, \Delta_j\varphi]=B[\Delta_{-j}(\alpha u), \varphi]+G_j[u, \varphi],$$

then†

$$\begin{aligned}G_j[u, \varphi]=&\sum_{i=1}^{n}((\Delta_{-j}E_i)\,(\alpha u)\,(x-h_j), E_i\varphi)\\&-\sum_{i=1}^{n}(E_i\alpha u, (\Delta_jE_i)\,\varphi(x+h_j))+((\Delta_{-j}F)\,(\alpha u)\,(x-h_j), \varphi)\\&+\int\Delta_{-j}(\rho(x'))\,(\alpha u)\,(x'-h_j)\,\overline{\varphi(x')}\,\mathrm{d}x',\end{aligned} \tag{3.97}$$

where Δ_{-j} is obtained by replacing h with $-h$ in its definition.

Summing up, we have

$$B[\Delta_{-j}(\alpha u), \varphi]=(j(x)\,(\alpha f)\,(x), \Delta_j\varphi)-C[u, \Delta_j\varphi]-G_j[u, \varphi]. \tag{3.98}$$

In this case, if we let $h\to 0$, then the first term of the right hand side converges to $(j(x)\,(\alpha f)\,(x), \partial\varphi/\partial x_j)$ because $f\in L^2(\Sigma_\delta)$. This convergence is

† In general, for differential operators $E(x, \mathrm{D})$, $F(x, \mathrm{D})$, we have
$(Eu, F(\Delta_j\varphi))=(Eu, \Delta_j(F\varphi)-(\Delta_jF)\varphi(x+h_j))=(\Delta_{-j}(Eu), F\varphi)-(Eu, (\Delta_jF)\varphi(x+h_j))$
$=(E(-\Delta_ju), F\varphi)+((\Delta_{-j}E)u(x-h_j), F\varphi)-(Eu, (\Delta_jF)\varphi(x+h_j)).$

uniform on the unit sphere of $\mathscr{E}_{L^2}{}^1(\Omega)$ $(\ni\varphi)$. It is easy to see that similar results can be established for $C[u, \Delta_j\varphi]$ and $G_j[u, \varphi]$. So that the left hand term of (3.98) is uniformly convergent on the unit sphere $\{\varphi: \|\varphi\|_1 = 1\}$. Therefore by (3.94), $\Delta_{-j}(\alpha u)$ is convergent by the topology of $\mathscr{E}_{L^2}{}^1(\Sigma_\delta)$ as $h \to 0$. Hence, it follows that

$$\frac{\partial}{\partial x_j}(\alpha u) \in \mathscr{E}_{L^2}{}^1(\Sigma_\delta) \qquad (j=1, 2, \ldots, n-1).$$

Furthermore, if $f \in \mathscr{E}_{L^2}{}^2(\Sigma_\delta)$, we repeat the previous argument to obtain

$$\left(\frac{\partial}{\partial x'}\right)^\gamma (\alpha u) \in \mathscr{E}_{L^2}{}^1(\Sigma_\delta) \qquad (|\gamma| \leqslant s+1)$$

for $\alpha \in \mathscr{D}(B_\delta)$. Finally, applying the same argument as at the beginning half of this section, we have $u \in \mathscr{E}_{L^2}{}^{2+s}(\Sigma_{\delta-\varepsilon})$ for an arbitrary $\varepsilon(>0)$.

Hence, we have the following:

THEOREM 3.25. Let Ω be an interior domain surrounded by a sufficiently smooth closed hypersurface S. If $f \in L^2(\Omega)$ in (3.91), then $u \in \mathscr{E}_{L^2}{}^2(\Omega)$. Accordingly $u(x)$ is a solution of (3.91)′. Furthermore, if $f \in \mathscr{E}_{L^2}{}^s(\Omega)$ $(s=0, 1, 2, \ldots)$, then $u \in \mathscr{E}_{L^2}{}^{2+s}(\Omega)$. In particular, if $s = [\frac{1}{2}n]+1$, then u is twice continuously differentiable, including the boundary S. Therefore, u is a genuine solution of (3.91)′.

Note 1. We have not yet mentioned the smoothnesses of the boundary S, the coefficients of (3.91)′, and $\sigma(x)$. For instance, for the first part of the theorem 3.25 we can assume† that S is of the C^2-class $a_i(x)$ are of the C^1-class including the boundary, and $\sigma(x)$ is of the C^1-class on S.

Note 2. As for an exterior domain, if we assume the boundedness of $a_i(x)$ and $c(x)$, then the theorem is true. To see this, taking an $\alpha \in C^\infty$ such that $\alpha(x)=0$ in the neighbourhood of S and $\alpha(x)=1$ in the neighbourhood of ∞, and operating with $\alpha(x)$ on (3.91)′ from the left, we have

$$(-\Delta+1)(\alpha u) = g,$$

$$g = \alpha f - 2\sum \frac{\partial\alpha}{\partial x_i}\frac{\partial u}{\partial x_i} - \Delta\alpha\cdot u + \alpha u - \alpha\left(\sum a_i(x)\frac{\partial u}{\partial x_i} + c(x)\, u\right) \in L^2(R^n).$$

† We can assume $c(x)$ to be bounded.

This equation is valid in R^n, so that, by a Fourier transform, $\alpha u \in \mathscr{D}_{L^2}{}^2(R^n)$. Therefore, $u \in \mathscr{E}_{L^2}{}^2(\Omega)$.

13 *The interpolation theorem of* $\mathscr{E}_{L^2}{}^m(R_+{}^n)$

We shall now prove the interpolation theorem which we used in section 12 without proof. To do this we first prove the following.†

LEMMA 3.13. Let u be of C^2-class defined on $[0, \infty]$, If $u, u'' \in L^2(0, \infty)$, then $u' \in L^2(0, \infty)$ and

$$\rho \int_0^\infty |u'(t)|^2 \, dt \leqslant \int_0^\infty |u(t)|^2 \, dt + \rho^2 \int_0^\infty |u''(t)|^2 \, dt \tag{3.99}$$

where ρ is an arbitrary positive number. In general, if u is of C^j-class defined on $[0, \infty)$, and if $u, u^{(j)} \in L^2(0, \infty)$, then $u^{(i)} \in L^2(0, \infty)$ $(0<i<j)$ and

$$\rho^i \int_0^\infty |u^{(i)}(t)|^2 \, dt \leqslant c(i,j) \left[\int_0^\infty |u(t)|^2 \, dt + \rho^j \int_0^\infty |u^{(j)}(t)|^2 \, dt \right], \tag{3.100}$$

where $c(i,j)$ is a positive constant and ρ is an arbitrary positive number.

Proof.

(i) From our hypothesis, we show $u^{(i)} \in L^2(0, \infty)$ $(0 \leqslant i \leqslant j)$. In fact, if $\alpha(t)$ is a C^∞-function such that $\alpha(t)=0$ in $t \leqslant 1$, and $\alpha(t)=1$ in $t \geqslant 2$, then from our hypothesis, we have (αu), $(\alpha u)^{(j)} \in L^2(-\infty, +\infty)$, so that by a Fourier transform, we have $(1+|\xi|)^j \widehat{(\alpha u)}(\xi) \in L^2$. Therefore, we see that

$$(\alpha u)^{(i)}(t) \in L^2(-\infty, +\infty) \qquad (0 \leqslant i \leqslant j).$$

From this, it follows that $u^{(i)} \in L^2(0, \infty)$.

(ii) We prove (3.99) where we can assume u is a real-valued function. Now, we have

$$\{u''^2 - u'^2 + u^2 - (u'' + u' + u)^2\} = -\{(u'+u)^2\}'.$$

We see that $u'^2(X) \to 0$, $u(X)^2 \to 0$ $(X \to \infty)$ in $-[(u'+u)^2]_0^{X \to +\infty}$ $= [u'(0) + u(0)]^2$.

† For this lemma see Hardy, Littlewood & Pólya: *Inequalities*, p. 187, Cambridge, 1952.

By integrating over $(0, \infty)$, we have

$$\int_0^\infty \{u''^2 - u'^2 + u^2\}\, dt = \{u'(0) + u(0)\}^2 + \int_0^\infty \{u'' + u' + u\}^2\, dt \geqslant 0,$$

therefore, we have proved

$$\int_0^\infty u'(t)^2\, dt \leqslant \int_0^\infty u(t)^2\, dt + \int_0^\infty u''(t)^2\, dt.$$

In this inequality, we put $u(\delta t)$ instead of $u(t)$ where $\delta(>0)$ is an arbitrary positive number. Then, we have

$$\delta^2 \int_0^\infty u'(\delta t)^2\, dt \leqslant \int_0^\infty u(\delta t)^2\, dt + \delta^4 \int_0^\infty u''(\delta t)^2\, dt.$$

Now, we put $\delta t = t'$, and then $\delta^2 = \rho$, and we have (3.99).

(iii) Let us demonstrate (3.100). From the result (ii), if $0 < i < j$ we have

$$\int_0^\infty |u^{(i)}(t)|^2\, dt \leqslant c(i, j) \left[\int_0^\infty |u(t)|^2\, dt + \int_0^\infty |u^{(j)}(t)|^2\, dt\right].$$

We replace $u(t)$ by $u(\delta t)$, and then $\delta^2 = \rho$.

Q.E.D.

PROPOSITION 3.8. Let $R_+{}^n = \{x;\, x_n > 0\}$.

(1) If u, $(\partial/\partial x_n)^m u \in L^2(R_+{}^n)$, then $(\partial/\partial x_n)^i u \in L^2(R_+{}^n)$ $(0 < i < m)$ and

$$\left\|\left(\frac{\partial}{\partial x_n}\right)^i u(x)\right\|^2 \leqslant c(i, m)\left[\|u(x)\|^2 + \left\|\left(\frac{\partial}{\partial x_n}\right)^m u(x)\right\|^2\right].$$

(2) Let $(x_1, \ldots, x_{n-1}) = x'$. If $(\partial/\partial x')^\alpha u \in L^2(R_+{}^n)$ $(|\alpha| \leqslant m)$ and $(\partial/\partial x_n)^m u \in L^2(R_+{}^n)$, then $(\partial/\partial x)^\alpha u \in L^2(R_+{}^n)$ $(|\alpha| \leqslant m)$, i.e. $u \in \mathscr{E}_{L^2}{}^m(R_+{}^n)$ and

$$\|u(x)\|_m{}^2 \leqslant c(m)\left[\sum_{|\alpha| \leqslant m}\left\|\left(\frac{\partial}{\partial x'}\right)^\alpha u(x)\right\|^2 + \left\|\left(\frac{\partial}{\partial x_n}\right)^m u(x)\right\|^2\right].$$

Proof. (1) can be proved by applying the previous lemma. Nevertheless, we have to do this in a special way.

First, we put $x_n = y$. Using a mollifier ρ_ε, we observe $u_\varepsilon(x) = \rho_\varepsilon * u(x', y + 2\varepsilon)$.

In this case, we regard $u_\varepsilon(x)$ as a distribution defined on $R_+{}^n$. We have

$$D^\alpha u_\varepsilon(x) = \rho_\varepsilon * D^\alpha u(x', y+2\varepsilon),$$

where $D^\alpha u(x', y+2\varepsilon)$ can be regarded as a distribution defined on $y > -2\varepsilon$ by a parallel transformation of $D^\alpha u(x)$ with the y-axis of measure -2ε.

In general, $\rho_\varepsilon * T$ has no meaning for a distribution T defined on an open set Ω, but if Ω_1 (open) $\subset \Omega$, and the distance between Ω_1 and the boundary of Ω is greater than ε, then for $x \in \Omega_1$ we have

$$(\rho_\varepsilon * T)(x) = \langle T_\xi, \rho_\varepsilon(x-\xi)\rangle$$

which is meaningful, and which is a C^∞-function in Ω_1. Furthermore,

$$\begin{aligned} D^\alpha(\rho_\varepsilon * T)(x) &= \langle T_\xi, D_x{}^\alpha \rho_\varepsilon(x-\xi)\rangle = (-1)^{|\alpha|} \langle T_\xi, D_\xi{}^\alpha \rho_\varepsilon(x-\xi)\rangle \\ &= \langle D_\xi{}^\alpha T, \rho_\varepsilon(x-\xi)\rangle, \end{aligned}$$

i.e.

$$D^\alpha(\rho_\varepsilon * T)(x) = (\rho_\varepsilon * D^\alpha T)(x) \qquad (x \in \Omega_1).$$

Next, we note that if $T = f(x) \in L^1{}_{\text{loc}}(\Omega)$ then

$$(\rho_\varepsilon * f)(x) = \int \rho_\varepsilon(x-y) f(y)\, dy.$$

In our case $\Omega = \{x; y > -2\varepsilon\}$, $\Omega_1 = R_+{}^n$. The following facts are obvious.
(*a*) $u_\varepsilon(x) \to u(x)$ in $L^2(R_+{}^n)$. It follows from this that $D^\alpha u_\varepsilon(x) \to D^\alpha u(x)$ in $\mathscr{D}'(R_+{}^n)$.
(*b*) $D_y{}^m u_\varepsilon \in \mathscr{E}^\infty$ and $D_y{}^m u_\varepsilon(x) \to D_y{}^m u(x)$ in $L^2(R_+{}^n)$. Therefore, if we apply the previous lemma to $u_\varepsilon(x) \equiv u_\varepsilon(x', y)$, we have

$$\int_0^\infty |D_y{}^i u_\varepsilon(x', y)|^2\, dy \leqslant c(i, m) \left[\int_0^\infty |u_\varepsilon(x', y)|^2\, dy + \int_0^\infty |D_y{}^m u_\varepsilon(x', y)|^2\, dy\right]$$

for x' almost everywhere. We integrate this with respect to x' to get

$$\|D_y{}^i u_\varepsilon(x', y)\|^2 \leqslant c(i, m)\, [\|u_\varepsilon(x', y)\|^2 + \|D_y{}^m u_\varepsilon(x', y)\|^2].$$

If $\varepsilon \to 0$, then we have (1) because of facts (*a*) and (*b*).

(2) To prove the second half of the proposition we proceed as follows. From our assumption, (1) is valid, so that $(\partial/\partial x_n)^i u \in L^2(R_+{}^n)$ $(i = 0, 1, \ldots, m)$. By the same method which we used in proposition 3.4, we can extend $u(x)$ to

the lower half-space, i.e. if $y<0$ then

$$v(x', y)=\sum_{\nu=1}^{m} a_\nu u(x', -\nu y); \quad \sum_{\nu=1}^{m} a_\nu(-\nu)^j=1 \quad (j=0, 1, \cdots, m-1).$$

Let $U(x', y)=\tilde{u}(x', y)+\tilde{v}(x', y)$. Then, if $0\leqslant i\leqslant m$ we have

$$D_y{}^i U(x', y)=D_y{}^i(\tilde{u}+\tilde{v})=(D_y{}^i u)^\sim+(D_y{}^i v)^\sim \in L^2(R^n).$$

Also, by (3.49)

$$D_{x'}{}^\alpha U(x', y)=D_{x'}{}^\alpha(\tilde{u}+\tilde{v})=(D_{x'}{}^\alpha u)^\sim+(D_{x'}{}^\alpha v)^\sim \in L^2(R^n) \quad (|\alpha|\leqslant m).$$

Therefore, for the Fourier transform $\hat{U}(\xi', \eta)$ of $U(x', y)$, we have

$$(1+|\xi'|)^m\, \hat{U}(\xi', \eta), \quad (1+|\eta|)^m\, \hat{U}(\xi', \eta)\in L^2.$$

From

$$(1+|\xi'|+|\eta|)^m\leqslant 2^m(1+|\xi'|)^m+2^m|\eta|^m,$$

we have $(1+|\xi'|+|\eta|)^m\hat{U}(\xi', \eta)\in L^2$, i.e. $U(x', y)\in\mathscr{D}_{L^2}{}^m(R^n)$. Hence $u(x', y)\in\mathscr{E}_{L^2}{}^m(R_+{}^n)$. From this the inequality of the proposition is also clear.

Q.E.D.

Note. We have some remarks about the proposition.

Part (1). If we replace $u(x', y)$ by $u(x', (\sqrt{\rho})y)$ as we have done in lemma 3.13, then we have

$$\rho^i\|Df_{x_n}{}^i u\|^2\leqslant c(i, m)\,[\|u\|^2+\rho^m\|D_{x_n}{}^m u\|^2]$$

where ρ is an arbitrary positive number. In other words, if we put $\rho^{m-i}=\varepsilon$, then

$$\|D_{x_n}{}^i u\|^2\leqslant c(i, m)\,[\varepsilon^{-i/(m-i)}\|u\|^2+\varepsilon\|D_{x_n}{}^m u\|^2]$$

where $\varepsilon(>0)$ is arbitrary.

Part (2). If $D_{x'}{}^\alpha D_{x_n}{}^m u\in L^2(R_+{}^n)$ $(|\alpha|\leqslant s)$ and $D_{x'}{}^\gamma u\in L^2(R_+{}^n)$ $(|\gamma|\leqslant m+s)$, then

$$D_{x'}{}^\alpha\, D_{x_n}{}^i u\in L^2(R_+{}^n) \qquad (0\leqslant i\leqslant m,\ i+|\alpha|\leqslant m+s).$$

In fact, we can apply the previous result to $u(x)$ and also $D_{x'}{}^\alpha u(x)$ $(|\alpha|\leqslant s)$.

Finally, under the condition (2), for $|\alpha| \leqslant m-1$, we note that

$$\|D^\alpha u\|^2 \leqslant c(m) [\varepsilon^{-|\alpha|/(m-|\alpha|)}\|u\|^2 + \varepsilon(\sum_{|\nu|=m} \|D^\nu u\|^2)],$$

where $\varepsilon(>0)$ is arbitrary. We now have:

THEOREM 3.26. Let Ω be an interior or exterior domain surrounded by a closed hypersurface S of C^m-class. For an arbitrary $u \in \mathscr{E}_{L^2}{}^m(\Omega)$ we have

$$\sum_{|\alpha|=i} \|D^\alpha u(x)\|^2 \leqslant c(i, m, \Omega) [\varepsilon \sum_{|\alpha|=m} \|D^\alpha u(x)\|^2 + \varepsilon^{-i/(m-i)}\|u(x)\|^2] \tag{3.101}$$

where $\varepsilon(>0)$ is arbitrary.

The proof is left to the reader.

14 *Some remarks on Dirichlet problems*

We have treated Dirichlet problems which are somewhat different from the traditional Dirichlet problem. Historically speaking, the problem of obtaining a solution $u(x)$ for $A[u]=0$ was most important. Let us consider this classical problem. We assume that the value $\varphi(s)$ $(s \in S)$ is given on the boundary S of Ω. Let A satisfy

$$A = -\sum a_{ij}(x) \frac{\partial^2}{\partial x_i \partial x_j} + \sum a_i(x) \frac{\partial}{\partial x_i} + c(x) \tag{3.71}$$

with the same conditions as we treated in section 9.

PROBLEM. To obtain $u \in \mathscr{E}_{L^2}{}^1(\Omega)$ satisfying $A[u]=0$ in Ω and $\gamma u = \varphi(s)$.

Given $\varphi(s)$, we assume that there exists a certain $h \in \mathscr{E}_{L^2}{}^1(\Omega)$ satisfying $\gamma h = \varphi(s)$. We put $u = v + h$, then, the problem is equivalent to that of obtaining a solution $v \in \mathscr{D}_{L^2}{}^1(\Omega)$ satisfying

$$A[v] = -A \cdot h. \tag{3.102}$$

Now, let us assume that A has a Green's operator G. We see that $A \cdot h \in \mathscr{D}'_{L^2}{}^1(\Omega)$ so that $v = -GAh$ is a solution. Therefore,

$$u(x) = (I - GA)\, h(x) \tag{3.103}$$

is a desired solution to the problem.

In this case $u(x)$ does not depend on h, i.e. if $\gamma h_1=\varphi(s)$ $(h_1\in\mathscr{E}_{L^2}{}^1(\Omega))$, then $\gamma(h-h_1)=0$, so that $h-h_1\in\mathscr{D}_{L^2}{}^1(\Omega)$ (by theorem 3.24), therefore, by (3.13), we have

$$(I-GA)(h-h_1)=0.$$

If we assume sufficient smoothness of the hypersurface S and $\varphi(s)$, then $h(x)$ is also sufficiently smooth in $\bar{\Omega}$. Further, if we assume the smoothness of the coefficients of A, then by theorem 3.24, we have a function $u(x)$ which is smooth including the boundary of S. This is the classical result. Note that this is true for the interior and exterior of the hypersurface S.

Let us consider the connection between the representation of the solution (3.103) of the problem and the classical result. For simplicity, let Ω be an interior domain, also let $\psi\in\mathscr{D}(\Omega)$. We have

$$((I-GA)\,h,\psi)=(h,\psi)-(GAh,\psi)=(h,\psi)-(Ah,G^*\psi).$$

In general, if $u, v\in\mathscr{E}_{L^2}{}^2(\Omega)$, we have

$$(Au,v)-(u,A^*v)=\int_S\frac{\mathrm{d}u}{\mathrm{d}n_0}\,\bar{v}\,\mathrm{d}S-\int_S u\,\frac{\mathrm{d}\bar{v}}{\mathrm{d}n_0}\,\mathrm{d}S-\int_S C[u,\bar{v}]\,\mathrm{d}S, \qquad (3.104)$$

where $C[\bar{u},\bar{v}]$ is a bilinear form with respect to u, $\bar{v}$, and $\mathrm{d}/\mathrm{d}n_0$ is a derivative in the transversal direction with respect to A, i.e. it points inwardly so that

$$\frac{\mathrm{d}}{dn_0}u=\sum\beta_i\frac{\partial u}{\partial x_i},\qquad \beta_i=\sum a_{ij}\cos\gamma_j,$$

where γ_j is the angle between the inner normal and x_i-axis.

Now, as we have shown in section 9, G^* is a Green's operator for the Dirichlet problem of A^*, and from $\psi\in\mathscr{D}(\Omega)$, $G^*\psi\in\mathscr{D}_{L^2}{}^1(\Omega)\cap\mathscr{E}_{L^2}{}^2(\Omega)$. Therefore,

$$(Ah,G^*\psi)=(h,A^*G^*\psi)-\int_S u\,\frac{\mathrm{d}}{\mathrm{d}n_0}\,\overline{G^*\psi}\,\mathrm{d}S.$$

Also, from $A^*G^*\psi=\psi$ and $\gamma u=\varphi(s)$,

$$((I-GA)\,h,\psi)=\int_S\varphi(s)\frac{\mathrm{d}}{\mathrm{d}n_0}\,\overline{G^*\psi}\,\mathrm{d}S.$$

To obtain a concrete representation of the right hand side term we use

'Green's kernel' (in classical terminology, this is equivalent to the 'Green's function for the Dirichlet problem of A^*'), i.e.

$$(G^*\psi)(x) = \int_\Omega G'(x, \xi)\,\psi(\xi)\,d\xi.$$

Using this operator, by Fubini's theorem,

$$((I - GA)h, \psi) = \left\langle \int_S \varphi(s) \frac{d}{dn_{0,s}} \overline{G'(s, \xi)}\,dS, \overline{\psi(\xi)} \right\rangle.$$

Therefore, using the Green's kernel of A, $G(x, \xi) = \overline{G'(\xi, x)}$: $(Gf)(x) = \int G(x, \xi) f(\xi) d\xi$, we have

$$u(x) = \int_S \varphi(s) \frac{d}{dn_{0,s}} G(x, s)\,dS \qquad (x \in \Omega). \tag{3.105}$$

Next, we assume that A has no Green's operator, i.e. $A[u] = 0$ has a solution $u \in \mathscr{D}_{L^2}{}^1(\Omega)$, $u(x) \not\equiv 0$. In this case, we have the same property for $A^*[v] = 0$. (See theorem 3.18.) Let $f_1(x), \ldots, f_p(x)$ be a set of linearly independent eigenfunctions for $\lambda = 0$ of A^*. In this case, given a sufficiently smooth function $\varphi(s)$ on S, we consider the conditions for the existence of the solution for the Dirichlet problem in our sense (of this section). To observe this, we pick $h \in \mathscr{E}_{L^2}{}^1(\Omega)$ satisfying $\gamma h(x) = \varphi(s)$. Then, the necessary and sufficient condition for the existence of the solution is

$$(Ah, f_i(x)) = 0 \qquad (i = 1, 2, \ldots, p)$$

(by theorem 3.18 and (3.102)).

Now, if we assume that φ is smooth, then we can also assume $h \in \mathscr{E}_{L^2}{}^2(\Omega)$. By (3.104) and from the fact that $\gamma f_i = 0$, $A^* f_i = 0$ this condition can be written as

$$\int_S \varphi(s) \frac{d}{dn_0} \overline{f_i(s)}\,dS = 0 \qquad (i = 1, 2, \ldots, p),$$

where $df_i(s)/dn_0 \not\equiv 0$, and the p functions are linearly independent. In fact, if

$$\sum C_i \frac{d}{dn_0} f_i(s) \equiv 0$$

then

$$\sum C_i f_i(s) \equiv 0.$$

Therefore, $\Phi(x)=\sum C_i f_i(x)$, with its derivative in the direction of the normal, are zero on the boundary S of Ω.

From the unique continuation theorem[†] for an elliptic operator of the second degree, we have $\Phi(x)\equiv 0$, therefore, $C_i=0\,(i=1, 2, \ldots, p)$. If S and the coefficients (in $\bar{\Omega}$) are sufficiently smooth, then $\psi_i(s)=\mathrm{d}f_i(s)/\mathrm{d}n_0$ is also sufficiently smooth. Accordingly, this condition, $(\varphi, \psi_i)S=0$ $(i=1, 2, \ldots, p)$, is necessary and sufficient for the existence of a solution if we are given a smooth $\varphi(s)$. Hence, the co-dimension of the space of φ which has at least one solution and the dimension of the eigenspace of A are the same. Also, from our observation, given a smooth $\varphi(s)$, we see that the necessary and sufficient condition for the existence of a solution is equivalent to the solvability of the Dirichlet problem in the form which we have treated so far.

15 *The boundary value problem of the third kind*

We have already explained the manner of treating this problem in the last half of section 12. We make some remarks: our original equation was

$$\sum_{i=1}^{n}\left(\frac{\partial u}{\partial x_i}, \frac{\partial \varphi}{\partial x_i}\right)+\left(\sum a_i(x)\frac{\partial u}{\partial x_i}+c(x)\,u, \varphi\right)-\int_S \sigma(x)\,u\bar{\varphi}\,\mathrm{d}S=(f, \varphi), \tag{3.91}$$

for $\varphi\in\mathscr{E}_{L^2}{}^1(\Omega)$. If we assume that the boundary S is smooth, then the solution $u\in\mathscr{E}_{L^2}{}^2(\Omega)$ by theorem 3.25. Accordingly, we have

$$\left.\begin{aligned} &A[u]\equiv -\Delta u+\sum a_i(x)\frac{\partial u}{\partial_i x}+c(x)\,u=f,\\ &\frac{\mathrm{d}u}{\mathrm{d}n}+\sigma(x)\,u=0. \end{aligned}\right\} \tag{3.91$'$}$$

Which problem is in a conjugate relation to this one? Let us consider this question. First, as we did when we dealt with the Dirichlet problem, if we take sufficiently large $t>0$, then we can find a unique solution $u\in\mathscr{E}_{L^2}{}^2(\Omega)$ satisfying

$$(A+tI)\,[u]=f(x), \qquad \frac{\mathrm{d}u}{\mathrm{d}n}+\sigma(x)\,u=0,$$

i.e. a Green's operator $G_t f$ is obtained. In general, is λ is a complex parameter, then, in (3.91)′, the solution of $(A+\lambda I)\,[u]=f$ can be obtained by

[†] See theorem 4.3 and section 14, chapter 6.

the operation of G_t on this equation as follows:

$$u+(\lambda-t)\,G_t u=G_t f.$$

As in the case of the Dirichlet problem, solving this equation in $L^2(\Omega)$ is equivalent to solving our original problem.

In this case, let us consider the problem concerning the conjugate operator G_t^* of G_t. For (3.91), we consider the equation in $\mathscr{E}_{L^2}{}^1(\Omega)$ as defined by

$$\sum_{i=1}^{n}\left(\frac{\partial v}{\partial x_i},\frac{\partial \varphi}{\partial x_i}\right)+\left(v,\sum a_i(x)\frac{\partial \varphi}{\partial x_i}\right)+(\overline{c(x)}\,v,\varphi)-\int_S \overline{\sigma(x)}\,v\bar{\varphi}\,\mathrm{d}S$$
$$=(g,\varphi)\quad(\varphi\in\mathscr{E}_{L^2}{}^1(\Omega)). \tag{3.106}$$

This can be obtained by putting (3.91) as $B[u,\varphi]$, (3.106) as $B_1[v,\varphi]$, and $B_1[u,\varphi]=\overline{B[\varphi,u]}$. As in section 12, we take $t(>0)$ sufficiently large, and add $t(v,\varphi)$ to the equation (3.106). The resulting equation has a unique solution v in $\mathscr{E}_{L^2}{}^1(\Omega)$ for an arbitrary $g\in L^2(\Omega)$ where also $v\in\mathscr{E}_{L^2}{}^2(\Omega)$ by theorem 3.25. Therefore v is a solution of

$$\begin{cases}(A^*+tI)\,[v]\equiv-\Delta v-\sum\dfrac{\delta\varphi}{\partial x_i}\,(\overline{a_i(x)}\,v)+(\overline{c(x)}+t)\,v=g(x),\\[2ex] \dfrac{\mathrm{d}v}{\mathrm{d}n}+\sigma'(x)\,v=0;\quad \sigma'(x)=\overline{\sigma(x)}+\displaystyle\sum_{i=1}^{n}\overline{a_i(x)}\cos\gamma_i,\end{cases} \tag{3.106$'$}$$

where γ_i is the angle between the inner normal and x_i-axis.

Now, we write $\tilde{G}_t$ for a Green's operator for (3.106)′. Then we can see $\tilde{G}_t=G_t^*$. In fact, if we put

$$N=\left\{\psi;\,\psi(x)\in\mathscr{E}_{L^2}{}^2(\Omega),\ \frac{\mathrm{d}\psi}{\mathrm{d}n}+\sigma'(x)\,\psi=0\right\}$$

and $A_t^*=A^*+tI$, and if we consider $(G_t f, A_t^*\psi)$ for $f\in L^2(\Omega)$, $\psi\in N$, then

$$(A_t G_t f,\psi)-(G_t f, A_t^*\psi)$$
$$=\int_S \frac{\mathrm{d}}{\mathrm{d}n}\,G_t f\cdot\bar{\psi}-G_t f\cdot\frac{\mathrm{d}}{\mathrm{d}n}\,\bar{\psi}\,\mathrm{d}S-\int_S\left(\sum a_i(x)\cos\gamma_i\right)G_t f\cdot\bar{\psi}\,\mathrm{d}S, \tag{3.107}$$

because of the fact that $G_t f\in\mathscr{E}_{L^2}{}^2(\Omega)$ and the property of a Green's operator.

From the fact that $\psi \in N$ and

$$\frac{\mathrm{d}}{\mathrm{d}n} G_t f + \sigma(x)\, G_t f = 0,$$

the right hand side term of (3.107) is zero. Therefore, from $A_t G_t f = f$, we have

$$G_t^* A_t^* \psi = \psi \quad (\psi \in N).$$

On the other hand, for an arbitrary $\varphi \in L^2(\Omega)$, we have $\psi = \tilde{G}_t \varphi \in N$ so that $G_t{}^* A_t{}^* \tilde{G}_t \varphi = \tilde{G}_t \varphi (\varphi \in L^2(\Omega))$. From $A_t{}^* \tilde{G}_t \varphi = \varphi$, we have $\tilde{G}_t{}^* \varphi = \tilde{G}_t \varphi \ (\varphi \in L^2(\Omega))$.

Summing up our results, we establish the fact that if we put $(A^* + \lambda I)\,[v] = g$ in (3.106)′, then this is equivalent to considering

$$v + (\lambda - t)\, G_t{}^* v = G_t{}^* g$$

in $L^2(\Omega)$. From this we see that if Ω is an interior domain of a closed hypersurface S, then Fredholm's theorem 3.21 can be established for (3.91)′ and (3.106)′. In particular, if (3.91)′ has a solution $u(x) \not\equiv 0$, then (3.106)′ has a solution $v(x) \not\equiv 0$ when $A^*[v] = 0$. Let us write such independent solutions as $g_1(x), \ldots, g_p(x)$. Then, the necessary and sufficient condition for the existence of solutions of (3.91)′ when we replace the right hand side by f is

$$(f(x), g_i(x)) = 0 \quad (i = 1, 2, \ldots, p).$$

Next, we consider the following:

PROBLEM. Given $\varphi(s)$, find a solution $u \in \mathscr{E}_{L^2}{}^2(\Omega)$ satisfying $A[u] = 0$, and

$$\frac{\mathrm{d}}{\mathrm{d}n} u + \sigma(x)\, u = \varphi(s).$$

The answer to this problem is similar to that in the previous section, i.e. if A has a Green's operator G, then, for $h \in \mathscr{E}_{L^2}{}^2(\Omega)$ such that

$$\frac{\mathrm{d}h}{\mathrm{d}n} + \sigma(s)\, h = \varphi(s),$$

$u(x) = (I - GA)h$ is a solution for the problem. If A has no Green's operator, and Ω is an interior domain, then for the same $g_i(x)$, the necessary and sufficient condition for the existence of a solution is

$$(Ah, g_i(x))=0 \quad (i=1, 2, \cdots, p).$$

From (3.107), we have

$$(Ah, g_i)-(h, A^*g_i)=\int_S \left\{\frac{\mathrm{d}}{\mathrm{d}n} h\cdot\bar{g}_i - h\cdot\frac{\mathrm{d}}{\mathrm{d}n}\bar{g}_i - (\sum a_i \cos\gamma_i)\, h\cdot\bar{g}_i\right\} \mathrm{d}S.$$

Also, from $A^*g_i=0$, and

$$\frac{\mathrm{d}}{\mathrm{d}n} g_i + (\sum \bar{a}_i \cos\gamma_i + \bar{\sigma})\, g_i = 0,$$

we have

$$(Ah, g_i)=\int_S \left(\frac{\mathrm{d}}{\mathrm{d}n} h + \sigma h\right) \bar{g}_i \, \mathrm{d}S.$$

Finally, we see that the necessary and sufficient condition for the equation to have at least one solution is

$$\int_S \varphi(s)\, \overline{g_i(s)}\, \mathrm{d}S = 0 \quad (i=1, 2, \ldots, p). \tag{3.108}$$

Furthermore, the independence of $g_i(s)$ $(s=1, 2, \ldots, p)$ as functions defined on S can be proved as in the previous section.

16 *Extension of self-adjoint operators*

We have treated Dirichlet problems and boundary value problems of the third kind. Our methods rely on 'Friedrichs' extension of self-adjoint operators' if A is formally self-adjoint. In this case, we have seen that the Green's operator G_t for $(A+tI)$ is a bounded Hermitian operator, so that $(A+tI)$ is self-adjoint and A is the same. On the other hand, when we treat a Hilbert space, the uniqueness of the extension of a given A such that A is a symmetric operator and the extension is a self-adjoint operator must be examined. If it is unique, then we say that the operator A is *essentially self-adjoint*.

We now assume that the boundary S of Ω is a closed and smooth hypersurface (interior or exterior is immaterial). In this case we have theorems 3.24, and 3.25. From these, we can conclude that there exists an appropriate constant C for the solution of a Dirichlet problem or a boundary value problem of the third kind, satisfying

$$\|u\|_2 \leqslant C(\|Au\|_{L^2(\Omega)} + \|u\|_{L^2(\Omega)}) \tag{3.109}$$

where,

$$u \in \mathscr{D}_{L^2}{}^1(\Omega) \cap \mathscr{E}_{L^2}{}^2(\Omega)$$

for a Dirichlet problem and

$$u \in N,\ N = \left\{u;\ u \in \mathscr{E}_{L^2}{}^2(\Omega),\ \frac{\mathrm{d}}{\mathrm{d}n}u + \sigma u = 0\right\}$$

for a boundary value problem of the third kind. We note that this inequality can be established irrespective of A being formally self-adjoint or not.

We prove this fact for a boundary value problem of the third kind. If we take $t(>0)$ sufficiently large, then,

$$(A+tI)[u] = f, \quad \frac{\mathrm{d}}{\mathrm{d}n}u + \sigma u = 0, \tag{3.91$''$}$$

where A is given by (3.91)$'$, we have a unique solution $u \in N$ for an arbitrary $f \in L^2(\Omega)$. Conversely, $(A+tI)u \in L^2$ for $u \in N$. On the other hand, N is a closed subspace of $\mathscr{E}_{L^2}{}^2(\Omega)$, so that it itself is a Hilbert space. Therefore, we have a *linear mapping*, $u \to (A+tI)u$ which defines a bijective continuous mapping N onto $L^2(\Omega)$. From theorem 2.5, the inverse is also continuous, i.e. for a certain positive constant C, we have

$$\|u\|_2 \leqslant C\|(A+tI)u\|_{L^2} \quad (u \in N). \tag{3.110}$$

From this we have (3.109). If Ω is an interior domain, and if $Au = 0\,(u \in N)$ implies $u = 0$, then we can omit $\|u\|_{L^2}$ in (3.110) which is obvious from our previous argument.

Using (3.109), we see that the extension of self-adjointness is uniquely determined. Let us explain this. First we observe

$$A[u] \equiv -\Delta u + c(x)u = f, \quad \frac{\mathrm{d}u}{\mathrm{d}n} + \sigma(x)u = 0, \tag{3.111}$$

where $c(x)$ and $\sigma(x)$ take real values. Further, we impose a condition on the domain $\mathscr{D}(A)$ of A such that it is of $C^2(\bar{\Omega})$-class and a function has a bounded support if Ω is an exterior domain satisfying the boundary value condition.

In this case, we see that $\mathscr{D}(A)$ is dense in $L^2(\Omega)$, and $\bar{A}(=A^{**})$, the closure of the graph of A is in fact the minimum closed extension of self-adjointness. $\bar{A}$ is also symmetric. The problem now is to prove $\bar{A}^*=\bar{A}$. In fact

$$\mathscr{D}(\bar{A})=N\equiv\left\{u\in\mathscr{E}_{L^2}{}^2(\Omega);\ \frac{\mathrm{d}u}{\mathrm{d}n}+\sigma u=0\right\}.$$

To prove this we first note that $\mathscr{D}(\bar{A})\subset N$ from (3.109). Conversely, $u\in N$ implies $u\in\mathscr{D}(\bar{A})$ as follows. Let $u\in N$ and let us pick a sequence

$$f_j\in C^\infty(\bar{\Omega})\to(A+tI)\,u \quad \text{in} \quad L^2(\Omega).$$

On the other hand, we can define u_j uniquely such that $(A+tI)\,[u_j]=f_j\,(u_j\in N)$. In this case, u_j is also sufficiently smooth by theorem 3.25, and $u_j\in\mathscr{D}(A)$. By (3.110), $\{u_j\}$ is a convergent sequence of $\mathscr{E}_{L^2}{}^2(\Omega)$, so that $u_j\to u_0$ in $\mathscr{E}_{L^2}{}^2(\Omega)$ $(u_0\in N)$. Now, from this, we have

$$(A+tI)\,u_j\to(A+tI)\,u_0 \quad \text{in} \quad L^2(\Omega).$$

Accordingly, it follows that

$$(A+tI)\,u=(A+tI)\,u_0, \qquad u,\,u_0\in N$$

and $u=u_0$, i.e. $u\in\mathscr{D}(\bar{A})$.

We must prove $\bar{A}^*=\bar{A}$. $\bar{A}^*\supset\bar{A}$ is always true, so that we have to prove $\bar{A}^*\subset\bar{A}$. Let $f\in\mathscr{D}(\bar{A}^*)$. Then, if we take $t>0$ sufficiently large, there exists h such that

$$(\bar{A}^*+tI)\,f=(\bar{A}+tI)\,h \qquad (h\in N\equiv\mathrm{D}(\bar{A})).$$

Incidentally, for $u\in\mathscr{D}(\bar{A})$ we have

$$((\bar{A}+tI)\,u,f-h)=(u,\,(\bar{A}^*+tI)\,f)-(u,\,(\bar{A}+tI)\,h)=0.$$

If u runs through $\mathscr{D}(\bar{A})\equiv N$, then $(\bar{A}+tI)u$ runs through the whole of $L^2(\Omega)$, so that $f=h$. Accordingly we see that $f\in\mathscr{D}(\bar{A})$. For the Dirichlet problem, the situation is exactly similar. Let us state the result which we have obtained.

THEOREM 3.27. Let Ω be either the interior or the exterior domain of a closed hypersurface S. If S, the coefficients of the equation, and $\sigma(x)$

are all sufficiently smooth, then the extension of self-adjointness for A for the boundary value problem of the third kind given by (3.111) is unique.†

The same property holds for the Dirichlet problem.

17 *The Dirichlet problem for an elliptic operator of higher order*

We have defined a strongly elliptic operator in section 12 (definition 3.6). Let us first recall the definition; we assume that δ in (3.87) can be taken to be independent of $x \in \Omega$ (uniform ellipticity). In this case, if $a_\alpha \in \mathscr{B}^m(\Omega)$, then the following lemma can be established:

LEMMA 3.14 (*Gårding*). There exists an appropriate constant $C(>0)$ and γ, with

$$\mathrm{Re}\,\langle Au, \bar{u}\rangle \geqslant C\|u\|_m{}^2 - \gamma\|u\|^2$$

for all $u \in \mathscr{D}_{L^2}{}^m(\Omega)$.

Proof. For $u, v \in \mathscr{D}_{L^2}{}^m(\Omega)$, it is obvious that

$$|\langle Au, \bar{v}\rangle| \leqslant C\|u\|_m\|v\|_m.$$

From the fact that $\mathscr{D}(\Omega)$ is dense in $\mathscr{D}_{L^2}{}^m(\Omega)$, it is sufficient to prove the inequality for $u \in \mathscr{D}(\Omega)$. Now from the argument of (3.88) we see if that η is taken sufficiently small, then for an arbitrary $x_0 \in \Omega$, and for $u \in \mathscr{D}(B\eta(x_0) \cap \Omega)$,

$$\mathrm{Re}(Au, u) \geqslant \tfrac{1}{2}\delta\|u\|_m{}^2, \tag{3.112}$$

where $B_\eta(x_0)$ is an open sphere with centre x_0, radius η, which can be taken independent of the position of x_0. Let us fix such an η. Let $\eta/\sqrt{n} = \eta'$, and let us enumerate the set of lattice-points $(m_1\eta', m_2\eta', \ldots, m_n\eta')$ $(m_i = 0, \pm 1, \pm 2, \ldots)$ as $\{x^{(i)}\}_{i=1,2,\ldots}$.

† The smoothness of S and the coefficients of the equation must be high enough in a L^2-space according to the dimension n. For example, to check this property we may examine the regularity of the solution using the idea of a potential theory. Alternatively, assuming that they belong to $\mathscr{D}(A)$, we may consider that they are elements of $\mathscr{E}_{L^2}{}^2(\Omega)$ satisfying

$$\frac{du}{dn} + \sigma(x)u = 0$$

on S in the sense of trace. In the author's view the latter way is probably more natural than the former.

In this case we consider the following decomposition of unit $\{\alpha_i(x)\}$, satisfying

$$\sum_i \alpha_i(x)^2 \equiv 1 \qquad (\alpha_i(x) \geqslant 0, \alpha_i \in C^\infty),$$
$$\operatorname{supp}[\alpha_i(x)] \subset B_\eta(x^{(i)}),$$

and

$$\sup_x |D^\nu \alpha_i(x)| \leqslant M(\alpha) < +\infty \qquad (i=1, 2, \ldots),$$

for $|\nu| \leqslant 2m$.

It is obvious that such a decomposition exists.

Now, for $u \in \mathscr{D}(\Omega)$, we have

$$\operatorname{Re}(Au, u) = \operatorname{Re} \sum_i (\alpha_i(x)\, Au, \alpha_i(x)\, u).$$

On the other hand,

$$\alpha_i Au = A(\alpha_i u) - \sum_{|\nu| \geqslant 1} \frac{D^\nu \alpha_i}{\nu!} \cdot A^{(\nu)} u \quad \text{(Leibniz)},$$

where

$$A^{(\nu)}(x, \xi) = D_\xi{}^\nu A(x, \xi).$$

Therefore,

$$(\alpha_i Au, \alpha_i u) = (A(\alpha_i u), \alpha_i u) - \sum_{|\nu| \geqslant 1} \frac{1}{\nu!} (\alpha_i{}^{(\nu)}(x)\, A^{(\nu)} u, \alpha_i(x)\, u).$$

From (3.112), we have

$$\operatorname{Re}\,(\alpha_i Au, \alpha_i u) \geqslant \tfrac{1}{2}\delta \|\alpha_i u\|_m{}^2 - C \|u\|_{m-1(B_i)} \|u\|_{m(B_i)}$$

where C is independent of i and

$$\|u\|^2{}_{s(B_i)} = \sum_{|\nu| \leqslant s} \|D^\nu u(x)\|^2{}_{L^2(B_i)} \qquad (B_i = B_\eta(x^{(i)})).$$

So that

$$C \|u\|_{m-1(B_i)} \|u\|_{m(B_i)} \leqslant \varepsilon \|u\|^2{}_{m(B_i)} + \frac{C}{2\varepsilon} \|u\|^2{}_{m-1(B_i)}.$$

Hence,

$$\operatorname{Re}\,(Au, u) \geqslant \tfrac{1}{2}\delta \|u\|_m{}^2 - \varepsilon \|u\|_m{}^2 - \gamma(\varepsilon) \|u\|^2,$$

where $\gamma(\varepsilon)$ is taken sufficiently large as ε becomes smaller, and $\varepsilon (>0)$ is arbitrary. Therefore the lemma is proved.

Using this lemma, we can treat the Dirichlet problem for a $2m$-order strongly elliptic operator A:

$$A[u] = f \qquad u \in \mathscr{D}_{L^2}{}^m(\Omega)$$

for $f \in \mathscr{D}'_{L^2}{}^m(\Omega)$ in just the same way as we have done for the case which we have treated in section 9. In fact, in this case, the above Gårding's inequality means

$$\langle (A+A^*)\, u, \bar{u} \rangle = 2\ \mathrm{Re} \langle Au, \bar{u} \rangle \geqslant 2C \|u\|_m{}^2 - 2\gamma \|u\|^2 \quad (u \in \mathscr{D}_{L^2}{}^m(\Omega)).$$

Chapter 4

Initial value problems (Cauchy problems)

1 *Introduction*

In this section we consider the Cauchy problem for a linear partial differential operator

$$a\left(x, \frac{\partial}{\partial x}\right) = \sum_{|\nu| \leqslant m} a_\nu(x) \left(\frac{\partial}{\partial x}\right)^\nu. \tag{4.1}$$

Let S be a hypersurface defined in an n-dimensional space R^n, and in a neighbourhood of $x_0 \in S$, we have for $x_0 \in S$

$$\varphi(x) = 0; \quad \varphi_x(x) \equiv \left(\frac{\partial \varphi}{\partial x_1}(x), \ldots, \frac{\partial \varphi}{\partial x_n}(x)\right) \neq 0. \tag{4.2}$$

Let $u_0, \ldots, u_{m-1}$ be functions defined on S, in a neighbourhood of x_0. In this case, let us call $\Psi = (u_0, u_1, \ldots, u_{m-1})$ *the initial data* (*Cauchy data, or initial value*) or, more precisely, *the initial data on S for a differential operator of order m.*

Ψ is said† *to be in* $\mathscr{E}^l$ if all $u_0, \ldots, u_{m-1}$ are in $\mathscr{E}^l$ ($l = 0, 1, 2, \ldots, +\infty$), and Ψ is called *analytic* if $\varphi(x)$ is an analytic function, and $u_0, \ldots, u_{m-1}$ are analytic functions defined on S.

Now, given $f(x)$ in the neighbourhood U of $x_0 (\in S)$, and given Ψ in a certain neighbourhood of x_0 on S, by a *Cauchy problem* for a differential operator we mean the problem of seeking a solution $u(x)$ in an appropriate neighbourhood $U' (\subset U)$ of x_0 such that it satisfies the equation

$$a\left(x, \frac{\partial}{\partial x}\right) u(x) = f(x), \tag{4.3}$$

with an initial condition:

$$\left.\begin{aligned} u(x) &= u_0(x) \quad (x \in S), \\ \frac{\partial}{\partial n} u(x) = u_1(x), \ldots, \left(\frac{\partial}{\partial n}\right)^{m-1} u(x) &= u_{m-1}(x) \quad (x \in S), \end{aligned}\right\} \tag{4.4}$$

where n is a unit normal to S.

† Of course, we assume that S is of C^s-class ($s \geqslant l$).

The main problems concerning this are the following:

(1) Does a solution $u(x)$ exist, and if so, what is its range?

(2) If a solution exists at all, is it unique?

(3) Do the solutions $u(x)$ for the initial data Ψ and the right hand term f have some form of continuity?

(4) Is there a solution u in the large?

In what follows we shall see that in general, the answers to (3) and (4) are negative unless there are some strong relations between $(x, \partial/\partial x)$ and S and some conditions on f and Ψ.

This is an intrinsic difference between this case and that of an ordinary differential equation. To make the situation clear we make a 'reduction' as follows: From (4.2), without loss of generality, we can assume $\varphi_{x_n} \neq 0$. Let $x_1' = x_1, \ldots, x_{n-1}' = x_{n-1}, x_n' = \varphi(x)$.

We perform the transformation of variables, $(x_1, \ldots, x_n) \to (x_1', \ldots, x_n')$. This transformation† and its inverse are smooth in a sufficiently small neighbourhood of x_0. In this case, if we consider

$$\frac{\partial}{\partial x_j} = \frac{\partial}{\partial x_j'} + \varphi_{x_j} \frac{\partial}{\partial x_n'} \quad (j=1, 2, \ldots, n-1), \qquad \frac{\partial}{\partial x_n} = \varphi_{x_n} \frac{\partial}{\partial x_n'},$$

then

$$h(x, \varphi_x(x)) \left(\frac{\partial}{\partial x_n'} \right)^m u + \sum \cdots = f \tag{4.5}$$

where

$$\varphi_x = \left(\frac{\partial \varphi}{\partial x_1}, \ldots, \frac{\partial \varphi}{\partial x_n} \right)$$

and

$$h(x, \xi) = \sum_{|\nu|=m} a_\nu(x) \xi^\nu \quad (\xi = (\xi_1, \ldots, \xi_n)).‡ \tag{4.6}$$

We note that the terms of $\sum$ in (4.5) contain partial derivatives of at most $(m-1)$-degree with respect to x_n'. We separate the cases as follows.

(1) If $h(x, \varphi_x) \neq 0$ in the neighbourhood of $x = x_0$, then we divide both sides of (4.5) by $h(x, \varphi_x)$, i.e.

$$\left(\frac{\partial}{\partial x_n'} \right)^m u + \sum_{|\nu| \leq m,\, \nu_n \leq m-1} a_\nu'(x') \left(\frac{\partial}{\partial x'} \right)^\nu u = f/h(x, \varphi_x). \tag{4.7}$$

† We do not mention the differentiability of the transformation here, but we assume a degree of differentiability such that the following arguments are valid.

‡ I.e. $h(x, \varphi_x) = \sum_{|\nu|=m} a_\nu(x) \varphi_{x_1}^{\nu_1} \ldots \varphi_{x_n}^{\nu_n}$ $(|\nu| = \nu_1 + \cdots + \nu_n)$.

We call (4.7) a *normal form* of (4.3). In this case, the initial condition can be written as

$$\left(\frac{\partial}{\partial x_n'}\right)^j u(x_1', \dots, x_{n-1}', 0) = u_j(x_1', \dots, x_{n-1}') \quad (j=0, 1, \dots, m-1).$$

(2) If $h(x, \varphi_x)=0$ at $x=x_0$, our Cauchy problem is rather difficult to solve.

For the present, we confine ourselves to case (1). First we explain some new ideas related to this classification.

DEFINITION 4.1 *(Characteristic hypersurface, and characteristic direction).* If the surface S defined by $\varphi(x)=0$ $(\varphi_x \neq 0)$ satisfies

$$h(x, \varphi_x(x))=0 \quad (x \in S),$$

then we call S a *characteristic hypersurface*† or *characteristic variety* for an operator $a(x, \partial/\partial x)$. Also, in this case we call $\xi\,(\in R^n)$ satisfying $\xi \neq 0$, and $h(x, \xi)=0$ a *characteristic direction* at x for an operator $a(x, \partial/\partial x)$.

From this definition, we see that if S is a characteristic hypersurface then all normals of arbitrary points on S are characteristic directions. The converse of this statement is also true. Related to these concepts is that of a characteristic cone. This is defined as follows.

For a fixed $x \in R^n$, the set of ξ satisfying $h(x, \xi)=0$ forms a cone with its apex at the origin. In this case we call the cone a *characteristic cone* at x for $a(x, \partial/\partial x)$.

We are not concerned with the problem of the existence of a solution for a given initial value Ψ in case (2); we are only interested in case (1).

2 *The Cauchy–Kowalewski and Holmgren theorems*

Let us demonstrate the local existence of the solution for the Cauchy problem given by (4.3) with (4.4) in section 1 if $h(x, \varphi_x) \neq 0$, and if the coefficients Ψ, and f are all analytic. By our assumptions, we can transform the equation (4.3) into a normal form (see (4.7)), so that it is sufficient to demonstrate the existence of the solution for the normalized equation obtained in this way.

† For brevity, we say simply 'characteristic surface'.

First we modify our notation as follows. Let us represent a point belonging to R^{n+1} as $(x, t)=(x_1, \ldots, x_n, t)$ and a point belonging to R^n as $x=(x_1, \ldots, x_n)$. Also, let

$$L \equiv \left(\frac{\partial}{\partial t}\right)^m + \sum_{|\nu|+j \leqslant m,\, j \leqslant m-1} a_{\nu, j}(x, t) \left(\frac{\partial}{\partial x}\right)^{\nu} \left(\frac{\partial}{\partial t}\right)^{j} \tag{4.8}$$

be an operator in normal form.

THEOREM 4.1 *(Cauchy–Kowalewski).* Let the coefficients of L be well-defined in a neighbourhood U of the origin of the space (x, t). Let us assume that they are also analytic. Assume f is also analytic in U. Let Ψ be an initial value which is analytic in a certain neighbourhood V of the origin of x-space. Then, there exists a neighbourhood W of the origin and a unique analytic function $u(x, t)$ defined on W, and

$$\left.\begin{aligned} &Lu=f \quad (x, t) \in W, \\ &\left(\frac{\partial}{\partial t}\right)^j u(x, 0)=u_j(x, 0), \quad x \in W \cap \{t=0\} \ (j=0, 1, 2, \ldots, m-1). \end{aligned}\right\} \tag{4.9}$$

Proof.† We put

$$\tilde{u}(x, t)=u(x, t)-\sum_{j=0}^{m-1} \frac{t^j}{j!} u_j(x).$$

Then, we can write (4.9) as

$$L[\tilde{u}]=f-\sum_{j=0}^{m-1} L\left[\frac{t^j}{j!} u_j(x)\right].$$

Without loss of generality, we can assume that the initial datum of $u(x, t)$ is 0. First, if $u(x, t)$ is analytic in a neighbourhood of the origin, then $u(x, t)$ can be uniquely determined. In fact, we can prove that the Taylor expansion of $u(x, t)$ at the origin is uniquely determined.

To prove this we write (4.9)

$$\left(\frac{\partial}{\partial t}\right)^m u(x, t)=\sum_{j=0}^{m-1} \alpha_j\left(x, t; \frac{\partial}{\partial x}\right)\left(\frac{\partial}{\partial t}\right)^j u(x, t)+f(x, t), \tag{4.10}$$

where $\alpha_j(x, t; \xi)$ is a polynomial of $(m-j)$-degree whose coefficients are analytic in the neighbourhood U of the origin.

† See Leray: *Hyperbolic equations* (Princeton lecture note), 1954.

Because the initial datum of $u(x, t)$ is 0 by our assumption, if we consider

$$u(x, t)=\sum_{j\geqslant m;\, \nu} c_{\nu, j}x^{\nu}t^{j} \tag{4.11}$$

and the Taylor expansion of the coefficients of $\alpha_j(x, t; \xi)$, then $c_{\nu, j}$ is uniquely determined.

In this case $c_{\nu, j}$ can be represented as a polynomial with:

(1) The coefficients of the Taylor expansion of an analytic function which appear as the coefficients of the polynomials $\alpha_0(x, t; \xi), \ldots, \alpha_{m-1}(x, t; \xi)$.

(2) The coefficients of the Taylor expansion of $f(x, t)$.

(3) The $c_{\nu, l}$ $(l\leqslant j-1)$.

Therefore, by a recursive process with respect to j, we can conclude that $c_{\nu, j}$ can be represented as a polynomial with the positive coefficients in (1) and (2).

Next, we observe the convergence of the sequence (4.11), to do this, we use a majorant series, which we define as follows. By saying that $F(x, t)$ is a *majorant series* of $f(x, t)$ at the origin we mean that $F(x, t)$ is analytic at the origin and the coefficients $C_{\nu, j}$ of the Taylor expansion of $F(x, t)$ are greater than or equal to the absolute value of the corresponding coefficients $c_{\nu, j}$ of the Taylor expansion of $f(x, t)$, i.e. $C_{\nu, j}\geqslant |c_{\nu, j}|$.

We extend the definition to the case of a differential polynomial. By saying that $A(x, t; \xi)$ is a *majorant series* of $\alpha(x, t; \xi)$ we mean that for each function which appears as a coefficient of ξ^{ν}, the function corresponding to A is a majorant series of the function corresponding to α in the sense defined above. For (4.10), we define appropriate majorant series $A_j(x, t; \xi)$, $F(x, t)$ of $\alpha_j(x, t; \xi)$, $f(x, t)$ respectively.

Consider the equation

$$\left(\frac{\partial}{\partial t}\right)^{m} w=\sum_{i=0}^{m-1} A_j\left(x, t; \frac{\partial}{\partial x}\right)\left(\frac{\partial}{\partial t}\right)^{j} w+F(x, t). \tag{4.12}$$

Now our line of proof is as follows. If the solution w of the equations which have a positive Cauchy datum† is analytic in the neighbourhood of the origin, then w is a majorant series of u. So that $u(x, t)$ is analytic at the origin, therefore, the existence of the solution of the equation is proved.

† By saying that a *Cauchy datum is positive* we mean that all

$$w(x, 0), \frac{\partial}{\partial t} w(x, 0), \ldots, \frac{\partial^{m-1}}{\partial t^{m-1}} w(x, 0)$$

have Taylor expansions with non-negative coefficients.

Now, we assume that all the coefficients of $\alpha_j(x, t; \xi)$ and $f(x, t)$ are analytic in $|x_i| \leqslant r (i=1, 2, \ldots, n)$, $|t| < r$. Let all the coefficients of $\alpha_j(x, t; \xi)$ $< M$, and let $f(x, t) < d$. In this case all the coefficients of $\alpha_j(x, t; \xi)$ have majorant series

$$\frac{M}{(1-x_1/r)\cdots(1-x_n/r)\,(1-t/r)},$$

or

$$\frac{M}{1-(x_1+\cdots+x_n+t)/r}.$$

Introducing $\rho\,(>1)$, we write

$$\frac{M}{1-(x_1+\cdots+x_n+\rho t)/r}$$

as the majorant series. For f,

$$\frac{d}{1-(x_1+\cdots+x_n+\rho t)/r}$$

is one of the majorant series, where $\rho > 1$ (which we shall determine later). Therefore, there are appropriate polynomials $b(\xi_0, \xi_1, \ldots, \xi_n)$, $c(\xi_0, \xi_1, \ldots, \xi_n)$ and

$$\left(\frac{\partial}{\partial t}\right)^m w = \frac{1}{1-(x_1+\cdots+x_n+\rho t)/r} \times \left[b\left(\frac{\partial}{\partial t}, \frac{\partial}{\partial x_1}, \ldots, \frac{\partial}{\partial x_n}\right) w + c\left(\frac{\partial}{\partial t}, \ldots, \frac{\partial}{\partial x_n}\right) w + d \right] \tag{4.13}$$

satisfies (4.12), where b is a polynomial of mth degree and $b(1, 0, \ldots, 0) = 0$, and c is a polynomial of $(m-1)$th degree.†

Let us look for a solution w such that w is representable as a function of only one variable

$$s = (x_1+\cdots+x_n+\rho t)/r \quad (|s| < 1).$$

† For example, we can take $b(\xi_0, \ldots, \xi_n) = M \sum_{|\nu|=m} \xi_0^{\nu_0} \ldots \xi_n^{\nu_n} - M\xi_0^m$, $c(\xi_0, \ldots, \xi_n) = M \sum_{|\nu| \leqslant m-1} \xi^\nu$.

In this case (4.13) is an ordinary differential equation

$$\left(\frac{\rho}{r}\right)^m w^{(m)}(s)=\frac{1}{1-s}\left[\frac{b(\rho)}{r^m}w^{(m)}(s)+c\left(\frac{\rho}{r}\frac{\mathrm{d}}{\mathrm{d}s},\frac{1}{r}\frac{\mathrm{d}}{\mathrm{d}s},\ldots,\frac{1}{r}\frac{\mathrm{d}}{\mathrm{d}s}\right)w+d\right],$$

where $b(\rho)=b(\rho, 1,\ldots, 1)$ and b is at most of $(m-1)$th degree so that this can be written as

$$w^{(m)}(s)=\left(\frac{r}{\rho}\right)^m\frac{1}{1-b(\rho)\,\rho^{-m}-s}\left[c\left(\frac{\rho}{r}\frac{\mathrm{d}}{\mathrm{d}s},\ldots,\frac{1}{r}\frac{\mathrm{d}}{\mathrm{d}s}\right)w+d\right]. \qquad (4.14)$$

We now choose ρ sufficiently large such that $b(\rho)\rho^{-m}<1$.

As is well known in the theory of ordinary linear differential equations, (4.14) with the initial value $w^{(j)}(0)=0$ $(j=0, 1,\ldots, m-1)$, always has a unique regular solution in $|s|<1-b(\rho)\rho^{-m}$. Obviously $w(s)$ has a Taylor expansion with positive coefficients. Therefore, if we consider

$$w(x, t)\equiv w(s)=w((x_1+\cdots+x_n+\rho t)/r),$$

(4.13) has a solution $w(x, t)$ which has positive Taylor coefficients convergent in

$$\sum_{i=1}^{n}|x_i|+\rho|t|<r\{1-b(\rho)\,\rho^{-m}\}. \qquad (*)$$

Hence, (4.11) itself is an expansion which is convergent in $(*)$. Hence, it is a solution in the neighbourhood of the origin.

Q.E.D.

Note. Let us consider the problem 'How can we determine the range W of existing $u(x, t)$?' From $(*)$, it is easy to see that W depends on the radius of convergence r of the coefficients $a_{\nu, j}(x, t)$ of L and $f(x, t)$. Also, it depends on the maximum value M of the absolute value of $a_{\nu, j}$, but it *does not* depend on the maximum value d of $|f(x, t)|$. From these facts, we can see that W depends only on M and not on the radius r if the initial value Ψ is a polynomial and $L[u]=0$. This is important: we shall use this fact in order to prove Holmgren's theorem (theorem 4.2).

Let the coefficients of L be analytic (this includes the case of constant coefficients), and let Ψ and f be C^∞-functions instead of analytic functions. Can we still establish the above local existence theorem? In fact, in general the answer to this question is in the negative, except in the case when L is an operator of a certain type (hyperbolic operator) as we shall see later. If

L is hyperbolic, then we can proceed further. For example, adding some conditions, we can even prove a global existence theorem of solution (chapter 6).

Now, if u_1 and u_2 are analytic solutions of (4.9) for the same initial value Ψ, and f, then from theorem 4.1 $u_1 = u_2$ in some neighbourhood of the origin. On the other hand, if u_i is not analytic, theorem 4.1 gives no information. Holmgren has proved that even in this case if L has analytic coefficients, then the uniqueness of the solution can be estabilished. This is an interesting fact because in other words we can claim that for the equation $Lu = f$, if f, Ψ are not analytic, then $u(x)$ is not analytic, but, nevertheless, the solution of the equation is unique.

More precisely, we can state:

THEOREM 4.2 *(Holmgren)*. Let us write

$$D_\varepsilon = \{(x, t) \in R^{n+1};\ |x|^2 + |t| < \varepsilon\}.$$

Let all the coefficients $a_{\nu, j}(x, t)$ of L be analytic in the neighbourhood U of the origin. Then, there exists $\varepsilon_0\,(>0)$ such that for ε satisfying $0 < \varepsilon < \varepsilon_0$ if $u(x, t) \in \mathscr{E}^m(D_\varepsilon)$ satisfies

$$Lu = 0 \quad \text{in} \quad D_\varepsilon$$

$$\left(\frac{\partial}{\partial t}\right)^j u(x, 0) = 0 \quad (j = 0, 1, \ldots, m-1), \quad x \in D_\varepsilon \cap \{t = 0\},$$

then $u(x, t) \equiv 0$ in D_ε.

Note. The uniqueness of the solution is valid in a half-space. In this case, add an extra assumption that $u(x, t)$ is defined on $D_\varepsilon' = D_\varepsilon \cap \{t \geqslant 0\}$ and is m-times continuously differentiable including the boundary.

Proof. Let us define a special type of transformation which is useful in what follows. By a *Holmgren transformation* we mean a transformation which is defined by $x_j' = x_j\,(j = 1, 2, \ldots, n)$, $t' = t + x_1^2 + \cdots + x_n^2$, and which maps the half-space $t \geqslant 0$ into a domain $\Omega = \{(x', t') \in R^{n+1};\ t' - |x'|^2 \geqslant 0\}$. We note that the function $u(x', t')$ becomes 0 on the hypersurface, $t' - |x'|^2 = 0$ after the transformation including its mth derivative.

Therefore, if we extend this function u by putting its value equal to zero outside Ω, then the extended function has its support in Ω and is an $\mathscr{E}^m$-class function. The differential operators are transformed into other differential operators in a sufficiently small neighbourhood; we write (x, t) instead of (x', t'). Then, we have a new equation†

$$L[u] \equiv \left(\frac{\partial}{\partial t}\right)^m u + \sum a_{\nu,j}(x,t) \left(\frac{\partial}{\partial x}\right)^\nu \left(\frac{\partial}{\partial t}\right)^j u = 0$$

where the coefficients are analytic, and the support of u is contained in Ω.

Let us now define an important operator. By the *transposed operator* ${}^tL$ of L we mean an operator satisfying

$${}^tL[u] \equiv (-1)^m \left(\frac{\partial}{\partial t}\right)^m u + \sum (-1)^{|\nu|+j} \left(\frac{\partial}{\partial x}\right)^\nu \left(\frac{\partial}{\partial t}\right)^j [a_{\nu,j}(x,t)\, u]. \tag{4.15}$$

In general, if $v(x, t)$, an m-times continuously differentiable function, is defined in the neighbourhood of $D \equiv \Omega \cap \{0 \leqslant t \leqslant h\}$ and satisfies ${}^tL[v] = 0$, then

$$\int_D \{u\,{}^tL[v] - vL[u]\}\, \mathrm{d}x\, \mathrm{d}t = 0.$$

Furthermore, if $v(x, t)$ satisfies

$$v(x,h) = \frac{\partial}{\partial t} v(x,h) = \cdots = \left(\frac{\partial}{\partial t}\right)^{m-2} v(x,h) = 0$$

on the hyperplane $t = h$, then integrating by parts we have

$$\int_D \{u\,{}^tL[v] - vL[u]\}\, \mathrm{d}x\, \mathrm{d}t = \int_{t=h} (-1)^m u(x,t) \left(\frac{\partial}{\partial t}\right)^{m-1} v(x,t)\, \mathrm{d}x = 0.$$

We note that here ${}^tL$ is of the same form as (4.8), and the coefficients of $v(x, t)$ are analytic.

Therefore, from theorem 4.1 and the subsequent note, if we add the condition

$$\left(\frac{\partial}{\partial t}\right)^{m-1} v(x,h) = P(x),$$

where $P(x)$ is a polynomial, to the already imposed initial condition, then

† Of course, $a_{\nu,j}(x, t)$ is also to be understood as a new coefficient.

the solution of the equation $'L[v]=0$ exists in the neighbourhood of $D\equiv\Omega\cap\{0\leqslant t\leqslant h\}$ for all h which satisfy $0<h<h_0$, $h_0>0$, and h_0 is sufficiently small. h_0 can be chosen independent of the polynomial $P(x)$. From this, it follows that $\int u(x,h)\,P(x)\mathrm{d}x=0$ for an arbitrary polynomial $P(x)$, i.e. $u(x,h)$, a continuous function having a compact support with regard to x, is orthogonal with an arbitrary $P(x)$.

Then, we see that by the Weierstrass theorem (theorem 1.13) $u(x,h)=0$, so that $u(x,t)\equiv 0$, in $(x,t)\in\Omega\cap\{0\leqslant t\leqslant h_0\}$. If $t\leqslant 0$, then we replace t by $-t$.

Q.E.D.

In Holmgren's theorem we assume that the coefficients are all analytic. There were age-old conjectures about this condition, namely, we can establish the same theorem if the coefficients are of C^∞-class, or if the operator is of an elliptic type, but Pliś discovered counter examples to deny these conjectures.† Nevertheless, the study of sufficient conditions has been carried out by many mathematicians.

Before closing this section we mention without proof just one of the remarkable results which were obtained by Calderón in 1958. When (x,t) is two-dimensional, Carleman obtained the same results as early as 1938.

THEOREM 4.3 *(Calderón)*. Let all the coefficients of the highest order $a_{\nu,j}(x,t)$ $(|\nu|+j=m)$ of the operator L in (4.8) be real-valued and of $C^{1+\sigma}$ $(\sigma>0)$ in the neighbourhood U of the origin‡ (σ is an arbitrary positive number), and let the other coefficients be bounded in U. In this case if the characteristic equation of L at the origin

$$p(\lambda,\xi)=\lambda^m+\sum_{|\nu|+j=m} a_{\nu,j}(0,0)\,\xi^\nu\lambda^j=0 \tag{4.16}$$

has roots $\lambda_i(\xi)(i=1,2,\ldots,m)$ which are distinct for all ξ, where ξ is a real number $\neq 0$, then the conclusion of theorem 4.2 is valid.

For the proof see section 14, chapter 6.

3 *Notes on the solubility of the Cauchy problem*

In the previous section we posed the question whether the Cauchy–

† A. Pliš: 'The problem of uniqueness of the solution of a system of partial differential equations', *Bull. de l'Acad. Polonaise des Sci.* 2 (1954), 55–7.

‡ I.e. the first derivatives have the property of σ-Hölder continuity.

Kowalewski theorem could be established for the initial value Ψ, a C^∞-class function and for f. We shall examine this problem in this section. Let our equation be

$$L[u]=\left(\frac{\partial}{\partial t}\right)^m u+\sum_{\substack{|\nu|+j\leqslant m\\ j\leqslant m-1}} a_{\nu,j}(x,t)\left(\frac{\partial}{\partial x}\right)^\nu\left(\frac{\partial}{\partial t}\right)^j u=f. \tag{4.17}$$

For simplicity, we assume that the coefficients are defined in a neighbourhood of the origin, and also $a_{\nu,j}(x,t)\in\mathscr{E}(U)$.

Definition 4.2 *(Local solubility)*. By saying that the equation (4.17) is *locally soluble* at the origin in $\mathscr{E}$, we mean that for an arbitrary $f(x,t)\in\mathscr{E}(U)$ and for an initial value $\Psi\equiv(u_0(x),\ldots,u_{m-1}(x))\in\mathscr{E}_x$, there exists a neighbourhood $D_{(f,\Psi)}$ of the origin, $u(x,t)\in\mathscr{E}(D_{(f,\Psi)})$ satisfying $L[u]=f$ for $(x,t)\in D_{(f,\Psi)}$ and

$$\left(\frac{\partial}{\partial t}\right)^j u(x,0)=u_j(x),\quad x\in D_{(f,\Psi)}\cap\{t=0\}\quad (j=0,1,\ldots,m-1).$$

The condition we imposed above is in a rather strong form; we can employ a definition weaker than this as follows.

Definition 4.2′. By saying that the equation (4.17) is *locally soluble* at the origin in $\mathscr{E}^m$, we mean that for an arbitrary $f(x,t)\in\mathscr{E}(U)$ and for an initial value $\Psi\in\mathscr{E}_x$, there exists a neighbourhood $D_{(f,\Psi)}$ and a function u m-times continuously differentiable in $D_{(f,\Psi)}\cap\{t>0\}$ and satisfying $Lu=f$, and in $D_{(f,\Psi)}\cap\{t\geqslant 0\}$ u is $(m-1)$-times continuously differentiable satisfying the initial condition at $t=0$.

There are some examples of equations which are not locally soluble in the sense of definition 4.2′. The following example is due to Hadamard.

Example. The Cauchy problem for the equation

$$\Delta u(x,y,z)\equiv\frac{\partial^2}{\partial x^2}u+\frac{\partial^2}{\partial y^2}u+\frac{\partial^2}{\partial z^2}u=0$$

with an initial value Ψ on the plane $z=0$ is not locally soluble at the origin.

To show this, we consider as an initial value Ψ

$$\left.\begin{aligned} &u(x, y, 0)=u_0(x, y)\in\mathscr{E}, \\ &\frac{\partial}{\partial z}u(x, y, 0)=0. \end{aligned}\right\}$$

Further, we assume that a solution $u(x, y, z)$ exists for the equation in the sense of definition 4.2′, and u is defined on $B_\delta=\{(x, y, z);\ x^2+y^2+z^2<\sigma^2\}$ for a sufficiently small σ, and also in $z\geqslant 0$.

In this case, if we put

$$\tilde{u}(x, y, z)=\begin{cases} u(x, y, z) & (z\geqslant 0), \\ u(x, y, -z) & (z<0), \end{cases}$$

then $\tilde{u}$ is a function defined on B_δ satisfying $\Delta\tilde{u}=0$ in $B\sigma$ in the sense of distribution. Therefore, $\tilde{u}(x, y, z)$ is an analytic function of (x, y, z) in B_δ as we shall prove later. $\tilde{u}(x, y, 0)$ $(\equiv u_0(x, y))$ is also an analytic function of (x, y).

Finally, from this, for $u_0(x, y)$ if we pick a function such that its restriction on any neighbourhood of the origin is not analytic, then the corresponding Cauchy problem for this function has no local solution u at the origin.

Let us prove $\Delta\tilde{u}=0$ in B_δ. This is nothing but the classical theorem of the Schwartz principle of reflection. In the language of distribution, the meaning of this is: let $\varphi\in\mathscr{D}(B_\delta)$, and let

$$\begin{aligned} \langle\tilde{u}, \Delta\varphi\rangle &= \int \tilde{u}(x, y, z)\, \Delta\varphi(x, y, z)\, dx\, dy\, dz \\ &= \lim_{\varepsilon\to 0}\left\{-\int_{|z|\geqslant\varepsilon} \frac{\partial\tilde{u}}{\partial z}\frac{\partial\varphi}{\partial z}\, dx\, dy\, dz + \int_{|z|\geqslant\varepsilon}\left(\frac{\partial^2\tilde{u}}{\partial x^2}+\frac{\partial^2\tilde{u}}{\partial y^2}\right)\varphi\, dx\, dy\, dz\right\}. \end{aligned}$$

From

$$-\int_{|z|\geqslant\varepsilon} \frac{\partial\tilde{u}}{\partial z}\frac{\partial\varphi}{\partial z}\, dx\, dy\, dz = \int\left[\frac{\partial\tilde{u}}{\partial z}\cdot\varphi\right]_{z=-\varepsilon}^{z=\varepsilon} dx\, dy + \int_{|z|\geqslant\varepsilon}\frac{\partial^2\tilde{u}}{\partial z^2}\varphi\, dx\, dy\, dz,$$

we have

$$\begin{aligned} \langle\tilde{u}, \Delta\varphi\rangle = \lim_{\varepsilon\to 0}\bigg\{&\int \frac{\partial\tilde{u}}{\partial z}(x, y, \varepsilon)\, \varphi(x, y, \varepsilon)\, dx\, dy \\ &-\int \frac{\partial\tilde{u}}{\partial z}(x, y, -\varepsilon)\, \varphi(x, y, -\varepsilon)\, dx\, dy\bigg\}=0. \end{aligned}$$

Then we can see that $\tilde{u}(x, y, z)\in\mathscr{E}(B_\delta)$ by the corollary of theorem 3.22, chapter 3 (*interior regularity*), i.e. $\tilde{u}$ is an analytic and harmonic function.

To see this, we prove the following fact: an arbitrary harmonic function defined on an open $\Omega \subset R^n$ is analytic. First, as we have shown in (2.20), chapter 2, we have the fact that

$$E(x) = -\frac{1}{(n-2)\, S_n} \cdot \frac{1}{|x|^{n-2}},$$

satisfies $\Delta R(x) = \delta$ where S_n is the surface area of the unit sphere of n-dimensional space ($n > 2$). We fix a point taken from Ω. Let us take $\alpha \in \mathscr{D}(\Omega)$ such that $\alpha(x) \equiv 1$ in the neighbourhood U of this fixed point. Let us write

$$\Delta(\alpha u) = \alpha(\Delta u) + 2 \sum \frac{\partial \alpha}{\partial x_i} \frac{\partial u}{\partial x_i} + \Delta \alpha \cdot u$$

and the right hand side $= g(x)$. Then, $g \in \mathscr{D}(\Omega)$.

We convolve $E(x)$ on both sides of the equation to get

$$E(x) * \Delta(\alpha u) = \alpha u = E(x) * g(x) = -\frac{1}{(n-2)\, S_n} \int \frac{g(y)}{|x-y|^{n-2}}\, \mathrm{d}y$$

where we have used the property $E(x) * \Delta(\alpha u) = (\Delta E(x)) * (\alpha u) = \delta * (\alpha u)$ (proposition 2.14 or proposition 2.17)†

Now, we take $\delta\,(>0)$ sufficiently small and consider $\alpha(x)$ within $|x - x_0| < \delta$. Then, we see $\alpha(x) \equiv 1$. Also, $g(x)$ has no support in the same neighbourhood, so that

$$u(x) = -\frac{1}{(n-2)\, S_n} \int_{|y-x_0| \geqslant \delta} \frac{g(y)}{|x-y|^{n-2}}\, \mathrm{d}y \qquad (|x - x_0| < \delta).$$

From this we see that $u(x)$ is analytic in the neighbourhood of x_0, where x_0 is arbitrary. Hence, $u(x)$ is analytic in Ω. If $n = 2$, we use

$$E(x) = -\frac{1}{2\pi} \log \frac{1}{|x|}$$

to obtain a similar result.

We shall close this section with the following remark on the existence of a global solution. We give an example which shows the fact that if the coefficients of the operator L, f and Ψ are analytic, then by the Cauchy–Kowalewski theorem there exists a local solution, although, in general, the local solution can not be extended to a global one.

† Here δ is Dirac's δ-measure.

EXAMPLE *(Hadamard)*. Let $L\equiv(\partial^2/\partial x^2)+(\partial^2/\partial y^2)$ in R^2. In this case,

$$u(x, y)=\mathrm{Re}\,\frac{1}{z-a}=\frac{x-a}{(x-a)^2+y^2} \qquad (a>0)$$

satisfies $L[u]=0$. Obviously, the initial value $u(0, y)$ and $(\partial/\partial x)\,u(0, y)$ are analytic with respect to y. But, because the singularity of the solution at $(a, 0)$, we cannot regard the solution as a global one for $L[u]=0$ in $x\geqslant 0$. On the other hand, the solution of $\Delta u=0$ is analytic, so that the solution for $L[u]=0$ with the initial value given at $x=0$ is *not* a solution in a half-space, $x\geqslant 0$.

4 *Local solubility of a Cauchy problem*

In the previous section, we did show that the solubility of a Cauchy problem can not be established if $L=\Delta$. In this section, on the other hand, we shall obtain a necessary condition of the solubility from general view-point. In the sequel, 'local solubility' should be understood in the sense of definition 4.2. First, we give a result obtained by Lax.†

PROPOSITION 4.1. Let the coefficients of (4.17) be analytic. Let us assume that a Cauchy problem concerning L is locally soluble at the origin. In this case, there exists a $\delta(>0)$, and for an arbitrary initial value $\Psi\equiv(u_0(x),\dots, u_{m-1}(x))\in\mathscr{E}_x$, there exists a unique solution $u(x, t)\in\mathscr{E}^m(D_\delta)$ satisfying

$$Lu=0 \quad\text{in}\quad D_\delta, \left(\frac{\partial}{\partial t}\right)^j u(x, 0)=u_j(x), \quad x\in D_\delta\cap\{t=0\}.$$

Note. We have given the definition of D_δ before Holmgren's theorem, i.e. $D_\delta=\{(x, t)\in R^{n+1};\ |x|^2+|t|<\delta\}$. In the definition of local solubility, each range of an existing solution depends on an initial Ψ. On the other hand the above lemma says that all u corresponding to Ψ share a common domain.

Proof. By Holmgren's theorem (4.2), there exists $\varepsilon_0(>0)$, for $\varepsilon<\varepsilon_0$, a solution satisfying

$$u(x, t)\in\mathscr{E}^m(D_\varepsilon) \quad\text{and}\quad Lu=0, \quad \left\{\left(\frac{\partial}{\partial t}\right)^j u(x, 0)\right\}=\Psi(x) \quad\text{in}\quad x\in D_\varepsilon\cap\{t=0\}$$

† Lax: 'Asymptotic solutions of oscillatory initial value problems', *Duke Math. J.* 24 (1957), 627–646.

is uniquely determined. Now, we pick a sequence $\varepsilon_0>\varepsilon_1>\varepsilon_2>\cdots\to 0$ and separate $\Psi(x)\in\prod_{i=1}^{m}\mathscr{E}_x(R^n)$ as a sequense of sets $\{A_{k,p}\}_{k=0,1,2\ldots;\ p=1,2,3,}$ in the following manner: $\Psi(x)\equiv(u_0(x),\ldots,u_{m-1}(x))\in A_{k,p}$ if and only if:

(1) there exists $u(x,t)\in\mathscr{E}_{L^2}{}^{[\frac{1}{2}n]+1+m}(D_{\varepsilon_k})$ such that

$$\left\{\left(\frac{\partial}{\partial t}\right)^j u(x,0)\right\}=\Psi(x),\quad x\in D_{\varepsilon_k}\cap\{t=0\};\quad Lu=0\ \text{ in }\ D_{\varepsilon_k},$$

(2) $\|u(x,t)\|_{[\frac{1}{2}n]+1+m}\leqslant p$.

We see that $A_{k,p}$ is symmetric, i.e. if $\Psi\in A_{k,p}$ then $-\Psi\in A_{k,p}$, and convex. From the assumption of local solubility we have $\bigcup_{k,p} A_{k,p}=\prod\mathscr{E}_x(R^n)$. $A_{k,p}$ is also closed (we prove this below). Let $\Psi_j\in A_{k,p}\to\Psi_0$ in $\prod\mathscr{E}_x(R^n)$. $u_j(x,t)$ corresponding to Ψ_j form a bounded set of $\mathscr{E}_{L^2}{}^{[\frac{1}{2}n]+1+m}(D_{\varepsilon_k})$.

Therefore, there exists a subsequence u_{j_p} which is weakly convergent. Furthermore, by the corollary of Theorem 3.7, u_{j_p} can be assumed to be a convergent sequence of $\mathscr{E}_{L^2(\text{loc})}{}^{[\frac{1}{2}n]+m}(D_{\varepsilon_k})$ (we pick a subsequence if necessary). So that if we write the limit of the weak convergence of u_{j_p} as u_0, then we see that $u_0(x,t)\in\mathscr{E}_{L^2}{}^{[\frac{1}{2}n]+1+m}(D_{\varepsilon_k})$ and $\|u_0\|_{[\frac{1}{2}n]+1+m}\leqslant p$ (see lemma 2.5). Also $u_{j_p}(x,t)\to u_0(x,t)$ in $\mathscr{E}_{L^2(\text{loc})}{}^{[\frac{1}{2}n]+m}(D_{\varepsilon_k})$. To see that $A_{k,p}$ is closed, we proceed as follows: From $Lu_{j_p}=0$, for an arbitrary $\varphi\in\mathscr{D}(D_{\varepsilon_k})$ we have $\langle Lu_{j_p},\varphi\rangle=\langle u_{j_p},{}^tL\varphi\rangle=0$ so that $\langle u_0,{}^tL\varphi\rangle=0$. By Sobolev's lemma, we have

$$u_{j_p}(x,0)\to u_0(x,0)\ \text{ in }\ \mathscr{E}^{m-1}(D_{\varepsilon_k}\cap\{t=0\}).$$

Hence,

$$\left\{\left(\frac{\partial}{\partial t}\right)^j u_0(x,0)\right\}=\Psi_0(x),\quad x\in D_{\varepsilon_k}\cap\{t=0\},$$

i.e. $\Psi_0\in A_{k,p}$, so that $A_{k,p}$ is a closed set.

By Baire's category theorem (theorem 2.2), one of $A_{k,p}$, A_{k_0,p_0} contains an open sphere of $\prod\mathscr{E}_x(R^n)$. On the other hand, because A_{k_0,p_0} is symmetric and also convex (as a set), it contains a certain neighbourhood of the origin. Therefore, if we write Ψ as an initial value, then for a sufficiently small $\lambda(\neq 0)$, $\lambda\Psi$ belongs to A_{k_0,p_0}. Hence, a solution

$$u(x,t)\in\mathscr{E}_{L^2}{}^{[\frac{1}{2}n]+1+m}(D_{\varepsilon_{k_0}})$$

exists for $Lu=0$ with an initial value $\lambda\Psi$. From this fact, it follows that

$$u(x,t)\in\mathscr{E}^m(D_{\varepsilon_{k_0}}).$$

Hence, $D_\delta\,(\varepsilon_{k_0}=\delta)$ satisfies the condition of the proposition.

Q.E.D.

From Banach's closed graph theorem (theorem 2.6), we have:†

THEOREM 4.4. We assume that (4.17) all the coefficients are analytic, and a Cauchy problem for an L in the $\mathscr{E}$-class is soluble at the origin.‡

In this case, the solution is continuous for the initial value: if D_δ is an open set in the sense of proposition 4.1, the linear mapping

$$\Psi(x)\equiv(u_0(x),\dots,u_{m-1}(x))\to u(x,t)$$

is a continuous mapping from $\prod\mathscr{E}(R^n)$ to $\mathscr{E}^m(D_\delta)$.

Proof. The graph of the linear mapping $\Psi\to u$ is closed by the uniqueness of the solution. In fact, if $\Psi_j\to 0$ in $\prod\mathscr{E}(R^n)$ and $u_j\to u_0$ in $\mathscr{E}^m(D_\delta)$, then $Lu_0=0$, $(\partial/\partial t)^j u_0(x,0)=0$, and $x\in D_\delta\cap\{t=0\}$ $(j=0,\dots,m-1)$. Therefore, it follows that $u_0(x,t)\equiv 0((x,t)\in D_\delta$, i.e. $u_j\to 0$ in $\mathscr{E}^m(D_\delta)$ (see the note after theorem 2.6). $\prod\mathscr{E}(R^n)$ and $\mathscr{E}^m(D_\delta)$ are Fréchet spaces, so that in this case we can establish Banach's closed graph theorem.

Q.E.D.

Note. In the above theorem, we assume that the coefficients of L are analytic because Holmgren's theorem guarantees the uniqueness of the solution. On the other hand, if the coefficients are of the C^∞-class, in general, this uniqueness is not guaranteed, so that the theorem is not valid. If the uniqueness can be established, for example for the operator L in theorems 4.3 then 4.4 is valid.

Also, in the case when we deal with solubility in the halfspace (see definition 4.2′), theorem 4.4 is valid. We examine this point more carefully.

In definition 4.2′, we assumed that $u(x,t)$ was continuously differentiable $(m+[\frac{1}{2}n]+1)$ times in $D_{(f,\Psi)}\cap\{t\geqslant 0\}$. We make this condition a little

† A similar result can be found in Banach: *Théorie des opérations linéaires*, pp. 44–8, Warsaw, 1932, as an application of the closed graph theorem.

‡ See definition 4.2.

stronger. As we stated in the proof of lemma 3. 6. we consider the extension of $u(x, t)$ to the lower half-space.

Then, there exists an open $\tilde{D}_{\varepsilon_k}$ by an extension of $D_{\varepsilon_k} \cap \{t \geqslant 0\}$ to the lower half-space, where $u(x, t)$ is $(m+[\frac{1}{2}n]+1)$-times differentiable and satisfies $Lu=0$ in $t \geqslant 0$. Therefore proposition 4.1 is valid. Hence, $u(x, t) \in \mathscr{E}^m(D_\delta \cap \{t \geqslant 0\})$. That is, u is m-times continuously differentiable on the hyperplane $t=0$. From this we can prove theorem 4.4 directly.

Theorem 4.4 says that if L has local solubility then the continuity of the solution for the given initial value can be established. The question then arises: in what case does the continuity follow?

In fact, historically, the question arose in the reverse way; Hadamard took the view (in his famous *Le problème de Cauchy*), for physically meaningful Cauchy problems, assuming the uniqueness of a solution, the continuity of the solution for a given initial value can be established. He called a Cauchy problem having this property a '*well posed*' (Fr. *bien posé*) problem. In what follows we give various solutions to this problem.

PROPOSITION 4.2. Let L be the equation (4.17), and let its coefficient be of the C^∞-class. The necessary condition of the continuity† of the solution u of L with a given initial value Ψ is that the roots $\lambda_1(\xi), \ldots, \lambda_m(\xi)$ of the characteristic equation of L at the origin

$$p(\lambda, \xi) = \lambda^m + \sum_{|\nu|+j=m} a_{\nu,j}(0, 0)\, \xi^\nu \lambda^j = 0 \tag{4.16}$$

are all real-valued for all real ξ.

Proof. The proof is very complicated‡. We only give a proof when the coefficients of L are all constant. To do this, it is enough to show that there exist ξ^* and λ_i (at this point we suppose this is λ_1), if $\operatorname{Im} \lambda_1(\xi^*) \neq 0$, the continuity of the solution for the initial value is never valid. In fact,* we assume

$$-\operatorname{Im} \lambda_1(\xi^*) = c > 0 \tag{4.18}$$

† We assume the uniqueness of a solution.

‡ For example, see Mizohata: 'Some remarks on the Cauchy problem', *Journal of Math. of Kyoto Univ.* 1 (1961), 109–27.

* The following argument was due to Hadamard when $L=\Delta$, and Petrowsky recognized the usefulness of the method (plane wave method).

(replace ξ^* by $-\xi^*$ if necessary). For the root $\lambda^*(\xi)$ of

$$P(\lambda, \mathrm{i}\xi) \equiv p(\lambda, \mathrm{i}\xi) + \sum_{|\nu|+j \leqslant m-1} a_{\nu,j}(\mathrm{i}\xi)^\nu \lambda^j = 0, \tag{4.19}$$

we consider

$$u_\xi(x, t) = \exp(\lambda^*(\xi)\, t) \exp(\mathrm{i}\xi x) \quad (\xi x = \xi_1 x_2 + \cdots + \xi_n x_n). \tag{4.20}$$

These satisfy $Lu_\xi = 0$, where ξ is a real parameter.

Now, let $\xi = \tau\xi^*$ in (4.19), and let $\tau\,(>0)$ be sufficiently large. Let us demonstrate that among the roots $\lambda_i^*(\tau\xi^*)$ $(i = 1, 2, \ldots, m)$ of $P(\lambda, i\tau\xi^*) = 0$, there exists a $\lambda^*(\tau\xi^*)$ satisfying

$$\operatorname{Re} \lambda^*(\tau\xi^*) \geqslant \tfrac{1}{2} c\tau \tag{4.21}$$

as $\tau \to +\infty$. In fact, if we put $\lambda/\tau = \lambda'$ then we can write (4.19) as

$$\tau^m \left\{ p(\lambda', \mathrm{i}\xi^*) + \frac{1}{\tau} Q(\lambda', \tau) \right\} = 0,$$

where $Q(\lambda', \tau)$ is a polynomial of less than $(m-1)$-degree with respect to λ', and its coefficients are polynomials of $1/\tau$. By our assumption, $p(\lambda', \mathrm{i}\xi^*) = 0$ has a root $\mathrm{i}\lambda_1(\xi^*)$, therefore, as $\tau \to +\infty$, $\mathrm{i}\lambda_1(\xi^*)$ has a root $\mathrm{i}\lambda_1(\xi^*) + \varepsilon$, where ε tends to 0 as $1/\tau \to 0$.

Hence, if $\tau > \tau_0$, then the real part of the root $\mathrm{i}\lambda_1(\xi^*) + \varepsilon$ is greater than $-\frac{1}{2}\operatorname{Im} \lambda_1(\xi^*)$. So that if $\tau > \tau_0$, then (4.19) has $\lambda^*(\tau\xi)$ satisfying

$$\operatorname{Re} \lambda^*(\tau\xi^*) \geqslant -\tfrac{1}{2}\tau \operatorname{Im} \lambda_1(\xi^*) = \tfrac{1}{2} c\tau.$$

Therefore (4.21) is proved.

Next, if (4.21) is true, then we see easily that the continuity of the solution with the given initial value for L (in the sense of theorem 4.4) is not valid. In fact, if it is valid, then for an arbitrary compact set K of D_δ, there exists a positive real C and a positive integer l, and for $u(x, t) \in \mathscr{E}$ satisfying $Lu = 0$

$$\sup_{(x,t) \in K} |u(x, t)| \leqslant C \sum_{j=0}^{m-1} \left| \left(\frac{\partial}{\partial t} \right)^j u(x, 0) \right|_l \tag{4.22}$$

where $|\cdot|_l$ indicates a norm of $\mathscr{B}^l(R^n)$. Let us take K containing a point $(0, t_0) \subset D_\delta$ $(t_0 > 0)$. Then, $u_\tau(x, t) = \exp[\lambda^*(\tau\xi^*)t] \exp(i\tau\xi^* x)$ is a solution in $\mathscr{E}$ satisfying $Lu_\tau = 0$. Hence, from $|\lambda^*(\tau\xi^*)| \leqslant c'\tau\,(\tau > \tau_0)$, (4.21) and (4.22), we have $\exp(\frac{1}{2}c\tau t_0) \leqslant \exp[\mathrm{Re}\,\lambda^*(\tau\xi^*)t_0] = |u_\tau(0, t_0)| \leqslant C\tau^{m+l-1}$, where $\tau > \tau_0$. If $\tau \to +\infty$ then this inequality is false.

Q.E.D.

From this proposition and theorem 4.4 we have:

THEOREM 4.5. In (4.17) we assume that the coefficients are analytic.† The necessary condition that the Cauchy problem for L should be locally soluble in the class $\mathscr{E}$ is that the roots $\lambda_i(\xi)$ of

$$p(\lambda, \xi) = \lambda^m + \sum_{|\nu|+j=m} a_{\nu,j}(0, 0)\, \xi^\nu \lambda^j = 0 \tag{4.16}$$

are real-valued for an arbitrary $\xi \in R^n$.

COROLLARY. The Cauchy problem for an elliptic operator with analytic coefficients is not locally soluble in the class $\mathscr{E}$.

5 *The continuity of solutions for initial value problem*

Proposition 4.2 can be improved if the coefficients of L are constant, i.e. a necessary and sufficient condition for the continuity of the solution of L with respect to the initial value can be given.

Let L satisfy

$$L[u] \equiv \left(\frac{\partial}{\partial t}\right)^m u + \sum_{\substack{|\nu|+j \leqslant m \\ j \leqslant m-1}} a_{\nu,j} \left(\frac{\partial}{\partial x}\right)^\nu \left(\frac{\partial}{\partial t}\right)^j u = f. \tag{4.23}$$

In this case we note that the characteristic equation for L including the terms of lower degree has been treated in the previous section, i.e.

$$P(\lambda, \mathrm{i}\xi) \equiv \lambda^m + \sum_{|\nu|+j \leqslant m} a_{\nu,j}(\mathrm{i}\xi)^\nu \lambda^j = 0. \tag{4.19}$$

The following theorem is known as Hadamard's condition (Hadamard proved the theorem by using the same argument in the case when $L = \Delta$).

THEOREM 4.6. Let our equation be (4.23). In the sense of theorem 4.4, the

† We can relax this condition; see the note after theorem 4.4.

necessary condition for the solution to have continuity for a given initial value is that there exist appropriate positive C and p, and, for all $\xi \in R^n$,

$$|\mathrm{Re}\,\lambda_i(\xi)| \leqslant p\log(1+|\xi|)+C \qquad (i=1, 2, \ldots m), \tag{4.24}$$

where $\lambda_i(\xi)$ $(i=1, 2, \ldots, m)$ are the roots of (4.19).

Note. We shall that if we consider the case $t \geqslant 0$ then (4.24) becomes $\mathrm{Re}\,\lambda_i(\xi) \leqslant p\log(1+|\xi|)+C$. Gårding has shown that (4.24) is equivalent to

$$|\mathrm{Re}\,\lambda_i(\xi)| \leqslant C. \tag{4.25}$$

The proof is based on the facts of algebraic geometry.†

Proof (By contradiction). Let us assume that (4.24) is not true.‡ For an arbitrary $j(>0)$, there exists a certain ξ^* $(|\xi^*|>1)$, and there exists a root of (4.19) satisfying $\mathrm{Re}\,\lambda(\xi^*) \geqslant j\log(1+|\xi^*|)$. On the other hand, $u(x, t)=\exp\{\lambda(\xi^*)t + \mathrm{i}\xi^* x\}$ satisfies $L[u]=0$. Also

$$|u(0, t)| = \exp\{\mathrm{Re}\,\lambda(\xi^*)\,t\} \geqslant (1+|\xi^*|)^{jt} \qquad (t>0),$$

and

$$\sum_{j=0}^{m-1} \sum_{|\nu| \leqslant l} \left| \left(\frac{\partial}{\partial x}\right)^{\nu} \left(\frac{\partial}{\partial t}\right)^{j} u(x, 0) \right| \leqslant C(l)\,(1+|\xi^*|)^{m+l-1},$$

where j is arbitrary. So that from these inequalities we see that the continuity of the solution for the initial value is not available in the sense of theorem 4.4. Hence, (4.24) is a necessary condition.

Q.E.D.

Next, if (4.24) is true, then the existence of a solution and the continuity of the solution for a given initial value can be proved. We need some tools for proving this. Let us prove the following:

LEMMA 4.1 *(Petrowsky).* Let us assume that we are given a system of different equations with constant coefficients

$$\frac{\mathrm{d}}{\mathrm{d}t} v_i = \sum_{j=1}^{N} a_{ij} v_j \qquad (i=1, \ldots, N). \tag{4.26}$$

† Gårding: 'Linear hyperbolic partial differential equations with constant coefficients', *Acta Math.* 146 (1950).

‡ Without loss of generality we assume that $\mathrm{Re}\,\lambda_i(\xi) \leqslant p\log(1+|\xi|)+C$ is not true.

We can write this as

$$\frac{d}{dt}v=Av. \tag{4.27}$$

Then, there is a regular matrix $C=(c_{ij})$, satisfying (1) $CA=DC$ where D satisfies

$$D=\begin{bmatrix} a_{11}^* & & 0 \\ a_{21}^* & a_{22}^* & \\ \vdots & & \ddots \\ a_{N1}^* & \cdots & a_{NN}^* \end{bmatrix},$$

and a_{ii}^* on the diagonal are the roots $\lambda_1, \lambda_2, \ldots, \lambda_N$ of $\det(\lambda I-A)=0$ (allowing duplication), i.e. if we put $Cv=w$, then (4.27) is transformed into

$$\frac{d}{dt}w=Dw, \tag{4.28}$$

where

(2) $|\det C|=1 \quad (|c_{ij}|\leqslant 1)$.
(3) $|a_{ij}^*|\leqslant (N-1)!2^N (\max_{i,j}|a_{ij}|) \quad (j<i)$.

Proof. Compared with the reduction of matrices to a Jordan form, this lemma is not precise. However, these estimations (2) and (3) are important in what follows, we shall examine them rather closely. As is wellknown, we can take $(b_1, \ldots, b_N)\neq 0$ satisfying

$$\sum_{i=1}^{N} b_i a_{ij}=\lambda b_j \quad (j=1, 2, \ldots, N),$$

where λ is necessarily an eigenvalue of A. Let us write $\lambda=\lambda_1$ in this case.

Of course, if $(b_1, \ldots, b_N)$ is a solution, then $(cb_1, \ldots, cb_N)$ also satisfies the above condition. Therefore, there exists i_0 satisfying $b_{i_0}=1, |b_j|\leqslant 1$. Now, let us write†

$$\left.\begin{aligned} v_1^{(1)} &= \sum_{j=1}^{N} b_j v_j, \\ v_i^{(1)} &= v_i \quad (i\geqslant 2, i\neq i_0), \\ v_{i_0}^{(1)} &= v_1. \end{aligned}\right\}$$

† If $i_0=1$, we put $v_i^{(1)}=v_i \quad (i\geqslant 2)$.

Then (4.26) is transformed into

$$\left.\begin{aligned} &\frac{\mathrm{d}}{\mathrm{d}t} v_1{}^{(1)} = \lambda_1 v_1{}^{(1)}, \\ &\frac{\mathrm{d}}{\mathrm{d}t} v_i{}^{(1)} = a_{i1}{}^{(1)} v_1{}^{(1)} + \sum_{j=2}^{N} a_{ij}{}^{(1)} v_j{}^{(1)} \quad (i \geqslant 2). \end{aligned}\right\}$$

Obviously, a_{ij} satisfies $|a_{ij}{}^{(1)}| \leqslant 2K$, $K = \max |a_{ij}|$, and properties (2) and (3).

Next, if we observe $\sum_{j=2}^{N} a_{ij}{}^{(1)} v_j{}^{(1)}$ in this way, then by putting $v_1{}^{(2)} = v_1{}^{(1)}$ we have

$$\left.\begin{aligned} &\frac{\mathrm{d}}{\mathrm{d}t} v_1{}^{(2)} = \lambda_1 v_1{}^{(2)}, \\ &\frac{\mathrm{d}}{\mathrm{d}t} v_2{}^{(2)} = a_{21}{}^{(2)} v_1{}^{(2)} + \lambda_2 v_2{}^{(2)}, \\ &\frac{\mathrm{d}}{\mathrm{d}t} v_i{}^{(2)} = a_{i1}{}^{(2)} v_1{}^{(2)} + a_{i2}{}^{(2)} v_2{}^{(2)} + \sum_{j=3}^{N} a_{ij}{}^{(2)} v_j{}^{(2)} \quad (i \geqslant 3). \end{aligned}\right\}$$

It is easy to see that the transformation $(v_1, \ldots, v_N) \rightarrow (v_1{}^{(2)}, \ldots, v_N{}^{(2)})$ satisfies the property (2). From

$$|a_{21}{}^{(2)}| \leqslant (N-1)\, 2K \quad \text{and} \quad |a_{ij}{}^{(2)}| \leqslant 2^2 K \quad (i > 2),$$

we see that $|a_{ij}{}^{(2)}| \leqslant (N-1) 2^2 K$. Continuing this process we can prove the lemma.

Q.E.D.

One of the methods of solving (4.23) is the one using a Fourier transform in x-space. Let us consider this. Suppose we transform (4.23) by a Fourier transform in x-space. Then we have an equation

$$\tilde{L}[v] \equiv \frac{\mathrm{d}^m}{\mathrm{d}t^m} v + \sum a_{\nu, j} (2\pi \mathrm{i} \xi)^\nu \frac{\mathrm{d}^j}{\mathrm{d}t^j} v = 0. \tag{4.29}$$

This is an ordinary differential equation subjected to a parameter ξ. In this case the solution $v(\xi, t)$ of (4.29) exists in $-\infty < t < +\infty$ and is uniquely determined for a given initial value at $t = 0$.

We consider the estimation of $v(\xi, t)$ as $|\xi| \rightarrow +\infty$. First we write $v_j(\xi, t)$

as the solution of (4.29) satisfying an initial condition

$$\frac{d^i}{dt^i} v(\xi, t)\bigg|_{t=0} = \delta_j^{\ i}, \tag{4.30}$$

for $i, j = 0, 1, \ldots, m-1$, where $\delta_j^{\ i}$ is Kronecker's symbol.

LEMMA 4.2. Let us assume that Hadamard's condition (4.24) is given. In this case, if we fix $T(>0)$, then there exist an integer $l(>0)$ and a positive constant C satisfying

$$|v_j(\xi, t)| \leqslant C(1+|\xi|)^l \quad (\xi \in R^n) \tag{4.31}$$

for $v_j(\xi, t)$ in $|t| \leqslant T$, where C and l in general depend on T.

Proof. We write one of v_j as v, putting

$$v^{(0)} = v, \quad v^{(1)} = \frac{d}{dt} v, \quad \ldots, \quad v^{(m-1)} = \frac{d^{m-1}}{dt^{m-1}} v.$$

Then, $\tilde{L}[v] = 0$ becomes

$$\frac{d}{dt}\begin{bmatrix} v^{(0)} \\ v^{(1)} \\ \vdots \\ v^{(m-1)} \end{bmatrix} = \begin{bmatrix} 0 & 1 & & & \\ & 0 & 1 & & \\ & & \ddots & \ddots & \\ & & & 0 & 1 \\ \alpha_m(\xi) & & & & \alpha_1(\xi) \end{bmatrix} \begin{bmatrix} v^{(0)} \\ v^{(1)} \\ \vdots \\ v^{(m-1)} \end{bmatrix}.$$

Obviously we have

$$|\alpha_i(\xi)| \leqslant C_0(1+|\xi|)^m \quad (i = 1, 2, \ldots, m).$$

If we let the right hand term in the above matrix equation be $A(\xi)$, then the roots $\lambda_i(\xi)$ of $\det(\lambda I - A(\xi)) = 0$ satisfy the following property: $\lambda = \lambda_i(\xi/2\pi)$ are the roots of (4.19).

Let us apply lemma 4.1 considering this fact. First we fix $\xi \in R^n$. We see that there exists a regular matrix C satisfying the conditions (2) and (3) of the previous lemma. In this case

$$\frac{d}{dt} w_i(\xi, t) = \lambda_i(\xi)\, w_i(\xi, t) + \sum_{j<i} a_{ij}^{\ *} w_j(\xi, t) \tag{4.32}$$

where we put

$$C\begin{bmatrix} v^{(0)} \\ \vdots \\ v^{(m-1)} \end{bmatrix} = \begin{bmatrix} w_1 \\ \vdots \\ w_m \end{bmatrix}.$$

Therefore $|a_{ij}^*| \leqslant c_1 (1+|\xi|)^m$ where c_1 can be chosen independent of ξ. Hence from the condition imposed on C there exists a constant c_2 independent of ξ, and $|w_j(\xi, 0)| \leqslant c_2$ $(j=1, 2, \ldots, m)$. From (4.32) we have

$$w_1(\xi, t) = w_1(\xi, 0) \exp[\lambda_1(\xi)\, t].$$

From this fact, and by (4.24), we have $|w_1(\xi, t)| \leqslant c_3 (1+|2\pi\xi|)^{pt}$.

Next, using this inequality we have an estimate for $w_2(\xi, t)$. We repeat the same process. Finally, for $w_i(\xi, t)$ we have an estimate in the form of (4.31) considering $|t| \leqslant T$, i.e. as $|\xi| \to \infty$ the solution has at most the increasing order of a polynomial of $|\xi|$. A similar argument holds for $v^{(j)}(\xi, t)$. Consequently, we have

$$\left| \left(\frac{\mathrm{d}}{\mathrm{d}t}\right)^i v_j(\xi, t) \right| \leqslant C(1+|\xi|)^l \quad (0 \leqslant i, j \leqslant m-1,\ |t| \leqslant T). \tag{4.33}$$

Q.E.D.

Note. Considering (4.29), (4.33) holds for $i=m, m+1, \ldots$, i.e. the partial derivative of arbitrary degree with respect to t of $v_j(\xi, t)$ has the increasing order of a polynomial in $|t| \leqslant T$. On the other hand

$$v_j(\xi, t) \in (\mathscr{S}')_\xi,$$

so that we write its Fourier inverse image as $R_j(x, t)$ where t is a parameter. By lemma 2.15, we have $R_j(x, t) \in \mathscr{D}'_{L^2}$.

LEMMA 4.3. Let us write $v(\xi, t)$ as one of $v_j(\xi, t)$. Let $\mathscr{F}[v(\xi, t)] = R(x, t)$ and let

$$\left| \left(\frac{\partial}{\partial t}\right)^i v(\xi, t) \right| \leqslant C(1+|\xi|)^p \quad (i=0, 1, 2, \cdots, m, \quad a \leqslant t \leqslant b).$$

For $u(x) \in \mathscr{D}_{L^2}{}^k$ $(k=0, \pm 1, \pm 2, \ldots)$ we put $f(x, t) = R(x, t) \underset{(x)}{*} u(x)$. Then, we have

(1) $t \to f(x, t) \in \mathscr{D}_{L^2}{}^{k-p}$ is m-times continuously differentiable, i.e. $f(x, t)$ is m-times continuously differentiable with respect to t and has its value in $\mathscr{D}_{L^2}{}^{k-p}$.

(2) $L[f]=0$.

(3) If $f_j(x, t) = R_j(x, t) * u(x)$ then

$$\left(\frac{\partial}{\partial t}\right)^i f_j(x, t) \underset{t \to 0}{\to} \delta_j{}^i \cdot u(x) \quad \text{in} \quad \mathscr{D}_{L^2}{}^{k-p} \quad (0 \leqslant i, j \leqslant m-1).$$

Proof. From lemma 2.14 we have

$$\hat{f}(\xi, t) = v(\xi, t)\, \hat{u}(\xi), \quad (D_x{}^\alpha f(x, t))\hat{}\,(\xi) = (2\pi i \xi)^\alpha\, v(\xi, t)\, \hat{u}(\xi)$$

Also, from $\| f(x, t)\|_{k-p} = \|(1+|\xi|)^{k-p} v(\xi, t) \hat{u}(\xi)\|_{L^2}$, $f(x, t) \in \mathscr{D}_{L^2}{}^{k-p}$ where f is continuous with respect to t.

In fact,

$$\|f(x, t) - f(x, t')\|_{k-p} = \|(1+|\xi|)^{k-p}\, [v(\xi, t) - v(\xi, t')]\, \hat{u}(\xi)\|_{L^2}.$$

So that, from Lebesgue's theorem, the right hand side of this equation tends to 0 as $t' \to t$.

Next, we see that

$$\lim_{t' \to t} \frac{f(x, t') - f(x, t)}{t' - t} = \frac{\partial R}{\partial t} \underset{(x)}{*} u(x) \quad \text{in} \quad \mathscr{D}_{L^2}{}^{k-p},$$

where $\partial R / \partial t$ is the inverse Fourier image of $\partial v(\xi, t)/\partial t$. Using the property

$$\frac{v(\xi, t') - v(\xi, t)}{t' - t} - \frac{\partial}{\partial t} v(\xi, t) = \frac{1}{t' - t} \int_t^{t'} \left\{ \frac{\partial}{\partial s} v(\xi, s) - \frac{\partial}{\partial t} v(\xi, t) \right\} ds$$

we can prove this quite easily. (2) and (3) follow immediately from the above argument.

Q.E.D.

From lemma 4.3, we have the inverse of theorem 4.6:

THEOREM 4.7.† For an operator with constant coefficients defined in (4.23)

† This theorem is due to Petrowsky. See Petrowsky: 'Über das Cauchysche Problem für ein System linearer partieller Differentialgleichungen im Gebiete der nicht-analytischen Funktionen', *Bull. de l'Univ. État, Moscow* (1938), 1–74.

assume Hadamard's condition (4.24). In this case, for $\Psi \equiv (u_0(x), u_1(x), \ldots, u_{m-1}(x)) \in \mathscr{D}$,

$$u(x,t) = \sum_{j=0}^{m-1} R_j(x,t) \underset{(x)}{*} u_j(x) \tag{4.34}$$

belongs to $\mathscr{E}(R^{n+1})$, $L[u(x,t)]=0$, and it takes an initial value Ψ at $t=0$. At the same time, in $\mathscr{E}$ the continuity of a solution for the initial value which we mentioned in theorem 4.4 is valid.

Proof. Let $\Psi \in \mathscr{D}_{L^2}{}^{\infty}$, and let us fix $T(>0)$. We consider (4.34) in $t \in [-T, T]$. As we have said in the note after lemma 4.2, in this interval, $(\partial/\partial t)^l v_j(\xi, t)$ has at most the increasing order of a polynomial. From the previous lemma, $t \to u(x,t) \in \mathscr{D}_{L^2}{}^{(\frac{1}{2}n)+1+m'}$ is m'-times continuously differentiable in $[-T, T]$ with respect to t. In fact, we can put $k = [\frac{1}{2}n]+1+m'+p$, $\Psi \in \mathscr{D}_{L^2}{}^k$. On the other hand, from Sobolev's lemma $\mathscr{D}_{L^2}{}^{(\frac{1}{2}n)+1+m'} \subset \mathscr{B}^{m'}$. Therefore,

$$t \to \left(\frac{\partial}{\partial t}\right)^j u(x,t) \in \mathscr{B}^{m'} \quad (0 \leqslant j \leqslant m)$$

is continuous. Hence, $u(x,t) \in \mathscr{E}^{m'}$. Nevertheless, m' is arbitrary, so that $u(x,t) \in \mathscr{E}$.

Next, we demonstrate that (4.34) has continuity in the sense of theorem 4.4 with respect to a solution u for an initial value Ψ. Let $\alpha \in \mathscr{D}$ be a function which is identically 1 in a neighbourhood of $x \in D_\delta \cap \{t=0\}$. For $\Psi \in \mathscr{E}$, we associate

$$\alpha\Psi \equiv (\alpha(x)\, u_0(x), \ldots, \alpha(x)\, u_{m-1}(x)) \in \mathscr{D}. \tag{4.35}$$

For an initial value Ψ, we pick (4.34) for which $\alpha\Psi$ is an initial value. Then, this is a solution of the problem if we restrict ourselves to D_δ of theorem 4.4. D_δ is uniquely chosen, so that we need only consider an initial value in the form of (4.35).

From this, we have

$$|u(x,t)|_{\mathscr{B}^m(D_\delta)} \leqslant C \sum_{j=0}^{m-1} \|\alpha(x) u_j(x)\|_{[\frac{1}{2}n]+1+m+p} \leqslant C' \sum_{j=0}^{m-1} |u_j(x)|_{\mathscr{B}^k(K)}$$

where K is the support of $\alpha(x)$, and $k = [\frac{1}{2}n]+1+m+p$.

Q.E.D.

Corollary. Under the same conditions as theorem 4.7, for $f(x, t) \in \mathscr{D}$,

$$u(x, t) = \sum_{j=0}^{m-1} R_j(x, t) \underset{(x)}{*} u_j(x) + \int_0^t R_{m-1}(x, t-\tau) \underset{(x)}{*} f(x, \tau)\, d\tau \tag{4.36}$$

takes an initial value of Ψ at $t=0$, and is a solution of $L[u]=f$ where $u(x, t) \in \mathscr{E}$.

Proof. The Fourier transform of

$$\int_0^t R_{m-1}(x, t-\tau) \underset{(x)}{*} f(x, \tau)\, d\tau \tag{4.37}$$

with respect to x is $g(\xi, t) = \int_0^t v_{m-1}(\xi, t-\tau)\hat{f}(\xi, \tau) d\tau$. To see this, we recall the definition of the integral of (4.37). The integrand is a continuous function of τ which takes its value in L^2 when we fix t. We note that $\mathscr{F}$ is a linear continuous mapping from L^2 to L^2. So

$$\mathscr{F}\Big[\int_0^t R_{m-1}(x, t-\tau) \underset{(x)}{*} f(\tau)\, d\tau\Big] = \int_0^t \mathscr{F}[R_{m-1}(x, t-\tau) \underset{(x)}{*} f(x, \tau)]\, d\tau.$$

Also, it is a well-known fact in the theory of ordinary differential equations that $g(\xi, t)$ satisfies $\tilde{L}[g(\xi, t)] = \hat{f}(\xi, t)$.

Q.E.D.

6 *Dependence domain*

In the previous section we stated a solution of the existence theorem but we did not discuss the uniqueness of the solution. We consider this now. Throughout this section, we assume all equations have an operator (4.23) with constant coefficients. Also, we assume Hadamard's condition (4.24). In this case, from the proof of lemma 4.2, all the roots $\lambda_i(\xi)$ of

$$p(\lambda, \xi) = \lambda_m + \sum_{|\nu|+j=m} a_{\nu, j} \xi^\nu \lambda^j = 0 \tag{4.16}$$

are obviously real. Now, we apply Holmgren's theorem (theorem 4.2) as follows.

Let us consider a surface S defined for $\varphi(x, t) = 0$. Let $\varphi(x, t)$ be analytic, and let $\varphi_t \neq 0$. In this case, as we demonstrated in section 1, if we perform a transformation of (4.23) by

$$\left.\begin{aligned} t' &= \varphi(x, t) \\ x_i' &= x_i \end{aligned}\right\},$$

then, we have

$$\left(\varphi_t^m + \sum_{|\nu|+j=m} a_{\nu,j} \varphi_x^\nu \varphi_t^j\right) \left(\frac{\partial}{\partial t'}\right)^m u + \sum \cdots = f, \tag{4.38}$$

where the second term of the left hand side is at most of degree $(m-1)$ with respect to $\partial/\partial t'$.

Also, φ is analytic, so all the coefficients of φ are analytic with respect to (x, t). Therefore, they are also analytic with respect to (x', t'). By the transformation, S is transformed into a hyperplane $t'=0$, so that if

$$\varphi_t^m + \sum_{|\nu|+j=m} a_{\nu,j} \varphi_x^\nu \varphi_t^j \neq 0 \tag{4.39}$$

Holmgren's theorem can be applied.

On the other hand, if we put

$$\lambda_{\max} = \max_{i=1,\dots,m;|\xi|=1} \lambda_i(\xi), \tag{4.40}$$

then from $\lambda_i(-\xi) = -\lambda_j(\xi)$ we have $\lambda_{\max} \geq 0$. In what follows, we assume $\lambda_{\max} > 0$. Now we have:

Lemma 4.4. Assume that an analytic surface S is defined by $\varphi(x, t)=0$ in the neighbourhood of $(x_0, t_0) \in S$.

At the same time we assume that $\varphi(x, t)$ satisfies

$$\varphi_t^2 > \lambda_{\max}^2 \sum_{i=1}^n \varphi_{x_i}^2. \tag{4.41}$$

On the other hand, if $u(x, t) \in \mathscr{E}^m$ is defined in a certain neighbourhood of (x_0, t_0) and if $L[u]=0$ with the Cauchy datum 0 on S, then $u(x, t) \equiv 0$ in a certain neighbourhood of (x_0, t_0).

Proof. It is enough to prove that (4.39) follows from (4.41). From

$$p(\lambda, \xi) = \prod_{i=1}^m (\lambda - \lambda_i(\xi))$$

and the definition of λ_{max}, the value of the term is positive if $\lambda > \lambda_{max}|\xi|$. Also, from $\lambda_{min} \equiv \inf \lambda_i(\xi) = -\lambda_{max}$, $p(\lambda, \xi) \neq 0$ if $\lambda < -\lambda_{max}|\xi|$. Therefore, $p(\lambda, \xi) \neq 0$ if $\lambda^2 > \lambda_{max}^2 |\xi|^2$. Putting $\xi = \varphi_x$, $\lambda = \varphi_t$ we have (4.41), so that (4.39) holds.

Q.E.D.

From the lemma we have the following theorem:

THEOREM 4.8. Let $t_0 > 0$, and let D be the interior of a backward cone: $\{(x, t): |x - x_0| = \lambda_{max}(t_0 - t)\}$ having (x_0, t_0) as its apex satisfying $t \geqslant 0$.

We see that if $u(x, t) \in \mathscr{E}^m$ is defined in D satisfying $L[u] = 0$, and if

$$\left(\frac{\partial}{\partial t}\right)^j u(x, 0) = 0, \; x \in D \cap \{t = 0\} \quad (j = 0, 1, \ldots, m-1),$$

i.e. the initial value of u is 0 on $t = 0$, then $u(x, t) \equiv 0$ in D.

In particular, if $u(x, t)$ is continuous on $\bar{D}$, then $u(x_0, t_0) = 0$.

Proof. Let $S_\theta (0 < \theta \leqslant \lambda_{max}^2 t_0^2)$ be an analytic hypersurface defined by the equation $\varphi(x, t; \theta) \equiv \lambda_{max}^2 (t - t_0)^2 - |x - x_0|^2 - \theta = 0$ $(t < t_0)$ belonging to θ. Obviously, $\bigcup S_\theta \supset D$, i.e. D is 'swept out' by S_θ. Also, we have

$$\frac{\varphi_t^2}{\sum \varphi_{x_i}^2} = \frac{\lambda_{max}^2 (t - t_0)^2}{|x - x_0|^2} = \lambda_{max}^2 \frac{|x - x_0|^2 + \theta}{|x - x_0|^2} > \lambda_{max}^2$$

on S_θ. Therefore, in this case the condition (4.41) is satisfied. Hence, for each S_θ, we can apply lemma 4.4.

More precisely, if the Cauchy datum of u on a certain S_{θ_0} is 0, then, for a sufficiently small ε, it is also 0 throughout on S_θ providing $|\theta - \theta_0| < \varepsilon$ (in this case we can consider $u(x, t)$ inside a backward cone extending from 0 if $t \leqslant 0$). Therefore, for a fixed $\delta (> 0)$, the Cauchy datum on S_θ is 0 where $\delta \leqslant \theta \leqslant \lambda_{max}^2 t_0^2$.

Hence, all the derivatives up to degree m are 0 on S_θ. The collection of such θ is an open set. On the other hand, obviously it is a closed set too, so that the set coincides with the set† of all $[\delta, \lambda_{max}^2 t_0^2]$. But δ is arbitrarily small, so that $u(x, t) = 0$ throughout on D.

Q.E.D.

† This is the same argument as the following. We choose the first point θ from $\lambda_{max}^2 t_0^2$ which corresponds to S_θ such that the Cauchy datum on S_θ is not 0. Then, we produce a contradiction.

Note. If $\lambda_{max}=0$, the theorem is true for the backward cone, $\{(x,t)\in R^{n+1};\ \varepsilon(t_0-t)\geqslant|x-x_0|\}$ where $\varepsilon(>0)$ is arbitrary. The uniqueness of the solution has been demonstrated by theorem 4.8. In fact, theorem 4.8 is even stronger. We illustrate the situation in what follows. Given a 'disturbance' in the neighbourhood of the origin, the behaviour of the transition of the disturbance with respect to t can be represented by (4.34) (theorem 4.7). The above theorem says that the speed of propagation of the disturbance is at most λ_{max}. To see this, we proceed as follows.

Let the support of the initial value Ψ lie within V, the neighbourhood of the origin.† The value of $u(x,t)$ at (x_0, t_0) $(t_0>0)$ is determined by the value of Ψ in the interior of the sphere, $|x-x_0|<\lambda_{max}t_0$, i.e. the interior of the intersection of the 'initial' plane and the backward cone having (x_0, t_0) as its apex. If the sphere, $|x-x_0|<\lambda_{max}t_0$ does not intersect V, then $\Psi(x)\equiv 0$ in the interior of the sphere, so that $u(x_0, t_0)=0$. From this we see that the support of $u(x,t)$ lies within the area which is swept by the forward cone $\check{C}$, $\check{C}=\{(x,t);\ (t\geqslant 0),\ \lambda_{max}t\geqslant|x|\}$, when the apex runs through the support of $\Psi(x)$, i.e. in the language of vector analysis, it lies within the set

$$\operatorname{supp}[\Psi]+\check{C}.$$

Therefore, the support of $u(x,t)$ lies in the neighbourhood of $\check{C}$. This shows that the speed of propagation of the disturbance is at most λ_{max}. Summing up, we have:

THEOREM 4.9. Under the same assumption as theorem 4.7, let the support of the initial value Ψ be a bounded set of R^n. In this case, the support of the solution u of $L[u]=0$ at $t=t(>0)$, taking the initial value Ψ at $t=0$ is contained in $\{x;\ \bigcup_{\xi\in\operatorname{supp}[\Psi]}|x-\xi|\leqslant\lambda_{max}t\}$, i.e. intuitively speaking, the phenomena governed by the operator L propagate with a finite speed which does not exceed λ_{max}, where λ_{max} should be undersood in the sense of (4.40).

7 *The existence theorem of solutions*

Let us recall the results we have obtained so far. First, from theorem 4.5, the necessary condition that the Cauchy problem of the equation (4.17)

† The support of $\Psi=(u_0(x),\dots,u_{m-1}(x))$ should be understood as $\bigcup_{j=0}^{m-1}\operatorname{supp}[u_j(x)]$.

should be locally soluble is that the characteristic root $\lambda_i(x, t; \xi)$ is real-valued for all ξ at $(x, t)=(0, 0)$. In particular, if the operator L has constant coefficients, the necessary condition is (4.24), Hadamard's condition. Also, in this case, this condition is also sufficient (theorem 4.7), and, moreover, the global existence of a solution can be demonstrated. Then, by theorem 4.9, the phenomena governed by this equation have a finite speed of propagation, i.e. propagation in a wave form. We state a definition which is important in what follows.

DEFINITION 4.3. By a *hyperbolic operator* we mean an operator with constant coefficients L such that

$$L[u]\equiv\left(\frac{\partial}{\partial t}\right)^m u+\sum_{|\nu|+j\leqslant m,\, j\leqslant m-1} a_{\nu,j}\left(\frac{\partial}{\partial x}\right)^{\nu}\left(\frac{\partial}{\partial t}\right)^{j} u \tag{4.23}$$

satisfies Hadamard's condition (4.24).

Now, we re-examine Hadamard's condition. Could we express the condition in the term of the principal part of (4.23) as

$$L_P[u]\equiv\left(\frac{\partial}{\partial t}\right)^m u+\sum_{|\nu|+j=m} a_{\nu,j}\left(\frac{\partial}{\partial x}\right)^{\nu}\left(\frac{\partial}{\partial t}\right)^{j} u \quad ? \tag{4.42}$$

The answer is, unfortunately, 'no', as we shall see below. For this reason we modify the problem as follows. What are the necessary and sufficient conditions for the operator L_P (4.42) and, at the same time, for the operator (4.23) with arbitrary terms of lower degrees which are added to L_P, to be hyperbolic?

To simply our terminology, we define the new operator as follows.

DEFINITION 4.4. An operator L_P with constant coefficients of mth homogeneous degree is a *strongly hyperbolic operator* if and only if an operator L which we obtain as a result of adding arbitrary terms of lower than mth degree to L_P is hyperbolic.

THEOREM 4.10. The necessary and sufficient condition for L_P to be strongly hyperbolic is that the roots $\lambda_i(\xi)$ of

$$p(\lambda, \xi) \equiv \lambda^m + \sum_{|\nu|+j=m} a_{\nu,j}\xi^\nu \lambda^j = 0 \tag{4.43}$$

are all distinct reals for an arbitrary $\xi (\neq 0)$.

Proof (Sufficiency). Let the roots of (4.43) all be distinct reals. In this case, for an arbitrary $a_{\nu,j}(|\nu|+j \leqslant m-1)$, the roots $\tilde{\lambda}_i(\xi)$ of

$$p(\lambda, \mathrm{i}\xi) + \sum_{|\nu|+j\leqslant m-1} a_{\nu,j}(\mathrm{i}\xi)^\nu \lambda^j = 0 \tag{4.44}$$

satisfy Hadamard's condition. To see this we let a point on the unit sphere be $\xi^0 = \xi/|\xi|$. Write $\lambda/|\xi| = \lambda'$. Then, (4.44) becomes

$$\prod_{j=1}^{m} (\lambda' - \mathrm{i}\lambda_j(\xi^0)) + \frac{Q(\lambda', \xi)}{|\xi|} = 0, \tag{4.45}$$

where $Q(\lambda', \xi)$ is a polynomial in the form of

$$a_1(\xi)\lambda'^{m-1} + a_2(\xi)\lambda'^{m-2} + \cdots + a_m(\xi)$$

and every coefficient $a_i(\xi)$ is bounded for $|\xi| \geqslant 1$.

From our hypothesis we have $d = \min_{i\neq j,\, \xi^0\in\Omega} |\lambda_i(\xi^0) - \lambda_j(\xi^0)| > 0$. Furthermore, we put

$$K = \max_{i,\, \xi^0\in\Omega} |\lambda_i(\xi^0)|, \quad M = \sup_{|\xi|\geqslant 1,\, |\lambda'|\leqslant K+1} |Q(\lambda', \xi)|.$$

Then, we choose $c(>0)$ satisfying

$$c(\tfrac{1}{2}d)^{m-1} \geqslant 2M.$$

Let $\Gamma_1, \ldots, \Gamma_m$ be circles in the λ'-plane with centres $\mathrm{i}\lambda_1(\xi^0), \ldots, \mathrm{i}\lambda_m(\xi^0)$ and radius $c|\xi|^{-1}$.

Without loss of generality, we can restrict our argument to the case when $|\xi|$ is sufficiently large, i.e. the radius of these circles is less than $\frac{1}{2}d$ and 1. If $\lambda' \in \Gamma_k$, we have

$$\left| \prod_{j=1}^{m} (\lambda' - \mathrm{i}\lambda_j(\xi^0)) \right| \geqslant \frac{c}{|\xi|} (\tfrac{1}{2}d)^{m-1} \geqslant \frac{2M}{|\xi|},$$

also $|\lambda'| \leqslant K+1$. Therefore, $|Q(\lambda', \xi)|/|\xi| \leqslant M/|\xi|$. Hence, if $\lambda' \in \Gamma_k$

$(k=1, 2, \ldots, m)$, we see

$$|\prod_j (\lambda' - \mathrm{i}\lambda_j(\xi^0))| > \frac{|Q(\lambda', \xi)|}{|\xi|}.$$

By Rouché's theorem, we see that in each Γ_k there exists the root $\lambda_k'(\xi)$ of (4.45), i.e.

$$|\lambda_j'(\xi) - \mathrm{i}\lambda_j(\xi^0)| < c/|\xi| \quad (j=1, 2, \ldots, m).$$

From this, the root $\tilde{\lambda}_j(\xi)$ of (4.44) satisfies

$$|\tilde{\lambda}_j(\xi) - \mathrm{i}\lambda_j(\xi)| < c \quad (j=1, 2, \ldots, m)$$

where $\lambda_j(\xi)$ is real. So that $\tilde{\lambda}_j(\xi)$ obviously satisfies Hadamard's condition.

(*Necessity*) From proposition 4.2, we see that '$\lambda_j(\xi)$ is real' is a necessary condition. Therefore, it is enough to show that the $\lambda_j(\xi)$ are distinct. We show this by contradiction.

Let us assume that there exists $\xi^* \neq 0$ and at least two of the roots of (4.43) for ξ^* are equal, i.e.

$$p(\lambda, \xi^*) = (\lambda - \lambda_1(\xi^*))^p \prod_{j=2}^{m-p+1} (\lambda - \lambda_j(\xi^*)) \quad (p \geqslant 2),$$

where $\lambda_2(\xi^*), \ldots, \lambda_{m-p+1}(\xi^*)$ are different from $\lambda_1(\xi^*)$. We pick $\xi = \tau\xi^*$ where τ is a real parameter. Let us consider

$$p(\lambda, \mathrm{i}\tau\xi^*) + c\tau^{m-1} = 0 \tag{4.46}$$

where c is a constant which we shall determine later. We put $\lambda' = \lambda/\tau - \mathrm{i}\lambda_1(\xi^*)$. Then, (4.46) can be written as

$$\begin{aligned}\lambda'^p &\prod_{j=2}^{m-p+1} \{\lambda' + \mathrm{i}(\lambda_1(\xi^*) - \lambda_j(\xi^*))\} + c/\tau \\ &= \lambda'^p \{a_0(\xi^*) + a_1(\xi^*)\lambda' + \cdots + a_{m-p-1}(\xi^*)\lambda'^{m-p+1} + \lambda'^{m-p}\} + c/\tau = 0\end{aligned}$$

where $a_0(\xi^*) \neq 0$.

Now, we consider the Puiseux expansion of the equation in the neighbourhood of $\tau = \infty$. We have

$$\lambda_k'(\tau) = \left(\frac{-c}{a_0(\xi^*)}\right)^{1/p} \exp\left(\frac{2\pi \mathrm{i}k}{p}\right) \tau^{-1/p} + O(\tau^{-2/p}) \quad (k=1, 2, \ldots, p).$$

From $p \geqslant 2$, we can choose c such that at least one of $\lambda'(\tau)$ has a positive real part, i.e. for a certain k_0, $\mathrm{Re}\,\lambda_{k_0}'(\tau) \geqslant c_0 \tau_p^{-1}$ is true for $\tau > \tau_0$. Therefore $\mathrm{Re}\,\lambda_{k_0}(\tau) \geqslant c_0 \tau_p^{-1}$ is true for $\tau > \tau_0$.

On the other hand, we can choose b_ν such that $c = \sum_{|\nu|=m-1} b_\nu (i\xi^*)^\nu$. So that the roots of the equation (4.44) for

$$L = L_p + \sum_{|\nu|=m-1} b_\nu \left(\frac{\partial}{\partial x}\right)^\nu$$

do not satisfy Hadamard's condition. This is a contradiction, the statement '$\lambda_i(\xi)$ are distinct' is a necessary condition for L_p to be strongly hyperbolic.

Q.E.D.

8 *Phenomena with a finite speed of propagation*

So far our argument has been based on the equation (4.43). But, as we shall observe in the next chapter, there are other types of equations of evolution. We re-examine the results of section 6 in greater generality. Let our operator be

$$L[u] \equiv \left(\frac{\partial}{\partial t}\right)^m u + \sum_{j<m} a_{\nu,j} \left(\frac{\partial}{\partial x}\right)^\nu \left(\frac{\partial}{\partial t}\right)^j u = 0 \tag{4.47}$$

with constant coefficients, where we do not impose any restriction on the degree of the derivatives in the x-direction which appear in the left hand side terms.

In this case, we can consider a type of Cauchy problem and we shall see that if there are some derivatives which satisfy $|\nu| + j \geqslant m+1$, then the phenomena governed by (4.47) can never have a finite speed of propagation. In fact, let (4.47) have a unique solution $u(x, t) \in \mathscr{E}^\infty$ $(t \geqslant 0)$ for an arbitrary initial value $\Psi \equiv (u_0(x), \ldots, u_{m-1}(x)) \in \mathscr{D}$. Furthermore, we assume that (4.47) has a finite speed of propagation. From our observation in section 6, we see that, for an arbitrary $\Psi \in \mathscr{E}^\infty(R^n)$, there exists a unique solution $u(x, t) \in \mathscr{E}^\infty$ $(t \geqslant 0)$. To see this, we assume that (4.47) has a speed of propagation less than v.

Let (x_0, t_0) $(t_0 > 0)$ be an arbitrarily chosen point, and let D be the intersection of the backward cone C: $\{(x, t);\ v(t_0 - t) \geqslant |x - x_0|\ (t \leqslant t_0)\}$ with the apex (x_0, t_0) and the hyperplane $t = 0$. If we choose $\alpha \in \mathscr{D}$ such that $\alpha(x) = 1$ in the neighbourhood of D, we see that for an arbitrary initial value $\Psi \in \mathscr{E}^\infty(R^n)$, the restriction $u(x, t)$ of the solution $\tilde{u}(x, t)$ for the initial value

$\alpha(x)\Psi \equiv (\alpha(x)u_0(x), \ldots, \alpha(x)\, u_{m-1}(x)) \in \mathscr{D}$ to the interior of C is a 'local' representation of the solution for Ψ in the interior of C. Then, Banach's closed graph theorem says that the mapping $\Psi(x) \to u(x, t)$ is a continuous mapping from $\mathscr{E}^\infty(R^n)$ to $\mathscr{E}^\infty(t \geqslant 0)$.

On the other hand, this property is not available if $|v|+j \geqslant m+1$. In fact, let us write

$$\left(\frac{\partial}{\partial t}\right)^m u + \sum a_j\left(\frac{\partial}{\partial x}\right)\left(\frac{\partial}{\partial t}\right)^j u = 0$$

where $\alpha_j(\xi)$ is a polynomial, and its order is m_j. By our hypothesis, there exists a certain j_0 and $m_{j_0}+j_0 \geqslant m+1$. In this case, there exists a certain rational number $k(>1)$ and $kj+m_j \leqslant km$ $(j=0, 1, 2, \ldots, m-1)$. Also, for a certain j the above equality holds. If this is so, then we can write the homogeneous part of degree $k(m-j)$ of $\alpha_j(\xi)$ as $h_j(\xi)$. Consider

$$p(\lambda, \zeta) = \lambda^m + \sum h_j(\zeta)\lambda^j = 0 \tag{4.48}$$

where ζ is a complex vector, i.e. $\zeta \in C^n$. Obviously, there exists $\zeta_0(|\zeta_0|=1)$ such that a root $\lambda_1(\zeta_0)$ of (4.48) is positive.

Next, consider the equation

$$P(\lambda, \zeta) \equiv \lambda^m + \sum \alpha_j(\zeta)\lambda^j = p(\lambda, \zeta) + \sum(\alpha_j(\zeta) - h_j(\zeta))\lambda^j = 0 \tag{4.49}$$

where τ is a real parameter. Consider $\lambda^*(\tau\zeta_0)$ corresponding to $\zeta = \tau\zeta_0$. For $\lambda_1(\zeta_0)$ we put $\lambda/\tau^k = \lambda'$ taking into account the fact that $\lambda_1(\tau\zeta_0) = \tau^k\lambda_1(\zeta_0)$. Then, (4.49) becomes

$$p(\lambda', \zeta_0) + \sum_{j=0}^{m-1} \tau^{-k(m-j)}[\alpha_j(\tau\zeta_0) - h_j(\tau\zeta_0)]\lambda'^j = 0.$$

As $\tau \to +\infty$ we see $\alpha_j(\tau\zeta_0) - h_j(\tau\zeta_0) = o(\tau^{k(m-j)})$. There exists $\tau_0(>0)$ and if $\tau > \tau_0$, then, for the root $\lambda^*(\tau\zeta_0)$ of (4.49),

$$\operatorname{Re}\lambda^*(\tau\zeta_0) \geqslant c\tau^k, \tag{4.50}$$

where c is an appropriate positive constant. Then, we observe

$$u_\tau(x, t) = \exp\{x \cdot \tau\zeta_0 + t\lambda^*(\tau\zeta_0)\}. \tag{4.51}$$

Take an arbitrary R, from $|\lambda^*(\tau\zeta_0)| \leqslant \to M \cdot \tau^k$ we see

$$\sup_{|x| \leqslant R}\left|\left(\frac{\partial}{\partial t}\right)^j\left(\frac{\partial}{\partial x}\right)^v u_\tau(x, 0)\right| \leqslant M'\tau^{|v|+kj}\exp(\tau R) \quad (j=0, 1, 2, \ldots, m-1).$$

On the other hand, for $t_0(>0)$, by (4.50) we see that

$$|u_\tau(0, t_0)| \geqslant \exp(ct_0\tau^k) \quad (k>1).$$

This fact shows that as $\tau \to +\infty$, the mapping from $\Psi(x) \in \mathscr{E}^\infty(R^n)$ to $u(x, t) \in \mathscr{E}^\infty$ $(t \geqslant 0)$ is not continuous for the family of the solutions of (4.47).

DEFINITION 4.5. In (4.47) (we allow variable coefficients), if the degree of the derivatives which appear on the right hand side satisfies $|\nu| + j \leqslant m$, then we call these derivatives *kowalewskians*. Using this new terminology, we have

THEOREM 4.11. If the phenomena which are governed by equation (4.47) with constant coefficients have finite speeds of propagation, then the derivatives which appear in (4.47) are kowalewskian.

9 *Solution of wave equations*

In this section we examine some specific examples related to our argument. We look at concrete representations of solutions of wave equations

$$\frac{\partial^2}{\partial t^2} u - \Delta u = 0, \tag{4.52}$$

and

$$\frac{\partial^2}{\partial t^2} u - \Delta u = f \tag{4.53}$$

using Fourier transforms.

First, from (4.30), we shall obtain a solution of

$$\frac{d^2}{dt^2} v(\xi, t) + 4\pi^2 |\xi|^2 v(\xi, t) = 0$$

satisfying

$$v(\xi, 0) = 0, \qquad \frac{d}{dt} v(\xi, 0) = 1.$$

The solution is given by

$$v(\xi, t) = \frac{\sin 2\pi|\xi| t}{2\pi|\xi|}.$$

We regard t as a parameter putting

$$E_x(t)=\mathscr{F}\left[\frac{\sin 2\pi|\xi|t}{2\pi|\xi|}\right]. \tag{4.54}$$

From theorem 4.7, if $u_1 \in \mathscr{E}$, then

$$u_1(x, t)=E_x(t) \underset{(x)}{*} u_1(x)$$

is the solution of (4.52) satisfying

$$u_1(x, 0)=0, \quad \frac{\partial}{\partial t} u_1(x, 0)=u_1(x).$$

Next, we see that $\mathrm{d}E_x(t)/\mathrm{d}t$ satisfies

$$\frac{\mathrm{d}^2}{\mathrm{d}t^2} E_x(t)=\Delta E_x(t)\to 0 \quad (t\to 0),$$

so that

$$u_0(x, t)=\frac{\mathrm{d}}{\mathrm{d}t} E_x(t) \underset{(x)}{*} u_0(x)$$

is the solution of (4.52) satisfying

$$u_0(x, 0)=u_0(x), \quad \frac{\partial}{\partial t} u_0(x, 0)=0.$$

We use these results to prove the following:

THEOREM 4.12. Given an initial datum

$$\left(u(x, 0), \frac{\partial}{\partial t} u(x, 0)\right)=(u_0(x), u_1(x))\in \mathscr{E},$$

and the right hand side term $f(x, t)\in \mathscr{E}$, there exists a unique solution

$$\begin{aligned} u(x, t)=&\frac{\mathrm{d}}{\mathrm{d}t} E_x(t) \underset{(x)}{*} u_0(x)+E_x(t) \underset{(x)}{*} u_1(x)\\ &+\int_0^t E_x(t-\tau) * f(x, \tau)\,\mathrm{d}\tau \end{aligned} \tag{4.55}$$

for (4.53), where $E_x(t)$ is defined in (4.54).

More precisely, if $t>0$, then the solution is as follows.

(1) When $n=1$,

$$\left.\begin{aligned} &E_x(t)=\begin{cases}\frac{1}{2} & (|x|<t),\\ 0 & (|x|>t),\end{cases}\\ &\frac{\mathrm{d}}{\mathrm{d}t}E_x(t)=\tfrac{1}{2}(\delta_t+\delta_{-t}).\end{aligned}\right\} \tag{4.56$_1$}$$

(2) When $n=2$,

$$E_x(t)=\begin{cases}\dfrac{1}{2\pi(t^2-|x|^2)^{\frac{1}{2}}} & (|x|<t),\\ 0 & (|x|>t).\end{cases} \tag{4.56$_2$}$$

(3) When $n=3$,

$$E_x(t)=\frac{1}{4\pi t}\,\delta_{|x|-t}. \tag{4.56$_3$}$$

Proof. Let us prove $(4.56)_1$. Let $\Psi(\rho)=\sin 2\pi\rho t/2\pi\rho$. Using $(2.111)_1$ we have

$$\begin{aligned}\Phi(r)&=2\int_0^{\to\infty}\Psi(\rho)\cos(2\pi\rho r)\,\mathrm{d}\rho=2\int_0^{\to\infty}\frac{\sin 2\pi\rho t}{2\pi\rho}\cos(2\pi\rho r)\,\mathrm{d}\rho\\ &=\frac{1}{2\pi}\int_0^{\to\infty}\left\{\frac{\sin 2\pi(t+r)\,\rho}{\rho}+\frac{\sin 2\pi(t-r)\,\rho}{\rho}\right\}\mathrm{d}\rho=\begin{cases}\frac{1}{2} & (r<t),\\ 0 & (r>t).\end{cases}\end{aligned}$$

$(4.56)_2$ is obviously nothing but a special case of the example (2.109) when $a=2\pi t$. Let us prove $(4.56)_3$. To do this we first observe that from $J_{\frac{1}{2}}(x)=\sqrt{(2/\pi x)}\sin x$, we have

$$\begin{aligned}E_x(t)&=\frac{1}{\pi r}\lim_{A\to+\infty}\int_0^A\sin(2\pi\rho t)\sin(2\pi\rho r)\,\mathrm{d}\rho\\ &=\frac{1}{2\pi r}\lim_{A\to+\infty}\left[\frac{\sin 2\pi(r-t)\,A}{2\pi(r-t)}-\frac{\sin 2\pi(r+t)\,A}{2\pi(r+t)}\right].\end{aligned}$$

By the Riemann–Lebesgue theorem, the second term tends to 0 as $A\to+\infty$. The first term tends to $\frac{1}{2}\delta_{r-t}$ by the fundamental equation of a Dirichlet integral where $t>0$. Therefore, we have

$$E_x(t)=\frac{1}{4\pi t}\,\delta_{r-t}.$$

Q.E.D.

Using these results $(4.56)_{1-3}$, we can re-write the solution (4.55) in a more concrete form as follows ($t>0$, and for simplicity, we put $f=0$):

$$u(x,t)=\tfrac{1}{2}[u_0(x+t)+u_0(x-t)]+\tfrac{1}{2}\int_{x-t}^{x+t} u_1(\xi)\,d\xi \quad (n=1), \tag{4.57$_1$}$$

$$u(x,t)=\frac{1}{2\pi}\frac{\partial}{\partial t}\iint_{|x-\xi|\leqslant t}\frac{u_0(\xi)\,d\xi}{(t^2-|x-\xi|^2)^{\frac{1}{2}}}+\frac{1}{2\pi}\iint_{|x-\xi|\leqslant t}\frac{u_1(\xi)\,d\xi}{(t^2-|x-\xi|^2)^{\frac{1}{2}}} \quad (n=2), \tag{4.57$_2$}$$

$$u(x,t)=\frac{\partial}{\partial t}\left(\frac{1}{4\pi t}\iint_{|x-\xi|=t} u_0(\xi)\,dS_\xi\right)+\frac{1}{4\pi t}\iint_{|x-\xi|=t} u_1(\xi)\,dS_\xi \quad (n=3). \tag{4.57$_3$}$$

Next, we consider the problem of seeking a concrete expression for $E_x(t)$ in a multi-dimensional case. In this case $E_x(t)$ becomes a distribution.

In fact, if n is an odd number, we put $n=2p+1$ ($p\geqslant 1$). Then,

$$E_x(t)=2^{-(p+1)}\pi^{-p}\left(\frac{1}{t}\frac{d}{dt}\right)^{p-1}\left(\frac{\delta_{r-t}}{t}\right) \tag{4.58}$$

from (2.113) and (4.56)$_3$. We note that the differential in (4.58) should be understood as being with respect to a parameter t. More precisely, we can write

$$E_x(t)*\psi(x)=2^{-(p+1)}\pi^{-p}\left(\frac{1}{t}\frac{d}{dt}\right)^{p-1}\left[t^{n-2}\int_{|\xi|=1}\psi(x-t\xi)\,dS_\xi\right].$$

Similarly, if n is an even number, we put $n=2p$. Then, we have

$$\begin{aligned}E_x(t)&=(2\pi)^{-p}\left(\frac{1}{t}\frac{d}{dt}\right)^{p-1}\{[(t^2-|x|^2)^+]^{-\frac{1}{2}}\}\\&=[2\pi^{\frac{1}{2}(n-1)}\,\Gamma(\tfrac{3}{2}-\tfrac{1}{2}n)]^{-1}\,\mathrm{Pf.}\,[(t^2-|x|^2)^+]^{\frac{1}{2}-\frac{1}{2}n}\end{aligned} \tag{4.59}$$

where $a^+=\max(a,0)$, and Pf. stands for the finite part in Hadamard's sense. We can write

$$E_x(t)*\psi(x)=(2\pi)^{-p}\left(\frac{1}{t}\frac{d}{dt}\right)^{p-1}\left[t^{n-1}\int_{|\xi|\leqslant 1}\frac{\psi(x-t\xi)}{(1-|\xi|^2)^{\frac{1}{2}}}\,d\xi\right].$$

From the corollary of theorem 4.7, in general, it is enough to obtain $R_0,\ldots,R_{m-1}$ for the representation of the solution of the Cauchy problem of a hyperbolic equation $L[u]=0$ with constant coefficients. We note that $R_0,\ldots,R_{m-2}$ can be easily obtained from $E_x(t)$ if we have $R_{m-1}(x,t)=E_x(t)$. This is obvious from the argument in the case of wave equations.

We call $E_x(t)$ $(t \geqslant 0)$ an *elementary solution* (or fundamental solution) for the Cauchy problem for $L[u]=0$, where $E_x(t)$ $(t \geqslant 0)$ satisfies

$$E_x(0)=\frac{\mathrm{d}}{\mathrm{d}t}E_x(0)=\cdots=\left(\frac{\mathrm{d}}{\mathrm{d}t}\right)^{m-2}E_x(0)=0,\quad \left(\frac{\mathrm{d}}{\mathrm{d}t}\right)^{m-1}E_x(0)=\delta,$$

in the sense of the solution for the Cauchy problem ($E_x(t)$ is determined uniquely in $\mathscr{S}'$).

10 *Systems of hyperbolic first-order equations*

We have treated a single high-order equation. We now briefly consider a system of equations of the following type:

$$\frac{\partial}{\partial t}u=\sum_{k=1}^{n}A_k\frac{\partial u}{\partial x_k}+Bu+f \tag{4.60}$$

where u is a vector having $u_1(x,t),\ldots,u_N(x,t)$ as its N-components, f is the same type of vector as u, and A_k, B are matrices of Nth degree.

We shall consider the case when A_k and B have variable coefficients in chapter 6. In this section let us assume they have constant coefficients. We write the roots of a characteristic equation

$$P(\lambda;\xi)=\det(\lambda I-\mathrm{i}\sum A_k\xi_k-B)=0 \tag{4.61}$$

as $\lambda_1(\xi),\ldots,\lambda_N(\xi)$. In this case we note that theorems 4.5 and 4.6 are true (the reader should verify this). Accordingly, we can define a system of 'hyperbolic type' equations similar to definition 4.3.

Also, the definition of a 'strongly hyperbolic' system is similar to definition 4.4. (B corresponds to terms of lower degree.)

For the equation including a principal part

$$M[u]=\frac{\partial}{\partial t}u-\sum A_k\frac{\partial}{\partial x_k}u=0 \tag{4.62}$$

we consider the roots $\lambda_i(\xi)$ of

$$p(\lambda;\xi)=\det(\lambda I-\sum A_k\xi_k)=0 \tag{4.63}$$

where we note that $\lambda_i(\xi)$ are homogeneous functions of the first degree. We have:

THEOREM 4.13. For (4.62) to be a strongly hyperbolic system it is necessary that the roots $\lambda_i(\xi)$ of (4.63) are real, and for an arbitrary $\xi \in R^n$, $\sum A_k\xi_k$ is diagonalizable.

Proof. The first half of the proof is almost the same as that of proposition 4.2. We prove only the rest.

If $\sum A_k\xi_k$ is not diagonalizable, there exists a regular matrix N such that

$$N(\sum A_k\xi_k{}^*)\,N^{-1} = \begin{bmatrix} \lambda_1 & 0 & 0 & \cdots & 0 \\ 1 & \lambda_1 & 0 & \cdots & 0 \\ & & * & & \end{bmatrix}.$$

We define B as satisfying

$$NBN^{-1} = \begin{bmatrix} 0 & 1 & 0 & \cdots & 0 \\ 0 & 0 & 0 & \cdots & 0 \\ & & \cdots & & \\ 0 & & \cdots & & 0 \end{bmatrix}.$$

For simplicity, we write

$$A\cdot\xi = \sum A_k\xi_k. \tag{4.64}$$

If we put $\xi = \tau\xi^*(\tau > 0)$ in (4.61), we have

$$\begin{aligned} \det(\lambda I - iA\cdot\tau\xi^* - B) &= \det(\lambda I - i\tau N(A\cdot\xi^*)\,N^{-1} - NBN^{-1}) \\ &= \begin{vmatrix} \lambda - i\tau\lambda_1(\xi^*) & -1 & 0 & \cdots & 0 \\ -i\tau & \lambda - i\tau\lambda_1(\xi^*) & 0 & \cdots & 0 \\ & * & & & \end{vmatrix} = 0. \end{aligned}$$

So that it has a root $(\lambda - i\tau\lambda_1(\xi^*))^2 - i\tau = 0$.

In this case, $\lambda_\pm(\tau\xi^*) = i\tau\lambda_1(\xi^*) \pm \sqrt{(i\tau)}$, where $\lambda_1(\xi^*)$ is real, so that there exists some $\mathrm{Re}\,\lambda_\pm(\tau\xi^*)$ which behaves like $\tau^{\frac{1}{2}}$ as $\tau \to +\infty$. This shows Hadamard's condition (4.24) is not satisfied. Therefore (4.60) is not hyperbolic. In other words, the Cauchy problem is not well-posed in this case.

Q.E.D.

The following problem is interesting for some applications.

THEOREM 4.14. For (4.62), if either of the following two conditions is valid, then the equation is strongly hyperbolic.

(1) A_k is an Hermitian matrix, i.e. $A_k{}^*(= {}^t\bar{A}_k) = A_k (k = 1, 2, \ldots, n)$.

(2) The roots of (4.63), $\lambda_i(\xi)$ $(i=1, 2,\ldots, N)$, are all real and distinct, i.e. $\lambda_i(\xi)\neq\lambda_j(\xi)$ $(i\neq j,\ \xi\neq 0)$.

Proof.

(1) Let A_k be Hermitian. We put $f=0$ in (4.60). Let a fundamental system of solutions of

$$\frac{\mathrm{d}}{\mathrm{d}t}v(\xi, t)=(2\pi\mathrm{i}A\cdot\xi+B)\,v(\xi, t) \tag{4.65}$$

be $v^1(\xi, t),\ldots, v^N(\xi, t)$ where the equation is obtained from (4.60) by a Fourier transform with respect to x, i.e.

$$v^j(\xi, 0)=\begin{bmatrix}0\\ \vdots\\ 1\\ \vdots\\ 0\end{bmatrix},$$

where 1 is in the jth row. Let us write $v_j(\xi, t)$ as $v(\xi, t)$. Then,

$$\begin{aligned}\frac{\mathrm{d}}{\mathrm{d}t}|v(\xi, t)|^2&=\frac{\mathrm{d}}{\mathrm{d}t}v(\xi, t)\cdot\overline{v(\xi, t)}+v(\xi, t)\cdot\frac{\mathrm{d}}{\mathrm{d}t}\overline{v(\xi, t)}\\ &=(2\pi\mathrm{i}A\xi+B)\,v\cdot\bar{v}+v\cdot\overline{(2\pi\mathrm{i}A\xi+B)\,v}\,.\end{aligned}$$

A is Hermitian, so that

$$\frac{\mathrm{d}}{\mathrm{d}t}|v(\xi, t)|^2=2\,\mathrm{Re}\,Bv\cdot\bar{v}\leqslant c|v(\xi, t)|^2$$

and

$$|v(\xi, t)|^2\leqslant \mathrm{e}^{ct}|v(\xi, 0)|^2=\mathrm{e}^{ct}\,. \tag{4.66}$$

From the above, if we put $R_x{}^j(t)=\mathscr{F}_\xi[v^j(\xi, t)]$ and let $u_j\in\mathscr{D}_{L^2}{}^1$, then from lemma 4.3 we see that

$$u(x, t)=\sum_{j=1}^{N} R_x{}^j(t)*u_j(x) \tag{4.67}$$

is a solution of (4.60) such that $u(t)=u(x, t)\in\mathscr{E}_t{}^1(L^2)$ $(t\geqslant 0)$ (in case $f=0$).

We omit the proof of (2) because it is almost the same as that of theorem 4.10.

Q.E.D.

In (4.60) if we assume A_k is Hermitian, then we see that the uniqueness of the solution is guaranteed by the theorem. Furthermore, in this case, an energy inequality is also valid as we shall see in chapter 6. In fact, let $u(t)\in\mathscr{E}_t^1(L^2)$ $(t\geqslant 0)$ and let $u(t)\in\mathscr{E}_t^0(\mathscr{D}_{L^2}^1)$, $f(t)\in\mathscr{E}_t^0(L^2)$. Then, we calculate

$$\begin{aligned}\frac{\mathrm{d}}{\mathrm{d}t}\|u(t)\|^2&=\left(\frac{\mathrm{d}}{\mathrm{d}t}u(t),u(t)\right)+\left(u(t),\frac{\mathrm{d}}{\mathrm{d}t}u(t)\right)\\&=\left(\frac{\mathrm{d}}{\mathrm{d}t}\hat{u}(\xi,t),\hat{u}(\xi,t)\right)+\left(\hat{u}(\xi,t),\frac{\mathrm{d}}{\mathrm{d}t}\hat{u}(\xi,t)\right)\\&=((2\pi\mathrm{i}A\cdot\xi+B)\,\hat{u},\hat{u})+(\hat{u},(2\pi\mathrm{i}A\xi+B)\,\hat{u})+2\,\mathrm{Re}\,(\hat{u},\hat{f})\\&=2\,\mathrm{Re}\,(Bu,u)+2\,\mathrm{Re}\,(u,f)\leqslant 2\|u\|(\gamma\|u\|+\|f\|),\end{aligned}$$

i.e.

$$\frac{\mathrm{d}}{\mathrm{d}t}\|u(t)\|\leqslant\gamma\|u(t)\|+\|f(t)\|. \tag{4.68}$$

Therefore

$$\|u(t)\|\leqslant \mathrm{e}^{\gamma t}\|u(0)\|+\int_0^t e^{\gamma(t-s)}\|f(s)\|\,\mathrm{d}s. \tag{4.69}$$

Chapter 5

Evolution equations

1 *Introduction*

In the previous chapter, starting from the Cauchy–Kowalewski theorem we considered the local solubility of a partial differential equation. Nevertheless, there are actually many equations in physics which do not fit in this framework. For instance, a heat equation such as

$$\frac{\partial}{\partial t} u = \Delta u, \quad \left(\Delta = \sum_{i=1}^{n} \frac{\partial^2}{\partial x_i^{\,2}}\right) \tag{5.1}$$

is an example. We seek a solution $u(x, t)$ satisfying (5.1) in $t \geqslant 0$ for a given initial datum $u(x, 0)$ at $t=0$. In this case, the hyperplane $t=0$ is a characteristic hypersurface of an operator $L = \partial/\partial t - \Delta$. (see definition 4.1.) Therefore, this is the case we excluded in the observation of the previous chapter.

In section 1, chapter 3, we treated the problem of finding solutions for given boundary conditions and initial data. This problem also lies outside the scope of chapter 4. Now, for simplicity, we consider a single equation

$$L[u] \equiv \left(\frac{\partial}{\partial t}\right)^m u + \sum_{j<m} a_{\nu, j}(x, t) \left(\frac{\partial}{\partial x}\right)^{\nu} \left(\frac{\partial}{\partial t}\right)^{j} u = f(x, t) \tag{5.2}$$

where we do not impose any restriction (unlike chapter 4) on the order of the derivatives in the x-direction appearing in the equation.

2 *The Cauchy problem*

Let us consider a Cauchy problem for (5.2). In this case, an important point is that a 'local uniqueness' property (for example, Holmgren's theorem) is no longer available. So we have to impose some restrictions on the behaviour of a solution $u(x, t)$ at the infinity of x-space. In other words, we appoint a function space which contains $u(x, t)$ by regarding t as a parameter in $u(x, t)$.

In this section, let us pick $\mathscr{D}_{L^2}{}^{\infty}$ as a function space for our purpose. In chapter 2 we have already given the definition of $\mathscr{D}_{L^2}{}^{\infty}$, but, for the reader's convenience, we repeat the definition here. By $f \in \mathscr{D}_{L^2}{}^{\infty}$ we mean $f \in C^{\infty}$, all its derivatives $\in L^2$ and so, clearly, $\mathscr{D}_{L^2}{}^{\infty}$ is a Fréchet space with semi-norms

$$p_m(f) = \|f\|_m = \Big(\sum_{|\alpha| \leqslant m} \|\mathrm{D}^{\alpha} f\|_{L^2}{}^2\Big)^{\frac{1}{2}} \quad (m = 1, 2, 3, \ldots),$$

i.e. $f_j(x) \to 0$ in $\mathscr{D}_{L^2}{}^{\infty}$ if and only if for an arbitrary derivative $\mathrm{D}^{\alpha} f_j(x)$, $\mathrm{D}^{\alpha} f_j(x) \to 0$ in L^2. It is a rather strong assumption, but we assume that the coefficients $a_{\nu,j}(x, t) \in \mathscr{B}(R^n)$, and $t \to a_{\nu,j}(x, t) \in \mathscr{B}$ are continuous.

Then we regard (5.2) as an equation of evolution as follows: (5.2) can be written as a system of equations

$$\begin{cases} \dfrac{\mathrm{d}}{\mathrm{d}t} u_0 = u_1, \\ \dfrac{\mathrm{d}}{\mathrm{d}t} u_1 = u_2, \\ \vdots \\ \dfrac{\mathrm{d}}{\mathrm{d}t} u_{m-1} = \alpha_0\left(x, t; \dfrac{\partial}{\partial x}\right) u_0 + \alpha_1\left(x, t; \dfrac{\partial}{\partial x}\right) u_1 + \cdots \\ \qquad + \alpha_{m-1}\left(x, t; \dfrac{\partial}{\partial x}\right) u_{m-1} + f, \end{cases} \tag{5.3}$$

where

$$u = u_0, \quad \frac{\partial}{\partial t} u = u_1, \quad \left(\frac{\partial}{\partial t}\right)^2 u = u_2, \ldots, \left(\frac{\partial}{\partial t}\right)^{m-1} u = u_{m-1}.$$

We say that $u(t) = (u_0(t), \ldots, u_{m-1}(t))$ is a solution of (5.3) in $\mathscr{D}_{L^2}{}^{\infty}$ if and only if $u_i \in \mathscr{E}_t{}^1(\mathscr{D}_{L^2}{}^{\infty})$ $(t \geqslant 0)$ $(i = 0, 1, 2, \ldots, m-1)$, i.e. we regard each component $u_i(t)$ as a function taking its value in $\mathscr{D}_{L^2}{}^{\infty}$ where for $t \geqslant 0$, $u_i(t)$ is continuously differentiable to first order satisfying (5.3).

In general, if $v(t) \in \mathscr{E}_t{}^1(\mathscr{D}_{L^2}{}^{\infty})$ $(T \geqslant t \geqslant 0)$, we consider a semi-norm

$$\max_{0 \leqslant t \leqslant T} p_m(v(t)) + \max_{0 \leqslant t \leqslant T} p_m\left(\frac{\mathrm{d}}{\mathrm{d}t} v(t)\right) (m = 1, 2 \ldots).$$

This also gives a Fréchet space. Using Banach's closed graph theorem as we did in theorem 4.4, we have:

THEOREM 5.1 *(The continuity of a solution for an initial value).* In (5.2) we put $f=0$. We assume that there exists a constant $T(>0)$, and for an arbitrary initial value $\Psi=(u_0(x),\ldots,u_{m-1}(x))\in\prod\mathscr{D}_{L^2}{}^\infty$ there exists a solution (in the sense which we have just described). In this case, a linear mapping from the given initial value Ψ to $(u(t),(\mathrm{d}/\mathrm{d}t)u(t),\ldots,(\mathrm{d}/\mathrm{d}t)^{m-1}u(t))$ is continuous if we consider it as a mapping from $\prod\mathscr{D}_{L^2}{}^\infty$ to $\prod\mathscr{E}_t^1(\mathscr{D}_{L^2}{}^\infty)$ $(0\leqslant t\leqslant T)$.

Proof. It is easy to see that the mapping is a closed operator.† Therefore, we can apply Banach's closed graph theorem.

Q.E.D.

Let us now assume that the coefficients of (5.2) are functions of t only, i.e.

$$L[u]\equiv\left(\frac{\partial}{\partial t}\right)^m u+\sum_{j<m}a_{\nu,j}(t)\left(\frac{\partial}{\partial x}\right)^\nu\left(\frac{\partial}{\partial t}\right)^j u=f \tag{5.4}$$

where $a_{\nu,j}(t)$ are continuous functions. We seek the unique solubility condition for a solution of (5.4).

For this purpose we introduce the following notion:

DEFINITION 5.1 *(Uniformly well-posed Cauchy problem).* Put $f=0$ in (5.4). In this case, by saying a Cauchy problem is *uniformly well posed,* we shall mean that, for an arbitrary initial datum $\Psi=(u_0(x),\ldots,u_{m-1}(x))\in\prod\mathscr{D}_{L^2}{}^\infty$ and an arbitrary initial time $t^0(0\leqslant t_0<T)$ there exists a unique solution $u(t;t_0)\in\mathscr{E}_t^m(\mathscr{D}_{L^2}{}^\infty)$ $(t_0\leqslant t\leqslant T)$ and the continuity of the solution for the given initial datum is uniform with respect to t_0.

More precisely, for arbitrary l, $\varepsilon(>0)$, there exist p, $\delta(>0)$ independent of t_0 such that if $\sum_{j=0}^{m-1}\|u_j(x)\|_p<\delta$ then

$$\sum_{j=0}^{m-1}\max_{t_0\leqslant t\leqslant T}\left\|\left(\frac{\mathrm{d}}{\mathrm{d}t}\right)^j u(x,t;t_0)\right\|_l<\varepsilon.$$

where $\|\cdot\|_s$ indicates $\mathscr{D}_{L^2}{}^s(R^n)$-norm. Then, from (5.3), we can write (5.4) as

$$M[u]\equiv\frac{\mathrm{d}}{\mathrm{d}t}u-A\left(t;\frac{\partial}{\partial x}\right)u=g;\quad u=\begin{bmatrix}u_0(x,t)\\ \vdots\\ u_{m-1}(x,t)\end{bmatrix},\quad g=\begin{bmatrix}0\\ \vdots\\ f(x,t)\end{bmatrix}. \tag{5.5}$$

† Translator's note: The original says (literally) 'the mapping is a closed graph'.

Then, we put $g=0$. In this case, we perform a Fourier transform on (5.5) with respect to the variables of the space. We have

$$\tilde{M}[v(\xi, t)] \equiv \frac{\mathrm{d}}{\mathrm{d}t} v(\xi, t) - A(t; 2\pi \mathrm{i}\xi)\, v(\xi, t) = 0. \tag{5.6}$$

This is an ordinary differential equation containing a parameter ξ.

Now, let the system of the fundamental solutions be $v^0(\xi, t; t_0), \ldots, v^m(\xi, t; t_0)$ where $t=t_0$ is the initial time, i.e.

$$v^j(\xi, t; t_0) = \begin{bmatrix} v_0{}^j(\xi, t; t_0) \\ \vdots \\ v_{m-1}{}^j(\xi, t; t_0) \end{bmatrix} \qquad (v_i{}^j(\xi, t_0; t_0) = \delta_i{}^j)$$

($\delta_i{}^j$ is Kronecker's symbol). Furthermore, we define $|v^j(\xi, t; t_0)|^2 = \sum_{i=1}^{m} |v_i{}^j(\xi, t; t_0)|^2$.

The following theorem is valid:

THEOREM 5.2 *(Petrowsky).* The necessary and sufficient condition for a Cauchy problem to be uniformly well posed for $t \in [0, T]$ with respect to the future direction is that there exist positive constants C and p and

$$|v^j(\xi, t; t_0)| \leqslant C(1+|\xi|)^p \quad (0 \leqslant t_0 \leqslant t \leqslant T, \quad j=1, 2, \ldots, m), \tag{5.7}$$

where C and p are independent of t_0.

Proof (*Necessity*). Let us assume that (5.7) is not true. In this case, for an arbitrary positive integer j, there exist $\xi^*(|\xi^*| \geqslant 2)$, t^*, $t_0{}^*$ and k, such that

$$|v^k(\xi, t^*; t_0{}^*)| \geqslant j(1+|\xi|)^j \quad (t_0{}^* < t^* < T)$$

is true in a certain neighbourhood U of ξ^*. Without loss of generality, we can assume that $\mathrm{dis}(0, U) \geqslant \frac{1}{2}|\xi^*|$ and $\max_{\xi \in U} \mathrm{dis}(0, \xi) \leqslant 2|\xi^*|$. If $\hat{f}(\xi)$ is a function such that $\|\hat{f}(\xi)\|_{L^2} = 1$ and its support lies in $\bar{U}$, then

$$u(x, t) = \int \mathrm{e}^{2\pi \mathrm{i} x\xi} v^k(\xi, t; t_0{}^*) \hat{f}(\xi)\, \mathrm{d}\xi \in \mathscr{E}_t{}^1(\mathscr{D}_{L^2}{}^\infty)$$

is a solution of $M[u]=0$ for $t \geqslant t_0$, and, from Plancherel's theorem,

$$\begin{aligned} \|u(x, t^*)\| &= \left(\int |v^k(\xi, t^*; t_0{}^*)|^2 |\hat{f}(\xi)|^2\, \mathrm{d}\xi\right)^{\frac{1}{2}} \\ &\geqslant j(1+\tfrac{1}{2}|\xi^*|)^j. \end{aligned} \tag{5.8}$$

On the other hand, we have

$$\sum_{|\alpha|\leqslant l}\left\|\left(\frac{\partial}{\partial x}\right)^{\alpha} u(x, t_0{}^*)\right\| = \sum \left(\int |(2\pi i\xi)^{\alpha} v^k(\xi, t_0{}^*; t_0{}^*)|^2 |\hat{f}(\xi)|^2 \,d\xi\right)^{\frac{1}{2}}$$
$$\leqslant c(l)\,(1+2|\xi^*|)^l\,, \tag{5.9}$$

where $c(l)$ is a constant related only to l.

We note that, in (5.8), j can be taken arbitrarily large. So that, comparing (5.9) with (5.8), we see that this is a contradiction to our hypothesis that the Cauchy problem is uniformly well posed. Therefore, the necessity of (5.7) is proved.

(Sufficiency) Let† $\mathscr{F}[v_0{}^j(\xi, t; t_0)] = R_x{}^j(t; t_0)$

We define

$$u(x, t; t_0) = \sum_{j=0}^{m-1} R_x{}^j(t; t_0) \underset{(x)}{*} u_j(x) + \int_{t_0}^{t} R_x{}^{m-1}(t, \tau) \underset{(x)}{*} f(x, \tau)\, d\tau. \tag{5.10}$$

Under the condition $f(x, t) \in \mathscr{E}_t{}^0(\mathscr{D}_{L^2}{}^\infty)$ $(0 \leqslant t \leqslant T)$, we see, by lemma 4.3, that (5.10) is a solution for (5.4) satisfying the given conditions. The uniqueness of the solution follows from the fact that the Fourier transform $\hat{u}(\xi, t)$ of the solution $u(x, t)$ is a solution of $\tilde{M}[\hat{u}(\xi, t)] = \hat{g}(\xi, t)$ in (5.6).

Q.E.D.

If $a_{\nu, j}$ are constant coefficients, we write $\lambda_j(\xi)$ $(j = 1, 2, \ldots, m)$ for the roots of

$$P(\lambda; \xi) = \lambda^m + \sum_{j<m} a_{\nu, j}(i\xi)^\nu \lambda^j = 0\,. \tag{5.11}$$

Then, we have:

THEOREM 5.3 *(Hadamard)*. The necessary and sufficient condition for the Cauchy problem of (5.4) to be uniformly well posed in the future direction in $t \in [0, T]$ (T is arbitrary) is that there exist certain constants C and p and

$$\operatorname{Re} \lambda_j(\xi) \leqslant p \log(1 + |\xi|) + C \quad (\xi \in R^n, \quad j = 1, 2, \ldots, m)$$

where C and p are dependent on T, but independent of ξ.

† $v_0{}^j(\xi, t; t_0)$ $(j = 0, 1, \ldots, m-1)$ satisfies $(d/dt)^i v_1{}^j(\xi, t; t_0)_{t \to t_0} = \delta_i{}^j$, and is a solution of $\tilde{L}[v] \equiv \left(\frac{d}{dt}\right)^m v + \sum a_{\nu, j}(t)\,(2\pi i\xi)^\nu \left(\frac{d}{dt}\right)^j v = 0.$

Proof. The basic principle of the proof is the same as that of theorem 4.6. Therefore, we leave it to the reader.

3 *Laplace transforms and semi-groups*

We have used a Fourier transform with respect to a space variable. Another method is the one using a Laplace transform with respect to a time variable, which is a familiar method in classical mathematical physics. Using this method, we can argue the case when the coefficients of a partial differential operator are variables similarly to the case when they are constants. Also, at least formally, even when a boundary condition is given, we can argue the case within the framework of a Cauchy problem. But, one disadvantage is that it is difficult to formulate a problem in a rigorous manner.

To overcome this shortcoming we employ a 'semi-group' method which lies in the same conceptual sphere as a Laplace transform. Let our equation be

$$\frac{\mathrm{d}}{\mathrm{d}t}u(t)=Au(t), \tag{5.12}$$

where A is a linear operator defined on a Banach space E, and in general, its boundedness is not presupposed. Nevertheless, if A is not bounded, then we assume that the domain of the definition of A is a dense subspace of E, and A is a closed operator.

By saying that $u(t)\,(t\geqslant 0)$ is a *solution* of (5.12) we mean that $t\to u(t)\,(t\geqslant 0)$ is a once continuously differentiable function of t having its value in E, i.e. $u(t)\in\mathscr{E}_t^1(E)$ $(t\geqslant 0)$, $u\in\mathscr{D}(A)$ and u satisfies (5.12). First, we shall find a solution of (5.12) by a constructive method. For this purpose, let us assume that there exists a solution $u(t)$ for a given initial datum $u_0\in D(A)$, and

$$\|u(t)\|\leqslant C\,\mathrm{e}^{\beta t} \tag{5.13}$$

is true as $t\to+\infty$, where C and β are appropriate positive constants.

In this case, as we have said in section 10, chapter 2, the Laplace transform of the solution $u(t)$ (more exactly, if $t<0$, we put $u(t)=0$ to obtain $Y(t)u(t)$) exists when $\mathrm{Re}\,p>\beta$,

$$U(p)=\int_0^\infty \mathrm{e}^{-pt}u(t)\,\mathrm{d}t \qquad (\mathrm{Re}\,p=\xi>\beta) \tag{5.14}$$

is a holomorphic function of p having its value in E, and

$$AU(p)=pU(p)-u(0). \tag{5.15}$$

In fact, if we put $U_n(p)=\int_0^n e^{-pt}u(t)dt$, then obviously, $U_n(p)\to U(p)$. Also, $AU_n(p)=\int_0^n e^{-pt}Au(t)dt$, so that

$$AU_n(p)=\int_0^n e^{-pt}\frac{d}{dt}u(t)\,dt=[e^{-pt}u(t)]_0^n+p\int_0^n e^{-pt}u(t)\,dt$$

converges to $-u(0)+p\int_0^\infty e^{-pt}u(t)dt$ as $n\to+\infty$, where A is a closed operator.

Therefore, (5.15) is true and can be written as

$$(pI-A)U(p)=u(0). \tag{5.16}$$

In this case, if resolvent$(pI-A)^{-1}$ exists Re$p>\beta$, then, because $u(t)$ is once continuously differentiable, a Laplace's inversion formula (theorem 2.19) can be established, and for $t>0$,

$$\begin{aligned} u(t) &= \lim_{A\to+\infty}\frac{1}{2\pi i}\int_{\xi-iA}^{\xi+iA} e^{pt}U(p)\,dp \\ &= \lim_{\omega\to+\infty}\frac{1}{2\pi i}\int_{\xi-i\omega}^{\xi+i\omega} e^{pt}(pI-A)^{-1}u(0)\,dp. \end{aligned} \tag{5.17}$$

This is a representation of the solution using the initial value $u(0)\in\mathscr{D}(A)$ and a resolvent set (Laplace representation), and is important when we examine the properties of solutions.

Now we come back to our original problem; making sure that $u(t)$ defined in (5.17) with some appropriate conditions imposed on the resolvent set is in fact a solution of (5.12). We note that it is difficult to discuss the problem in a general form. To ease this difficulty we use the notion of 'semi-group' to derive the existence theorem of solution.† We have assumed the existence of a resolvent $(pI-A)^{-1}$. We examine this point more closely. Suppose the following conditions are given.

(1) There exists a unique solution $u(t)$ for an arbitrary initial value $u_0\in\mathscr{D}(A)$.

(2) $\|u(t)\|\leqslant Ce^{\beta t}\|u_0\|$.

In this case, if we put $u(t)=T_tu_0$ $(t\geqslant0)$, then, from (2), T_t can be regarded as a bounded operator defined on E. Furthermore, we see that

† These results are mainly due to Hille and Yosida.

$$\left.\begin{array}{l}\text{(1) } T_t \text{ is a linear operator satisfying } \|T_t\| \leqslant Ce^{\beta t} \text{ where } C, \beta \\ \text{are appropriate constants.} \\ \text{(2) } T_{t+s}=T_tT_s\ (t, s\geqslant 0) \text{ and } \lim\limits_{\eta\to+0} T_\eta u_0=u_0\ (u_0\in E).\end{array}\right\} \quad (5.18)$$

For $u_0\in\mathscr{D}(A)$,

$$\frac{\mathrm{d}}{\mathrm{d}t}(T_t u_0)_{t=0}=\lim_{\eta\to+0}\frac{T_\eta-I}{\eta}u_0=Au\,. \tag{5.19}$$

Now, in general, we call $\{T_t\}_{t\geqslant 0}$ *semi-groups*, if they satisfy the conditions (1) and (2) of (5.18), and if $T_t u_0$ $(u_0\in E)$ is a continuous function of t.

Instead of (5.12) we define an operator A as satisfying

$$\lim_{\eta\to+0}\frac{T_\eta-I}{\eta}u=Au\,, \tag{5.20}$$

i.e. the domain of the definition of A is a set of u which satisfies (5.20) (lim exists for u), and the set forms a subspace of E. Generally speaking, A is not a bounded operator. We call A an *infinitesimal generator* of T_t. Now, for the Laplace transform of T_t,

$$T_t(\mathrm{e}^{-pt})x=\int_0^\infty \mathrm{e}^{-pt}T_t x\,\mathrm{d}t \qquad (x\in E, \operatorname{Re} p=\xi>\beta)\,, \tag{5.21}$$

the following proposition can be established.

Proposition 5.1.

(1) For an arbitrary $x\in E$, $T_t(\mathrm{e}^{-pt})x\in\mathscr{D}(A)$ and $(pI-A)T_t(\mathrm{e}^{-pt})x=x$.

(2) For $x\in\mathscr{D}(A)$, $T_t(\mathrm{e}^{-pt})(pI-A)x=x$.

(3) A is a closed operator and its domain of definition is dense.

Proof.

(1) Let us recall that the infinitesimal generator A is defined by (5.20). We put $A_\eta=(T_\eta-I)/\eta$ $(\eta>0)$, where A_η is a bounded operator. We see

$$\begin{aligned} A_\eta T_t(\mathrm{e}^{-pt})x &= \frac{T_\eta-I}{\eta}T_t(\mathrm{e}^{-pt})x=\frac{1}{\eta}\int_0^\infty \mathrm{e}^{-pt}(T_{t+\eta}-T_t)x\,\mathrm{d}t \\ &= \frac{1}{\eta}\int_\eta^\infty (\mathrm{e}^{-p(t-\eta)}-\mathrm{e}^{-pt})T_t x\,\mathrm{d}t-\frac{1}{\eta}\int_0^\eta \mathrm{e}^{-pt}T_t x\,\mathrm{d}t \\ &= \frac{\mathrm{e}^{p\eta}-1}{\eta}\int_\eta^\infty \mathrm{e}^{-pt}T_t x\,\mathrm{d}t-\frac{1}{\eta}\int_0^\eta \mathrm{e}^{-pt}T_t x\,\mathrm{d}t\,. \end{aligned}$$

We note that

$$A_\eta T_t(e^{-pt})x \to p\int_0^\infty e^{-pt}T_t x\, dt - x$$

as $\eta \to +0$.

(2) First note that for $x \in \mathscr{D}(A)$, $T_t x \in \mathscr{D}(A)$ and

$$\frac{d}{dt}(T_t x) = AT_t x = T_t Ax. \tag{5.22}$$

In fact, $(1/\eta)(T_{t+\eta} - T_t)x = A_\eta T_t x = T_t A_\eta x$ where $\eta > 0$. So that if $x \in \mathscr{D}(A)$ then the last term tends to $T_t Ax$ as $\eta \to +0$. If $t > 0$, $\eta > 0$, then we write $(1/-\eta)(T_{t-\eta} - T_t)x = T_{t-\eta}A_\eta x$ and, from this, we see that $T_{t-n}A_\eta x \to T_t Ax$ as $\eta \to +0$.

Assume $x \in \mathscr{D}(A)$, then

$$\begin{aligned}
T_t(e^{-pt})Ax &= \int_0^\infty e^{-pt}T_t Ax\, dt = \lim_{L\to+\infty}\int_0^L e^{-pt}T_t Ax\, dt \\
&= \lim_{L\to\infty}\int_0^L e^{-pt}\frac{d}{dt}(T_t x)\, dt \\
&= \lim_{L\to+\infty}[e^{-pt}T_t x]_0{}^L + p\lim_{L\to\infty}\int_0^L e^{-pt}T_t x\, dt \\
&= -x + pT_t(e^{-pt})x.
\end{aligned}$$

(3) We have for $\lambda > \beta$

$$\lambda\int_0^\infty e^{-\lambda t}T_t x\, dt \to x \in E \quad (\lambda \to +\infty),^\dagger$$

and from the result (1), we have $\int_0^\infty e^{-\lambda t}T_t x dt \in \mathscr{D}(A)$. Hence $\mathscr{D}(A)$ is dense in E. Next, from (1) and (2) we see that $T_t(e^{-pt})$ is nothing but the resolvent of A. So that $(pI - A)^{-1}$ is bounded, hence $pI - A$ is a closed operator, therefore A is also a closed operator.

Q.E.D.

Summing up, we have:

† To see this, choose a small $\delta(>0)$ and divide the integrand into $[0, \delta]$ and $[\delta, \infty)$, then note that $\int_0^\delta \lambda e^{-\lambda t}\, dt = 1 - e^{-\lambda\delta} \to 1 \quad (\lambda \to +\infty)$ and $\lambda\int_\delta^\infty e^{-\lambda t}T_t x\, dt \to 0 \quad (\lambda \to +\infty)$.

THEOREM 5.4. Let us write the infinitesimal generator of a semi-group T_t satisfying (5.18) as A. There exists a resolvent $(pI-A)^{-1}$ of A for $\mathrm{Re}\,p>\beta$. It is equal to the Laplace transform of T_t, i.e.

$$T_t x \sqsupset (pI-A)^{-1}x \quad (x\in E,\ \xi>\beta). \tag{5.23}$$

In this case, we also have

$$\|(pI-A)^{-m}\| \leqslant \frac{C}{(\mathrm{Re}\,p-\beta)^m} \quad (m=1, 2, 3, \ldots). \tag{5.24}$$

Proof. (5.23) is nothing but the previous proposition. For (5.24), we observe

$$(pI-A)^{-1}x = \int_0^\infty e^{-pt}T_t x\,dt \quad \mathrm{Re}\,p>\beta.$$

Differentiating this m times with respect to the parameter p, we have

$$(-1)^m m!\,(pI-A)^{-m-1}x = \int_0^\infty e^{-pt}(-t)^m T_t x\,dt.$$

From (5.18), we have (5.24).

Q.E.D.

We justify (5.17) by the following:

COROLLARY. Under the same condition as theorem 5.4, if $x\in\mathscr{D}(A)$, then for $t>0$,

$$T_t x = \lim_{\omega\to+\infty} \frac{1}{2\pi i}\int_{\xi-i\omega}^{\xi+i\omega} e^{pt}(p-A)^{-1}x\,dp \quad (\xi>\beta) \tag{5.25}$$

where the convergence of the right hand side term is uniform in t in the interval $[\varepsilon, 1/\varepsilon]$ ($\varepsilon>0$ is arbitrary).

Proof. $x\in\mathscr{D}(A)$, so that by (5.22), $T_t x$ is of bounded variation in an arbitrary finite interval, and it is not continuous only at $t=0$. Therefore, from theorem 2.19, we have (5.25).

Q.E.D.

It is important to note that the converse of theorem 5.4 is also true; it gives the existence theorem of the evolution equation (5.12).

THEOREM 5.5 *(Hille–Yosida)*. For a closed operator A, having a dense domain of definition, if there exists β such that for $\lambda>\beta$, there exists a resolvent $(\lambda I-A)^{-1}$ of A satisfying

$$\|(\lambda I-A)^{-m}\| \leqslant \frac{C}{(\lambda-\beta)^m} \quad (\lambda>\beta,\ m=1,2,3,\ldots), \tag{5.26}$$

then there exists a unique semi-group T_t having an infinitesimal generator, A. In this case T_t satisfies (5.18).

COROLLARY.† In the above theorem, if

$$\|(\lambda I-A)^{-1}\| \leqslant \frac{1}{\lambda-\beta} \quad (\lambda>\beta), \tag{5.27}$$

then the conclusions of theorem 5.5 are valid. (In fact, in this case (5.26) is satisfied.)

Proof of theorem 5.5. Let $A_1=A-\beta I$. From (5.26), we have

$$\|(\lambda I-A_1)^{-m}\| \leqslant \frac{C}{\lambda^m} \quad (\lambda>0). \tag{5.28}$$

On the other hand, if we can prove the existence of a semigroup S_t having an infinitesimal generator A_1, and if $\|S_t\| \leqslant C$, then $T_t=e^{\beta t}S_t$ obviously satisfies the conditions of the theorem.

Therefore, without loss of generality, we can assume $\beta=0$ in (5.26), i.e.

$$\|(\lambda I-A)^{-m}\| \leqslant \frac{C}{\lambda^m} \quad (\lambda>0,\ m=1,2,3,\ldots). \tag{5.26$'$}$$

Let

$$J_\lambda=\left(I-\frac{A}{\lambda}\right)^{-1} \quad (\lambda>0). \tag{5.29}$$

Then, we shall see:

(1) $\|J_\lambda{}^m\| \leqslant C \quad (m=1,2,3,\ldots)$

(2) For $x\in\mathscr{D}(A)$, $AJ_\lambda x=J_\lambda Ax=\lambda(J_\lambda-I)x$. From this we have for $x\in E$,

$$AJ_\lambda x=\lambda(J_\lambda-I)x, \tag{5.30}$$

$$J_\lambda x\to x \quad (\lambda\to+\infty) \quad (x\in E). \tag{5.31}$$

† Some authors call this result the Hille–Yosida theorem.

In fact, for $x \in \mathscr{D}(A)$, we have $(J_\lambda - I)x = (1/\lambda)J_\lambda Ax \to 0$ $(\lambda \to +\infty)$. So that from $\|J_\lambda\| \leqslant C$, $\mathscr{D}(A)$ is dense, therefore, (5.31) follows.

In general, when A and B are bounded operators, we define

$$\exp(A) = \sum_{j=0}^{\infty} \frac{A^j}{j!}.$$

In this case we have

$$\|\exp(A)\| \leqslant \exp(\|A\|).$$

If A and B are commutative, then we see

$$\exp(A+B) = \exp(A)\exp(B),$$

$$\frac{\mathrm{d}}{\mathrm{d}t}\exp(tA) = A\exp(tA) = \exp(tA)A.$$

On the other hand, $AJ_\lambda = \lambda(J_\lambda - I)$ is a bounded operator, so that, from

$$\exp(tAJ_\lambda) = \exp\{t\lambda(J_\lambda - I)\} = \exp(t\lambda J_\lambda)\exp(-t\lambda I)$$

and from the property of J_λ (see (1), (2)), for $t \geqslant 0$, we have

$$\|\exp(tAJ_\lambda)\| \leqslant C\exp(t\lambda)\exp(-t\lambda) = C. \tag{5.32}$$

Furthermore,

$$\frac{\mathrm{d}}{\mathrm{d}t}\exp(tAJ_\lambda) = AJ_\lambda\exp(tAJ_\lambda) = \exp(tAJ_\lambda)AJ_\lambda = \exp(tAJ_\lambda)J_\lambda A.$$

Note that the last equality holds if $x \in \mathscr{D}(A)$.

It is obvious that J_λ, J_μ $(\lambda, \mu > 0)$ are commutative, so that $AJ_\mu (= \mu(J_\mu - I))$ and $\exp(tAJ_\lambda)$ are commutative. We write $T_t^{(\lambda)} = \exp(tAJ_\lambda)$. For $x \in \mathscr{D}(A)$, we have

$$T_t^{(\lambda)} - T_t^{(\mu)} = x\int_0^t \frac{\mathrm{d}}{\mathrm{d}s}\{T_{t-s}{}^{(\mu)}T_s{}^{(\lambda)}x\}\,\mathrm{d}s = \int_0^t T_{t-s}{}^{(\mu)}T_s{}^{(\lambda)}(J_\lambda - J_\mu)Ax\,\mathrm{d}s.$$

From (5.31) and (5.32), this equation converges uniformly in a finite interval of $t \geqslant 0$ as $\lambda, \mu \to +\infty$.

Furthermore, from the fact that $\mathscr{D}(A)$ is dense and (5.32), we see that $T_t^{(\lambda)}x - T_t^{(\mu)}x$ converges uniformly in the finite interval $t \geqslant 0$ for an arbitrary $x \in E$.

We write the limit of the convergence as $T_t x$. Now, $T_t x$ is continuous for $t \geqslant 0$, and

(1) $\|T_t x\| \leqslant C\|x\|$,

(2) $T_{t+s} = T_t \cdot T_s$.

To prove (2) we observe $T_{t+s}{}^{(\lambda)}x = T_t^{(\lambda)} T_s^{(\lambda)} x$ as $\lambda \to +\infty$. We see also $T_0 x = x$, so that

(3) $T_t x \to x$, as $t \to +0$.

Finally, let us prove that the infinitesimal generator of T_t is A. To do this, we write the infinitesimal generator of T_t as A', and show $A' \supset A$. In fact, from proposition 5.1, for $\lambda > 0$, $(\lambda I - A')$ is a bijection of $\mathscr{D}(A')$ onto E. Therefore, by our assumption, $(\lambda I - A)$ is also a bijection of $\mathscr{D}(A')$ onto E. Hence, from $\mathscr{D}(A) \subset \mathscr{D}(A')$, we have $\mathscr{D}(A) = \mathscr{D}(A')$. To prove $A' \supset A$ we proceed as follows. For $x \in \mathscr{D}(A)$ we have

$$T_t^{(\lambda)} x - x = \int_0^t T_t^{(\lambda)} J_\lambda A x \, dt,$$

so that

$$T_t x - x = \int_0^t T_t A x \, dt$$

as $\lambda \to +\infty$. Hence,

$$\lim_{t \to +0} (T_t x - x)/t = Ax,$$

Proof of uniqueness. Let T_t be an arbitrary semi-group having an infinitesimal generator A. In this case, we do not assume (1) in (5.18). We have

$$T_t x - \exp(tAJ_\lambda)x = \int_0^t T_{t-s} \exp(sAJ_\lambda)(A - AJ_\lambda)x \, ds \quad (x \in \mathscr{D}(A)),$$

where we have used $AJ_\lambda \supset J_\lambda A$ to establish the equality. From this, $\exp(tAJ_\lambda)x \to T_t x$ as $\lambda \to +\infty$.

Q.E.D.

We give an existence theorem for an equation of evolution

$$\frac{d}{dt} u(t) = Au(t) + f(t), \tag{5.33}$$

in a Banach space E.

THEOREM 5.6 *(Existence theorem)*. Let A be a closed operator having a dense domain of definition. We assume either (5.26) or (5.27). We also assume that $t \to f(t)$ and $t \to Af(t)$ are continuous for $t \in [0, T]$ for the right hand side $f(t)$. In this case, we can obtain a unique solution $u(t) \in \mathscr{E}_t^1(E)$ $(0 \leqslant t \leqslant T)$ for an arbitrary initial value $u_0 \in \mathscr{D}(A)$, and

$$u(t) = T_t u_0 + \int_0^t T_{t-s} f(s)\,\mathrm{d}s, \tag{5.34}$$

where T_t is a semi-group which is known to exist by the previous theorem.

Proof. To show (5.34) is in fact a solution, it is enough to prove

$$\frac{\mathrm{d}}{\mathrm{d}t}\psi(t) = A\psi(t) + f(t)$$

where

$$\psi(t) = \int_0^t T_{t-s} f(s)\,\mathrm{d}s$$

(see the previous theorem).

We note that $\psi \in \mathscr{D}(A)$ and

$$A\psi(t) = \int_0^t T_{t-s} Af(s)\,\mathrm{d}s.$$

To prove this, first recall the definition of integration and note that A is a closed operator. Therefore, $A\psi(t)$ is also continuous. On the other hand, for $\eta > 0$,

$$\frac{\psi(t+\eta) - \psi(t)}{\eta} = \frac{1}{\eta}\int_t^{t+\eta} T_{t+\eta-s} f(s)\,\mathrm{d}s + \frac{T_\eta - I}{\eta}\int_0^t T_{t-s} f(s)\,\mathrm{d}s.$$

So that as $\eta \to +0$, $\psi_+'(t) = f(t) + A\psi(t)$, where the right hand side terms are continuous functions of t. Therefore, $\psi'(t) = f(t) + A\psi(t)$. We now prove uniqueness. We use J_λ which we defined in the previous theorem. We see easily that the solution $u_\lambda(t)$ of

$$\frac{\mathrm{d}}{\mathrm{d}t} u_\lambda(t) = (AJ_\lambda) u_\lambda(t) + f(t)$$

is uniquely determined for a given initial value u_0 because AJ_λ is a bounded operator. We put $u(t)-u_\lambda(t)=v_\lambda(t)$ for the solution $u(t)$ of (5.33). Then, we have

$$\frac{\mathrm{d}}{\mathrm{d}t}v_\lambda(t)=(AJ_\lambda)v_\lambda(t)+(A-AJ_\lambda)u(t).$$

Therefore, using the same symbol as in the previous theorem, we have

$$v_\lambda(t)=\int_0^t T_{t-s}{}^{(\lambda)}(A-AJ_\lambda)u(s)\,\mathrm{d}s \qquad (u\in\mathscr{D}(A))$$

(note that $v_\lambda(0)=0$), and this is equal to

$$\int_0^t T_{t-s}{}^{(\lambda)}(I-J_\lambda)Au(s)\,\mathrm{d}s.$$

$\|T_{t-s}{}^{(\lambda)}(I-J_\lambda)Au(s)\|$ is uniformly bounded for λ as s runs through a bounded interval. Also, if we fix s and make $\lambda\to+\infty$, we see that this value tends to 0. By Lebesgue's theorem, we see $v_\lambda(t)\to 0$, i.e. $u(t)$ is given by the limit of $u_\lambda(t)$. Hence, $u(t)$ is uniquely determined.

Q.E.D.

4 *Parabolic semi-groups*

In section 3, when we described the Laplace representation (5.17), we said that it is difficult to obtain an existence theorem of the solution starting from the representation. In some cases, the situation is different and we can reach the theorem in a natural way. For example, in the case of the heat equation (5.1), we see that $A=\Delta$ is an elliptic operator, and if we impose an initial condition, the resolvent of A has a good property. In fact, in theorem 3.19 a Green's operator $G_\lambda=(\lambda I-A)^{-1}$ satisfies (3.82). In the next section, we consider an operator A which has a resolvent satisfying (3.82).

The condition for a resolvent

Let us consider a and $\theta(0<\theta<\frac{1}{2}\pi)$. Consider a fan-shaped domain Σ (see the diagram) bounded by two lines which radiate from the point $(a, 0)$ and make angles θ, $-\theta$ measured from the negative direction with the real axis of the p-plane. There exists a resolvent $(pI-A)^{-1}$ in the complementary set of Σ, and

$$\|(pI-A)^{-1}\|\leqslant\frac{C}{|p|+1}, \tag{5.35}$$

where A is a closed operator having a dense domain of definition.

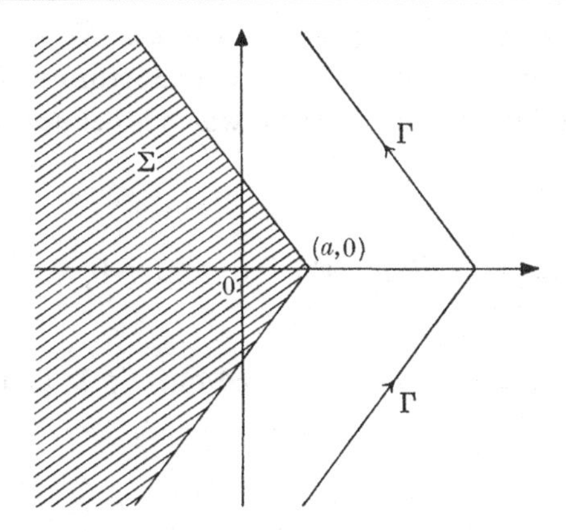

Now, let us assume that there exists a semi-group T_t having an infinitesimal generator A. In this case, we can write the Laplace representation (5.25) in the following form. First we note that in (4.25) the Laplace axis can be any line providing it satisfies $\xi > a$. If $\xi > 0$ and $t > 0$, we have

$$\lim_{\omega\to+\infty} \frac{1}{2\pi i}\int_{\xi-i\omega}^{\xi+i\omega} e^{pt}\frac{1}{p}x\,dp = \frac{1}{2\pi i}\int_{\xi-i\infty}^{\xi+i\infty}\frac{1}{t}e^{pt}\frac{1}{p^2}x\,dp = \frac{1}{2\pi i}\int_\Gamma \frac{1}{t}e^{pt}\frac{dp}{p^2}x$$

where Γ are fan-shaped lines (see the diagram) parallel to the boundary of Σ and situated on the right hand side of the origin.

Obviously,

$$\frac{1}{2\pi i}\int_\Gamma \frac{1}{t}e^{pt}\frac{dp}{p^2} = \operatorname*{Res}_{p=0}\left[\frac{1}{t}e^{pt}\frac{1}{p^2}\right] = 1,$$

where Res indicates a residue. Therefore, from (5.25) and (5.35)

$$\begin{aligned} T_t x - x &= \frac{1}{2\pi i}\int_{\xi-i\infty}^{\xi+i\infty} e^{pt}(pI-A)^{-1}Ax\frac{dp}{p} \qquad (x\in\mathscr{D}(A),\ \xi>a,\ t>0)\\ &= \frac{1}{2\pi i}\int_\Gamma e^{pt}(pI-A)^{-1}Ax\frac{dp}{p}. \end{aligned}$$

From

$$\frac{1}{2\pi i}\int_\Gamma e^{pt}\frac{dp}{p} = 1$$

we have

$$T_t x = \frac{1}{2\pi i}\int_\Gamma e^{pt}(pI-A)^{-1}x\,dp \qquad (t>0) \tag{5.36}$$

for $x\in E$. In fact, for $x\in\mathscr{D}(A)$, (5.36) is obvious. We know that $\mathscr{D}(A)$ is dense, so that, by the continuity of T_t we have (5.36) for $x\in E$.

Conversely, we write S_t for an operator which is defined by the right hand side terms of (5.36), i.e.

$$S_t x=\frac{1}{2\pi i}\int_{\Gamma} e^{pt}(pI-A)^{-1}x\,dp \quad (t>0,\, x\in E) \tag{5.37}$$

where we assume that $(pI-A)^{-1}$ is defined outside Σ, and (5.35) is valid.†

Now, we have:

PROPOSITION 5.2.

(1) $\|S_t\|\leqslant C\,e^{\beta t}, \quad \lim_{t\to+0} S_t x=x,$

(2) $\lim_{t\to+0}((S_t x-x)/t)=Ax \quad$ for $x\in\mathscr{D}(A)$,

(3) $S_{s+t}=S_s S_t \quad (s,\,t\geqslant 0)$.

Proof.

(1) For the estimate of $\|S_t\|$, if $0<t\leqslant 1$, we proceed as follows (when $t\geqslant 1$ it is obvious). First, in (5.37) Γ moves depending on t. We pick a (standard) Γ_1. Let Γ_λ $(\lambda\geqslant 1)$ be the trace of $\{\lambda p\}$ as p moves along Γ_1. Obviously, Γ_λ lies within the complementary set of Σ. We consider the integral of (5.37) along $\Gamma_{1/t}$ depending on t.

Furthermore, by the change of variable $pt=p'$, we have a mapping from $\Gamma_{1/t}$ to Γ_1 so that

$$S_t x=\frac{1}{2\pi i}\cdot\frac{1}{t}\int_{\Gamma_1} e^{p}\left(\frac{p}{t}I-A\right)^{-1} x\,dp.$$

From

$$\left\|\left(\frac{p}{t}-A\right)^{-1}\right\|\leqslant\frac{C}{\dfrac{|p|}{t}+1}\leqslant\frac{t}{|p|}C,$$

we have

$$\|S_t\|\leqslant\frac{C}{2\pi}\int_{\Gamma_1} e^{\operatorname{Re} p}\frac{1}{|p|}\,d|p|=C'<+\infty.$$

---

† We take for Γ fixed, fan-shaped lines which lie on the right hand side of the origin, parallel with the boundary of Σ.

Next, if $t>0$, $x\in\mathscr{D}(A)$, then

$$S_t x - x = \frac{1}{2\pi i}\int_\Gamma e^{pt}\left\{(pI-A)^{-1} - \frac{I}{p}\right\} x\, dp$$

$$= \frac{1}{2\pi i}\int_\Gamma e^{pt}(pI-A)^{-1}Ax\,\frac{dp}{p}$$

$$= \frac{1}{2\pi i}\int_{\xi-i\infty}^{\xi+i\infty} e^{pt}(pI-A)^{-1}\,Ax\frac{dp}{p} \quad (\xi>a).$$

Obviously, the representation of the right hand side shows $S_t x - x$ is continuous for $t\geqslant 0$.

Therefore,

$$\lim_{t\to+0}(S_t x - x) = \frac{1}{2\pi i}\int_{\xi-i\infty}^{\xi+i\infty}(pI-A)^{-1}Ax\,\frac{dp}{p}.$$

If $\xi\to+\infty$ then the right hand side $\to 0$, so that the right hand side is 0. From the boundedness of $\|S_t\|$, for $x\in E$, $\lim_{t\to+0} S_t x = x$.

(2) If $t>0$, then,

$$\frac{d}{dt}(S_t x) = S_t{}'x = \frac{1}{2\pi i}\int_\Gamma p e^{pt}(pI-A)^{-1}x\,dp \quad (x\in E).$$

For $x\in\mathscr{D}(A)$, the right hand side can be written as

$$\frac{1}{2\pi i}\int_\Gamma e^{pt}\,dp\cdot x + \frac{1}{2\pi i}\int e^{pt}(pI-A)^{-1}Ax\,dp.$$

The first term is 0, therefore, $dS_t x/dt = S_t Ax$ $(x\in\mathscr{D}(A))$, and $S_t(Ax)$ is continuous.

(3) Let t, $t'>0$, and let $x\in E$. We have

$$S_t\cdot S_{t'}x = \frac{1}{(2\pi i)^2}\int_\Gamma \exp(pt)\,(pI-A)^{-1}dp\int_{\Gamma'}\exp(qt')(qI-A)^{-1}x\,dq,$$

where Γ' is obtained by displacing Γ along the real-axis in the positive direction. By Fubini's theorem, the right hand side is

$$\left(\frac{1}{2\pi \mathrm{i}}\right)^2 \int_\Gamma \int_{\Gamma'} \exp(pt)\exp(qt')(pI-A)^{-1}(qI-A)^{-1}x \, \mathrm{d}p \, \mathrm{d}q \, .$$

By the formula of resolvent, this is equal to

$$\left(\frac{1}{2\pi \mathrm{i}}\right)^2 \int_\Gamma \int_{\Gamma'} \exp(pt)\exp(qt') \frac{1}{q-p} [(pI-A)^{-1}-(qI-A)^{-1}]x \, \mathrm{d}p \, \mathrm{d}q \, .$$

We note that

$$\int_{\Gamma'} \exp(qt') \frac{\mathrm{d}q}{q-p} = 2\pi \mathrm{i} \exp(pt'), \qquad \int_\Gamma \exp(pt) \frac{\mathrm{d}p}{q-p} = 0 \, .$$

So that, again by using Fubini's theorem, we have

$$\frac{1}{2\pi \mathrm{i}} \int_\Gamma \exp[p(t+t')](pI-A)^{-1}x \, \mathrm{d}p = S_{t+t'}x \, .$$

Q.E.D.

By proposition 5.2, we see that S_t is a semi-group satisfying the condition (5.18), so that theorem 5.4 is true. Let us write A' for the infinitesimal generator of S_t. By (2) of proposition 5.2, we have $A' \supset A$. If $\mathrm{Re}\, p$ is sufficiently large, $(pI-A')$ and $(pI-A)$ are bijective linear mappings from $\mathscr{D}(A')$, $\mathscr{D}(A)$ respectively onto E. Hence, $\mathscr{D}(A)' = \mathscr{D}(A)$, and $A' = A$.

Now we have:

THEOREM 5.7. Under the assumption (5.35), the semi-group S_t which has an infinitesimal generator A is uniquely determined; S_t is given by (5.37). This semi-group S_t has the property that for $x \in E$, $t > 0$, $S_t x \in \mathscr{D}(A)$, and

$$AS_t x = S_t' x = \frac{1}{2\pi \mathrm{i}} \int_\lambda \mathrm{e}^{pt} p(pI-A)^{-1} x \, \mathrm{d}p \, . \tag{5.38}$$

Also, we have

$$\|S_t'\| \leqslant \frac{C'}{t} \mathrm{e}^{\beta t} \, . \tag{5.39}$$

Proof. The uniqueness of S_t follows from theorem 5.5. In fact, by theorem 5.4, the Laplace transform $(pI-A)^{-1}$ of S_t satisfies the assumption (5.26) of theorem 5.5. In the definition (5.37) of S_t, we observe that the term is differentiable with respect to t under the sign of the integration. So that (5.38) is valid. To see (5.39), we consider the representation of the estimation of $\|S_t\|$ in proposition 5.2, i.e. for $0<t\leqslant 1$ we have

$$S_t'x=\frac{1}{2\pi i t^2}\int_{\Gamma_1} e^p p\left(\frac{p}{t}I-A\right)^{-1} x\,dp,$$

so that

$$\|S_t'\|\leqslant\frac{C}{2\pi t}\int_{\Gamma_1} e^{\operatorname{Re} p}\,d|p|\leqslant\frac{C'}{t}.$$

Q.E.D.

We rephrase the result in the form of an existence theorem which is compatible with theorem 5.6. In the present case the theorem is concerned with a parabolic semi-group in contrast with the previous case which is concerned with any semi-group satisfying (5.26) or (5.27).

THEOREM 5.8. *(Existence theorem for a parabolic equation).* Let A be a closed operator satisfying (5.35) and having a dense domain of operand. For an arbitrary initial value $u_0\in\mathscr{D}(A)$ and for an arbitrary $f(t)$ having Hölder's continuity with respect to t, $\|f(t)-f(t')\|\leqslant C|t-t'|^\alpha (0<\alpha\leqslant 1)$, there is a uniquely determined solution $u\in\mathscr{E}_t^1(E)$ $(t\geqslant 0)$ of (5.33) given by

$$u(t)=S_t\cdot u_0+\int_0^t S_{t-s}f(s)\,ds \tag{5.40}$$

where S_t should be understood as in the previous theorem 5.7.

Proof. As we have seen in the proof of theorem 5.6, in order to see that (5.40) is a solution it is sufficient to show that if we put $\psi(t)=\int_0^t S_{t-s}f(s)ds$ then $\psi'(t)=A\psi(t)+f(t)$ and $A\psi(t)$ is continuous. So we have to show $\psi\in\mathscr{D}(A)$.
First, for $\varepsilon(>0)$, we consider

$$\psi_\varepsilon(t)=\int_0^{t-\varepsilon} S_{t-s}f(s)\,ds=\int_0^{t-\varepsilon} S_{t-s}[f(s)-f(t)]\,ds+\int_0^{t-\varepsilon} S_{t-s}f(t)\,ds.$$

Obviously $\psi_\varepsilon(t) \to \psi(t)(\varepsilon \to +0)$. On the other hand we have

$$A\psi_\varepsilon(t) = \int_0^{t-\varepsilon} AS_{t-s}[f(s)-f(t)]\,\mathrm{d}s + \int_0^{t-\varepsilon} AS_{t-s}f(t)\,\mathrm{d}s$$

$$= \int_0^{t-\varepsilon} S'_{t-s}[f(s)-f(t)]\,\mathrm{d}s + \int_0^{t-\varepsilon} S'_{t-s}\,\mathrm{d}sf(t),^\dagger$$

where the second term $=(S_t - S_\varepsilon)f(t) = (S_t - I)f(t) + (I - S_\varepsilon)f(t)$.

Therefore, as $\varepsilon \to +0$, from Hölder's continuity of $f(t)$ and (5.39), $A\psi_s(t)$ tends to $\int_0^t S'_{t-s}[f(s) - f(t)]\mathrm{d}s + (S_t - I)f(t)$. So that, by the fact that A is a closed operator, we have $\psi(t) \in \mathscr{D}(A)$, and this term is equal to $A\psi(t)$. Because $A_\varepsilon\psi(t)$ is uniformly convergent to $A\psi(t)$ in the bounded interval of t as $\varepsilon \to +0$, we see that $A\psi(t)$ is continuous. Hence, from $\psi_+'(t) = A\psi(t) + f(t)$ (see the proof of theorem 5.6), we have $\psi'(t) = A\psi(t) + f(t)$. Hence, $u(t)$ defined in (5.40) belongs to $\mathscr{E}_t^1(E)$ $(t \geqslant 0)$. We do not give the proof of uniqueness here because the proof is exactly the same as that of theorem 5.6.

Q.E.D.

Note. In this theorem we assume the initial value $u_0 \in \mathscr{D}(A)$, but, in some cases, we assume simply $u_0 \in E$ to obtain a solution of a parabolic equation, then, we understand that u_0 as $u(t) \to u_0(t \to +0)$ and the equation is satisfied when $t > 0$.

It is easy to see that, by theorem 5.7, $u(t)$ given by (5.40) is qualified as a unique solution.

5 *Semi-groups for self-adjoint operators*

Let H be a self-adjoint operator in a Hilbert space $\mathscr{H}$. The semi-group obtained from H with $\mathrm{i}H$ as an infinitesimal generator is well known and has been used for various applications. (Some properties of this type of semi-group were not, however, rigorously established at an early stage of development.)

Let E_λ be a spectral resolution of H. In this case, a semi-group with $\mathrm{i}H$ as an infinitesimal generator (in fact, a group) is given by

$$\mathrm{e}^{\mathrm{i}tH}u = \int_{-\infty}^{+\infty} \mathrm{e}^{\mathrm{i}t\lambda}\,\mathrm{d}E_\lambda u \quad (u \in \mathscr{H}). \tag{5.41}$$

† This follows from $S'_{t-s} = \mathrm{d}S_{t-s}/\mathrm{d}t = -\mathrm{d}S_{t-s}/\mathrm{d}s \quad (t > s)$.

Let H be a self-adjoint operator bounded above, i.e. the spectrum of H is bounded above. In this case, the semi-group having H as an infinitesimal generator is given by

$$e^{tH}u = \int_{-\infty}^{m} e^{t\lambda}\, dE_\lambda u \quad (u \in \mathscr{H}) \tag{5.42}$$

where m is an upper bound of the spectrum of H. We note that (5.41) is a unitary operator, and that the converse of this statement is also true, i.e. we claim that continuous one-parameter unitary group can be represented in the form of (5.41) (Stone's theorem). Furthermore, (5.42) shows that the semi-group is of parabolic type which we have treated in the previous section.

Although it may be obvious to the reader, we explain the situation concerning (5.41). Let $u \in \mathscr{D}(H)$, i.e. $\int_{-\infty}^{+\infty} \lambda^2\, d\|E_\lambda u\|^2 < +\infty$. In this case, (5.41) can be differentiated with respect to t under the integral sign we have:

$$\begin{aligned}(e^{itH}u)_t' &= \int_{-\infty}^{+\infty} i\lambda\, e^{it\lambda}\, dE_\lambda u = iH \int_{-\infty}^{+\infty} e^{it\lambda}\, dE_\lambda u \\ &= iH\, e^{itH}u \quad (u \in \mathscr{D}(H))\end{aligned} \tag{5.43}$$

This can be interpreted in terms of the operational calculus of self-adjoint operators.† To be exact, we have

$$\begin{aligned}\left\| \frac{(e^{i(t+h)H} - e^{itH})\, u}{h} - \int_{-\infty}^{+\infty} i\lambda e^{it\lambda}\, dE_\lambda u \right\|^2 &= \int_{-\infty}^{+\infty} \left| \frac{e^{i(t+h)\lambda} - e^{it\lambda}}{h} - i\lambda e^{it\lambda} \right|^2 d\|E_\lambda u\|^2, \\ &= \int_{-\infty}^{\infty} \left| \frac{e^{ith} - 1}{h} - i\lambda \right|^2 d\|E_\lambda u\|^2 \leqslant M \int_{-\infty}^{\infty} |\lambda|^2 d\|E_\lambda u\|^2 < +\infty,\end{aligned}$$

and this tends to 0 at every point as $h \to 0$. So that, by Lebesgue's theorem, we see that the integral itself tends to 0.

By (5.43), we see that the infinitesimal generator of the semi-group $T_t = e^{itH}$ is in fact an extension of iH. We show that this extension coincides with iH. We use a Laplace transform, i.e. when $\mathrm{Re}\, p > 0$ we have

$$\int_0^\infty e^{-pt}\, e^{itH}u\, dt = \int_0^\infty e^{-pt}\, dt \int_{-\infty}^{\infty} e^{it\lambda}\, dE_\lambda u$$

† Additivity: $e^{itH}\, e^{i\delta H} = e^{i(t+\delta)H}$ follows from the following fact. In general, if $w_1(\lambda)$, $w_2(\lambda)$ are bounded continuous functions, and if $w_i(H) = \int w_i(\lambda)\, dE_\lambda$ $(i = 1, 2)$, then $w_1(H)w_2(H) = \int w_1(\lambda)w_2(\lambda)\, dE_\lambda$.

using Fubini's theorem.

$$= \int_{-\infty}^{\infty} \mathrm{d}E_\lambda u \int_0^{\infty} \mathrm{e}^{-pt}\,\mathrm{e}^{\mathrm{i}t\lambda}\,\mathrm{d}t$$

$$= \int_{-\infty}^{+\infty} \frac{1}{p-\mathrm{i}\lambda}\,\mathrm{d}E_\lambda u = (pI-\mathrm{i}H)^{-1}\,u\,.$$

That is

$$\mathrm{e}^{\mathrm{i}tH}u \sqsupset (pI-\mathrm{i}H)^{-1}\,u \qquad (\mathrm{Re}\,p>0).$$

From this, and theorem 5.4, the infinitesimal generator of $\mathrm{e}^{\mathrm{i}tH}$ is in fact $\mathrm{i}H$.

6 *Two examples of parabolic equations*

We shall treat hyperbolic equations extensively in the next chapter. In this section we mention merely two examples of simple parabolic equations. Let us consider a parabolic equation

$$\frac{\partial}{\partial t}\,u = \Delta u + \sum a_i(x)\,\frac{\partial}{\partial x_i}\,u + c(x)\,u + f(x, t) \equiv Au + f \qquad (5.44)$$

defined in Ω, a domain lying in the interior or the exterior of a C^2-class hypersurface S which is compact in R^n. Our problem is: given an initial value $u_0(x)$, find a solution of (5.44) satisfying $\mathrm{d}u/\mathrm{d}n+\sigma(x)u=0$ $(x\in S)$.

We assume $a_i(x)$ is bounded and continuously once differentiable in $\bar{\Omega}$, $c(x)$ is bounded in Ω, and $\sigma(x)$ is of C^1-class defined on S. In general, $a_i(x)$, $c(x)$ and $\sigma(x)$ may be complex-valued.

We see that the following properties follow from the results in chapter 3.

(1) There exists a Green's operator $(pI-A)^{-1}=G_p$ in a certain region of p, and it satisfies the condition (5.35). The proof is similar to (3.82) (the Dirichlet's problem).

(2) We have

$$\mathscr{D}(A)=\{u;\, u\in\mathscr{E}_{L^2}{}^2(\Omega),\ \mathrm{d}u/\mathrm{d}n+\sigma u=0\}$$

and for $u\in\mathscr{D}(A)$,

$$\|u\|_2 \leqslant C(\|Au\|_{L^2(\Omega)} + \|u\|_{L^2(\Omega)})\,. \qquad (3.109)$$

Let us apply theorem 5.8 with the following assumptions:

$$\left.\begin{array}{ll}(1) & u_0\in\mathscr{D}(A),\\ (2) & \|f(x,t)-f(x,t')\|_{L^2(\Omega)} \leqslant C|t-t'|^\alpha \quad (0<\alpha\leqslant 1).\end{array}\right\} \qquad (5.45)$$

In this case, the theorem says that there exists a solution $u(x,t)\in\mathscr{D}(A)$. But, from

$$Au(t)=\frac{\mathrm{d}}{\mathrm{d}t}u(t)-f(t),$$

we see that for $t\geqslant 0$, $Au(t)$ is continuous in $L^2(\Omega)$. Therefore, from (3.109) we see that u itself is continuous in $\mathscr{E}_{L^2}{}^2(\Omega)$. Hence, we have

PROPOSITION 5.3. Under the assumption (5.45), a solution of (5.44) such that

$$t\to u(x,t)\in\mathscr{E}_{L^2}{}^2(\Omega),\quad t\to\frac{\partial}{\partial t}u(x,t)\in L^2(\Omega)$$

are continuous for $t\geqslant 0$ and $u(x, t)$ satisfies the boundary condition, exists, and is unique.

Note. For example, if $n=2, 3$, then by theorem 3.15, we have $\mathscr{E}_{L^2}{}^2(\Omega)\subset\mathscr{B}^0(\Omega)$, so that $u(x, t)$ is continuous everywhere including the boundary. Nevertheless to look at the smoothness, and the estimation of a solution, this method is not very efficient. Although it may be difficult, a generalized method based on classical potential theory is more accurate. The only definite advantage of the former method is that it gives more perspective to the argument of the existence and the uniqueness of a solution.

Next, let us demonstrate that the formal solution (3.9) given in section 1, chapter 3 is in fact qualified as a solution. The assumptions are the same as the previous case. Let Ω be an interior domain of S. In this case, the problem reduces to an example of a semi-group of self-adjoint operators as in the previous section, i.e. we have

$$f(x)=\sum_{i=1}^{\infty}(f,u_i)\,u_i(x)\quad(f\in L^2(\Omega)),$$

$$Af(x)=\sum_{i=1}^{\infty}\lambda_i(f,u_i)\,u_i(x)\quad(f\in\mathscr{D}(A)),\tag{3.41}$$

where

$$A=-\Delta+c(x),\quad\mathscr{D}(A)=\mathscr{D}_{L^2}{}^1(\Omega)\cap\mathscr{E}_{L^2}{}^2(\Omega).$$

By (5.42), for $f\in\mathscr{D}(A)$ we see that

$$u(x,t)=\exp(-tA)f=\sum\exp(-\lambda_i t)(f,u_i)u_i(x)\in\mathscr{D}(A)$$

is a $\mathscr{E}_{L^2}{}^2(\Omega)$-valued continuous function of $t\geqslant 0$ and represents a unique solution which is continuously once differentiable with respect to t by the topology of $L^2(\Omega)$.

Also, by a property of a parabolic semi-group, even if we simply assume $f \in L^2(\Omega)$ $u(x, t)$, defined above, satisfies the same conditions as above for $t>0$, and as $t \to +0$, $u(x, t) \to f(x)$ in $L^2(\Omega)$. Such a solution is unique.

Remark. Similarly, if we consider the solution (3.8) of a wave-equation in section 1, we see that if $f_0, f_1 \in \mathscr{D}(A)$, then $u(x, t) \in \mathscr{D}(A)$ and the unique solution† belongs to $\mathscr{E}_t^2(L^2(\Omega))$ $(-\infty < t < +\infty)$.

† See section 7, chapter 8.

Chapter 6

Hyperbolic equations

1 *Introduction*

In chapter 4 we discussed an initial-value problem, which led us to the definition of a hyperbolic equation in a natural way. Nevertheless, because we used a Fourier transform to deal with equations with constant coefficients in the most general form, we did not tough upon the case in which equations have variable coefficients. In the present chapter we shall discuss the initial value problem of hyperbolic equations with variable coefficients.

In chapter 4 we discussed exclusively higher-order single equations and in section 10 we considered briefly a system of equations of the first order. This time we reverse the order; first we extensively discuss systems of first-order equations and then we treat single higher-order equations as a special case. We follow Friedrichs' argument of a 'symmetric hyperbolic system'

2 *Energy inequality for a symmetric hyperbolic system*

Let us consider a system of first-order equations:

$$M[u] \equiv \frac{\partial}{\partial t} u - \sum_{k=1}^{n} A_k(x, t) \frac{\partial}{\partial x_k} u - B(x, t)u = f \tag{6.1}$$

where $A_k(x, t)$ $(k=1, 2, \ldots, n)$ are Hermitian matrices of Nth-degree, i.e. if we write $A_k(x, t) = (a_{ij}^{(k)}(x, t))$ we have

$$\overline{a_{ij}^{(k)}(x, t)} = a_{ji}^{(k)}(x, t). \tag{6.2}$$

Furthermore, for differentiability, we assume

$$\left.\begin{array}{l} t \to A_k(x, t) \in \mathscr{B}^1(R^n),\ t \to \partial/\partial t\, A_k(x, t) \in \mathscr{B}^0(R^n) \text{ are continuous, i.e.} \\ a_{ij}^{(k)}(x, t),\ \partial/\partial x_\nu\, a_{ij}^{(k)}(x, t),\ (\partial/\partial t)\, a_{ij}^{(k)}(x, t) \text{ are bounded and con-} \\ \text{tinuous and } |a_{ij}^{(k)}(x, t) - a_{ij}^{(k)}(x, t')| \text{ tends to 0 uniformly as } t' \to t. \\ \text{The same assumption is set up for other functions.} \end{array}\right\} \tag{6.3}$$

For $B(x, t)$, we assume

$$t \to B(x, t) \in \mathscr{B}^0(R^n) \tag{6.4}$$

is continuous. Also, we put

$$u(x, t) \equiv u(t) = \begin{bmatrix} u_1(x, t) \\ \vdots \\ u_N(x, t) \end{bmatrix} \tag{6.5}$$

and write

$$\|u(t)\|^2 = \sum_{j=1}^{N} \|u_j(x, t)\|^2_{L^2(R^n)}. \tag{6.6}$$

In (6.1) we regard $\partial u(x, t)/\partial x_k$ as a derivative of a distribution $u(x, t)$ of x-space (t is a parameter) unless we state the contrary. Under the above assumptions, we have the following:

PROPOSITION 6.1. Let $f(t) \in \mathscr{E}_t^0(L^2)$ $(t \geq 0)$ in (6.1), and let $u(t) \in \mathscr{E}_t^0(\mathscr{D}_{L^2}^1)$ $(t \geq 0)$ for a solution $u(t)$. Further, we assume $u(t)$ is continuously differentiable in L^2; i.e. $u(t) \in \mathscr{E}_t^1(L^2)$ $(t \geq 0)$. Then, for $0 \leq t \leq T$, we have an energy inequality:

$$\|u(t)\| \leq e^{\gamma t} \|u(0)\| + \int_0^t e^{\gamma(t-s)} \|f(s)\| \, ds \tag{6.7}$$

where the constant γ may depend upon T, but is independent of $u(t)$ and $f(t)$.

Proof. Clearly.

$$\frac{d}{dt} \|u(t)\|^2 = \left(\frac{d}{dt} u(t), u(t)\right) + \left(u(t), \frac{d}{dt} u(t)\right).$$

On the other hand

$$\left(u(t), \frac{d}{dt} u(t)\right) = \sum_k \left(u, A_k(x, t) \frac{\partial}{\partial x_k} u\right) + (u, Bu + f).$$

Because A_k are Hermitian, this is equal to

$$-\sum_k \left(\frac{\partial}{\partial x_k} (A_k(x, t)u), u\right) + (u, Bu + f)$$

where we use the following fact: for $u, v \in \mathscr{D}_{L^2}{}^1$,

$$\left(u, \frac{\partial}{\partial x_k} v\right) = -\left(\frac{\partial}{\partial x_k} u, v\right). \tag{6.8}$$

To see this, we recall the note of definition 2.10; for $u \in \mathscr{D}_{L^2}{}^1$, we can pick a sequence from $\mathscr{D}$ such that $\varphi_j(x) \to u(x)$ in $\mathscr{D}_{L^2}{}^1$. Obviously,

$$\left(u, \frac{\partial}{\partial x_k} v\right) = \lim_{j\to\infty} \left(\varphi_j(x), \frac{\partial}{\partial x_k} v\right) = -\lim_{j\to\infty} \left(\frac{\partial}{\partial x_k} \varphi_j(x), v\right) = -\left(\frac{\partial}{\partial x_k} u, v\right).$$

Now,

$$\frac{\mathrm{d}}{\mathrm{d}t} \|u(t)\|^2 = -\sum \left(\frac{\partial A_k}{\partial x_k} u, u\right) + 2 \operatorname{Re}(Bu, u) + 2 \operatorname{Re}(u, f),$$

so that

$$\frac{\mathrm{d}}{\mathrm{d}t} \|u(t)\| \leqslant \gamma \|u(t)\| + \|f(t)\| \qquad (0 \leqslant t \leqslant T).$$

Hence, we have (6.7).

Q.E.D.

Notice that we assume $u \in \mathscr{D}_{L^2}{}^1$ but in (6.7) only the L^2-norm of $u(t)$ appears. In fact, for $u \in L^2$, we can derive the inequality (6.7) by using Friedrichs' mollifier (See chapters 1–2). The key lemma for this purpose is:

Lemma 6.1 *(Friedrichs)*. Let $a \in \mathscr{B}^1$, $u \in L^2$. For a commutator C_δ which is defined as

$$C_\delta u = \varphi_\delta * \left(a(x) \frac{\partial u}{\partial x_j}\right) - a(x) \left(\varphi_\delta * \frac{\partial u}{\partial x_j}\right)$$
$$= \left(\varphi_\delta *, a \frac{\partial}{\partial x_j}\right) u, \tag{6.9}$$

we have:

(1) $\|C_\delta u\| \leqslant C \|u\|$, where C (const.) depends only on φ.
(2) $C_\delta u \to 0$ in L^2 $\quad (\delta \to +0)$.

Proof.

$$C_\delta u(x) = -\int \varphi_\delta(x-y) \, [a(x) - a(y)] \frac{\partial u}{\partial y_j}(y) \, \mathrm{d}y, \tag{6.10}$$

where the integral is regarded as a symbolic operation in the sense of distribution. Now, we have

$$C_\delta u=\int \frac{\partial}{\partial y_j}\{\varphi_\delta(x-y)(a(x)-a(y))\}u(y)dy \tag{6.11}$$

(We prove this later).

We consider the right hand side. We can write this as

$$-\int \frac{\partial a}{\partial y_j}(y)\,\varphi_\delta(x-y)\,u(y)\,dy-\int (a(x)-a(y))\frac{\partial \varphi_\delta}{\partial y_j}(x-y)\cdot u(y)\,dy.$$

Note that

$$|a(x)-a(y)|\leqslant |a(x)|_{\mathscr{B}^1}\,|x-y|,\qquad \int_{|x-y|\leqslant\delta} |x-y|\left|\frac{\partial\varphi_\delta}{\partial y_j}(x-y)\right|dx<C.$$

By the Hausdorff–Young inequality (1.73), we see that the function represented by the above integral can be assessed by $C'|a(x)|_{\mathscr{B}^1}\|u(x)\|$ with the L^2-norm. Therefore, part (1) of the lemma has been proved.

The transition from (6.10) to (6.11)–an integration by parts in a wider sense–can be understood as follows:

For $u\in\mathscr{D}$ (6.11) is true. Therefore, we pick $\{u_j(x)\}$ of $\mathscr{D}$ which is convergent to $u(x)$ in the sense of L^2. Obviously $C_\delta u_j$ tends to the right hand side term of (6.11) in the sense of L^2. On the other hand, $C_\delta u_j$ tends to $C_\delta u$ in the sense of a distribution.† Hence, (6.11) is true.

Let us prove (2). If we regard $(a(x)-a(y))\varphi_\delta(x-y)$ as a function of y for fixed x, then it has a compact support. Therefore, we have

$$\int \frac{\partial}{\partial y_j}\{(a(x)-a(y))\,\varphi_\delta(x-y)\}\,dy=0.$$

Hence, we can write

$$\begin{aligned}(C_\delta u)\,(x)&=\int \frac{\partial}{\partial y_j}\{(a(x)-a(y))\,\varphi_\delta(x-y)\}\,(u(y)-u(x))\,dy\\ &=-\int \frac{\partial a}{\partial y_j}(y)\,\varphi_\delta(x-y)(u(y)-u(x))\,dy\\ &\quad-\int (a(x)-a(y))\,\frac{\partial\varphi_\delta}{\partial y_j}(x-y)(u(y)-u(x))\,dy\\ &\equiv\Phi_1(x)+\Phi_2(x).\end{aligned}$$

†From (6.9), we see that $C_\delta u_j\to C_\delta u$ in $\mathscr{D}'_{L^2}{}^1$.

From this, and by the method of the proof of lemma 1.3, (*d*), we have $\|\Phi_i(x)\|\to 0$ $(\delta\to 0, i=1,2)$.

In fact, for instance, we have an estimation

$$|\Phi_2(x)|\leqslant |a(x)|_{\mathscr{B}^1}\int_{|x-y|\leqslant\delta}|x-y|\left|\frac{\partial\varphi_\delta}{\partial x_j}(x-y)\right||u(y)-u(x)|\,dy.$$

On the other hand, we have

$$\int |x|\left|\frac{\partial\varphi_\delta}{\partial x_j}\right|dx\leqslant c$$

where c is independent of δ. Now, we see that we can use the same method that we used in lemma 1.3.

Q.E.D.

COROLLARY. Under the same conditions as were imposed on the lemma, if $u\in\mathscr{D}_{L^2}{}^1$, then:

(1) $\|C_\delta u\|_1\leqslant C_1\|u\|_1$ $\quad(\|u\|_1=\|u\|_{1,L^2})$,

(2) $C_\delta u\to 0$ in $\mathscr{D}_{L^2}{}^1$.

Proof. Let $C_\delta u=u_\delta(x)$. Then we have

$$\frac{\partial}{\partial x_k}u_\delta(x)=\varphi_\delta*\left(\frac{\partial a}{\partial x_k}\frac{\partial u}{\partial x_j}\right)-\frac{\partial a}{\partial x_k}\left(\varphi_\delta*\frac{\partial u}{\partial x_j}\right)+\left[\varphi_\delta*,\,a\frac{\partial}{\partial x_j}\right]\frac{\partial u}{\partial x_k}.$$

By lemmas 1.3 and 6.1 we conclude (1) and (2) as desired.

Q.E.D.

THEOREM 6.1 *(Friedrichs)*. Let $u(t)$ be a solution of (6.1) such that $u(t)\in\mathscr{E}_t{}^0(L^2)$ $(t\geqslant 0)$ and $u(t)\in\mathscr{E}_t{}^1(\mathscr{D}'_{L^2})$ $(t\geqslant 0)$. Then the energy inequality (6.7) is true.

Proof. We operate with $\varphi_\delta(x)$ on $u(t)$. Then, we see easily that $u_\delta(t)=\varphi_\delta\underset{(x)}{*}u(t)\in\mathscr{E}_t{}^1(\mathscr{D}_{L^2}{}^1)$ for $t\geqslant 0$. We have

$$\frac{\partial}{\partial t}u_\delta(t)=\frac{\partial}{\partial t}(\varphi_\delta\underset{(x)}{*}u(t))=\varphi_\delta\underset{(x)}{*}\frac{\partial u}{\partial t}.$$

So that from (6.1) we obtain an equation which is satisfied by u_δ as follows.

First, we operate with $\varphi_\delta *$ from the left of (6.1),

$$\frac{\partial}{\partial t} u_\delta = \sum_k \varphi_\delta \underset{(x)}{*} \left(A_k \frac{\partial}{\partial x_k} u \right) + \varphi_\delta \underset{(x)}{*} Bu + \varphi_\delta \underset{(x)}{*} f,$$

so that

$$\frac{\partial}{\partial t} u_\delta = \sum_k A_k \frac{\partial}{\partial x_k} u_\delta + Bu_\delta + f_\delta + C_\delta u, \tag{6.12}$$

where

$$C_\delta u = \sum \left\{ \varphi_\delta \underset{(x)}{*} \left(A_k \frac{\partial}{\partial x_k} u \right) - A_k \left(\varphi_\delta \underset{(x)}{*} \frac{\partial}{\partial x_k} u \right) \right\} + \{ \varphi_\delta \underset{(x)}{*} Bu - B(\varphi_\delta \underset{(x)}{*} u) \}$$
$$= \sum_k \left[\varphi_\delta *, A_k \frac{\partial}{\partial x_k} \right] u + [\varphi_\delta *, B] \, u .$$

Next, we apply proposition 6.1 to (6.12).

$$\|u_\delta(t)\| \leqslant e^{\gamma t} \|u_\delta(0)\| + \int_0^t e^{\gamma(t-s)} \{ \|f_\delta(s)\| + \|C_\delta u(s)\| \} \, ds .$$

On the other hand, by lemma 6.1 we have $\|C_\delta u(s)\| \leqslant c \|u(s)\|$, where c is independent of δ and $\|C_\delta u(s)\| \to 0$ $(\delta \to 0)$. We now apply Lebesgue's theorem to the result of the above calculation; the integral on the right hand side tends to

$$\int_0^t e^{\gamma(t-s)} \|f(s)\| \, ds$$

as $\delta \to 0$. Thus, we have proved (6.7) by a limiting process.

Q.E.D.

3 *Remarks on energy inequalities*

In the previous section we have demonstrated the fact that the L^2-norm of a solution of (6.1) can be estimated by the L^2-norm of the initial value and the right hand terms of the equations. A question arises here: could we achieve such an estimation in the case of an L^p-norm $(p \neq 2)$? In general the answer to the question is in the negative.† Furthermore, we may ask if an energy inequality can be established in the sense of maximum-norm. Later,

† W. Littman: 'The non-existence of certain estimates for the wave equation', *Proceedings of International Conference*, (*M.R.C.*), Wisconsin, Madison (1960).

we shall show that, in general, except in the one-dimensional case (as we shall demonstrate with an example) the answer is also negative.

From the view-point of an existence theorem, we may ask if $f=0$ in (6.1) and the initial value is of C^m-class, then is a solution $u(x, t)$ of C^m-class, too? Let us assume that this is true and also that we can establish the uniqueness of the solution. Then, by Banach's closed graph theorem, we must have an energy inequality in the sense of the B^m-norm

$$\max_{0 \leqslant t \leqslant T} |u(x, t)|_{\mathscr{B}^m} \leqslant c(T)|u(x, 0)|_{\mathscr{B}^m}.$$

But, this is not generally true (except in the one-dimensional case) as we mentioned just now. We may say that except when $n=1$ the classical method is not sufficiently powerful to include hyperbolic equations in general. We note that if $n=1$ we have the following:

Haar's inequality

We consider a system of first-order equations in the one-dimensional space.

$$\frac{\partial}{\partial t} u - A(x, t) \frac{\partial}{\partial x} u - B(x, t)\, u = f \tag{6.13}$$

where $A(x, t)$ satisfies the condition: all the roots of $\det(\lambda I - A(x, t)) = 0$ are real and distinct. In this case,

$$|u(x_0, t_0)| \leqslant c(T)\{|u(x, 0)|_0 + |f(x, t)|_0\} \quad (0 \leqslant t_0 \leqslant T) \tag{6.14}$$

where the maximum-norm is taken in an appropriate domain, as we shall explain below.

Proof.† Let $\lambda_1(x, t), \ldots, \lambda_N(x, t)$ be the roots of $\det(\lambda I - A) = 0$, and let $\lambda_1 < \cdots < \lambda_N$. We note that $A(x, t)$ is diagonalizable. So that we have a regular matrix $N(x, t)$ satisfying

$$N(x, t)\, A(x, t) = D(x, t)\, N(x, t) \qquad D(x, t) = \begin{bmatrix} \lambda_1(x, t) & & 0 \\ & \ddots & \\ 0 & & \lambda_N(x, t) \end{bmatrix}.$$

Furthermore, we can assume $|N(x, t)| \geqslant \delta > 0$. Operating with N on (6.13) we

† See I. G. Petrowsky: *Lectures on partial differential equations*, pp. 67–73. Interscience, 1955.

have

$$\frac{\partial}{\partial t}(Nu)-N_t'u-D\frac{\partial}{\partial x}(Nu)+DN_x'u-NBu=Nf,$$

i.e. if we write $Nu=v$, then

$$\frac{\partial}{\partial t}v=D\frac{\partial}{\partial x}v+B_1v+Nf, \tag{6.15}$$

$B_1=(N_t'-DN_x'+NB)N^{-1}$. The principal part has then been diagonalized. Therefore, (6.15) can be transformed into a Volterra-type integral equation.

Let (x_0, t_0) be an arbitrary point such that $t_0>0$. Through this point we draw a characteristic curve

$$L_i:\frac{\mathrm{d}x}{\mathrm{d}t}=-\lambda_i(x,t)$$

in the direction of decreasing t.

We write D as a domain (including its boundary) which is bounded by L_1, L_N, and $t=0$, and so contains the characteristic curves which we have just drawn. We also write D_0 as $D\cap\{t=0\}$. Then we have

$$|u(x_0,t_0)|\leqslant c\{\sup_{x\in D_0}|u(x,0)|+\sup_{(x,t)\in D}|f(x,t)|\}$$

where c is independent of u and f.

Q.E.D.

Next, we examine a wave equation when $n\geqslant 2$ to show that an energy equation cannot be established in the sense of a maximum-norm.

EXAMPLE. As is well known, if

$$t\to u(t)\in\mathscr{D}_{L^2}{}^2,\quad t\to\frac{\partial}{\partial t}u\in\mathscr{D}_{L^2}{}^1,$$

are continuous, a solution of

$$\Box u\equiv\frac{\partial^2}{\partial t^2}u-\Delta u=0 \tag{6.16}$$

satisfies

$$E(t;u)=\tfrac{1}{2}\int_{t=t}\left\{\left(\frac{\partial u}{\partial t}\right)^2+\sum_{i=1}^{n}\left(\frac{\partial u}{\partial x_i}\right)^2\right\}\mathrm{d}x=\text{constant} \tag{6.17}$$

where $u(x, t)$ is a real-valued function. In fact, we have

$$\frac{\mathrm{d}}{\mathrm{d}t} E(t; u) = \int_{t=t} \{u_t u_{tt} + \sum u_{x_i} u_{x_i t}\} \, \mathrm{d}x .$$

By (6.8) this is equal to

$$\int_{t=t} \{u_t u_{tt} - \sum u_{x_i x_i} u_t\} \, \mathrm{d}x = 0 .$$

Now, we write

$$E_1(t; u) = \sup_x \left\{ \left| \frac{\partial u}{\partial t} \right| + \sum_i \left| \frac{\partial u}{\partial x_i} \right| \right\} . \tag{6.18}$$

For simplicity, we consider the case $n=3$, and we demonstrate that for any c the inequality

$$E_1(t_0; u) \leqslant c E_1(0; u)$$

is not valid. This fact shows that the differentiability of a solution does not propagate in the direction of t.

As we have seen in theorem 4.12, $E(x, t) = \delta_{|x|-t}/4\pi t$ is an *elementary solution* for a wave equation, i.e. for $t>0$ this satisfies

$$\square E = 0, \quad E(x, 0) = 0, \quad \frac{\partial}{\partial t} E(x, 0) = \delta .$$

From this, if we define

$$\tilde{E}(x, t) = \begin{cases} E(x, t) & (t \geqslant 0), \\ -E(x, -t) & (t \leqslant 0), \end{cases}$$

then we can easily see that $\tilde{E}(x, t)$ is well defined in $R^3 \times \{-\infty < t < +\infty\}$ satisfying $\square \tilde{E}(x, t) = 0$. This is a solution in the sense of distribution.

On the other hand, if $\varphi \in \mathscr{D}$ and if we write $\varphi^{(\varepsilon)}(x) = \varphi(x/\varepsilon)$ for $\varepsilon > 0$, then $u_\varepsilon(x, t) = \tilde{E}(x, t - t_0) \underset{(x)}{*} \varphi^{(\varepsilon)}(x) \in C^\infty$ satisfies $\square u_\varepsilon(x, t) = 0$. We see that

$$\frac{\partial}{\partial t} u_\varepsilon(x, t_0) = \frac{\partial E}{\partial t}(x, 0) \underset{(x)}{*} \varphi^{(\varepsilon)}(x) = \delta * \varphi^{(\varepsilon)}(x) = \varphi^{(\varepsilon)}(x),$$

$$\frac{\partial}{\partial x_j} u_\varepsilon(x, t_0) = E(x, 0) \underset{(x)}{*} \frac{\partial}{\partial x_j} \varphi^{(\varepsilon)}(x) = 0,$$

so that

$$E_1(t_0; u_\varepsilon) = \sup_x |\varphi^{(\varepsilon)}(x)| = \sup_x |\varphi(x)| . \tag{6.19}$$

In (6.19) we take a fixed $\varphi(x)$ as satisfying the conditions that $\varphi(0)=1$, the support of $\varphi(x)$ lies in $|x|\leqslant 1$, and $|\varphi(x)|\leqslant 1$. In this case we have

$$\left|\frac{\partial}{\partial x_j}\varphi^{(\varepsilon)}(x)\right|\leqslant\frac{\gamma}{\varepsilon},$$

so that at $t=0$ we have

$$\frac{\partial}{\partial t}u_\varepsilon(x,0)=\frac{\partial}{\partial t}E(x,t_0)\underset{(x)}{*}\varphi^{(\varepsilon)}(x)=\left[\frac{\partial}{\partial t}\left(\frac{t}{4\pi}\int_{|\xi|=1}\varphi^{(\varepsilon)}(x-t\xi)\,\mathrm{d}S_\xi\right)\right]_{t=t_0}.$$

On the other hand,

$$\int_{B_\varepsilon(x_0)\cap\{|\xi|=1\}}\mathrm{d}S_\xi=O(\varepsilon^2)$$

where $B_\varepsilon(x_0)$ is a sphere having a radius ε, centred at x. From this it follows

$$\left|\frac{\partial}{\partial t}u_\varepsilon(x,0)\right|=O(\varepsilon).$$

Similarly, we see

$$\left|\frac{\partial}{\partial x_j}u_\varepsilon(x,0)\right|=O(\varepsilon).$$

From (6.19) we have $E_1(t_0\,;u_\varepsilon)=1$, which shows that $E_1(t_0\,;u_\varepsilon)\leqslant cE_1(0\,;u_\varepsilon)$ is not valid in this case.

Note. We have examined the case R^3, but by (4.58) and (4.59), using $u_\varepsilon(x,t)$, we can give a counter example in the case of R^2.

4 *Existence theorem 1 for a solution of a system of symmetric hyperbolic equations (the case in which the coefficients are independent of t)*

Let us assume that in (6.1) all coefficients are not dependent upon t. Of course, we also assume (6.2), (6.3) and (6.4). From (6.3) and (6.4), we have

$$A_k\in\mathscr{B}^1,\quad B\in\mathscr{B}^0. \tag{6.20}$$

Let us write

$$A=\sum A_k(x)\frac{\partial}{\partial x_k}+B(x), \tag{6.21}$$

then theorem 5.6 (Hille–Yosida) is used as follows:

We take

$$\mathscr{D}(A)=\{u;\, u\in L^2,\ Au\in L^2\} \tag{6.22}$$

as the domain of the definition of A. We note that $u\in L^2$, so that $Au\in\mathscr{D}'_{L^2}{}^1$. Let λ be a real number and let $u\in\mathscr{D}_{L^2}{}^1(\subset\mathscr{D}(A))$. In this case, we have

$$\begin{aligned}\|(I-\lambda A)\,u\|^2&=((I-\lambda A)\,u,\,(I-\lambda A)\,u)\\&=\|u\|^2-\lambda\{(Au,u)+(u,Au)\}+\lambda^2\|Au\|^2.\end{aligned}$$

Considering the symmetry of A_k and (6.8), we have

$$(u,Au)=\sum_k\left(u,\,A_k\frac{\partial}{\partial x_k}u\right)+(u,Bu)=-\sum_k\left(\frac{\partial}{\partial x_k}(A_k(x)\,u),\,u\right)+(u,Bu).$$

So that there exists a certain positive constant β with

$$\|(I-\lambda A)\,u\|^2\geqslant(1-\beta|\lambda|)\,\|u\|^2.$$

For a sufficiently small ε_0, we have

$$\|(I-\lambda A)\,u\|\geqslant(1-\beta|\lambda|)\|u\|\qquad(|\lambda|<\varepsilon_0). \tag{6.23}$$

Next, we demonstrate that (6.23) is valid for an arbitrary $u\in\mathscr{D}(A)$. Using Friedrichs' mollifier φ_δ, we have

$$(I-\lambda A)(\varphi_\delta*u)=\varphi_\delta*(I-\lambda A)\,u-\lambda\{A(\varphi_\delta*u)-\varphi_\delta*(Au)\}.$$

By lemma 6.1, we have

$$C_\delta u=A(\varphi_\delta*u)-\varphi_\delta*(Au)\to0\quad\text{in}\quad L^2\qquad(\delta\to+0).$$

Because $\varphi_\delta*u\in\mathscr{D}_{L^2}{}^1$, we observe

$$\|(I-\lambda A)(\varphi_\delta*u)\|\geqslant(1-\beta|\lambda|)\|\varphi_\delta*u\|.$$

If $\delta\to+0$, both sides of this inequality tend to the terms on both sides of (6.23) respectively. Finally, we conclude (6.23) is valid for $u\in\mathscr{D}(A)$. From this, we have:

PROPOSITION 6.2. $(I-\lambda A)$ ($\lambda\neq0$, $|\lambda|<\varepsilon$, where ε is a sufficiently small positive number) realizes a bijection from $\mathscr{D}(A)$ onto L^2.

Proof.

(1) We note that $(I-\lambda A)\mathscr{D}(A)$ is closed in L^2. To see this, we note that A is a closed operator. In fact, from $u_n \to 0$ it follows that $Au_n \to 0$ in $\mathscr{D}'_{L^2}{}^1$, and if $Au_n \to v_0$ in L^2, then $v_0=0$ because the natural injection from L^2 into $\mathscr{D}'_{L^2}{}^1$ is in fact a bijection.

Now, let $(I-\lambda A)u_n \to v_0$. Then, by (6.23), $\{u_n\}$ is a Cauchy sequence. Therefore, $u_n \to u_0 \in L^2$. But $\{Au_n\}$ is also a Cauchy sequence. From the fact that A is a closed operator we have $u_0 \in \mathscr{D}(A)$ and $Au_n \to Au_0$, i.e. $v_0 = (I-\lambda A)u_0$.

(2) $\mathrm{Im}((I-\lambda A)u_0)$ is dense in L^2, for if not, there exists $\psi \neq 0 (\in L^2)$ and $((I-\lambda A)u, \psi)=0$ $(u \in \mathscr{D}(A))$. This is of course true for $u \in \mathscr{D}$, therefore,

$$(I-\lambda A^*)\,\psi=0 \qquad A^*\psi=-\sum_k \frac{\partial}{\partial x_k}(A_k\psi)+B^*\psi\,.$$

On the other hand, $A^*\psi \in L^2$. From

$$A^*\psi=-\sum A_k \frac{\partial}{\partial x_k}\psi-\sum \frac{\partial A_k}{\partial x_k}\psi+B^*\psi\,,$$

we have $\psi \in \mathscr{D}(A^*)$. For $(I-\lambda A^*)$, obviously an inequality in the same form as (6.23) can be established in $|\lambda|<\varepsilon'$. Hence, $\psi=0$, which is a contradiction.

Therefore, if we put $\varepsilon=\min(\varepsilon_0, \varepsilon')$ in Proposition 6.2, then it follows that $(I-\lambda A)\mathscr{D}A=L^2$ from our results (1) and (2).

Q.E.D.

Problem in the space $\mathscr{D}_{L^2}{}^1$

Let us re-examine the previous result in $\mathscr{D}_{L^2}{}^1$. As the domain of the definition of A we take

$$\mathscr{D}(A)=\{u\,;\, u \in \mathscr{D}_{L^2}{}^1,\ Au \in \mathscr{D}_{L^2}{}^1\}\,. \tag{6.24}$$

On the other hand, for the smoothness of the coefficients, we assume

$$A_k \in \mathscr{B}^{1+\sigma} \qquad (\sigma>0,\ B \in \mathscr{B}^1) \tag{6.25}$$

where $\mathscr{B}^{1+\sigma}$ is a function belonging to $\mathscr{B}^1$ and whose derivatives of first order have a uniform Hölder continuity of degree σ.

In this case, from the corollary of lemma 6.1 (similar to the case L^2) for $u \in \mathscr{D}(A)$ we have

$$\|(I-\lambda A)\,u\|_1 \geq (1-\beta_1|\lambda|)\|u\|_1 \qquad (\beta_1>0,\ |\lambda|<\varepsilon_1) \tag{6.26}$$

with regard to the $\mathscr{D}_{L^2}{}^1$-norm. We prove the part which corresponds to proposition 6.2. Although it is easy to see that $(I-\lambda A)\,\mathscr{D}(A)$ is closed in $\mathscr{D}_{L^2}{}^1$, we have to elaborate a little further in order to prove the denseness of the image in $\mathscr{D}_{L^2}{}^1$.

To do this, we introduce a concept, that of 'singular integral operator' which we shall employ extensively later. We use the notations which are employed by Calderón and Zygmund.†

DEFINITION 6.1.

(1) Operator Λ. For $u\in\mathscr{D}'_{L^2}$ we define

$$(\Lambda u)\hat{}\,(\xi)=|\xi|\,\hat{u}(\xi).$$

From this we have $\Lambda^2=-\Delta/4\pi^2$.

(2) Riesz operator R_j. For $u\in\mathscr{D}'_{L^2}$, we define

$$(R_j u)\hat{}\,(\xi)=\frac{1}{2\pi i}\frac{\xi_j}{|\xi|}\hat{u}(\xi).$$

Obviously, R_j is a bounded operator in L^2-space. Considering (2.79) we define

$$\text{v.p. } R_j(x)=c_n \text{ v.p. } x_j/|x|^{n+1} \quad \text{where} \quad c_n=\frac{1}{2\pi}\cdot\frac{\Gamma\{\frac{1}{2}(n+1)\}}{\pi^{\frac{1}{2}(n+1)}}. \tag{6.27}$$

We can write

$$R_j u=\text{v.p. } R_j(x)*u=c_n \text{ v.p. } \frac{x_j}{|x|^{n+1}}*u(x). \tag{6.28}$$

Also we have

$$\Lambda u=\sum_{j=1}^{n}\frac{\partial}{\partial x_j}R_j u=\sum_{j=1}^{n}R_j\frac{\partial}{\partial x_j}u. \tag{6.29}$$

LEMMA 6.2. Let $a\in\mathscr{B}^{1+\sigma}(\sigma>0)$. Then the commutator Cu for $a(x)$ and Λ, $Cu=(a(x)\Lambda-\Lambda a(x))u$ is a bounded operator of L^2.

† See A. P. Calderón and A. Zygmund: 'Singular integral operators and differential equations', *Amer. J. Math.* 79 (1957), 901–21.

Proof. It is sufficient to prove the case $u \in \mathscr{D}$. We can write

$$(Cu)(x) = \sum_j a(x) R_j \frac{\partial}{\partial x_j} u - R_j \frac{\partial}{\partial x_j}(a(x) u)$$
$$= \sum_j (a(x) R_j - R_j a(x)) \frac{\partial}{\partial x_j} u - \sum_j R_j \frac{\partial a}{\partial x_j} u.$$

The last term is obviously a bounded operator, so we examine the first term. Considering (6.28), we write

$$v_\varepsilon(x) = \int_{|x-y| \geq \varepsilon} \{a(x) - a(y)\} R_j(x-y) \frac{\partial}{\partial y_j} u(y)\, dy$$
$$= \int_{|x-y| = \varepsilon} [a(x) - a(y)] R_j(x-y) u(y) \cos \gamma_j \, dS_\varepsilon$$
$$+ \int_{|x-y| \geq \varepsilon} \frac{\partial a}{\partial y_j}(y) R_j(x-y) u(y)\, dy$$
$$+ \int_{|x-y| \geq \varepsilon} [a(x) - a(y)] \left(\frac{\partial}{\partial x_j} R_j\right)(x-y) u(y)\, dy.$$

If we let $\varepsilon \to +0$, then the first term tends to the function which can be estimated by $C|a(x)|_1 |u(x)|$. Therefore, the L^2-norm of the first term can be estimated by $C|a(x)|_1 \cdot \|u\|$. Next, we look at the second term. We can write this as v.p.$R_j(x) * a_{x_j}(x) u(x)$, so that this gives us no problem.

Finally, we estimate the third term. We write the integral separately as follows

$$\int_{\varepsilon \leq |x-y| \leq 1} dy + \int_{|x-y| \geq 1} dy$$

where the second integrand cannot exceed

$$2|a(x)|_0 \left| \left[\frac{\partial}{\partial x_j} R_j(x) \right]_{|x| \geq 1} \right| * |u(x)|$$

in its absolute value. On the other hand,

$$\frac{\partial}{\partial x_j} R_j(x) = O(|x|^{-n-1}),$$

so that it is summable. By the Hausdorff–Young theorem, we can estimate this by $C'|a(x)|_0 \|u\|$ with the L^2-norm. If we put $a(x) - a(y)$ $= \sum a_{x_i}(x)(x_i - y_i) + b(x, y)$, then from the fact that $|b(x, y| \leq |_{1+\sigma} |x-y|^{1+\sigma}$, we have an estimation

$$\left|\sum_i a_{x_i}(x) \int_{\varepsilon \leqslant |x-y| \leqslant 1} (x_i - y_i) \frac{\partial R_j}{\partial x_j}(x-y) u(y)\, dy\right|$$
$$+ |a|_{1+\sigma} \int_{|x-y|\leqslant 1} |x-y|^{1+\sigma} \left|\frac{\partial R_j}{\partial x_j}(x-y)\right| |u(y)|\, dy.$$

The second term of this expression also gives us no problem (use the Hausdorff–Young inequality). We examine only the remaining first term. Obviously, $x_i \partial R_j(x)/\partial x_j$ is a homogeneous function of degree $(-n)$, and its integral mean is 0. We have to show that the Fourier transform of v.p.$[x_i \partial R_j(x)/\partial x_j]_{|x| \leqslant 1}$ is in fact a bounded function. To see this, we put $x_i \partial R_j/\partial x_j = K(x)$. Because the spherical mean of $K(\xi)$ is 0, we have

$$\lim_{\varepsilon \to +0} \int_{\varepsilon \leqslant |x| \leqslant 1} e^{-2\pi i x \xi} K(x)\, dx = \lim_{\varepsilon \to +0} \int_{\Omega} K(\omega)\, d\omega \int_{\varepsilon}^{1} \frac{\exp(-2\pi i r \rho \cos \gamma)}{r}\, dr,$$
$$= \int_{\Omega} K(\omega)\, d\omega \int_{0}^{1} \frac{\exp(-2\pi i r \rho \cos \gamma)}{r}\, dr.$$

Obviously, from this, we see that it is bounded for $|\xi| = \rho < 1$.

Let us prove the boundedness as $|\xi| \to +\infty$. To do this, we may refer to proposition 2.12 and its proof. $\mathscr{F}$[v.p. $K(x)$] is bounded, consequently, we have to show that $\mathscr{F}[(K(x))_{|x| \geqslant 1}]$ is bounded in $|\xi| \geqslant 1$. To this end, we first consider a function $\alpha(r) \in C^\infty$ which is 0 in $r \leqslant 2$ and is 1 in $r \geqslant 3$. We see that

$$\mathscr{F}[(K(x))_{|x|\geqslant 1}] = \mathscr{F}[\alpha(r)K(x)] + \mathscr{F}[(1-\alpha(r))\,(K(x))_{|x|\geqslant 1}].$$

Obviously the second term is a bounded function of ξ. On the other hand, the first term is bounded because

$$\mathscr{F}\left[\frac{\partial}{\partial x_j}\{a(r)K(x)\}\right]$$

is a bounded function, being a Fourier transform of a function of L^1, so that $2\pi i \xi_j \times \mathscr{F}[\alpha(r)K(x)]$ $(j = 1, 2, \ldots, n)$ is bounded. Finally, we see that it is bounded for $|\xi| \geqslant 1$.

Q.E.D.

We now return to the original problem; we shall show that a similar result to that of proposition 6.2 can be established in $\mathscr{D}_{L^2}{}^1$.

PROPOSITION 6.3. Let the domain of A be (6.24), and let us further assume (6.25). Then, $(I-\lambda A)$ $(|\lambda|<\varepsilon_1)$ is a bijection from $\mathscr{D}(A)$ onto $\mathscr{D}_{L^2}{}^1$.

Proof. We can easily demonstrate that the image $(I-\lambda A)\mathscr{D}(A)$ is closed, as we have done in proposition 6.2, so that we have to show the denseness. If it is not dense, then there exists a certain $\psi_1 \in \mathscr{D}_{L^2}{}^1(\psi_1 \neq 0)$ with

$$((\Lambda+1)(I-\lambda A)u, (\Lambda+1)\psi_1)=0 \quad (u \in \mathscr{D}(A)).$$

Let us write $(\Lambda+1)\psi_1=\psi \in L^2$, so that $\psi \neq 0$. From this,

$$(I-\lambda A^*)(\Lambda+1)\psi=0.$$

This relation can be written as

$$(\Lambda+1)\left[\psi+\lambda \sum \frac{\partial}{\partial x_k}(A_k\psi)\right]+\lambda(C\psi-B^*(x)\Lambda\psi)=0,$$

$$-C=\sum_k \frac{\partial}{\partial x_k}(\Lambda A_k(x)-A_k(x)\Lambda),$$

i.e.

$$\psi+\lambda \sum \frac{\partial}{\partial x_k}(A_k(x)\psi)+\lambda K\psi=0, \tag{6.30}$$

where $K=(\Lambda+1)^{-1}[C-B^*(x)(\Lambda+1)]$, and $((\Lambda+1)^{-1}u)\hat{}(\xi)=(|\xi|+1)^{-1}\hat{u}(\xi)$.

We note that K is a bounded operator of L^2. In fact, $(\Lambda+1)^{-1}C$ is bounded because from lemma 6.2 $(\Lambda+1)^{-1}\partial/\partial x_k$, $(\Lambda A_k(x)-A_k(x)\Lambda)$ are all bounded operators. For u, $v \in L^2$, we have

$$\begin{aligned}|(u, (\Lambda+1)^{-1}B^*\Lambda v)| &= |(B(\Lambda+1)^{-1}u, \Lambda v)| \leqslant \|B(\Lambda+1)^{-1}u\|_1 \|\Lambda v\|_{-1} \\ &\leqslant \|B(x)\| \, \|(\Lambda+1)^{-1}u\|_1 \|v\|_{L^2} \\ &\leqslant C\|u\|_{L^2}\|v\|_{L^2},\end{aligned}$$

where $\|B(x)\|$ is a norm of an operator $B(x)$ which carries $\mathscr{D}_{L^2}{}^1$ into $\mathscr{D}_{L^2}{}^1$, so that $(\Lambda+1)^{-1}B^*(x)\Lambda$ is also bounded. From this and (6.30) we can apply the inequality (6.23). The left hand term of (6.30) in the L^2-norm is larger than

$$(1-\beta'|\lambda|)\|\psi\|-|\lambda|\|K\|\|\psi\|=\{1-(\beta'+\|K\|)|\lambda|\}\|\psi\|.$$

Therefore, if $|\lambda|<(\beta'+\|K\|)^{-1}$ then $\psi=0$, which is a contradiction.

Q.E.D.

Next, we show that under the assumption $u\in\mathscr{E}_t^0(\mathscr{D}_{L^2}^1)$ and also $u\in\mathscr{E}_t^1(L^2)$ we can establish an energy inequality in the sense of the $\mathscr{D}_{L^2}^1$-norm, where in (6.3) we impose a stronger condition on $B(x, t)$: $t\to B(x, t)\in\mathscr{B}^1(R^n)$ is continuous.

First we also impose a stronger condition on $u(t)$: $u\in\mathscr{E}_t^0(\mathscr{D}_{L^2}^2)$ and $u\in\mathscr{E}_t^1(\mathscr{D}_{L^2}^1)$.

If we operate with $\partial/\partial x_i = \mathrm{D}_i$ on

$$\frac{\mathrm{d}}{\mathrm{d}t}u(t)=A\cdot u(t)+f(t), \tag{6.1}$$

we get

$$\frac{\mathrm{d}}{\mathrm{d}t}\mathrm{D}_i u(t)=A(\mathrm{D}_i u)+(\mathrm{D}_i A)u+\mathrm{D}_i f \quad (i=1,2,\ldots,n), \tag{6.31}$$

where

$$\mathrm{D}_i A=\sum(\mathrm{D}_i A_k)\frac{\partial}{\partial x_k}+\mathrm{D}_i B.$$

If $f\in\mathscr{E}_t^0(\mathscr{D}_{L^2}^1)$, then we can apply proposition 6.1 to $\mathrm{D}_i u(t)$.

We now have

$$\frac{\mathrm{d}}{\mathrm{d}t}\|\mathrm{D}_i u(t)\|^2\leqslant\gamma'\|\mathrm{D}_i u(t)\|^2+2\|\mathrm{D}_i u(t)\|(\|\mathrm{D}_i f(t)\|+C\|u(t)\|_1).$$

With this result in mind we consider

$$\frac{\mathrm{d}}{\mathrm{d}t}\|u(t)\|^2\leqslant\gamma'\|u(t)\|^2+2\|u(t)\|\cdot\|f(t)\|$$

and

$$\|u(t)\|_1^2=\|u(t)\|^2+\sum\|\mathrm{D}_i u(t)\|^2.$$

We have

$$\frac{\mathrm{d}}{\mathrm{d}t}\|u(t)\|_1^2\leqslant 2\gamma_1\|u(t)\|_1^2+2\|u(t)\|_1\cdot\|f(t)\|_1.$$

From this, for $0\leqslant t\leqslant T$ we have

$$\|u(t)\|_1\leqslant\exp(\gamma_1 t)\|u(0)\|_1+\int_0^t\exp\{\gamma_1(t-s)\}\|f(s)\|_1\,\mathrm{d}s, \tag{6.32}$$

where γ_1 is determined according to T. We note that the proof of the same fact under the original condition for $u(t)$ can be carried out by the same method as we have done when we proved theorem 6.1. (Use lemma 6.1.)

We can state our result as follows:

THEOREM 6.2 *(Existence theorem).* Let us assume that for (6.1) $A_k(x)$ is Hermitian, and also that it satisfies (6.25). In this case, for an arbitrary initial value $u_0 \in \mathscr{D}_{L^2}^1$, and for $f(t) \in \mathscr{E}_t^0(\mathscr{D}_{L^2}^1)$ $(t \geqslant 0)$ (the right hand term of (6.1)), there exists a unique solution of (6.1) such that it satisfies $u(t) \in \mathscr{E}_t^0(\mathscr{D}_{L^2}^1)$ $(t \geqslant 0)$ and $u \in \mathscr{E}_t^1(L^2)$ $(t \geqslant 0)$.† Also, we have that the energy inequality (6.32) is true.

Proof. From the inequality (6.26) and proposition 6.3, we can apply theorem 5.6 (Hille–Yosida). (We have $E = \mathscr{D}_{L^2}^1$.) We see that for $u_0 \in \mathscr{D}(A)$ $(\subset \mathscr{D}_{L^2}^1)$ and $f(t)$, $Af(t) \in \mathscr{E}_t^0(\mathscr{D}_{L^2}^1)$ $(t \geqslant 0)$, there exists a unique solution $u(t) \in \mathscr{E}_t^1(\mathscr{D}_{L^2}^1)$ $(t \geqslant 0)$. We do not yet know the properties of $\mathscr{D}(A)$.

Let us assume that $u_0 \in \mathscr{D}_{L^2}^2$ $(\subset \mathscr{D}(A))$ and $f(t) \in \mathscr{E}_t^0(\mathscr{D}_{L^2}^2)$ $(t \geqslant 0)$. Of course, in this case, the hypothesis of theorem 6.2 is compatible with these assumptions.

Now, on any u_0 and $f(t)$ which satisfy the hypothesis of the theorem, we can operate with a Friedrichs' mollifier φ_δ to get $u_0^{(\delta)} = \varphi_\delta \underset{(x)}{*} u_0$, and $f_\delta(t) = \varphi_\delta * f(t)$. These also satisfy the same hypothesis. So that there exists a solution $u_\delta(t) \in \mathscr{E}_t^1(\mathscr{D}_{L^2}^1)$ $(t \geqslant 0)$. But, if we apply (6.32) to $u_\delta(t) - u_{\delta'}(t)$, then for a fixed $T (>0)$, we have

$$\max_{0 \leqslant t \leqslant T} \|u_\delta(t) - u_{\delta'}(t)\|_1 \leqslant C(T) \left[\|u_0^{(\delta)} - u_0^{(\delta')}\|_1 + \int_0^T \|f_\delta(t) - f_{\delta'}(t)\|_1 \, dt\right].$$

Therefore, from Lebesgue's theorem, as $\delta \to +0$, $\{u_\delta(t)\}$ form a Cauchy sequence in $\mathscr{E}_t^0(\mathscr{D}_{L^2}^1)$ $(0 \leqslant t \leqslant T)$, i.e.

$$u_\delta(t) \to u(t) \in \mathscr{E}_t^0(\mathscr{D}_{L^2}^1) \quad (t \geqslant 0).$$

On the other hand, if we apply the limiting process to

$$u_\delta(t) = u_0^{(\delta)} + \int_0^t \{Au_\delta(s) + f_\delta(s)\} \, ds,$$

we have

$$u(t) = u_0 + \int_0^t \{Au(s) + f(s)\} \, ds,$$

† I.e. $u(t)$ is continuously differentiable to first order with respect to t by the topology of L^2.

where integration is carried out under the topology of L^2. From this

$$\frac{\mathrm{d}}{\mathrm{d}t}u(t)=Au(t)+f(t)$$

is true in the sense of L^2-topology.

Q.E.D.

Note 1. Under the same assumption, there also exists a solution for a Cauchy problem for $t\leqslant 0$. An energy inequality can be established in this case. To be more precise, if we change t into $-t$, the equation (6.1) becomes

$$\frac{\mathrm{d}u}{\mathrm{d}t}=-Au-f.$$

Therefore, after replacing A by $-A$, the same existence theorem (theorem 5.6) can be established.

In fact, the right hand side of (6.26) is valid for $(I-\lambda A)$ irrespective of the sign of λ. Hence, the theorem can be paraphrased as: for $u_0\in\mathscr{D}_{L^2}{}^1$, $f(t)\in\mathscr{E}_t{}^0(\mathscr{D}_{L^2}{}^1)$ $(-\infty<t<+\infty)$, there exists a solution for (6.1) such that $u(t)\in\mathscr{E}_t{}^0(\mathscr{D}_{L^2}{}^1)$ $(-\infty<t<+\infty)$ and $u\in\mathscr{E}_t{}^1(L^2)$ $(-\infty<t<+\infty)$. In this case we have an energy inequality

$$\|u(t)\|_1\leqslant\exp(\gamma_1|t|)\|u_0\|_1+\left|\int_0^t\exp\{\gamma_1|t-s|\}\ \|f(s)\|_1\,\mathrm{d}s\right|.$$

Note 2. If $u_0\in L^2, f(t)\in\mathscr{E}_t{}^0(L^2)$ $(t\geqslant 0)$, there exists a unique solution for (6.1) such that $u(t)\in\mathscr{E}_t{}^0(L^2)$ $(\geqslant 0)$ and $u\in\mathscr{E}_t{}^1(\mathscr{D}'_{L^2}{}^1)$ $(t\geqslant 0)$. The proof is exactly the same as the previous case (use proposition 6.2). In this case, we note that $A_k\in\mathscr{B}^1$, $B\in\mathscr{B}^0$.

5 *Existence theorem* 2 *for a system of symmetric hyperbolic equations (general case)*

In the previous section, we assumed that A was independent of t in order to apply the theory of semi-groups. If A depends on t, the situation becomes more complicated: the generalization of the theory of semi-groups presents some difficulty. We explain this in what follows. Let us consider

$$\frac{\partial}{\partial t}u=a(t)\frac{\partial}{\partial x}u$$

where $a(t)$ is a real-valued continuous function. The reader should note that this is the simplest hyperbolic equation. By a Fourier transform, we have

$$\frac{\mathrm{d}}{\mathrm{d}t}v(\xi, t)=a(t)2\pi\mathrm{i}\xi v(\xi, t),$$

where the condition $v(\xi, 0)=1$ is satisfied by

$$v(\xi, t)=\exp\{2\pi\mathrm{i}\xi\sigma(t)\},\quad \sigma(t)=\int_0^t a(s)\,\mathrm{d}s.$$

Therefore,

$$E_x(t)=\mathscr{F}[\exp\{2\pi\mathrm{i}\xi\sigma(t)\}]=\delta_{-\sigma(t)}.$$

This is obvious from the proof of proposition 2.11. Therefore $u(x, t)=E_x(t)\underset{(x)}{*}u_0(x)$, so that $u(x, t)=u_0(x+\sigma(t))$ is a solution corresponding to the initial value $u_0(x)$.

From the stand-point of the theory of semi-groups, by (6.22), we have

$$\mathscr{D}(A(t))=\begin{cases}\mathscr{D}_{L^2}{}^1 & (a(t)\neq 0),\\ L^2 & (a(t)=0),\end{cases}$$

where $A(t)=a(t)\partial/\partial x$.

If $a(t)=0$, $a(t)\not\equiv 0$, we note that the domain of the definition of $\mathscr{D}(A(t))$ varies sharply, so that it is difficult to deal with the problem generally. This is the reason why we stated theorem 6.2 without giving a specific domain of definition. Now, let us apply Cauchy's 'zigzag line' method in the theory of ordinary differential equations to the existence problem of solutions. In fact, this method became widely known after Peano used it for deriving his famous theorem in ordinary differential equations. For the convenience of the reader, we shall explain it briefly.

Let us examine the equation

$$\frac{\mathrm{d}}{\mathrm{d}t}u(t)=f(t, u),$$

where $f(t, u)$ is continuous and bounded in $[0, 1]\times(-\infty<u<+\infty)$. In this case, we claim that there exists at least one solution $u(t)$ $(0\leqslant t\leqslant 1)$ for the differential equation, such that it takes an arbitrarily given value u_0 at $t=0$ (Peano's theorem). The proof is as follows. Let us consider a partition $\Delta: 0=t_0<t_1<\cdots<t_n=1$. Let $u_1(t)=u_0+f(t_0, u_0)\cdot(t-t_0)$ in $t_0\leqslant t\leqslant t_1$, and

let $u_2(t)=u_1(t_1)+f(t_1, u_1(t_1))\,(t-t_1)$ in $t_1 \leqslant t \leqslant t_2$, etc. We now have a function $u_\Delta(t)$ thus obtained.

Now, consider a sequence of partitions $\Delta_1, \Delta_2, \ldots$ where Δ_n has the maximum length of a piece of the partition of Δ_n which tends to 0 as $n \to +\infty$. Obviously, we can write

$$u_\Delta(t)-u_0=\int_0^t f(t, u_\Delta(t))\,\mathrm{d}t+\zeta_\Delta(t)$$

where $\zeta_\Delta(t)$ tends to 0 uniformly with respect to t as the maximum length of the partitions tends to 0 uniformly. On the other hand $\{u_{\Delta_n}(t)\}$ is uniformly bounded and also equi-continuous, so that, by the Ascoli–Arzela theorem, some subsequence of the sequence $\{u_{\Delta_{n_p}}(t)\}$ is uniformly convergent. Therefore, if we write this limit as $u(t)$, we have

$$u(t)=u_0+\int_0^t f(t, u(t))\,\mathrm{d}t.$$

Hence, there exists a solution.

Q.E.D.

Now, we write (6.1) as

$$\frac{\mathrm{d}}{\mathrm{d}t}u(t)=A(t)u(t)+f(t) \tag{6.34}$$

where for $A(t)$ we assume that $A_k(x, t)$ $(k=1, 2, \ldots, n)$ is Hermitian, and for the differentiability of $A(t)$ we assume that

$$t \to A_k(x, t) \in \mathscr{B}^{1+\sigma} \quad (\sigma>0); \quad t \to B(x, t) \in \mathscr{B}^1 \tag{6.35}$$

are continuous for $t \geqslant 0$.

Let us fix $T(>0)$ and let Δ be a partition of $[0, T]$, $0=t_0<t_1<\cdots<t_n=T$. Furthermore, we assume that

$$u_0 \in \mathscr{D}_{L^2}{}^1, \quad f(t) \in \mathscr{E}_t{}^0(\mathscr{D}_{L^2}{}^1) \quad (t \geqslant 0). \tag{6.36}$$

By theorem 6.2 we define $u_\Delta(t)$ as follows. First, for $t_0 \leqslant t \leqslant t_1$, we define $u_1(t)$ by

$$\frac{\mathrm{d}}{\mathrm{d}t}u_1(t)=A(t_0)u_1(t)+f(t) \quad (u_1(t_0)=u_0).$$

In general, for $t_{j-1} \leqslant t \leqslant t_j$, we define $u_j(t)$ by

$$\frac{\mathrm{d}}{\mathrm{d}t} u_j(t) = A(t_{j-1}) u_j(t) + f(t) \quad (u_j(t_{j-1}) = u_{j-1}(t_{j-1})).$$

Obviously $u_\Delta(t) \in \mathscr{E}_t^0(\mathscr{D}_{L^2}^1)$ $(0 \leqslant t \leqslant T)$.

For example, if we consider Δ_n as an equal, n-partition of $[0, T]$, then there exists M satisfying

$$\max_{0 \leqslant t \leqslant T} \|u_{\Delta_n}(t)\|_1 \leqslant M \quad (n = 1, 2, \ldots). \tag{6.37}$$

On the other hand, if $A_\Delta(t) = A(t_j)$, $t \in [t_j, t_{j+1})$ $(j = 0, 1, 2, \ldots, n-1)$ then

$$\frac{\mathrm{d}}{\mathrm{d}t} u_\Delta(t) = A_\Delta(t) u_\Delta(t) + f(t) \quad (t \neq t_j). \tag{6.38}$$

We write $u_{\Delta_n}(t) = u_n(x, t)$. Then, from (6.37), and (6.38), $\{u_n(x, t)\}_{n=1,2,\ldots}$ is a bounded sequence of $\mathscr{E}_{L^2}^1(\Omega)$ where $\Omega = R^n \times (0, T)$. Therefore, from theorem 2.12, we can choose a subsequence $\{u_{n_p}(x, t)\}_{p=1,2,\ldots}$ which is weakly convergent in $\mathscr{E}_{L^2}^1(\Omega)$. We write the limit as $u(x, t) \in \mathscr{E}_{L^2}^1(\Omega)$. By proposition 2.6, we see that

$$u_{n_p}(x, t) \to u(x, t), \quad \frac{\partial}{\partial t} u_{n_p}(x, t) \to \frac{\partial}{\partial t} u(x, t), \quad \frac{\partial}{\partial x_j} u_{n_p}(x, t) \to \frac{\partial}{\partial x_j} u(x, t)$$

in the sense of the weak topology of $L^2(\Omega)$. From this we have

$$\frac{\partial}{\partial t} u - \sum A_k(x, t) \frac{\partial}{\partial x_k} u - B(x, t) u = f(x, t) \tag{6.39}$$

in the sense of distribution of Ω. Therefore, it is also valid in $L^2(\Omega)$. In fact, from (6.38), for $\varphi(x, t) \in \mathscr{D}(\Omega)$

$$\left\langle \frac{\partial}{\partial t} u_n - \sum A_k^{(n)}(x, t) \frac{\partial}{\partial x_k} u_n - B^{(n)}(x, t) u_n, \varphi \right\rangle = \langle f, \varphi \rangle.$$

Now, we put $n = n_p$ and make $p \to \infty$, then, it is easy to see that

$$\left\langle \frac{\partial}{\partial t} u - \sum A_k(x, t) \frac{\partial}{\partial x_k} u - B(x, t) u, \varphi \right\rangle = \langle f, \varphi \rangle.$$

Next, because $u(x, t) \in \mathscr{E}_{L^2}{}^1(\Omega)$, if we write $\lim_{\varepsilon \to +0} u(x, \varepsilon) = \gamma u$, then, by lemma 3.5 (see also theorem 3.9),

$$(\gamma u)(x) = -\int_0^T \{\alpha(t) u(x, t)\}_t{}' \, \mathrm{d}t$$

is true almost everywhere, where α is a C^∞-function such that $\alpha(t) = 1$ for $t \in [0, \frac{1}{2}T]$ and $\alpha = 0$ in a neighbourhood $t = T$. Therefore, for $\varphi \in \mathscr{D}$.

$$\int (\gamma u)(x) \varphi(x) \, \mathrm{d}x = -\int \int_\Omega \{\alpha'(t) \varphi(x) u(x, t) + \alpha(t) \varphi(x) u_t{}'(x, t)\} \, \mathrm{d}t \, \mathrm{d}x. \qquad (6.40)$$

On the other hand,

$$\frac{\mathrm{d}}{\mathrm{d}t} \langle u_\Delta(t), \varphi(x) \rangle = \left\langle \frac{\mathrm{d}}{\mathrm{d}t} u_\Delta(t), \varphi(x) \right\rangle \quad (t \neq t_j)$$

and $u_\Delta(t)$ is continuous in L^2. From $u_\Delta(0) = u_0$, we have

$$\langle u_0, \varphi(x) \rangle_{R^n} = -\int_0^T \langle (\alpha(t) u_\Delta(t))', \varphi(x) \rangle_{R^n} \, \mathrm{d}t,$$

i.e.

$$\int u_0(x) \varphi(x) \, \mathrm{d}x = -\int \int_\Omega \{\alpha'(t) \varphi(x) u_n(x, t) + \alpha(t) \varphi(x) \, (u_n(x, t))_t{}'\} \, \mathrm{d}x \, \mathrm{d}t.$$

Now, we put $n = n_p$, and make $p \to \infty$. Then, the integral of the right hand side tends to the term on the right hand side of (6.40). Hence, it follows that $(\gamma u)(x) = u_0(x)$.

Next we shall demonstrate that $t \to u(x, t) \in L^2(R^n)$ is continuous (we adjust the value on sets having measure 0 in Ω, if necessary). If we consider theorem 2.7, for almost all x, we have

$$u(x, t) - u(x, t') = \int_{t'}^t \frac{\partial}{\partial t} u(x, t) \, \mathrm{d}t.$$

So that, from the Schwartz inequality,

$$|u(x, t) - u(x, t')|^2 \leqslant |t - t'| \int_{t'}^t \left| \frac{\partial}{\partial t} u(x, t) \right|^2 \mathrm{d}t,$$

and integration over R^n,

$$\|u(x,t)-u(x,t')\|_{L^2(R^n)} \leqslant |t-t'|^{\frac{1}{2}} \left\|\frac{\partial}{\partial t} u\right\|_{L^2(\Omega)}.$$

We define $u(x, 0)=\gamma u(x)=u_0(x)$, also, we define $u(x, T)$ such that $u(x,T)=\lim_{t\to T} u(x,t)$ in L^2 for $t=T$. We see that $t\to u(x,t)\in L^2(R^n)$ is continuous for $t\in[0, T]$. Finally, we show $u(x,t)\equiv u(t)$ is a solution in the sense of theorem 6.2.

To see this we use a mollifier; we write

$$\varphi_\delta \underset{(x)}{*} u(t)=u_\delta(t), \quad \varphi_\delta \underset{(x)}{*} f(t)=f_\delta(t).$$

Operating with the mollifier from the left in (6.39), we get

$$\frac{\partial}{\partial t} u_\delta(t)=\sum A_k(x,t)\frac{\partial}{\partial x_k} u_\delta(t)+B(x,t)u_\delta(t)+f_\delta(t)+C_\delta u(t), \tag{6.41}$$

where

$$C_\delta u=\sum_k \left[\varphi_\delta \underset{(x)}{*}, A_k\frac{\partial}{\partial x_k}\right]u+[\varphi_\delta \underset{(x)}{*}, B]u.$$

From lemma 6.1, we have $\|C_\delta u(s)\|_1\to 0$ $(\delta\to 0)$ almost everywhere with respect to s. Also, we have $\|C_\delta u(s)\|_1\leqslant C\|u(s)\|_1$. Therefore, applying the energy inequality (6.32) to $u_\delta(t)-u_{\delta'}(t)$, we have

$$\begin{aligned}&\|u_\delta(t)-u_{\delta'}(t)\|_1\\ &\quad\leqslant C\Big[\|u_\delta(0)-u_{\delta'}(0)\|_1+\int_0^T \{\|f_\delta(s)-f_{\delta'}(s)\|_1+\|C_\delta u(s)-C_{\delta'}u(s)\|_1\}\,\mathrm{d}s\Big]\end{aligned}$$

for $t\in[0, T]$.

By Lebesgue's theorem (note that $\int_0^T\|u(s)\|_1\,\mathrm{d}s<+\infty$) the right hand term tends to 0 as $\delta,\delta'\to 0$. Therefore, $\{u_\delta(t)\}$ is a Cauchy sequence as $\delta\to 0$ in $\mathscr{E}_t^0(\mathscr{D}_{L^2}{}^1)$ $(0\leqslant t\leqslant T)$, hence, $u(t)\in\mathscr{E}_t^0(\mathscr{D}_{L^2}{}^1)$ $(0\leqslant t\leqslant T)$. Also, from (6.41), we have

$$u(t)-u_0=\int_0^t \{A(s)u(s)+f(s)\}\,\mathrm{d}s,$$

so that we see $u(t)\in\mathscr{E}_t^1(L^2)$ $(0\leqslant t\leqslant T)$. We state our result in the following way:

THEOREM 6.3 *(Existence theorem)*. Let A_k be Hermitian, and let us assume (6.35). In this case, for an arbitrary $u_0 \in \mathscr{D}_{L^2}{}^1$ and for an arbitrary $f(t) \in \mathscr{E}_t{}^0(\mathscr{D}_{L^2}{}^1)$ $(t \geqslant 0)$, we have a unique solution of (6.1) such that $u(t) \in \mathscr{E}_t{}^0(\mathscr{D}_{L^2}{}^1)$ $(t \geqslant 0)$ and $u \in \mathscr{E}_t{}^1(L^2)$ $(t \geqslant 0)$.

In theorem 6.3, if we impose the smoothness not only on the coefficients A_k and B but also on u_0 and f, then we have an improved form of the theorem. Let us explain this situation. We assume that

$$t \to A_k(x, t) \in \mathscr{B}^2, \quad t \to B(x, t) \in \mathscr{B}^2; \quad u_0 \in \mathscr{D}_{L^2}{}^2, \quad f(t) \in \mathscr{E}_t{}^0(\mathscr{D}_{L^2}{}^2) \quad (t \geqslant 0).$$

By theorem 6.3, we see that $u(t) \in \mathscr{E}_t{}^0(\mathscr{D}_{L^2}{}^1)$ $(t \geqslant 0)$ is a solution. Now we differentiate both sides of

$$M[u] = f(t). \tag{6.1}$$

Writing $\partial/\partial x_i = \mathrm{D}_i$, we have

$$M[\mathrm{D}_i u] - \sum_k (\mathrm{D}_i A_k)[\mathrm{D}_k u] = \mathrm{D}_i f + (\mathrm{D}_i B)[u] \quad (i = 1, 2, \ldots, n) \tag{6.42}$$

where we note that

$$g_i = \mathrm{D}_i f + (\mathrm{D}_i B)[u] \in \mathscr{E}_t{}^0(\mathscr{D}_{L^2}{}^1) \quad (t \geqslant 0).$$

On the other hand, (6.42) is a system of hyperbolic equations having a principal part M whose unknown functions are $(\mathrm{D}_1 u, \ldots, \mathrm{D}_n u)$. Therefore, the system has a solution such that $(\mathrm{D}_1 u, \ldots, \mathrm{D}_n u) \in \mathscr{E}_t{}^0(\mathscr{D}_{L^2}{}^1)$ $(t \geqslant 0)$ and $\in \mathscr{E}_t{}^1(L^2)$ $(t \geqslant 0)$, if we use almost the same method as in the previous theorem and if we note that $\mathrm{D}_i u(0) = D_i u_0(x) \in \mathscr{D}_{L^2}{}^1$ $(i = 1, 2, \ldots, n)$. It is easy to see that for

$$M[v_i] - \sum_k (\mathrm{D}_i A_k)\,[v_k] = g_i,$$

an energy inequality can be established in L^2-space, so that the system has a unique solution in L^2-space. From this fact, we have $u(t) \in \mathscr{E}_t{}^0(\mathscr{D}_{L^2}{}^2)$ $(t \geqslant 0)$. More generally, we have:

THEOREM 6.4. In theorem 6.3, if we assume $t \to A_k(x, t) \in \mathscr{B}^{\max[1+\sigma, m]}$ $(0 < \sigma < 1)$ and $t \to \mathscr{B}(x, t) \in \mathscr{B}^m$ are continuous, and $u_0 \in \mathscr{D}_{L^2}{}^m$, $f(t) \in \mathscr{E}_t{}^0(\mathscr{D}_{L^2}{}^m)$ $(t \geqslant 0)$ $(m = 0, 1, 2, \ldots)$, then there exists a unique solution such that $u(t) \in \mathscr{E}_t{}^0(\mathscr{D}_{L^2}{}^m)$ $(t \geqslant 0)$, where we take the differentiation with respect to t in the sense of the topology of $\mathscr{D}_{L^2}{}^{m-1}$.

Note. If we assume the smoothness of the coefficients and f with respect to t, then we can conclude that u is also smooth with respect to t. The proof is quite similar to the argument above, so it is left as an exercise.

COROLLARY. If $m \geqslant [\frac{1}{2}n]+2$ in theorem 6.4, then $u(x, t)$ is a genuine solution (use Sobolev's lemma).

6 *Non-symmetric hyperbolic systems*

Recall the strongly hyperbolic system which we treated in theorem 4.14. That was that in

$$\frac{\partial}{\partial t} u - \sum_{k=1}^{n} A_k \frac{\partial}{\partial x_k} u - Bu = f \tag{6.40}$$

if we write $\sum A_k \xi_k = A \cdot \xi$, then the roots of $\det(\lambda I - A \cdot \xi) = 0$ satisfy $\lambda_1(\xi) < \lambda_2(\xi), < \cdots < \lambda_N(\xi)$, where $\lambda_1(\xi)$ is a homogeneous function of the first degree with respect to ξ. Obviously, in this case, there exists $N(\xi)$ satisfying the following conditions.

(1) $N(\xi)$ is a homogeneous function of zero-th degree, and is bounded on the unit sphere with respect to ξ. Also, there exists δ such that $|\det N(\xi)| \geqslant \delta > 0$.

(2) $N(\xi) A \cdot \xi = \mathscr{D}(\xi) N(\xi)$ where

$$D(\xi) = \begin{bmatrix} \lambda_1(\xi) & & 0 \\ & \cdot & \\ 0 & & \lambda_N(\xi) \end{bmatrix}.$$

Now we consider a solution $u(x, t) \in \mathscr{E}_t{}^0(L^2)$. By a Fourier transform of (4.60) with respect to x, we have

$$\frac{\mathrm{d}}{\mathrm{d}t} \hat{u}(\xi, t) = (2\pi \mathrm{i} A \cdot \xi + B) \hat{u}(\xi, t) + \hat{f}(\xi, t).$$

Operating with $N(\xi)$ on the left, and putting $N(\xi)\hat{u}(\xi, t) = v(\xi, t)$, we have

$$\frac{\mathrm{d}}{\mathrm{d}t} v(\xi, t) = 2\pi \mathrm{i} D(\xi) v(\xi, t) + B'(\xi) v(\xi, t) + N(\xi) \hat{f}(\xi, t) \tag{6.43}$$

where $B'(\xi) = N(\xi) B N(\xi)^{-1}$ and $B'(\xi)$ is bounded. $D(\xi)$ is a real-valued function, so that

$$\frac{\mathrm{d}}{\mathrm{d}t}|v(\xi,t)|^2=\frac{\mathrm{d}}{\mathrm{d}t}v\cdot\bar{v}+v\cdot\frac{\mathrm{d}}{\mathrm{d}t}\bar{v}=(2\pi \mathrm{i}D(\xi)+B')v\cdot\bar{v}+v\cdot(-2\pi\mathrm{i}D(\xi)+\bar{B}')\bar{v}$$
$$+2\,\mathrm{Re}\,N\hat{f}\cdot\bar{v}=2\,\mathrm{Re}\,\{B'(\xi)v\cdot\bar{v}+N\hat{f}\cdot\bar{v}\}.$$

Therefore, integrating with respect to ξ, and applying Plancherel's theorem we have

$$\frac{\mathrm{d}}{\mathrm{d}t}\|\mathcal{N}\underset{(x)}{*}u(t)\|^2\leqslant 2\gamma\|\mathcal{N}\underset{(x)}{*}u(t)\|^2+2\|\mathcal{N}\underset{(x)}{*}f(t)\|\cdot\|\mathcal{N}\underset{(x)}{*}u(t)\|,$$

where $\mathcal{N}(x)$ is a Fourier inverse image of $N(\xi)$ and is a distribution. From this fact, we have

$$\|\mathcal{N}*u(t)\|\leqslant \mathrm{e}^{\gamma t}\|\mathcal{N}*u(0)\|+\int_0^t \mathrm{e}^{\gamma(t-s)}\|\mathcal{N}*f(s)\|\,\mathrm{d}s, \tag{6.44}$$

where we note that the operator $\mathcal{N}*$ and its inverse are continuous in L^2. In fact, $\|\mathcal{N}*u\|=\|N(\xi)\hat{u}(\xi)\|$. There exists $\mathcal{N}^{-1}(\xi)$.

Hence, it is bounded with respect to ξ.

Therefore by (6.44) we have

$$\|u(t)\|\leqslant C[\|u(0)\|+\int_0^t \mathrm{e}^{\gamma(t-s)}\|f(s)\|\,\mathrm{d}s].$$

The important point of the above argument is that if we write $\mathcal{N}*u=w(t)$, then, by the operator Λ (see definition 6.1), we have the principal part of (6.43) as

$$\frac{\mathrm{d}}{\mathrm{d}t}w(t)=2\pi\mathrm{i}\Delta(x)*\Lambda w(t)$$

$$\mathcal{F}[\Delta(x)]=\begin{bmatrix}\lambda_1\left(\frac{\xi}{|\xi|}\right) & & 0\\ & \ddots & \\ 0 & & \lambda_N\left(\frac{\xi}{|\xi|}\right)\end{bmatrix}.$$

We write $\mathcal{H}=\Delta(x)*$, where $\mathcal{H}$ is a bounded Hermitian operator of L^2 satisfying the following relation:

$$\mathcal{H}\Lambda u=\Lambda\mathcal{H}u \quad \text{for} \quad u\in\mathcal{D}_{L^2}{}^1.$$

Therefore, if we rewrite the above equation as

$$\frac{d}{dt}w(t)=2\pi i\mathscr{H}\Lambda w(t), \tag{6.45}$$

then we see that $i\mathscr{H}\Lambda$ is an *anti-Hermitian operator* i.e.

$$(i\mathscr{H}\Lambda)^*=-i\mathscr{H}\Lambda.$$

Our final object is to show that a similar property holds in the case of variable coefficients. To this end, we need the 'Calderón–Zygmund singular integral operator', which we shall explain in the following section.

7 *Singular integral operator*†

As we have shown in sections 6–7, chapter 2, if $N(x)$ is a homogeneous function of $(-n)$th degree, smooth in $R^n\setminus\{0\}$, and its mean over a sphere is 0, then we see that

$$\text{v.p. } N(x)*f(x)=\lim_{\varepsilon\to 0}\int_{|x-y|\geq\varepsilon} N(x-y)f(y)\,dy$$

defines a bounded operator of L^2.

Conversely, given a smooth homogeneous function of degree zero which has a mean of zero over a sphere, then the Fourier inverse image $N(x)$ satisfies the condition just mentioned.

Now given a differential operator

$$L\left(x, \frac{\partial}{\partial x}\right)=\sum_j a_j(x)\frac{\partial}{\partial x_j},$$

for $u\in\mathscr{S}$, we have

$$(Lu)(x)=\int e^{2\pi i x\xi}(2\pi i\sum a_j(x)\xi_j)\hat{u}(\xi)\,d\xi.$$

We write $h(x, \xi)=2\pi i\sum a_j(x)\xi_j/|\xi|$, and define

$$Hu(x)=\int e^{2\pi i x\xi}h(x, \xi)\hat{u}(\xi)\,d\xi. \tag{6.46}$$

We see that H is a bounded operator of L^2. In fact, if we use a Riesz

† See A. P. Calderón and A. Zygmund: 'Singular integral operators and differential equations', *Amer. J. Math.* 79 (1957), 901–21.

operator R_j (see definition 6.1), we have $Hu=(2\pi i)^2\sum a_j(x)(R_j*u)$. From this we have

$$Lu=H\Lambda u. \tag{6.47}$$

We study a more general case; replace $h(x, \xi)$ in (6.46) by a new function defined as follows. If we regard $h(x, \xi)$ as a function of ξ then it is homogeneous of degree zero, and belongs to $\mathscr{E}^\infty$ in $R^n\setminus\{0\}$. We wish to measure the dependence on $x(x\in R^n)$ of a function $(\in\mathscr{E}_\xi^\infty(R^n\setminus\{0\}))$ in this case.

For this purpose, we introduce semi-norms

$$p_s(h)=\sum_{|\nu|\leqslant s}\sup_{|\xi|\geqslant 1}\left|\left(\frac{\partial}{\partial\xi}\right)^\nu h(\xi)\right| \quad (s=0, 1, 2, \ldots).$$

in $\mathscr{E}_\xi^\infty(R^n\setminus\{0\})$. By $h(x, \xi)\in C_\beta^\infty(\beta\geqslant 0)$ we mean:

(1) If $\beta=0$ then $x\to h(x, \xi)\in\mathscr{E}_\xi(R^n\setminus\{0\})$ is bounded and continuous.

(2) If $0<\beta<1$, then $h(x, \xi)\in C_0^\infty$ and $x\to h(x, \xi)\in\mathscr{E}_\xi(R^n\setminus\{0\})$ has a uniform β-Hölder continuity in the following sense:

$$\sup_{|\xi|\geqslant 1}\left|\left(\frac{\partial}{\partial\xi}\right)^\nu h(x, \xi)-\left(\frac{\partial}{\partial\xi}\right)^\nu h(x', \xi)\right|\leqslant c_\nu|x-x'|^\beta \quad (\nu\geqslant 0).$$

(3) If $\beta\geqslant 1$, then $(\partial/\partial x)^\alpha h(x, \xi)\in C_0^\infty(|\alpha|\leqslant\beta)$, and if $|\alpha|=[\beta]$, then $(\partial/\partial x)^\alpha h(x, \xi)\in C_{\beta-[\beta]}^\infty$.

Now, let $h(x, \xi)\in C_0^\infty$. If we regard x as a parameter, then $h(x, \xi)$ is a C^∞-function defined on the unit sphere $|\xi|=1$. Therefore, it has an expansion by spherical functions which is uniformly convergent, i.e. if we write a normalized spherical function of order l as $Y_l(\xi)$ then $\int_{|\xi|=1} Y_l(\xi)^2 d\sigma_\xi=1$.

Let us write a base of the spherical functions of lth order as $\{Y_{lm}(\xi)\}_{m=1,2,\ldots,n(l)}$. As is well known, we have an expansion by spherical functions

$$h(x, \xi)=a_0(x)+\sum_{l\geqslant 1, m} a_{lm}(x)\,Y_{lm}(\xi), \tag{6.48}$$

where

$$a_{lm}(x)=\int_{|\xi|=1} h(x, \xi)\,Y_{lm}(\xi)\,d\sigma_\xi.$$

Then the following estimations are true.

(*a*) $|Y_{lm}(\xi)|\leqslant Cl^{\frac{1}{2}(n-2)}$.

(*b*) The number of the base Y_{lm} of the number of spherical functions of lth order has the order of l^{n-2}, i.e. $O(l^{n-2})$

(*c*) $|a_{lm}(x)| \leqslant CMl^{-\frac{3}{2}n}$ where

$$M = \sum_{|\nu| \leqslant 2n} \sup_{|\xi| \geqslant 1,\, x \in R^n} \left| \left(\frac{\partial}{\partial \xi}\right)^{\nu} h(x, \xi) \right|. \tag{6.49}$$

(*d*) In general, we have $|a_{lm}(x)| \leqslant c(n, r) M_{2r} l^{-2r+\frac{1}{2}n}$, where

$$M_{2r} = \sum_{|\nu| \leqslant 2r} \sup_{|\xi| \geqslant 1,\, x \in R^n} \left| \left(\frac{\partial}{\partial \xi}\right)^{\nu} h(x, \xi) \right|. \tag{6.50}$$

(*e*)
$$\sup_{|\xi| \geqslant 1} \left| \left(\frac{\partial}{\partial \xi}\right)^{\nu} Y_{lm}(\xi) \right| \leqslant c(v, n)\, l^{\frac{1}{2}(n-2)+|\nu|}.$$

From this, if $h(x, \xi_j) \in C_0^{\infty}$ then (6.48) is uniformly convergent. Furthermore, we can write the operator of (6.46) as

$$(Hu)(x) = a_0(x)u(x) + \sum_{l,\, m} a_{lm}(x)(\tilde{Y}_{lm}(x) * u(x)) \tag{6.51}$$

where $\tilde{Y}_{lm}(x) = \mathscr{F}[Y_{lm}(\xi)]$. Hence, by Plancherel's theorem, we have

$$\|Hu\| \leqslant \{|a_0(x)|_0 + \sum |a_{lm}(x)|_0 |Y_{lm}(\xi)|_0\} \, \|u\|.$$

Using (*a*), (*b*), (*c*), we have

$$\sum |a_{lm}(x)|_0 |Y_{lm}(\xi)|_0 \leqslant CM \sum_l l^{-\frac{3}{2}n + \frac{1}{2}(n-2) + n - 2} = CM \sum_l l^{-3} < +\infty.$$

Hence,

$$\|H\| \leqslant CM \tag{6.52}$$

where M is to be understood in the sense of (6.49). We call H a *singular integral operator*, and $h(x, \xi)$ the *symbol* of H. Some properties of h are as follows.

If $h_1(x, \xi), h_2(x, \xi) \in C_\beta^{\infty}$, then $h_1(x, \xi) + h_2(x, \xi)$ and $h_1(x, \xi)\, h_2(x, \xi) \in C_\beta^{\infty}$. Also, if $|h_2(x, \xi)| \geqslant \delta > 0$, then $h_1(x, \xi)/h_2(x, \xi) \in C_\beta^{\infty}$.

Next, we give the same definition in the form which was given by Calderón and Zygmund. They defined

$$(Hu)(x) = a_0(x)\, u(x) + \text{v.p.} \int k(x, x-y)\, u(y)\, dy$$

where $k(x, z)$ is a homogeneous function of $(-n)$th degree with respect to

z, belongs to $\mathscr{E}(R^n\setminus\{0\})$, and has the spherical mean zero. The expansion of $k(x, z)$ is

$$k(x, z)=\sum a_{lm}(x)\frac{Y_{lm}(z')}{|z|^n}, \quad a_{lm}(x)=\int_{|z|=1} k(x, z)\,Y_{lm}(z)\,d\sigma_z \qquad (z'=z/|z|).$$

From $\mathscr{F}[Y_{lm}(z')|z|^{-n}]=\gamma_l Y_{lm}(\xi)$, we define a symbol of H as

$$\sigma(H)=a_0(x)+\sum a_{lm}(x)\,\gamma_l Y_{lm}(\xi).$$

Let us then define operators $H^{\#}$ and $H_1\circ H_2$ which will play an important rôle in the theory of singular integral operators.

DEFINITION 6.2. Let H be a singular integral operator defined by $h(x, \xi)$. In this case, we write $H^{\#}$ as a singular integral operator having $\overline{h(x, \xi)}$ as a symbol. Also, we write $H_1\circ H_2$ as a singular integral operator having a symbol, which is the product of $h_1(x, \xi)$ and $h_2(x, \xi)$, where $\sigma(H_1)=h_1(x, \xi)$, $\sigma(H_2)=h_2(x, \xi)$.

Let us demonstrate the power of these concepts. Consider the following two differential operators:

$$L=\sum a_j(x)\frac{\partial}{\partial x_j}, \quad M=\sum_j b_j(x)\frac{\partial}{\partial x_j} \qquad (a_j, b_j\in\mathscr{B}^1).$$

We have

$$LM=\sum_{j,k} a_j(x)b_k(x)\frac{\partial^2}{\partial x_j\partial x_k}+\sum_{j,k} a_j(x)\frac{\partial b_k}{\partial x_j}\frac{\partial}{\partial x_k}.$$

If we define

$$L\circ M=\sum_{j,k} a_j(x)b_k(x)\frac{\partial^2}{\partial x_j\partial x_k},$$

then $LM-L\circ M$ is a first-order differential operator, i.e. $LM\equiv L\circ M$ (modulo first order operators). Next, for the time being, we define

$$L^{\#}=-\sum \overline{a_j(x)}\frac{\partial}{\partial x_j},$$

Then we have

$$L^*\equiv L^{\#} \quad \text{(modulo bounded operators)}.$$

This shows that, in a sense, we can approximate the product of two singular integral operators H_1, H_2 and a conjugate operator H^* by singular integral operators $H_1 \circ H_2$ and $H^{\#}$ respectively. In fact, we have:

THEOREM 6.5. Let H be a singular integral operator, and let $\sigma(H) = h(x\,\xi) \in C_\beta{}^\infty (\beta > 1)$.

In this case for $f \in \mathscr{D}_{L^p}{}^1 (1 < p < +\infty)$ the following inequalities are valid:†

$$\|(H\Lambda - \Lambda H)f\|_{L^p} \leqslant A_p M \|f\|_{L^p}, \qquad \|(H^*\Lambda - \Lambda H^*)f\|_{L^p} \leqslant A_p M \|f\|_{L^p},$$
$$\|(H^* - H^{\#})\Lambda f\|_{L^p} \leqslant A_p M \|f\|_{L^p}, \qquad \|\Lambda(H^* - H^{\#})f\|_{L^p} \leqslant A_p M \|f\|_{L^p},$$

where M is the sum of (6.49) and the upper bounds of

$$\left(\frac{\partial}{\partial \xi}\right)^\nu \frac{\partial}{\partial x_i} h(x, \xi) \quad (|\nu| \leqslant 2n, \; i = 1, 2, \ldots, n),$$

and their Hölder constants of degree $\beta - [\beta]$ for $x \in R^n, |\xi| \geqslant 1$; and A_p is a positive constant depending on p, n, and β.

If we assume that the symbols of H_1 and H_2 belong to $C_\beta{}^\infty (\beta > 1)$, then

$$\|(H_1 H_2 - H_1 \circ H_2)\Lambda f\|_{L^p} \leqslant A_p M_1 M_2 \|f\|_{L^p},$$
$$\|\Lambda(H_1 H_2 - H_1 \circ H_2)f\|_{L^p} \leqslant A_p M_1 M_2 \|f\|_{L^p},$$

where M_1, M_2 are Ms which are defined with respect to $\sigma(H_1)$ and $\sigma(H_2)$ respectively.

Note. We have considered a singular integral operator in L^2-space. In fact an operator of this type is a bounded operator and

$$\|Hf\|_{L^p} \leqslant A_p M \|f\|_{L^p}.$$

We note that the same fact which we stated in theorem 6.5 can be written as

$$\Lambda H \equiv H\Lambda, \qquad H^*\Lambda \equiv H^{\#}\Lambda, \qquad H_1 H_2 \Lambda \equiv (H_1 \circ H_2)\Lambda,$$

where $\equiv$ means 'modulo bounded operators in L^p'. That is, if we disregard bounded operators, both sides of each equation are equal.

† The proof of theorem 6.5 is lengthy, so we shall omit it. Interested readers should refer to the Calderón–Zygmund joint paper which was mentioned at the beginning of section 7.

For example, in this sense, $H_1 H_2$ is no longer a singular integral operator (the property of this object is difficult to characterize), but at least it can be approximated by $(H_1 \circ H_2)$.

8 *Properties of singular integral operators*

In the previous section we defined a singular integral operator and gave the result by Calderón–Zygmund in relation to this type of operator. In this section, we shall give their further properties from a different view-point.

For a differential operator L, the support of $(Lu)(x)$ is contained in that of $u(x)$, i.e. $\operatorname{supp}[Lu] \subset \operatorname{supp}[u]$. This property is called a *local* property of L. On the other hand, a singular integral operator has no such property.

Nevertheless, we have:

LEMMA 6.3 *(Quasi-local property)*. Let H be a singular integral operator such that $\sigma(H) \in C_\beta{}^\infty (\beta > 0)$. Let us denote $B_{2\eta}(x_0)$, a sphere having a radius 2η centred at $x_0 \in R^n$, by Ω. In this case, for an arbitrary $u \in \mathscr{D}_{L^2}{}^1$ such that its support is contained in $B_\eta(x_0)$, we have

$$\|H\Lambda u\|_{L^2(C\Omega)} \leqslant c(n, \eta) M' \|u\|,$$

where

$$M' = \sum_{|\nu| \leqslant 3n+1} \sup_{x \in R^n, |\xi| \geqslant 1} \left| \left(\frac{\partial}{\partial \xi}\right)^\nu h(x, \xi) \right|,$$

$c(n, \eta)$ is a positive constant depending only on n and η, and $C\Omega$ is to be understood as $R^n \setminus \Omega$.

Proof. Let us decompose $\Lambda = \Lambda_1 + \Lambda_2$, where $\hat{\Lambda}_1(\xi) = \alpha(\xi)|\xi|$, $\hat{\Lambda}_2(\xi) = (1 - \alpha(\xi))|\xi|$ and $\alpha(\xi) \in \mathscr{D}$ is a function such that $\alpha(\xi) = 1$ for $|\xi| \leqslant 1$, and $\alpha(\xi) = 0$ for $|\xi| > 2$. Because $\hat{\Lambda}_1(\xi)$ is bounded, Λ_1 is a bounded operator of L^2, and it is enough to show

$$\|H\Lambda_2 u\|_{L^2(C\Omega)} \leqslant c(n, \eta) M' \|u\|. \tag{6.53}$$

We define $Y_{lm}{}'(x)$ as

$$Y_{lm}{}'(x) \underset{\mathscr{F}}{\rightarrow} \hat{Y}_{lm}(\xi) \equiv Y_{lm}(\xi) \Lambda_2(\xi),$$

then we have

$$(H\Lambda_2 u)(x) = a_0(x) \Lambda_2 u + \sum a_{lm}(x)(Y_{lm}{}'(x) * u). \tag{6.54}$$

First, we shall demonstrate, for $2p \geqslant n+2$,

$$|Y_{lm}'(x)| \leqslant |x|^{-2p} c(p, n) |\hat{Y}_{lm}(\xi)|_{2p} \tag{6.55}$$

where

$$|\hat{Y}_{lm}(\xi)|_{2p} = \sum_{|\nu| \leqslant 2p} \sup_{|\xi| \geqslant 1} \left| \left(\frac{\partial}{\partial \xi}\right)^{\nu} \hat{Y}_{lm}(\xi) \right|.$$

To do this, we note that $Y_{lm}'(x) = |x|^{-2p}\{|x|^{2p} Y_{lm}'(x)\}$ and the Fourier image of $|x|^{2p} Y_{lm}'(x)$ is $c\Delta_{\xi}^{p} \hat{Y}_{lm}(\xi)$. From this we have

$$\begin{aligned} |Y_{lm}'(x)| &\leqslant |x|^{-2p} \left(\frac{1}{2\pi}\right)^{p} \int |\Delta_{\xi}^{p}((1-\alpha(\xi)) Y_{lm}(\xi)|\xi|)| \, d\xi \\ &\leqslant |x|^{-2p} c(n, p) |\hat{Y}_{lm}(\xi)|_{2p} \quad (2p \geqslant n+2), \end{aligned}$$

i.e. (6.55) is true.

Next, if u satisfies the conditions of the lemma, we shall demonstrate

$$\| |x|^{-2p} * u \|_{L^2(C\Omega)} \leqslant C(n, p, \eta) \|u\| \tag{6.56}$$

for $4p > n$. In fact, for $x \in C\Omega$ we have

$$|(|x|^{-2p} * u)(x)| = \left| \int_{\omega} \frac{u(y)}{|x-y|^{2p}} \, dy \right| \leqslant \|u\| \left(\int_{\omega} \frac{dy}{|x-y|^{4p}} \right)^{\frac{1}{2}} \quad (\omega = B_{\eta}(x_0)).$$

This value is bounded by

$$\|u\| (\operatorname{vol} \omega)^{\frac{1}{2}} (\operatorname{dis}(x, \omega))^{-2p} \quad (x \in C\Omega).$$

From this we have

$$\| |x|^{-2p} * u \|_{L^2(C\Omega)} \leqslant (\operatorname{vol} \omega)^{\frac{1}{2}} \|u\| \left(\int_{|x| \geqslant 2\eta} \frac{dx}{(|x| - \eta)^{4p}} \right)^{\frac{1}{2}},$$

i.e. (6.56) is true. From (6.55) and (6.56) we have

$$\|HA_2 u\|_{L^2(C\Omega)} \leqslant (\operatorname{vol} \omega)^{\frac{1}{2}} c(p, n) c(p, n, \eta) \left[\sum_{l \geqslant 0, m} |a_{lm}(x)|_0 |\hat{Y}_{lm}(\xi)|_{2p} \right] \|u\|.$$

In this equality we pick the minimum p satisfying $2p \geqslant n+2$. We see that the statement of the lemma is true.

Q.E.D.

When we proved lemma 3.14 (Gårding's lemma), we used a decomposition of the unity $\{\alpha_j(x)\}$, $\sum \alpha_j^2(x) \equiv 1 (\alpha_j \geqslant 0)$ in the sense of L^2. Note that the following lemma is true for the decomposition (we omit the proof because it is lengthy†).

LEMMA 6.4. For a singular integral operator H such that $\sigma(H) \in C_\beta^\infty (\beta > 0)$, we have

$$\sum_j \|[(H\Lambda)\alpha_j(x) - \alpha_j(x)(H\Lambda)]u\|^2 \leqslant \gamma \|u\|^2. \tag{6.57}$$

In particular, for $\sigma(H) = 1$, we have

$$\sum_j \|[\Lambda, \alpha_j(x)]u\|^2 \leqslant \gamma \|u\|^2. \tag{6.58}$$

Note. There is a problem in estimating γ. We satisfy ourselves with making the following remark about this problem. For a certain s, we give a function $x \to h(x, \xi) \in \mathscr{E}_{\xi^s}(R^n \setminus \{0\})$, and if the β-norm of the function is bounded, then γ may be taken as constant. To clarify the situation we give the following:

DEFINITION 6.3. By saying that $\{H\} \in C_\beta^\infty (0 < \beta < 1)$ is a *bounded set* we mean that for an arbitrary integer s, there exists a constant C_s, and the following inequalities are true:

$$p_s[h(x, \xi)] \leqslant C_s,$$
$$p_s[h(x, \xi) - h(x', \xi)] \leqslant C_s |x - x'|^\beta \quad (|x - x'| \leqslant 1, s = 0, 1, 2, \ldots).$$

(We give the definition of the symbols used here in section 7.) From lemmas 6.3 and 6.4, we have the following theorem (a generalization of Gårding's lemma (lemma 3.14) to the case of a singular integral operator):

THEOREM 6.6. Let H be a singular integral operator such that $\sigma(H) = h(x, \xi) \in C_\beta^\infty (\beta > 0)$. If

$$|h(x, \xi)| \geqslant \delta > 0 \quad ((x, \xi) \in R^n \times R^n), \tag{6.59}$$

† See Mizohata: 'Systèmes hyperboliques', *J. Math. Soc. Japan* 11 (1959), 205–33.

then there exists a certain $\delta'>0$, with

$$\|H\Lambda u\|^2 \geqslant \delta'\|\Lambda u\|^2 - \gamma\|u\|^2 \quad (u \in \mathscr{D}_{L^2}{}^1), \tag{6.60}$$

where γ is a positive number which we can take as a constant when H moves over a bounded set (in the sense of definition 6.3) satisfying (6.59).

Proof. We use the idea of 'decomposition of unity'. The degree of the decomposition is decided as follows. Let $x_0 \in R^n$ be an arbitrary point. Let $H(x_0)$ be a singular integral operator (in fact, a convolution operator) with respect to $h(x_0, \xi)$. If we take $\varepsilon(>0)$ sufficiently small, then for an arbitrary $u \in L^2$, we have

$$\|(H - H(x_0))u\|^2{}_{B_{2\varepsilon}(x_0)} \leqslant \tfrac{1}{4}\delta^2\|u\|^2 \tag{6.61}$$

where ε can be taken independent of x_0.

In fact, from (6.51) we have

$$[H - H(x_0)]u(x) = \{a_0(x) - a_0(x_0)\}u(x) + \sum\{a_{lm}(x) - a_{lm}(x_0)\}(\tilde{Y}_{lm}(x) * u).$$

The L^2-norm of the function on the right hand side, in $B_{2\varepsilon}(x_0)$, can be estimated by

$$c\|u\| \sum_{|\nu| \leqslant 2n} \sup_{x \in B_{2\varepsilon}(x_0),\ |\xi| \geqslant 1} \left|\left(\frac{\partial}{\partial \xi}\right)^{\nu} [h(x, \xi) - h(x_0, \xi)]\right| \qquad \text{(use (6.52))}.$$

By our hypothesis, $h(x, \xi) \in C_\beta{}^\infty$, so that this can be taken sufficiently small as $\varepsilon \to 0$. Let us fix such an ε.

We write a decomposition of unity as $\{\alpha_j(x)\}$ which we can decide by ε, i.e. $\sum \alpha_j(x)^2 \equiv 1 (\alpha_j \in \mathscr{D}, \geqslant 0)$ and the support of $\alpha_j(x)$ is in $B_\varepsilon(x^{(j)})$. We have $\|H\Lambda u\|^2 = \sum \|\alpha_j H\Lambda u\|^2$. Now, we consider $\alpha_j H\Lambda u$. Obviously,

$$\begin{aligned}\|\alpha_j H\Lambda u\|^2 &\geqslant \tfrac{1}{2}\|H\alpha_j\Lambda u\|^2 - \|(H\alpha_j - \alpha_j H)\Lambda u\|^2 \\ &\geqslant \tfrac{1}{2}\|H\alpha_j\Lambda u\|^2 - 2\|H(\Lambda\alpha_j - \alpha_j\Lambda)u\|^2 - 2\|\{(H\Lambda)\alpha_j - \alpha_j(H\Lambda)\}u\|^2\end{aligned}$$

so that, from lemma 6.4,

$$\begin{aligned}\sum \|H(\Lambda\alpha_j - \alpha_j\Lambda)u\|^2 &\leqslant \sum_j \|H\|^2\|(\Lambda\alpha_j - \alpha_j\Lambda)u\|^2 \\ &\leqslant c_1\|H\|^2\|u\|^2 \leqslant c_1\|u\|^2.\end{aligned}$$

Also,

$$\sum \|\{(H\Lambda)\alpha_j - \alpha_j(H\Lambda)\}u\|^2 \leqslant c_2\|u\|^2,$$

therefore,

$$\|H\Lambda u\|^2 \geqslant \tfrac{1}{2}\sum \|H\alpha_j\Lambda u\|^2 - c_3\|u\|^2. \tag{6.62}$$

It is enough to estimate $\|H\alpha_j\Lambda u\|^2$. To do this, we first write $H(x^{(j)})$ for a singular integral operator having $h(x^{(j)}, \xi)$ as its symbol. We have

$$\|H\alpha_j\Lambda u\|^2 \geqslant \tfrac{1}{2}\|H(x^{(j)})\alpha_j\Lambda u\|^2 - \|\{H - H(x^{(j)})\}\alpha_j\Lambda u\|^2.$$

From $|h(x, \xi)| \geqslant \delta$, we have

$$\tfrac{1}{2}\|H(x^{(j)})\alpha_j\Lambda u\|^2 \geqslant \tfrac{1}{2}\delta^2\|\alpha_j\Lambda u\|^2,$$

(here we use a Fourier transform) and, as in the proof of lemma 6.3, we write Ω_j as $B_{2\varepsilon}(x^{(j)})$. Then, we have a decomposition

$$\|\{H - H(x^{(j)})\}\alpha_j\Lambda u\|^2 = \|\{H - H(x^{(j)})\}\alpha_j\Lambda u\|_{\Omega_j}{}^2 + \|\{H - H(x^{(j)})\}\alpha_j\Lambda u\|^2{}_{C\Omega_j}.$$

From (6.61), we see that the first term of the right hand side can be estimated by $\tfrac{1}{4}\delta^2\|\alpha_j\Lambda u\|^2$. If we use lemma 6.3 in

$$\begin{aligned}\|\{H - H(x^{(j)})\}\alpha_j\Lambda u\|^2{}_{C\Omega_j} \leqslant\ & 2\|\{H - H(x^{(j)})\}\,(\alpha_j\Lambda - \Lambda\alpha_j)u\|^2{}_{C\Omega_j}\\ &+ 2\|\{H - H(x^{(j)})\}\Lambda\alpha_j u\|^2{}_{C\Omega_j},\end{aligned}$$

then we have

$$\|\{H - H(x^{(j)})\}\Lambda\alpha_j u\|^2{}_{C\Omega_j} \leqslant c(n, \varepsilon)^2 M'^2\|\alpha_j u\|^2.$$

Considering the fact that $\{H - H(x^{(j)})\}$ is a singular integral operator, and using lemma 6.4, we have

$$\sum \|\{H - H(x^{(j)})\}\,(\alpha_j\Lambda - \Lambda\alpha_j)u\|^2{}_{C\Omega_j} \leqslant c_4\|u\|^2.$$

Therefore,

$$\begin{aligned}\|H\Lambda u\|^2 &\geqslant \tfrac{1}{4}\delta^2 \sum \|\alpha_j\Lambda u\|^2 - c(n, \varepsilon)^2 M'^2\|u\|^2 - c_5\|u\|^2\\ &= \tfrac{1}{4}\delta^2\|\Lambda u\|^2 - \gamma\|u\|^2.\end{aligned}$$

Q.E.D.

We need the following result for later applications. We have, however, omitted the proof of the theorem because it is exactly the same as that of theorem 6.6.

COROLLARY 1. Let $\mathscr{H}=(H_{jk})$ be a matrix of Nth degree, and let $H_{jk}(j, k=1, 2, \ldots, N)$ be a singular integral operator. We write $\sigma(H_{jk}) =h_{jk}(x, \xi)\in C_\beta{}^\infty(\beta>0)$.

Further, we assume that there exists a positive constant δ, and for an arbitrary complex vector $\alpha=(\alpha_1, \ldots, \alpha_N)$, we have

$$|\sigma(\mathscr{H})(x, \xi)\cdot\alpha|\geqslant\delta|\alpha| \quad ((x, \xi)\in R^n\times R^n). \tag{6.63}$$

In this case, there exist certain $\delta'(>0)$ and $\gamma(\geqslant 0)$, and for $u(x)=(u_1(x), \ldots, u_N(x))\in\prod \mathscr{D}_{L^2}{}^1$, we have

$$\|\mathscr{H}\Lambda u\|^2\geqslant\delta'\|\Lambda u\|^2-\gamma\|u\|^2 \tag{6.64}$$

where $\|u\|^2=\sum\|u_j(x)\|^2$. Also, if $\mathscr{H}$ moves over a bounded set in the sense of definition 6.3 while satisfying the inequality (6.63), then δ' and γ in (6.64) can be taken as a certain identical value.

Let us give another result which can be proved by the same method as theorem 6.6.

COROLLARY 2. Let $\sigma(H)=h(x, \xi)\in C_\beta{}^\infty(\beta>0)$, and let $\operatorname{Re} h(x, \xi)\leqslant-\delta$ $(\delta>0)$. Then, there exist certain $\delta'(>0)$ and $\gamma(\geqslant 0)$, and for $u\in\mathscr{D}_{L^2}{}^1$ we have $((H+H^*)\Lambda u, \Lambda u)\leqslant-\delta'\|\Lambda u\|^2+\gamma\|u\|^2$.

9 *Energy inequality for a system of hyperbolic equations*

Let us demonstrate that the idea which we explained in section 6 is valid for the case of variable coefficients.

DEFINITION 6.4. Let us write $\Omega=R^n\times[0, T]$. We say that a differential operator

$$M=\frac{\partial}{\partial t}-\sum A_k(x, t)\frac{\partial}{\partial x_k} \tag{6.65}$$

defined on Ω, is *regularly hyperbolic* if it satisfies the following conditions.

(1) A_k is bounded.

(2) For arbitrary $(x, t)\in\Omega$ and $\xi\neq 0\in R^n$, the roots of $\det(\lambda I-\sum A_k(x, t)\xi_k)$

$=0$ are all real and distinct. Also, for $\lambda_1(x, t; \xi) < \cdots < \lambda_N(x, t; \xi)$,

$$\inf_{(x,t)\in\Omega,\ |\xi|=1,\ j\neq k} |\lambda_j(x, t; \xi) - \lambda_k(x, t; \xi)| > 0. \tag{6.66}$$

In this section, we consider only the hyperbolic system of this type. The meaning of 'regular' in our definition refers to the fact that (6.66) has a positive lower bound for (x, t) in Ω. Now, we write $\sum A_k(x, t)\, \partial/\partial x_k = \mathrm{i}\mathscr{H}(t)\, \Lambda$ using a singular differential operator, where

$$\sigma(\mathscr{H}(t)) = 2\pi \sum A_k(x, t) \frac{\xi_k}{|\xi|}, \tag{6.67}$$

and t is regarded as a parameter. M can be rewritten as

$$M = \frac{\partial}{\partial t} - \mathrm{i}\mathscr{H}(t)\Lambda. \tag{6.65}$$

Proposition 6.4. Let M be regularly hyperbolic in Ω. Assume $A_k(x, t) \in \mathscr{B}^{1+\sigma}(0 \leqslant t \leqslant T)$ and also assume A_k is real. Then, there exists a matrix $\sigma(\mathscr{N}(t)) = \sigma(\mathscr{N})(x, t; \xi)$ satisfying the following conditions.

(1) $\sigma(\mathscr{N}(t))\sigma(\mathscr{H}(t)) = \sigma(\mathscr{D}(t))\sigma(\mathscr{N}(t))$ where†

$$\sigma(\mathscr{D}(t)) = 2\pi \begin{bmatrix} \lambda_1(x, t; \xi') & & 0 \\ & \ddots & \\ 0 & & \lambda_N(x, t; \xi') \end{bmatrix} \qquad (\xi' = \xi/|\xi|).$$

(2) $\sigma(\mathscr{N}(t)) = \sigma(\mathscr{N})(x, t; \xi) \in C_{1+\sigma}^{\infty}$ where $\sigma > 0$ and t is regarded as a parameter.

(3) $|\det \sigma(\mathscr{N}(t))| \geqslant \delta' > 0 \quad ((x, t) \in \Omega,\ \xi \in R^n)$.

(4) $t \to \sigma(\mathscr{N})(x, t; \xi) \in C_{1+\sigma}^{\infty}(\sigma > 0)$ and $(\partial/\partial t)\sigma(\mathscr{N})(x, t; \xi) \in C_0^{\infty}$ are both continuous in $[0, T]$.

Proof. $\lambda_j(x, t; \xi')$, $(\xi' = \xi/|\xi|)$ must be single-valued functions because they are real and distinct on the unit sphere in R_ξ^n. Next, we shall show

† If we write this $N(x, t; \xi) \cdot A(x, t)\xi/|\xi| = D(x, t; \xi)N(x, t; \xi)$, the meaning may be more explicit, but we choose this form so as to suggest that $N(x, t: \xi)$ has the properties of the symbol of a singular integral equation.

$\lambda_j(x, t; \xi') \in C_{1+\sigma}^{\infty}$. To do this we note that λ_j is a root of

$$P(\lambda; x, t, \xi) \equiv \det(\lambda I - \sum A_k(x, t)\xi_k) = 0.$$

By the theorem of implicit functions, we have

$$\frac{\partial \lambda_j}{\partial x_k} = -\left(\frac{\partial P}{\partial x_k} \bigg/ \frac{\partial P}{\partial \lambda}\right)_{\lambda=\lambda_j}.$$

Considering

$$\left|\left(\frac{\partial P}{\partial \lambda}\right)_{\lambda=\lambda_j}\right| \geqslant d^{N-1}, \quad d = \inf_{(x,t)\in\Omega,\ |\xi|=1} |\lambda_j - \lambda_k|,$$

we see that $\partial\lambda_j(x, t; \xi')/\partial x_k$ have a σ-Hölder continuity with respect to x.

If we observe successive derivatives with respect to x, we see that $t \to \lambda_j(x, t; \xi') \in C_{1+\sigma}^{\infty}$ is continuous. Similarly, we see that

$$t \to \partial\lambda_j(x, t; \xi')/\partial t \in C_0^{\infty}$$

is continuous. Now, we observe the structure of $\sigma(\mathcal{N})$. If $\sigma(\mathcal{N}) = (n_{ij})\sigma(\mathcal{H}) = (a_{ij})$ then from an equation $\sigma(\mathcal{N})\sigma(\mathcal{H}) = \sigma(\mathcal{D})\sigma\mathcal{N})$ we must find the solutions for $\lambda_i n_{ij} = \sum_{s=1}^{N} n_{is} a_{sj}$, i.e. $(n_{i1}, n_{i2}, \ldots, n_{iN})$ must be an eigen vector corresponding to λ_i which is an eigenvalue of $A = (a_{ji})$. Let $\lambda_i = \lambda_1$. Then, if we write M_{ij} for the (i, j)-cofactor of $\lambda_1 I - A$, then $(M_{ij}, M_{2j}, \ldots, M_{Nj})$ $(j = 1, 2, \ldots, N)$ is the representation of an eigenvector. Among these N-components, we see that at least one component will not be trivial. So that, in this case;

$$\left(\frac{\pm M_{1j}}{\sqrt{(M_{1j}^2 + \cdots + M_{Nj}^2)}}, \ldots, \frac{\pm M_{Nj}}{\sqrt{(M_{1j}^2 + \cdots + M_{Nj}^2)}}\right)$$

represents a real eigenvector having unit length. Of course, in order to achieve this, we may need to change the sequence of the representing cofactors according to the value of $(x, t; \xi)$ in an appropriate way.

Note that the property of $\sum_{i,j} |M_{ij}(x, t; \xi)|$, namely, the existence of a positive largest lower bound follows from the hypothesis of regular hyperbolicity. Now, we fix $(x, t) \in \Omega$. Let ξ_0 be a point of the unit sphere. Let us find the direction of a real unit eigenvector $\hat{e}_1(x, t; \xi), \ldots, \hat{e}_N(x, t; \xi)$ at this point. If $n \geqslant 3$, then the monodromy principle gives the distribution of continuous $(\check{e}_1, \ldots, \check{e}_N)$ over $\Omega \times \{|\xi|=1\}$. By the definition of regular hyperbolicity we see that a matrix consisting of these vectors has a positive largest lower bound at its absolute value. From this, (3) and (4) follow easily.

In the case when $n=2$, in general, continuous distributions will not result. To meet this situation, we proceed as follows. Let us write $\theta(0 \leqslant \theta < 2\pi)$ for the point on the circle $|\xi|=1$. Then, we put $\check{e}_i(0, 0; 0)$ at $\theta=0$. If we let θ run round the circle up to 2π, then $\check{e}_i(0, 0; 2\pi-0) = \pm \check{e}_i(0, 0)$. If the sign is positive there is no problem. On the other hand, if it is negative, we use a complex vector; instead of $\check{e}_i$ we take $e^{\pm i\theta}\check{e}_i(0, 0; \theta)$. So we are able to define a continuous vector field on the circle when $\theta \to 2\pi$.

In this way we see that the statement is valid.

Q.E.D.

LEMMA 6.5 *(Energy inequality)*. Let M be regularly hyperbolic† in Ω. Let

$$t \to A_k(x, t) \in \mathscr{B}^{1+\sigma}(\sigma > 0), \qquad t \to \frac{\partial}{\partial t} A_k(x, t) \in \mathscr{B}^0$$

and also $t \to B(x, t) \in \mathscr{B}^0$, $t \to f(t) \in L^2$ be continuous. In this case, for the solution $u(t) \in \mathscr{E}_t^0(L^2)$ $(0 \leqslant t \leqslant T)$ of the equation

$$\frac{\partial}{\partial t} u - \sum A_k(x, t) \frac{\partial}{\partial x_k} u - B(x, t)u = f, \tag{6.68}$$

there exists an inequality relation

$$\|u(t)\| \leqslant c(T) \left[\|u(0)\| + \int_0^t \|f(s)\| \, ds\right]. \tag{6.69}$$

Proof. Let us assume $u(t) \in \mathscr{E}_t^0(\mathscr{D}_{L^2}^1)$, and $u(t) \in \mathscr{E}_t^1(L^2)$. We write (6.68) as

$$\frac{d}{dt} u(t) - i\mathscr{H}\Lambda u - B(t)u = f(t). \tag{6.68$'$}$$

We operate with $\mathscr{N}(t)$ (lemma 6.4) from the left to get

$$\frac{d}{dt}(\mathscr{N}u) - i\mathscr{N}\mathscr{H}\Lambda u - (\mathscr{N}B(t) + \mathscr{N}_t')u = \mathscr{N}f.$$

† In this case we assume that A_k is real.

Using the symbol of definition 6.2 we can re-write the condition of the previous lemma as $\mathscr{N}\circ\mathscr{H}=\mathscr{D}\circ\mathscr{N}$. By theorem 6.5 $\mathscr{N}\mathscr{H}\Lambda\equiv\mathscr{D}\mathscr{N}\Lambda$ (mod. bounded operators in L^2). In fact, $\mathscr{N}\mathscr{H}\Lambda\equiv(\mathscr{N}\circ\mathscr{H})\Lambda=(\mathscr{D}\circ\mathscr{N})\Lambda$ $\equiv\mathscr{D}\mathscr{N}\Lambda$. Furthermore, we have $\mathscr{D}\mathscr{N}\Lambda\equiv\mathscr{D}\Lambda\mathscr{N}$, so that if we put $\mathscr{N}u=v$, we have

$$\frac{\mathrm{d}}{\mathrm{d}t}v-\mathrm{i}\mathscr{D}\Lambda v-\mathscr{B}u=\mathscr{N}f, \tag{6.70}$$

where $\mathscr{B}$ is a bounded operator of L^2. We can describe this situation as 'a diagonalization of an operator'. (See (615)).

From the properties of $\mathscr{N}(t)$, we have $v(t)=\mathscr{N}(t)u(t)\in\mathscr{E}_t^0(\mathscr{D}_{L^2}{}^1)$ $\cap\mathscr{E}_t^1(L^2)$. Also, from (6.70), we have

$$\begin{aligned}\frac{\mathrm{d}}{\mathrm{d}t}(v,v)&=\left(\frac{\mathrm{d}}{\mathrm{d}t}v,v\right)+\left(v,\frac{\mathrm{d}}{\mathrm{d}t}v\right)\\&=(\mathrm{i}\mathscr{D}\Lambda v,v)+(v,\mathrm{i}\mathscr{D}\Lambda v)+2\mathrm{Re}(\mathscr{B}u+\mathscr{N}f,v).\end{aligned}$$

On the other hand $(\mathrm{i}\mathscr{D}\Lambda)^*=-\mathrm{i}\Lambda\mathscr{D}^*$, so that, if we consider the fact that $\mathscr{D}$ is in a diagonal form, then, from theorem 6.5, we have

$$-\mathrm{i}\Lambda\mathscr{D}^*\equiv-\mathrm{i}\Lambda\mathscr{D}^{\#}=-\mathrm{i}\Lambda\mathscr{D}\equiv-\mathrm{i}\mathscr{D}\Lambda\quad\text{(mod. bounded operators)},$$

where we used the fact $\mathscr{D}^{\#}=\mathscr{D}$. Because the symbol $\lambda_j(x,t;\xi')$ of $\mathscr{D}$ is real, we have

$$\frac{\mathrm{d}}{\mathrm{d}t}\|v\|^2\leqslant 2\gamma'\|v\|^2+2C\|u\|\,\|v\|+2\|\mathscr{N}f\|\,\|v\|,$$

i.e.

$$\frac{\mathrm{d}}{\mathrm{d}t}\|v(t)\|\leqslant\gamma'\|v(t)\|+C\|u(t)\|+\|(\mathscr{N}f)(t)\|. \tag{6.71}$$

We notice that $\|u(t)\|$ remains on the right hand side. In this case, in general, it seems to be difficult to expect to have an inequality in the following form

$$\|v\|=\|\mathscr{N}u\|\geqslant c\|u\|\quad(c>0).$$

So that we define a new norm $|||u|||$ as follows:

$$|||u|||=\|\mathscr{N}u\|+\beta\|(\Lambda+1)^{-1}u\|\quad(\beta>0) \tag{6.72}$$

where $(\Lambda+1)^{-1}$ is defined by $((\Lambda+1)^{-1}u(x))\hat{}(\xi)=(|\xi|+1)^{-1}u(\xi)$ for $u\in L^2$. By the corollary of theorem 6.6, there exists a positive constant c and

$$|||u|||\geqslant c\|u\|, \tag{6.73}$$

if we take β sufficiently large. This new norm can be defined by $\mathcal{N}(t)$, and c can be fixed for $t\in[0, T]$. From corollary 1 of Theorem 6.6, and from lemma 6.4, we have $\|\mathcal{N}(t)\Lambda u\|\geqslant\delta'\|\Lambda u\|-\beta'\|u\|$ for $u\in\mathscr{D}_{L^2}{}^1$ $(\delta',\beta'>0, t\in[0,T])$. From this fact we have

$$\begin{aligned}\|\mathcal{N}(t)u\| &= \|\mathcal{N}(t)\,\{\Lambda(\Lambda+1)^{-1}u+(\Lambda+1)^{-1}u\}\|\\ &\geqslant\|\mathcal{N}(t)\Lambda(\Lambda+1)^{-1}u\|-\|\mathcal{N}(t)\,(\Lambda+1)^{-1}u\|\\ &\geqslant\delta'\|\Lambda(\Lambda+1)^{-1}u\|-\beta''\|(\Lambda+1)^{-1}u\|\\ &\geqslant\delta'\|u\|-\beta\|(\Lambda+1)^{-1}u\|,\end{aligned}$$

i.e. $\|\mathcal{N}(t)u\|+\beta\|(\Lambda+1)^{-1}u\|\geqslant\delta'\|u\|$. Clearly $|||u|||\leqslant c'\cdot\|u\|$, so that $|||u|||$ and $\|u\|$ are uniformly equivalent norms for $t\in[0, T]$.

Operating with $(\Lambda+1)^{-1}$ on (6.68)$'$ we have

$$\frac{\mathrm{d}}{\mathrm{d}t}(\Lambda+1)^{-1}u(t)=\mathrm{i}(\Lambda+1)^{-1}\mathscr{H}\Lambda u(t)+(\Lambda+1)^{-1}(Bu+f).$$

Also, we see that

$$(\Lambda+1)^{-1}\mathscr{H}\Lambda\equiv(\Lambda+1)^{-1}\Lambda\mathscr{H}\ (\text{mod. bounded operator}).$$

The right hand side term is a bounded operator of L^2, so that

$$\frac{\mathrm{d}}{\mathrm{d}t}\|(\Lambda+1)^{-1}u\|\leqslant\delta_0\|u\|+\|(\Lambda+1)^{-1}f\|.$$

We add (6.71) and the product of this inequality with β, to get

$$\frac{\mathrm{d}}{\mathrm{d}t}|||u(t)|||\leqslant\gamma|||u(t)|||+|||f(t)|||, \tag{6.74}$$

(to see this, consider (6.73)), so that

$$|||u(t)|||\leqslant\mathrm{e}^{\gamma t}|||u(0)|||+\int_0^t\mathrm{e}^{\gamma(t-s)}|||f(s)|||\,\mathrm{d}s. \tag{6.75}$$

That is, we have a similar energy inequality to the case of a symmetric system (6.7). Note that

$$|||u(t)||| = \|\mathcal{N}(t)u(t)\| + \beta\|(\Lambda+1)^{-1}u(t)\|.$$

Therefore, by (6.73) we see that $\|u(t)\|$ and $|||u(t)|||$ are uniformly equivalent, hence, we have (6.69).

We use a mollifier φ_δ if there is a condition imposed on $u(t)$. We see this as follows. Let $\varphi_\delta * u(t) = u_\delta(t)$, then, from (6.68), we have $M[u_\delta] = f_\delta + C_\delta u$. From the fact that (6.69) is valid for u_δ and from $\int_0^T \|C_\delta u(s)\| \mathrm{d}s \to 0$ $(\delta \to 0)$, we complete the proof (see the proof of theorem 6.1).

Q.E.D.

We generalize this result a little; we have:

THEOREM 6.7. Suppose a regularly hyperbolic system in Ω is given. Let us assume

$$t \to A_k(x, t) \in \mathscr{B}^{\max[1+\sigma, m]} \ (\sigma > 0), \quad t \to \frac{\partial}{\partial t} A_k(x, t) \in \mathscr{B}^0 \quad (m = 0, 1, 2, \ldots)$$

are continuous. Furthermore, we assume $t \to B(x, t) \in \mathscr{B}^m$, $t \to f(t) \in \mathscr{D}_{L^2}{}^m$ are also continuous. In this case, for the solution $u(t) \in \mathscr{E}_t{}^0(\mathscr{D}_{L^2}{}^m)$ $(0 \leqslant t \leqslant T)$ of

$$M[u] \equiv \frac{\partial}{\partial t} u - \sum A_k(x, t) \frac{\partial}{\partial x_k} u - B(x, t)u = f \tag{6.68}$$

we have an energy inequality

$$\|u(t)\|_m \leqslant c(m, T)\left[\|u(0)\|_m + \int_0^t \|f(s)\|_m \,\mathrm{d}s\right] \quad (m = 0, 1, 2, \ldots). \tag{6.76}$$

Proof. If $m = 0$, then the situation is exactly the same as the previous lemma. Let $m = 1$.

If we operate with $\partial/\partial x_j$ on both sides of (6.68) and write $\partial u/\partial x_j = u^{(j)}$, then we have

$$M[u^{(j)}] - \sum_k \left(\frac{\partial}{\partial x_j} A_k\right) u^{(k)} = \frac{\partial}{\partial x_j} B \cdot u + \frac{\partial}{\partial x_j} f \quad (j = 1, 2, \ldots, n),$$

where $u^{(j)} \in \mathscr{E}_t^0(L^2)$. The right hand terms belong to L^2, therefore, the same argument as in the previous lemma can be applied to this case.

Let us write $\sum |||u^{(j)}(t)||| = \varphi_1(t)$. Then, we have†

$$\frac{\mathrm{d}}{\mathrm{d}t}\varphi_1(t) \leqslant \gamma_1 \varphi_1(t) + \sum_j \left|\left|\left| \frac{\partial}{\partial x_j} B \cdot u \right|\right|\right| + \sum_j \left|\left|\left| \frac{\partial}{\partial x_j} f \right|\right|\right|.$$

We integrate both sides of the inequality. Considering the previous lemma, we conclude (6.76) is true in the case $m=1$. We repeat the argument to obtain the lemma.

Q.E.D.

10 *Energy inequality for hyperbolic equations*

From the stand-point of the theory of singular integral operators, hyperbolic equations can be regarded as special cases of the hyperbolic systems, which we treated in section 9. To see this we proceed as follows. We write L_P as a homogeneous part of mth degree of a partial differential operator

$$L[u] \equiv \left(\frac{\partial}{\partial t}\right)^m u + \sum_{\substack{|\nu|+j \leqslant m \\ j<m}} a_{\nu,j}(x,t) \left(\frac{\partial}{\partial x}\right)^\nu \left(\frac{\partial}{\partial t}\right)^j u = g \tag{6.77}$$

defined on $\Omega = R^n \times [0, T]$, i.e.

$$\begin{aligned} L_p &= \left(\frac{\partial}{\partial t}\right)^m + \sum_{j=1}^m h_j\left(x, t; \frac{\partial}{\partial x}\right)\left(\frac{\partial}{\partial t}\right)^{m-j}, \\ h_j\left(x, t; \frac{\partial}{\partial x}\right) &= \sum_{|\nu|=j} a_{\nu, m-j}(x,t)\left(\frac{\partial}{\partial x}\right)^\nu. \end{aligned} \tag{6.78}$$

Now, let us write the roots of a characteristic equation

$$P(\lambda) = \lambda^m + \sum h_j(x, t; \xi)\, \lambda^{m-j} = 0, \tag{6.79}$$

obtained from L_P as $\lambda_1(x, t; \xi), \ldots, \lambda_m(x, t; \xi)$.

DEFINITION 6.5. We say that L_P is *regularly hyperbolic* in Ω if and only if

(1) $a_{\nu,j}(x,t)$ $(|\nu|+j=m)$ are bounded in Ω, and

(2) $$\inf_{(x,t)\in\Omega,\ |\xi|=1,\ j\neq k} |\lambda_j(x,t;\xi) - \lambda_k(x,t;\xi)| = k > 0.$$

† To be exact, we need a mollifier to get the result. We omit the argument involved in this idea because it is clear from the previous lemma.

As we did before, we make (6.77) a first-order equation with respect to $\partial/\partial t$. For this purpose, we define H_j as $\sigma(H_j(t))=h_j(x, t; 2\pi\xi/|\xi|)$ in L_P. Also, we add unknown functions

$$\{\mathrm{i}(\Lambda+1)\}^{m-j-1}\left(\frac{\partial}{\partial t}\right)^j u=v_j \quad (j=0, 1, 2, \ldots, m-1). \tag{6.80}$$

First, we note that

$$\frac{\partial}{\partial t} v_j=i(\Lambda+1)\, v_{j+1} \qquad (j=0, 1, 2, \ldots, m-2),$$

so that we can write L_P as

$$\frac{\partial}{\partial t} v_{m-1}+\mathrm{i}\sum_{j=0}^{m-1} H_{m-j}(t)\, \Lambda(\mathrm{i}\Lambda)^{m-j-1}\left(\frac{\partial}{\partial t}\right)^j u.$$

If we write

$$(\mathrm{i}\Lambda)^j=(1+S_j)\,\{\mathrm{i}(\Lambda+1)\}^j=\{\mathrm{i}(\Lambda+1)\}^j\,(1+S_j) \qquad (j\geqslant 1),$$

then

$$(S_j u)\hat{}(\xi)=\left\{\left(\frac{|\xi|}{1+|\xi|}\right)^j-1\right\}\hat{u}(\xi).$$

We now see S_j, $S_j\Lambda$, ΛS_j are all bounded operators of L^2, therefore,

$$L_P[u]=\frac{\partial}{\partial t} v_{m-1}+\mathrm{i}\sum_{j=0}^{m-1} H_{m-j}(t)\, \Lambda v_j+\mathrm{i}\sum_{j=0}^{m-1} H_{m-j}(t)(\Lambda S_{m-1-j})\, v_j$$

where the third term of the right hand side is a bounded operator (we put $S_0=0$). We put

$$\mathscr{H}(t)=\begin{bmatrix} 0 & 1 & & & \\ & \cdot & 1 & & \\ & & \cdot & \cdot & \\ & & & 0 & 1 \\ -H_m(t) & -H_{m-1}(t) & \cdots & & -H_1(t) \end{bmatrix}.$$

Then, (6.77) becomes

$$\frac{\mathrm{d}}{\mathrm{d}t} v(t)=\mathrm{i}\mathscr{H}(t)\, \Lambda v(t)+\mathscr{B}(t)\, v+f(t) \tag{6.81}$$

where

$$\sigma(\mathscr{H})=\begin{bmatrix} 0 & 1 & & & \\ & \ddots & \ddots & & \\ & & \ddots & \ddots & \\ & & & 0 & 1 \\ -h_m & -h_{m-1} & \cdots & & -h_1 \end{bmatrix}, \quad v=\begin{bmatrix} v_0 \\ v_1 \\ \vdots \\ v_{m-1} \end{bmatrix}, \quad f=\begin{bmatrix} 0 \\ \vdots \\ 0 \\ g \end{bmatrix}.$$

From $\det(\lambda I-\sigma(\mathscr{H}))=\lambda^m+\sum h_j(x, t; 2\pi\xi')\lambda^{m-j}$, $(\xi'=\xi/|\xi|)$, if L_P is regularly hyperbolic, we can apply the previous result which we obtained in section 9.

Therefore:

THEOREM 6.8. Let L be normally hyperbolic in Ω. We assume that if $|v|+j=m$ then

$$t\to a_{v,j}(x, t)\in\mathscr{B}^{\max[1+\sigma, k]}(\sigma\geqslant 0), \operatorname{Im} \quad t\to\frac{\partial}{\partial t}a_{v,j}(x, t)\in\mathscr{B}^0$$

are continuous. Furthermore, we assume if $|v|+j<m$ then $t\to a_{v,j}\in\mathscr{B}^k$, $t\to g(t)\in\mathscr{D}_{L^2}{}^k$ is continuous. Then, for a solution of (6.77) such that $(\partial/\partial t)^j u(t)\in\mathscr{E}_t{}^0(\mathscr{D}_{L^2}{}^{k+m-j-1})$, we have an energy inequality

$$\|u(t)\|_k'\leqslant c(T, k)\left[\|u(0)\|_k'+\int_0^t\|g(s)\|_k\,ds\right] \quad (k=0, 1, 2, \ldots)$$

$$\|u(t)\|_k'=\sum_{j=0}^{m-1}\left\|\left(\frac{\partial}{\partial t}\right)^j u(t)\right\|_{k+m-j-1}.$$

11 *Existence theorem for the solution of a system of hyperbolic equations*

In sections 4 and 5 we gave an existence theorem for the solution of a symmetric hyperbolic system (see theorems 6.2 and 6.4). On the other hand, in section 9, we proved an energy inequality, for a hyperbolic system. The existence theorem for the solution of a symmetric hyperbolic system is also true in the case of an asymmetric hyperbolic system by a method which we can prove as follows. To show this, we follow closely the argument which we gave in section 4 and section 5. We give a sketch of the proof. Let us consider a hyperbolic system whose coefficients are *not* dependent on t.

We see

$$M[u]=\frac{\partial}{\partial t}u-\sum_{k=1}^{n}A_k(x)\frac{\partial}{\partial x_k}u-B(x)u=f \tag{6.83}$$

where M is assumed to be regularly hyperbolic in the sense of definition 6.4 in $\Omega=R^n\times[0,T]$. Let

$$A=\sum A_k(x)\frac{\partial}{\partial x_k}+B(x). \tag{6.84}$$

From the previous argument which we used to prove an energy inequality, we see that if we write an unknown function as $v(t)=\mathcal{N}(t)u(t)$, then, by the diagonalization of the principal part, we have an equation

$$\frac{\mathrm{d}}{\mathrm{d}t}v-\mathrm{i}\mathcal{D}\Lambda v-\mathcal{B}u=\mathcal{N}f \tag{6.70}$$

the existence of whose solution we can demonstrate. Nevertheless, we shall employ a different approach as follows. We introduce an equivalent norm

$$(Lu,u)\equiv(\mathcal{N}\Lambda u,\mathcal{N}\Lambda u)+\beta(u,u) \tag{6.85}$$

in $\mathcal{D}_{L^2}{}^1$. As for the domain of the definition of A we consider

$$\mathcal{D}(A)=\{u;\,u\in\mathcal{D}_{L^2}{}^1,\ Au\in\mathcal{D}_{L^2}{}^1\} \tag{6.86}$$

as in the symmetric case.

LEMMA 6.6. Let M (6.83) be regularly hyperbolic in Ω. We assume $A_k\in\mathcal{B}^{1+\sigma}(\sigma>0)$ and $B\in\mathcal{B}^1$. In this case, if λ (real) is sufficiently small, then for $u\in\mathcal{D}(A)$, there exists a certain constant γ, and†

$$(L(I-\lambda A)u,(I-\lambda A)u)\geqslant(1-\gamma|\lambda|)(Lu,u).$$

$(I-\lambda A)$ is a bijection from $\mathcal{D}(A)$ onto $\mathcal{D}_{L^2}{}^1$.

Proof. For $u\in\mathcal{D}(A)$,

$$(L(I-\lambda A)u,(I-\lambda A)u)=(Lu,u)+\lambda^2(LAu,Au)-\lambda\{(LAu,u)+(Lu,Au)\}.$$

† I.e. if we write $\|u\|_H=\sqrt{(Lu,u)}$ then $\|(I-\lambda A)u\|_H\geqslant(1-\gamma'|\lambda|)\|u\|_H$ ($u\in\mathcal{D}(A)$) where $\gamma'>0$.

Therefore, if we prove

$$-\gamma(Lu, u) \leqslant (LAu, u) + (Lu, Au) \leqslant \gamma(Lu, u). \tag{6.87}$$

Then, the desired inequality is true.

Let $u \in \mathscr{D}_{L^2}{}^2$. Using $A = \mathrm{i}\mathscr{H}\Lambda + B$, we have

$$(LAu, u) + (Lu, Au) = \mathrm{i}\{\mathscr{N}\Lambda\mathscr{H}\Lambda u, \mathscr{N}\Lambda u) - (\mathscr{N}\Lambda u, \mathscr{N}\Lambda\mathscr{H}\Lambda u)\} + B[u, u],$$

and

$$|B[u, u]| \leqslant c\|u\|_1{}^2 \leqslant c'(Lu, u).$$

As we stated in lemma 6.5 in the previous section, we have

$$\mathscr{N}\Lambda\mathscr{H} \equiv \mathscr{N}\mathscr{H}\Lambda \equiv \mathscr{D}\mathscr{N}\Lambda \equiv \mathscr{D}\Lambda\mathscr{N} \quad \text{(mod. bounded operators in } L^2),$$

and

$$(\mathscr{D}\Lambda\mathscr{N}\Lambda u, \mathscr{N}\Lambda u) - (\mathscr{N}\Lambda u, \mathscr{D}\Lambda\mathscr{N}\Lambda u) = ((\mathscr{D}\Lambda - \Lambda\mathscr{D}^*)\mathscr{N}\Lambda u, \mathscr{N}\Lambda u).$$

Therefore,

$$\mathscr{D}\Lambda - \Lambda\mathscr{D}^* \equiv \mathscr{D}\Lambda - \Lambda\mathscr{D}^\# = \mathscr{D}\Lambda - \Lambda\mathscr{D} \equiv 0 \quad \text{(mod. bounded operators), i.e.}$$

$\mathscr{D}\Lambda - \Lambda\mathscr{D}^\#$ is a bounded operator of L^2. From these results, we have (6.87). Next, we shall show that (6.87) is also true for $u \in \mathscr{D}(A)$. In fact, for $\varphi_\delta * u = u_\delta$, we have

$$Au_\delta - Au = (Au_\delta - \varphi_\delta * (Au)) + (\varphi_\delta * (Au) - Au).$$

Because $Au \in \mathscr{D}_{L^2}{}^1$, we see that the second term tends to 0 by the topology of $\mathscr{D}_{L^2}{}^1$. We can say the same for the first term. Therefore, $Au_\delta \to Au$ in $\mathscr{D}_{L^2}{}^1$ $(\delta \to 0)$. Hence, the inequality holds.

Now, we shall prove the rest. A is obviously a closed operator. The inequality says that the image of $(I - \lambda A)\mathscr{D}(A)$ is closed in $\mathscr{D}_{L^2}{}^1$. Therefore, it is enough to show the image is dense in $\mathscr{D}_{L^2}{}^1$. If not, there exists $\psi_1 \in \mathscr{D}_{L^2}{}^1 (\psi_1 \neq 0)$ and

$$((\Lambda + 1)(I - \lambda A)u, (\Lambda + 1)\psi_1) = 0 \quad (u \in \mathscr{D}(A)).$$

So that, for $u \in \mathscr{D}$, we have

$$(I - \lambda A^*)(\Lambda + 1)\psi = 0 \quad (\psi = (\Lambda + 1)\psi_1 \in L^2). \tag{6.88}$$

Now, we use the representation of A^* by a singular integral operator.† From $A^* = -\mathrm{i}\Lambda\mathscr{H}^* + B^*$, we have

$$A^*(\Lambda+1)\psi = (-\mathrm{i}\Lambda\mathscr{H}^* + B^*)(\Lambda+1)\,\psi = (\Lambda+1)(-\mathrm{i}\Lambda\mathscr{H}^* + B^*)\psi + B_0\psi$$

where

$$B_0 = -\mathrm{i}\Lambda(\mathscr{H}^*\Lambda - \Lambda\mathscr{H}^*) + (B^*\Lambda - \Lambda B^*).$$

Therefore,

$$A^*(\Lambda+1)\psi = (\Lambda+1)(-\mathrm{i}\mathscr{H}^{\#}\Lambda + B^* + B_1)\psi + B_0\psi,$$

where $B_1 = (-\mathrm{i}\Lambda\mathscr{H}^* + \mathrm{i}\mathscr{H}^{\#}\Lambda)$, and $\tilde{B} = B_1 + (\Lambda+1)^{-1}B_0 + B^*$ is a bounded operator of L^2. From (6.88), we have $[I - \lambda(-\mathrm{i}\mathscr{H}^{\#}\Lambda + \tilde{B})]\psi = 0$, and more concretely,

$$\left[I - \lambda\left(-\sum {}^t\overline{A_k(x)}\frac{\partial}{\partial x_k} + \tilde{B}\right)\right]\psi = 0. \tag{6.89}$$

On the other hand,

$$\frac{\partial}{\partial t}v = -\sum_{k=1}^{n} {}^t\overline{A_k(x)}\frac{\partial}{\partial x_k}v + \tilde{B}v = \tilde{A}v \tag{6.90}$$

is also regularly hyperbolic‡ in Ω. In (6.89), from $\tilde{A}\psi \in L^2$, there exists a Hermitian form $(L_0 v, v)$, such that it defines a norm which is equivalent to a certain L^2-norm,* and

$$(L_0(I - \lambda\tilde{A})\,\psi, (I - \lambda\tilde{A})\,\psi) \geq (1 - \gamma'|\lambda|)(L_0\psi, \psi).$$

Hence, $(L_0\psi, \psi) = 0_1$, therefore $\psi = 0$. This is a contradiction to our hypothesis $\psi_1 = (\Lambda+1)^{-1}\psi \neq 0$.

Q.E.D.

The lemma shows that we can establish a similar result to theorem 6.2. Next, we note that if the coefficients are dependent on t, we can proceed as we did in theorem 6.3. We shall state the result. We observe

$$M[u] \equiv \frac{\partial}{\partial t}u - \sum_{k=1}^{n} A_k(x, t)\frac{\partial}{\partial x_k}u - B(x, t)\,u = f$$

† It is not necessary to use this, but it makes our argument easier.

‡ In this case A_k is real, so that ${}^t\bar{A}_k = {}^tA_k$. But, (6.90) is hyperbolic does not mean that A_k is real.

* We define $(L_0 v, v) = \|\mathscr{M}v\|^2 + \beta\|(\Lambda+1)^{-1}v\|^2$ where $\sigma(\mathscr{M}) = (\sigma(\mathscr{N})^*)^{-1}$.

where M is regularly parabolic in $\Omega = R^n \times [0, T]$ (see definition 6.4).
Corresponding to theorem 6.4, we have

THEOREM 6.9 *(Existence theorem).* Let

$$t \to A_k(x, t) \in \mathscr{B}^{\max[1+\sigma, m]} \quad (\sigma > 0), \qquad t \to \frac{\partial}{\partial t} A_k(x, t) \in \mathscr{B}^m$$

be continuous in $[0, T]$ where $m = 0, 1, 2, \ldots$. In this case, for an arbitrary $f(t) \in \mathscr{E}_t^0(\mathscr{D}_{L^2}^m)$ $(0 \leqslant t \leqslant T)$ (the right hand side of (6.68)) and an arbitrary initial value $u_0 \in \mathscr{D}_{L^2}^m$, there exists a unique solution $u(t) \in \mathscr{E}_t^0(\mathscr{D}_{L^2}^m)$ $\{0 \leqslant t \leqslant T\} \cap \mathscr{E}_t^1(\mathscr{D}_{L^2}^{m-1})$ $(0 \leqslant t \leqslant T)$† of (6.68). Also, an energy inequality (6.76) is true for the solution. If $m \geqslant [\frac{1}{2}n] + 2$, then u is a genuine solution.

12 *Dependence domain*

Let us demonstrate that the fact which we established in section 6 chapter 4 remains true in the case of a hyperbolic system with variable coefficients. To see this, we do not need a delicate argument such as Holmgren's theorem (essentially the theorem is based on the Cauchy–Kowalewski theorem); we can derive directly the existence of a finite dependence domain from an energy inequality for the hyperbolic system. Let us first prove the following proposition.

PROPOSITION 6.5.‡ Let $A_k (k = 1, 2, \ldots, n)$ be matrices of Nth order. If $\det (I - \sum A_k \xi_k) = 0$ has real and distinct roots $\lambda_1(\xi), \ldots, \lambda_N(\xi)$ for an arbitrary real vector $(\xi \neq 0)$, and if we write* $\lambda_{\max} = \sup_{|\xi|=1, 1 \leqslant j \leqslant N} \lambda_j(\xi)$, then if α is a non-zero real vector such that $|\alpha| < 1/\lambda_{\max}$, then $\det(\mu B - \sum A_k \xi_k) = 0$, $B = I - \sum A_k \alpha_k$, has also distinct real roots for an arbitrary non-zero real vector ξ. Furthermore, B has an inverse.

† I.e. $u(t)$ is continuously once differentiable by the topology of $\mathscr{D}_{L^2}^{m-1}$. If $m = 0$, then the topology is that of $\mathscr{D}'_{L^2}{}^1$.

‡ See H. F. Weinberger: 'Remarks on the preceding paper of Lax', *Comm. Pure App. Math.* 11 (1958), 197–216.

* From $\lambda_j(-\xi) = -\lambda_j(\xi)$ we can write $\lambda_{\max} = \sup |\lambda_j(\xi)|$.

Proof. Let us first demonstrate any eigenvalue ν_k of B is positive. In fact,

$$\det(\nu I - B) = \det(\nu I - (I - A\cdot\alpha)) = (-1)^N \det((1-\nu)\, I - A\cdot\alpha) = 0.$$

So that from $1-\nu_k = \lambda_k(\alpha) = |\alpha|\cdot\lambda_k(\alpha/|\alpha|)$, we have

$$\nu_k = 1 - |\alpha|\ \lambda_k(\alpha/|\alpha|) \geqslant 1 - |\alpha|\cdot\lambda_{\max} > 0. \tag{6.91}$$

Next, we consider an equation

$$\det(\mu B - \lambda I - A\cdot\xi) = (-1)^N \det((\lambda-\mu)\, I + A\cdot(\xi+\mu\alpha)) = 0,$$

where we regard ξ as a constant, μ as a parameter, and λ as an unknown. Let us write the roots of the equation as $\varphi_1(\mu), \ldots, \varphi_N(\mu)$. We have

$$\det(\mu B - \lambda I - A\cdot\xi) = (-1)^N(\lambda - \varphi_1(\mu))\cdots(\lambda - \varphi_N(\mu)),$$

so that $\det(\mu B - A\cdot\xi) = \varphi_1(\mu)\varphi_2(\mu)\ldots\varphi_N(\mu)$. We shall prove the following properties:

(1) $\varphi_j(\mu) \to \pm\infty$ $(\mu \to \pm\infty)$ (the signs being taken respectively) and

(2) $\varphi_j(\mu)$ is a monotone increasing function in the strict sense.

To prove (1) we proceed as follows. First, we see $\varphi_j(\mu) - \mu = \lambda_j(-\xi - \mu\alpha)$ so that $\varphi_j(\mu) = \mu - \lambda_j(\xi + \mu\alpha)$. Therefore, if ξ and α are mutually linearly independent, then, for an arbitrary μ, $\{\varphi_j(\mu)\}$ are real and all distinct.

From $\det(\mu B - \lambda I - A\cdot\xi) = 0$, we have $\det(B - (\lambda/\mu)I - (A\cdot\xi)/\mu) = 0$. ξ is fixed, so that $\varphi_j(\mu)/\mu$ tends to one of $\nu_j(>0)$, an eigenvalue of B as $\mu \to \pm\infty$. Therefore, $\varphi_j(\mu) \sim \mu\nu_j$. Hence (1) is true.

We prove (2) by contradiction. If (2) is not true, then there exist certain j_0 and $\mu_1 < \mu_2$ satisfying $\varphi_{j_0}(\mu_1) = \varphi_{j_0}(\mu_2) = \lambda_0$. In this case, μ satisfying $\varphi_j(\mu) = \lambda_0 (j \neq j_0)$ is one of the roots of det $(\mu B - \lambda_0 I - A\cdot\xi) = 0$. Therefore, the equation, having an unknown μ, has at least $(N+1)$ roots, which is a contradiction.

Therefore, the roots of $\det(\mu B - A\cdot\xi) = 0$ are given as the zero μ_j of $\varphi_j(\mu)$ $(j = 1, 2, \ldots, N)$. Finally, we see that if and only if ξ and α are linearly dependent and $\xi + \mu_0\alpha = 0$, then $\varphi_j(\mu)$ are all equal, and their common value is $\mu_0 (\neq 0)$. We consider the graph of $\varphi_j(\mu)$ in the interval $\mu \in (-\infty, \mu_0 - \varepsilon)$ if $\mu_0 > 0$, and the graph in the interval $\mu \in (\mu_0 + \varepsilon, +\infty)$ if $\mu_0 < 0$. We see that in this case the statement is also true. The existence of B^{-1} is obvious because all the eigenvalues ν_k of B are positive.

Q.E.D.

The following lemma is a generalization of Holmgren's theorem. We consider an equation

$$M[u] \equiv \frac{\partial}{\partial t} u - \sum A_k(x, t) \frac{\partial}{\partial x_k} u - B(x, t)\, u = 0. \tag{6.92}$$

LEMMA 6.7 *(Local uniqueness)*. Let M be regularly hyperbolic† in $\Omega = R^n \times [0, T]$. Let us assume that $A_k(x, t) \in \mathscr{B}_{x,t}^{1+\sigma} (\sigma > 0)$ and $B(x, t) \in \mathscr{B}_{x,t}^{0}$.

In this case, if $u(x, t) \in C^1$ satisfies $M[u] = 0$, and $u(x, 0) \equiv 0$ in the neighbourhood of the origin, then $u(x, t) \equiv 0$, $(x, t) \in V$ for a certain neighbourhood V of the origin.

Proof. Let $\mathscr{D}_\varepsilon = \{(x, t) \in \Omega; |x|^2 + t < \varepsilon, t \geqslant 0\} \subset \Omega$ where ε is sufficiently small. By a Holmgren transform,

$$t' = t + \sum x_j^2, \quad x_j' = x_j \quad (j = 1, 2, \cdots, n),$$

we have

$$(I - 2 \sum A_k x_k') \frac{\partial}{\partial t'} \tilde{u} = \sum A_k \frac{\partial}{\partial x_k'} \tilde{u} + B\tilde{u} \tag{6.93}$$

where we have written $\tilde{u}(x', t') = u(x, t)$.

We see that D_ε is mapped into $\tilde{D}_\varepsilon$ which is surrounded by $t' = \sum x_j'^2$ and $t' = \varepsilon$. $\tilde{u}$ is defined on $\tilde{D}_\varepsilon$. We put $\tilde{u} \equiv 0$ in the outside of $\tilde{D}_\varepsilon$, i.e. it is defined in $0 \leqslant t' \leqslant \varepsilon$. We write this new function as $\tilde{u}(x', t')$ which is defined on $R^n \times [0, \varepsilon]$, belonging to $\mathscr{E}^1$ there, and having its support in $\tilde{D}_\varepsilon$.

We look at (6.93): it is regularly hyperbolic in $\tilde{D}_\varepsilon$ because

$$(I - 2 \sum x_k' A_k)^{-1} \sum A_k \xi_k$$

has eigenvalues which are distinct reals. Note that we take ε smaller if necessary. Also we take $\alpha = 2x'$ by the previous lemma. Now, we can assume that by appropriate extensions of A_k, B to the outside of $\tilde{D}_\varepsilon$, (6.93) is valid in $\Omega' = R^n \times [0, \varepsilon]$, and

$$\frac{\partial}{\partial t'} \tilde{u} = \sum (I - 2 \sum x_k' A_k)^{-1} A_k \frac{\partial}{\partial x_k'} \tilde{u} + (I - 2 \sum \alpha_k' A_k)^{-1} B\tilde{u}$$

† It is not a necessary condition, we can simply assume that it is hyperbolic in the neighbourhood of the origin, i.e. $\lambda_j(x, t; \xi)$ are distinct reals.

is regularly hyperbolic in Ω'. From $\tilde{u}(x',0)\equiv 0$, and the energy equation (6.69) we have $\tilde{u}(x,t)\equiv 0$ in Ω'. Therefore, $u(x,t)\equiv 0$ is true in D_ε.

Q.E.D.

COROLLARY. Let M be normally hyperbolic in Ω, and let $\lambda_j(x,t;\xi)$ be the roots of $\det(\lambda I - A\cdot\xi)=0$. We write

$$\lambda_{\max} = \sup_{\substack{|\xi|=1,\,(x,t)\in\Omega\\ 1\leqslant j\leqslant N}} |\lambda_j(x,t;\xi)|, \tag{6.94}$$

and let S be a hypersurface defined by $\varphi(x,t)=0$ ($\varphi\in\mathscr{E}^3$) which passes through $(x_0,t_0)\in\Omega$ satisfying

$$\left(\frac{\partial}{\partial t}\varphi\right)^2 > \lambda_{\max}{}^2 \sum_{j=1}^{n}\left(\frac{\partial}{\partial x_j}\varphi\right)^2. \tag{6.95}$$

In this case, if $u(x,t)$ is a C^1-function defined in the neighbourhood of (x_0,t_0) satisfying $M[u]=0$ and $u(x,t)\equiv 0$ on S, then $u(x,t)\equiv 0$ is true in a certain neighbourhood of (x_0,t_0).

Proof. By $t'=\varphi(x,t)$, $x_j'=x_j$, (6.92) is transformed as

$$\left(\frac{\partial\varphi}{\partial t}I - \sum A_k\frac{\partial\varphi}{\partial x_k}\right)\frac{\partial}{\partial t'}\tilde{u} = \sum A_k(x,t)\frac{\partial}{\partial x_k'}\tilde{u} + B\tilde{u}$$

On the other hand, S is mapped into $t'=0$. If we write

$$\alpha = \left(\frac{\varphi_{x_1}}{\varphi_t},\ldots,\frac{\varphi_{x_n}}{\varphi_t}\right),$$

then, it satisfies the condition of lemma 6.5. Hence, by the operation of $(\varphi_t I - A\cdot\varphi x)^{-1}$ from the left, we have a new equation which is hyperbolic. Hence, by lemma 6.7, we have the desired result.

Q.E.D.

From this we have the following theorem which corresponds to theorem 4.8.

THEOREM 6.10 *(Theorem of dependence domain)*. Let M (defined in (6.92)) be regularly hyperbolic (see definition 6.4) in $\Omega=R^n\times[0, T]$. As for the coefficients we assume $A_k(x, t)\in\mathscr{E}_{x,t}^{1+\sigma}(\sigma>0)$,† and $B(x, t)\in\mathscr{E}_{x,t}^{0}$. Let us write D as the interior $t\geqslant 0$ of a backward cone $\{(x, t);\ |x-x_0| =\lambda_{\max}(t_0-t)\}$ having its apex $(x_0, t_0)\in\Omega$. If $u(x, t)\in\mathscr{E}^1$ is defined in D, satisfies $M[u]=0$ and $u(x, 0)=0$, $x\in D\cap\{t=0\}$, then $u(x, t)\equiv 0$ in D.

Proof. We omit the proof because it is completely analogous to that of theorem 4.8.

Q.E.D.

Note 1. Obviously, the result is also true in the case of $L[u]$ which is normally hyperbolic in Ω (see definition 6.5). The proof is left to the reader.

Note 2. The result is also true for a symmetric operator defined by (6.1). In this case, we cannot apply lemma 6.5 directly. We shall explain this situation very briefly. Let α be a vector satisfying the conditions of lemma 6.5. If A_k is a Hermitian matrix, then $B=I-\sum A_k\alpha_k$ is also a positive Hermitian matrix as we mentioned at the beginning of the proof of lemma 6.5. Now, we define

$$B^{\pm\frac{1}{2}}=\frac{1}{2\pi i}\int_\Gamma \lambda^{\pm\frac{1}{2}}(\lambda I-B)^{-1}\,d\lambda,$$

where the signs must be taken respectively, and Γ is a positive path of integration such that $\mathrm{Re}\,\lambda>0$, and it contains all point spectra of B in its interior.

We note that $B^{\pm\frac{1}{2}}$ is also a positive Hermitian matrix, and $B^{\pm\frac{1}{2}}B^{\pm\frac{1}{2}}=B^{\pm 1}$. Let

$$\tilde{M}[u]=(I-\sum A_k(x, t)\,\alpha_k(x, t))\frac{\partial}{\partial t}u-\sum A_k(x, t)\frac{\partial}{\partial x_k}u-C(x, t)\,u=0.$$

If we write $B(x, t)=I-A\cdot\alpha$, then from $A_k\in\mathscr{E}^1$ we have $B^{\pm\frac{1}{2}}\in\mathscr{E}^1$. Putting $B^{\frac{1}{2}}u=v$ we can write the equation as

$$\frac{\partial}{\partial t}v-\sum B^{-\frac{1}{2}}A_kB^{-\frac{1}{2}}\frac{\partial}{\partial x_k}v-\tilde{C}(x, t)\,v=0.$$

Because $\tilde{A}_k=B^{-\frac{1}{2}}A_kB^{-\frac{1}{2}}$ is also Hermitian, applying the energy inequality

† I.e. $A_k(x, t)$ are continuously differentiable to first order at Ω, and all their first partial derivatives have uniform σ-Hölder continuity on a bounded set of Ω. In the case of theorem 6.10, σ may vary with the bounded set of Ω.

(6.7) for v, we can repeat the same argument which we used in the hyperbolic case. Therefore, we have the local uniqueness of the solution.

In this case, as a necessary condition of theorem 6.10, we assume that $A_k(x, t)$ is bounded in Ω, but its first derivative and $B(x, t)$ may not be bounded; we assume only they are continuous.

Note 3. From the foregoing argument, we see that theorem 4.9 is true for symmetric hyperbolic or hyperbolic systems and hyperbolic equations. The phenomena which are governed by these equations/systems possess finite propagating speeds not exceeding $\lambda_{\max}$. Finally, we shall show the uniqueness of the solution of a semi-linear equation. For simplicity, we consider a system

$$M[u] \equiv \frac{\partial u}{\partial t} - \sum A_k(x, t) \frac{\partial u}{\partial x_k} + f(x, t; u) = 0 \tag{6.96}$$

which has a regularly hyperbolic principal part in Ω. We assume f is continuously once differentiable in $\Omega \times C^N$. As an application of theorem 6.10, we have

THEOREM 6.11. We consider the backward cone defined in theorem 6.10. We assume that $u_1(x, t), u_2(x, t) \in \mathscr{E}^1$ and $M[u]=0$ in D.

In this case, if $u_1(x, 0) = u_2(x, 0)$ in $x \in D \cap \{t=0\}$ and then $u_1(x, t) \equiv u_2(x, t)$ in D.

Proof. Let $v = u_1 - u_2$.

Using the mean-value theorem for

$$\frac{\partial}{\partial t} v - \sum A_k \frac{\partial}{\partial x_k} v + \{f(x, t; u_1) - f(x, t; u_2)\} = 0,$$

we have

$$f(x, t; u_1(x, t)) - f(x, t; u_2(x, t)) = B(x, t)\, v(x, t).$$

Now we see that the case is reduced to that of theorem 6.10.

Q.E.D.

13 *Existence theorem for the solution of a hyperbolic equation*

We give the existence theorem of the solution for the equation which we discussed in section 10. Let us consider

$$L[u]\equiv\left(\frac{\partial}{\partial t}\right)^m u+\sum_{\substack{|\nu|+j\leqslant m\\ j<m}} a_{\nu,j}(x,t)\left(\frac{\partial}{\partial x}\right)^\nu\left(\frac{\partial}{\partial t}\right)^j u=g \tag{6.77}$$

which is defined on $\Omega=R^n\times[0,T]$ and is regularly hyperbolic. Using a singular integral operator, we transform this into the form of (6.81). Then, we see that the argument of section 11 can be applied directly without any change. Therefore, the existence theorem for $\mathscr{D}_{L^2}{}^1$-space is valid.

Next, we operate with $D_i=\partial/\partial x_i$ from the left on $L[u]=g$. We have

$$L[D_iu]+(D_iL)[u]=D_ig \quad (i=1,2,\ldots,n).$$

We have a theorem which corresponds to theorem 6.4. To be more precise, we are considering $u(t)$ such that

$$u(t)\in\mathscr{D}_{L^2}{}^{m-1+k},\quad \frac{\partial}{\partial t}u(t)\in\mathscr{D}_{L^2}{}^{m-2+k},\ldots,\left(\frac{\partial}{\partial t}\right)^{m-1}u(t)\in\mathscr{D}_{L^2}{}^k, \tag{6.97}$$

are all continuous functions of t. We write

$$\tilde{u}(t)=\left(u(t),\frac{\partial}{\partial t}u(t),\ldots,\left(\frac{\partial}{\partial t}\right)^{m-1}u(t)\right), \tag{6.98}$$

and for (6.97),

$$\tilde{u}(t)\in\mathscr{E}_t{}^0(\tilde{\mathscr{D}}_{L^2}{}^k)\quad(0\leqslant t\leqslant T) \tag{6.99}$$

Also, we write

$$\|\tilde{u}(t)\|_k^2=\|u(t)\|_{m-1+k}^2+\left\|\frac{\partial}{\partial t}u(t)\right\|_{m-2+k}^2+\cdots+\left\|\left(\frac{\partial}{\partial t}\right)^{m-1}u(t)\right\|_k^2. \tag{6.100}$$

Then, we have:

Theorem 6.12 *(Existence theorem).* Under the same assumption as in theorem 6.8, for an arbitrary $g\in\mathscr{E}_t{}^0(\mathscr{D}_{L^2}{}^k)$ and for an arbitrary initial value

$$\tilde{u}(0)=\left(u(0),\frac{\partial}{\partial t}u(0),\ldots,\left(\frac{\partial}{\partial t}\right)^{m-1}u(0)\right)\in\tilde{\mathscr{D}}_{L^2}{}^k(k=0,1,2,\ldots),$$

there exists a unique solution $u(t)$, $\tilde{u}(t)\in\mathscr{E}_t{}^0(\tilde{\mathscr{D}}_{L^2}{}^k)$ $(0\leqslant t\leqslant T)$ and an energy inequality (6.82) is valid. If we use the same notation as (6.100), we have

$$\|\tilde{u}(t)\|_k \leqslant c(T, k)\,[\|\tilde{u}(0)\|_k + \int_0^t \|g(s)\|_k \,\mathrm{d}s]. \tag{6.101}$$

14 *Uniqueness of the solution of a Cauchy problem*

In chapter 4, we gave Holmgren's theorem. We remarked that theorem 4.3 is true for an equation with non-analytic coefficients. We shall give a proof of this. First, we must clarify the meaning of the problem. For example, let L be a second-order elliptic operator. Let us assume that the coefficients of the second order terms in the operator L are all real. In this case, we further assume that there exists a solution for $L[u]=0$ in a domain Ω. Also we assume $u \equiv 0$ for some open subset of Ω. Then, we ask if $u(x)=0$ for the entire Ω. The answer to this question is affirmative, i.e. if two solutions $u_1(x)$, $u_2(x)$ satisfying $L[u]=f$ are the same in a certain open set $\omega(\subset\Omega)$, then $u_1(x) \equiv u_2(x)$ in Ω. We call this fact the *unique continuation theorem.* It has important applications.

We note that if the solut on of $L[u]=0$ which is defined on a domain Ω surrounded by a surface S satisfies $u=\partial u/\partial n=0$ on a portion (open set) of S, then $u(x)\equiv 0\,(x\in\Omega)$. To prove this, it is enough to prove the uniqueness of the solution of a Cauchy problem for an elliptic equation of second order with real coefficients. In this sense we see that theorem 4.3 is very powerful.

Let us assume that we can make an equation be of first order and write it in terms of a singular integral operator:

$$\frac{\mathrm{d}u}{\mathrm{d}t} + \mathrm{i}\mathscr{H}\Lambda u = \mathscr{B}u. \tag{6.102}$$

Let us also assume that there exists $\mathscr{N}$ and $\mathscr{N}\circ\mathscr{H}=\mathscr{D}\circ\mathscr{N}$ i.e.

$$\frac{\mathrm{d}}{\mathrm{d}t}(\mathscr{N}u) + \mathrm{i}\mathscr{D}\Lambda(\mathscr{N}u) = \mathscr{B}_1 u \tag{6.103}$$

where $\mathscr{D}$ is in a diagonal form. We note that since the principal part is in a diagonalized form we consider a generalized weighted energy inequality for each component.†

† The proof can be found in A. P. Calderón: 'Uniqueness in the Cauchy problem for partial differential equations', *Amer. J. Math.* 80 (1958), 16–36.

Let us write

$$J=\int_0^h \varphi^2(t)\left\|\frac{\mathrm{d}}{\mathrm{d}t}u(t)+(P+\mathrm{i}Q)\Lambda u(t)\right\|^2 \mathrm{d}t, \tag{6.104}$$

where:

(1) $u(t)\in\mathscr{E}_t^1(L^2)$ $(0\leqslant t\leqslant h)$ and $u(t)\in\mathscr{E}_t^0(\mathscr{D}_{L^2}^1)$ $(0\leqslant t\leqslant h)$.

(2) $u(0)=0$, and $u(t)\not\equiv 0$ in all neighbourhoods of $t=0$.

(3) The symbols of $P(t)$ and $Q(t)$ belong to $C_{1+\sigma}^\infty$ $(\sigma>0)$.

They are real-valued continuously once differentiable functions with respect to t.

(4) $\varphi(t)>0$, $\varphi'(t)<0$.

The integrand of (6.104) is the integral of the square of

$$[(\varphi u)'+\mathrm{i}Q\Lambda\varphi u]+[P\Lambda\varphi u-\varphi' u],$$

where the first term is a hyperbolic operator. To estimate the integral, we proceed as follows. We divide the estimation into four parts. First,

$$\begin{aligned}&-\int_0^h \{[(\varphi u)', \varphi' u]+[\varphi' u, (\varphi u)']\}\,\mathrm{d}t\\ &\geqslant\int_0^h (\varphi\varphi''-\varphi'^2)\|u\|^2\,\mathrm{d}t,\end{aligned} \tag{6.105}$$

where [,] is an inner product.† Secondly,

$$\begin{aligned}&\int_0^h \{-\mathrm{i}[Q\Lambda\varphi u, \varphi' u]+\mathrm{i}[\varphi' u, Q\Lambda\varphi u]\}\,\mathrm{d}t\\ &\geqslant-\mathrm{const}\int_0^h \varphi|\varphi'|\|u\|^2\,\mathrm{d}t,\end{aligned} \tag{6.106}$$

because

$$-\mathrm{i}\{[Q\Lambda\varphi u, \varphi' u]-[\varphi' u, Q\Lambda\varphi u]\}=+\mathrm{i}\varphi\varphi'[(\Lambda Q^*-Q\Lambda)u, u],$$

so that

$$\Lambda Q^*-Q\Lambda\equiv 0 \text{ (mod. bounded operator).}$$

Thirdly,

$$\begin{aligned}&\mathrm{i}\int_0^h \{[Q\Lambda\varphi u, P\Lambda\varphi u]-[P\Lambda\varphi u, Q\Lambda\varphi u]\}\,\mathrm{d}t\\ &\geqslant-\mathrm{const}\int_0^h \varphi^2\|u\|\,\|\Lambda u\|\,\mathrm{d}t\end{aligned} \tag{6.107}$$

† We have used $-[(\varphi u)', \varphi' u]=-[\varphi u, \varphi' u]'+[\varphi u, (\varphi' u)']$.

(this is just the same result as (6.106)). Fourthly,

$$\int_0^h [(\varphi u)', PA\varphi u]+[PA\varphi u, (\varphi u)']\, dt \geqslant [\varphi u, PA\varphi u]_{t=h}$$
$$-\text{const}\int_0^h \varphi^2\|u\|\|Au\|\, dt-\text{const}\int_0^h \|\varphi u\|\|(\varphi u)'+iQA\varphi u\|\, dt, \tag{6.108}$$

because

$$[(\varphi u)', PA\varphi u]+[PA\varphi u, (\varphi u)']$$
$$=[\varphi u, PA\varphi u]'-[\varphi u, P'A\varphi u]-[\varphi u, PA(\varphi u)']+[PA\varphi u, (\varphi u)'],$$

So that when we consider the last two terms of this equality we have

$$[(PA-AP^*)\varphi u, (\varphi u)'+iQA\varphi u]-[(PA-AP^*)\varphi u, iQA\varphi u].$$

Summing up, we have

$$\left.\begin{aligned}
J\geqslant&\int_0^h \|(\varphi u)'+iQA\varphi u\|^2+\|PA\varphi u-\varphi' u\|^2\, dt\\
&+\int_0^h (\varphi\varphi''-\varphi'^2)\,\|u\|^2\, dt-\text{const}\int_0^h \varphi|\varphi'|\|u\|^2\, dt\\
&-\text{const}\int_0^h \varphi^2\|u\|\|Au\|\, dt\\
&-\text{const}\int_0^h \|\varphi u\|\|(\varphi u)'+iQA\varphi u\|\, dt-\varphi^2(h)|[u, PAu]_{t=h}|
\end{aligned}\right\} \tag{6.109}$$

where we take the second of the integrals as large as possible compared with the other terms.† We take φ as satisfying $\varphi\varphi''-\varphi'^2>0$ and larger than $\varphi|\varphi'|$. To do this, we note that $(\varphi\varphi''-\varphi'^2)/\varphi^2=(\log\varphi)''$. Then, as $n\to+\infty$, we define $\varphi(t)=(t+n^{-1})^{-n}$. Now we see that

$$\varphi\varphi''-\varphi'^2=n(t+n^{-1})^{-2}\varphi^2,$$

therefore,

$$\varphi\varphi''-\varphi'^2=(t+n^{-1})^{-1}\varphi|\varphi'|=n^{-1}\varphi'^2=n(t+n^{-1})^{-2}\varphi^2.$$

If we write an 'average' value as

$$I_n^2=\int_0^h \varphi'^2\|u\|^2\, dt, \quad \rho_n^2 I_n^2=\int_0^h \varphi^2\|PAu\|^2\, dt. \tag{6.110}$$

† If $P\equiv 0$, then an inequality holds when we replace the fourth and fifth integrals with 0.

Then, from the fact that $u(t)\not\equiv 0$ is valid in an arbitrary neighbourhood of $t=0$, we see that

$$J_n \geqslant (\rho_n - 1)^2 I_n{}^2 + I_n{}^2 n^{-1} - o(I_n{}^2 n^{-1}) - o(\rho_n I_n{}^2 n^{-1})$$

as $n\to\infty$, where J_n is obtained from (6.104) by replacing φ by φ_n, and $P(t)$ has the following restriction:

Assumption.

$$P(t)\equiv 0, \tag{6.111}$$

otherwise, P has a bounded inverse P^{-1}.

Summing up, we have

PROPOSITION 6.6.

$$J_n = \int_0^h \varphi_n{}^2(t)\left\|\frac{\mathrm{d}}{\mathrm{d}t}u(t) + (P+\mathrm{i}Q)\Lambda u(t)\right\|^2 \mathrm{d}t \gtrsim \frac{1}{2n}(I_n{}^2 + \rho_n{}^2 I_n{}^2)$$

as $n\to\infty$, where $\varphi_n(t) = (t+n^{-1})^{-n}$ and we assume (6.111). (Also, see (6.110) for $I_n{}^2$, $\rho_n{}^2 I_n{}^2$.) We return to the original problem. Let

$$\frac{\partial}{\partial t}u + \sum A_k(x,t)\frac{\partial}{\partial x_k}u + B(x,t)u = 0 \tag{6.112}$$

be a system of equations of first order, where we assume that u is defined in a certain V, a neighbourhood of the origin, and $u(x,0)=0$. We set up the following (in respect of (6.111)).

Assumption

$P(\lambda; x, t, \xi) = \det(\lambda I - \sum A_k(x,t)\xi_k)$ is a polynomial with real coefficients, and if $\xi\neq 0$, then the roots $\lambda_j(x, t; \xi)$ $(j=1, 2,\ldots, N)$ of $P=0$ are distinct. (6.113)

If we perform a Holmgren transformation, $t' = t + |x|^2$, $x' = x$ on (6.112), we see that our new assumption is still valid in a sufficiently small neighbourhood of the origin. Without loss of generality, we can assume that the support of u lies within $t\geqslant |x|^2$. Also, because the roots are all simple $\lambda_j(x, t; \xi)$ can be regarded as either

(1) always real-valued, or

(2) the imaginary part is never 0 on $|\xi|=1$. In this case, we already assume $\lambda_j(x, t; \xi)$ is a single-valued function defined on $V\times(R^n\backslash\{0\})$, but in general, it is not so. Here, we assume this, then we adjust our method to deal with general cases.

Now, we take h sufficiently small, and extend the value of $A_k(x, t)$ to $(t, x)\in[0, h]\times R^n$ without changing its value on the support of u in $0\leqslant t\leqslant h$ so as to make the amount of oscillation as small as possible. From this we can also take $\lambda_j(x, t; 2\pi\xi/|\xi|)\equiv\lambda_j(x, t; 2\pi\xi')$ such that for each (x, t) the amount of oscillation of its derivative up to the 2nth order with respect to ξ' tends to 0 uniformly on $|\xi'|=1$. Therefore, for λ_j which belongs to type (2) above, we have

$$\sigma(P_j)=-\operatorname{Im}\lambda_j(x, t; 2\pi\xi'), \quad \sigma(Q_j)=\operatorname{Re}\lambda_j(x, t; 2\pi\xi'),$$

so that $P_j(t)$ has a bounded inverse for $t\in[0, h]$. In fact, if we decompose $P_j=P_j{}^0+(P_j-P_j{}^0)$, $\sigma(P_j{}^0)=-\operatorname{Im}\lambda_j(0, 0; 2\pi\xi')$, then

$$\sigma(P_j-P_j{}^0)=-\operatorname{Im}[\lambda_j(x, t; 2\pi\xi')-\lambda_j(0, 0; 2\pi\xi')].$$

Therefore, $P_j{}^0$ has an inverse, and $\|P_j-P_j{}^0\|\leqslant\frac{1}{2}\|P_j{}^0\|$ where the norm is that of an L^2-operator. If we assume the existence of the inverse for $\mathcal{N}(t)$, then $\mathcal{N}(t)^{-1}$ is bounded.

Now, let us consider (6.103). Suppose that there is some component of $\mathcal{N}u=v$ which is $\not\equiv 0$ in all the neighbourhoods of $t=0$. If not, for a sufficiently small h, such a component $\equiv 0$ for $t\in[0, h]$. Therefore, we can apply lemma 6.6 to every component. We add these components to get

$$\int_0^h \varphi_n{}^2(t)\left\|\frac{\mathrm{d}}{\mathrm{d}t}(\mathcal{N}u)+\mathrm{i}\mathscr{D}\Lambda(\mathcal{N}u)\right\|^2 \mathrm{d}t$$

$$\geqslant\frac{1}{2n}\int_0^h \varphi_n{}'^2\|\mathcal{N}u\|^2\,\mathrm{d}t\geqslant\frac{n}{2}\int_0^h \varphi_n{}^2\|\mathcal{N}u\|^2\,\mathrm{d}t. \qquad (6.114)$$

From the boundedness of $\mathcal{N}^{-1}$, this is

$$\geqslant\text{const}\, n\int_0^h \varphi_n{}^2(t)\|u(t)\|^2\,\mathrm{d}t.$$

On the other hand

$$\int_0^h \varphi_n{}^2(t)\|\mathscr{B}_1u\|^2\,\mathrm{d}t\leqslant\text{const}\int_0^h \varphi_n{}^2(t)\|u(t)\|^2\,\mathrm{d}t,$$

so that as $n\to\infty$ we have a contradiction. Therefore, $\mathcal{N}u(t)\equiv 0$ in $0\leqslant t\leqslant h'(\leqslant h)$, i.e. $u(t)\equiv 0$. Summing up, we have:

THEOREM 6.3. In (6.112), let $A_k(x, t)\in C^{1+\sigma}(\sigma>0)$ and $B(x, t)$ be bounded. Furthermore, we assume (6.113). In this case we have the uniqueness of the solution of a Cauchy problem in a neighbourhood of the origin, i.e. if $u(x, t)\in C^1$ is a solution of (6.112) defined on $V\cap\{t\geqslant 0\}$ where V is a neighbourhood of the origin, and if $u(x, 0)=0, x\in V\cap\{t=0\}$, then $u(x, t)\equiv 0$ in a neighbourhood of the origin.

Although we have already given a rough sketch of the proof of the theorem, this time we present a detailed account of the argument. First, from the fact that the roots $\lambda_j(x, t; \xi)$ are distinct, and from the argument at the beginning of the proof of lemma 6.4, obviously, the construction of $(\sigma\mathcal{N})(x, t; \xi)=N(x, t; \xi)\in C_{1+\sigma}{}^{\infty}(\sigma>0)$ can be given on the unit sphere $\Omega=\{\xi; |\xi|=1\}$ in a local sense, i.e. for an arbitrary $\xi_0\in\Omega$, there exists a certain neighbourhood $V'\times W'$ of $(0, 0; \xi_0)$ in which there exists $N(x, t; \xi)\in C_{1+\sigma}{}^{\infty}$. Therefore, by the Heine–Borel theorem, there exists a finite family of open sets $V_i\times W_i$, and there is $N_i(x, t; \xi)$ for each i.

Let us write $V=\bigcap V_i$. Then, $\bigcup W_i=\Omega$. For W_i, we may assume that there is a certain $W_i'\supset W_i$ and the same property is satisfied by W_i'. Under this condition, we consider the decomposition of unity $\hat{\alpha}_i(\xi')\geqslant 0(\in C^{\infty})$, $\sum\hat{\alpha}_i(\xi')^2\equiv 1$, $\operatorname{supp}[\hat{\alpha}_i(\xi')]\subset W_i$ in the sense of L^2 depending on the covering $\{W_i\}$. We know

$$\sigma(\mathcal{H})=2\pi\sum_k A_k(x, t)\xi_k' \qquad (\xi_k'=\xi/|\xi|).$$

We shall adjust the value of this equation with respect to ξ'. To do this, we proceed as follows. For $\varphi_i(\xi')\in C^{\infty}$, $\varphi_i(\xi')=\xi'$ if $\xi'\in\bar{W}_i$, otherwise $\varphi_i(\xi')\in W_i'$.

Thus, we define $\mathcal{H}_i$ as

$$\sigma(\mathcal{H}_i)(x, t; \xi')=\sigma(\mathcal{H})(x, t; \varphi_i(\xi')). \tag{6.115}$$

In this case, if we write the inverse Fourier image of $\hat{\alpha}_i(\xi)\equiv\hat{\alpha}_i(\xi/|\xi|)$ as $\alpha_i(x)$, then we have

$$\mathcal{H}(\alpha_i \underset{(x)}{*} u)=\mathcal{H}_i(\alpha_i \underset{(x)}{*} u) \quad (u\in L^2). \tag{6.116}$$

In fact, in general,

$$K(\alpha_i * u)=\int e^{2\pi i x\xi}k(x, \xi)\hat{\alpha}_i(\xi)\hat{u}(\xi)\, d\xi, \tag{6.117}$$

where K is a singular integral operator, and $k(x, \xi)$ is the symbol of K. Consider the integral on the right hand side. The operator under the sign of the integral depends only on the value of $k(x, \xi)$ for $\xi \in W_i$ i.e. the operator remains the same even if we change the value of $k(x, \xi)$ at CW_i.

Knowing this special property of the operator, let us establish a correspondence between $i (i=1, 2,\ldots, p)$ and $\sigma(\mathcal{N}_i) \in C_{1+\sigma}{}^{\infty}$ $(\sigma>0)$. In this case, we have

$$\sigma(\mathcal{N}_i)\sigma(\mathcal{H}_i)=\sigma(\mathcal{D}_i)\sigma(\mathcal{N}_i) \tag{6.118}$$

where $\sigma(\mathcal{N}_i)(x, t; \xi')=N(x, t; \varphi_i(\xi'))$. A similar argument holds for $\sigma(\mathcal{D}_i)$. Thus, we can extend the process of local diagonalization to the entire unit sphere. Let us operate with $\alpha_i{}^*$ on

$$\frac{\mathrm{d}}{\mathrm{d}t}u(t)+\mathrm{i}\mathcal{H}\Lambda u(t)=\mathcal{B}u(t), \tag{6.102}$$

where $(\alpha_i{}^*)$ is a special type of a singular integral operator such that $\sigma(\alpha_i *)=\hat{\alpha}_i(\xi)$. We have

$$\frac{\mathrm{d}}{\mathrm{d}t}(\alpha_i * u)+\mathrm{i}\mathcal{H}\Lambda(\alpha_i * u)=\alpha_i * \mathcal{B}u-\mathrm{i}[\alpha_i *, \mathcal{H}\Lambda]u \quad (i=1, 2, \ldots, p),$$

where the last term is a bounded operator of L^2 (consider the properties of a singular integral operator). Therefore, by (6.116) we can write

$$\frac{\mathrm{d}}{\mathrm{d}t}(\alpha_i * u)+\mathrm{i}\mathcal{H}_i\Lambda(\alpha_i * u)=\mathcal{B}^{(i)}u \quad (i=1, 2, \ldots, p)$$

where $\mathcal{B}^{(i)}$ is a bounded operator.

From (6.118), operating with $\mathcal{N}_i$ on both sides of this equation, we have

$$\frac{\mathrm{d}}{\mathrm{d}t}\mathcal{N}_i(\alpha_i * u)+\mathrm{i}\mathcal{D}_i\Lambda\mathcal{N}_i(\alpha_i * u)=\mathcal{B}_0{}^{(i)}u \tag{6.119}$$

where $\mathcal{B}_0{}^{(i)}$ is also a bounded operator. From this we have

$$\frac{\mathrm{d}}{\mathrm{d}t}v_i(t)+\mathrm{i}\mathcal{D}_i\Lambda v_i(t)=\mathcal{B}_0{}^{(i)}u, \quad v_i(t)=\mathcal{N}_i(\alpha_i * u). \tag{6.120}$$

Therefore, from (6.114), we have

$$\int_0^h \varphi_n^2(t) \left\| \frac{\mathrm{d}}{\mathrm{d}t} v_i + \mathrm{i}\mathscr{D}_i \Lambda v_i \right\|^2 \mathrm{d}t \geqslant \tfrac{1}{2} n \int_0^h \varphi_n^2(t) \|v_i\|^2 \,\mathrm{d}t$$

$$\geqslant \text{const}\, n \int_0^h \varphi_n^2(t) \|\alpha_i * u(t)\|^2 \,\mathrm{d}t \qquad (\text{const} > 0).$$

Now, we sum up each side of the inequality with respect to i to get

$$\sum_{i=1}^{p} \int_0^h \varphi_n^2(t) \left\| \frac{\mathrm{d}}{\mathrm{d}t} v_i + \mathrm{i}\mathscr{D}_i \Lambda v_i \right\|^2 \mathrm{d}t \geqslant \text{const}\, n \int_0^h \varphi_n^2(t) \|u(t)\|^2 \,\mathrm{d}t.$$

In fact, we have $\sum \|\alpha_i * u\|^2 = \sum \|\hat{\alpha}_i(\xi)\hat{u}(\xi)\|^2 = \|\hat{u}(\xi)\|^2 = \|u\|^2$. On the other hand,

$$\sum_{i=1}^{p} \int_0^h \varphi_n^2(t) \|\mathscr{B}_0^{(i)} u(t)\|^2 \,\mathrm{d}t \leqslant \text{const} \int_0^h \varphi_n^2(t) \|u(t)\|^2 \,\mathrm{d}t.$$

Hence, from (6.120), these two inequalities are not compatible at $n \to \infty$. Hence, $u(t) \equiv 0$.

Note 1. From this argument, the proof of theorem 4.3 is easily obtained, i.e. if we represent a given equation in the form of (6.102) (as we did in the case of hyperbolic equations), then we see that the foregoing argument is completely valid.

Note 2. From theorem 4.3, we see that for the solution of a second-order elliptic equation such that its coefficients of the homogeneous second-order terms are real, a unique continuation theorem is valid. To be more precise, we assume here that the coefficient of the part of second order is $C^{1+\sigma}(\sigma > 0)$, and the solution is $u \in C^2$.

It had been conjectured that the unique continuation theorem might be valid for the solution of a general elliptic equation, but this conjecture is found to be false. In this connection, we mention the following fact which has been obtained by the author:

THEOREM 6.14. Let $L = P_1 P_2 +$ (a differential operator of order less than three), and let P_1 and P_2 be second-order elliptic operators having real coefficients which are also twice continuously differentiable. In this case, we have the unique continuation theorem for the solution of $L[u]=0$, $u \in C^4$. We omit the proof, although it is not difficult.†

Pliś gave counter examples‡ in the case of a sixth-order elliptic operator, and also a general fourth-order elliptic operator having fourth-order real coefficients.

† See S. Mizohata: 'Unicité du prolongement des solutions des équations elliptiques du quatrième ordre', *Proc. Jap. Acad.* 34 (1958), 687–92.

‡ See A. Pliś: 'A smooth linear elliptic differential equation without any solution in a sphere', *Comm. P. App. Math.* 14 (1961), 599–617.

Chapter 7

Semi-linear hyperbolic equations

1 *Introduction*

In this chapter we are concerned with a semi-linear hyperbolic equation

$$M[u] \equiv \frac{\partial}{\partial t} u - \sum_{k=1}^{n} A_k(x, t) \frac{\partial}{\partial x_k} u - f(x, t; u) = 0. \qquad (7.1)$$

We shall use the results which we obtained in the previous chapter in the case of linear hyperbolic equations to obtain an existence theorem for the solution of an equation of this type. A difficulty arises when we carry this out: suppose we wish to use an energy inequality. Unfortunately, however, this is not available in the sense of norms which are defined on spaces whose dimensions are more than two,† i.e. they are not valid in the sense of maximum-norms or L^p-norms ($p \neq 2$). Therefore, we are forced to use the energy inequality in the sense of L^2-norm which has already been established. To overcome this shortcoming, we use Sobolev's lemma in order to show the existence of a local solution of (7.1). Perhaps the reader may feel the following argument a little too elaborate for the purpose, but this is inevitable to establish a fundamental inequality which we shall need for the proof of the existence theorem.

2 *Estimation of a product of functions*

Let us consider the following lemma. The proof of the lemma is lengthy, so that interested readers should refer to the original paper written by Sobolev himself.‡

LEMMA 7.1 (*Sobolev*). Let p, q satisfy $p > 1, q > 1$ and $(1/p) + (1/q) > 1$. In

† See section 3 chapter 6.

‡ S. L. Sobolev: 'On a theorem of functional analysis', *Math. Sbornik*, 4 (1938), 279–82.

this case, if $g\in L^p(R^n)$, $h\in L^q(R^n)$, then

$$\left|\iint \frac{g(x)\,h(y)}{|x-y|^\lambda}\,\mathrm{d}x\,\mathrm{d}y\right| \leqslant K\|g\|_{L^p}\|h\|_{L^q} \tag{7.2}$$

where $\lambda=n(2-(1/p)-(1/q))$ and K is a positive constant depending only on p, q and n. Let us examine some facts which we can derive from the lemma.

We assume $u\in L^p(p>1)$, and let λ satisfy $0<\lambda<n$ and $\lambda/n>1-(1/p)$. We see that in this case $h(x)\to\int(u(y)*(1/|y|^\lambda))h(y)\,dy$ is a continuous linear form defined over L^q $(1/q=\{2-(1/p)\}-(\lambda/n))$.

Therefore, $u*(1/|x|^\lambda)\in L^{q'}$ where $1/q'=1-(1/q)=(\lambda/n)+(1/p)-1$.

COROLLARY 1. Let $u\in L^p(p>1)$ and let λ satisfy $0<\lambda<n$ and $\lambda/n>1-(1/p)$. Then, $(1/|x|^\lambda)*u\in L^{q'}$ $(1/q'=(\lambda/n)+(1/p)-1)$. Also, the mapping from $u\in L^p$ into $(1/|x|^\lambda)*u\in L^{q'}$ is continuous.

COROLLARY 2. Let $u\in L^2$, and let $\frac{1}{2}n<\lambda<n$. Then, $(1/|x|^\lambda)*u\in L^q(1/q=(\lambda/n)-\frac{1}{2})$, and $\|(1/|x|^\lambda)*u\|_{L^q}\leqslant K\|u\|_{L^2}$, where K is a constant depending on n and λ. Now we consider $\mathscr{D}_{L^2}{}^s(-\infty<s<+\infty)$ which was defined in definition 3.4. Recall

$$\|f(x)\|_s=\|(1+|\xi|)^s\hat{f}(\xi)\|_{L^2}. \tag{3.59}$$

PROPOSITION 7.1.

(1) If $u\in\mathscr{D}_{L^2}{}^s$ $(0\leqslant s<\frac{1}{2}n)$, then $u\in L^p$ $(1/p=\frac{1}{2}-(s/n)>0)$ and

$$\|u\|_{L^p}\leqslant c\,(s,n)\|u\|_s \tag{7.3}$$

where $c(s,n)$ is a constant depending on s and n.

(2) Let $u\in\mathscr{D}_{L^2}{}^s$ $(s>\frac{1}{2}n)$. Then, $u\in\mathscr{B}^\sigma$, and

$$\|u\|_{\mathscr{B}^\sigma}\leqslant c\,(s,n)\|u\|_s. \tag{7.4}$$

More precisely, for an arbitrary $\sigma\leqslant 1$ satisfying $0<\sigma<s-\frac{1}{2}n$,

$$\|u\|_{\mathscr{B}^\sigma}\leqslant c\,(s,n,\sigma)\|u\|_s. \tag{7.5}$$

Proof.

(1) This is obvious for $s=0$. We consider the case $0<s<\frac{1}{2}n$. Let $u\in\mathscr{D}_{L^2}{}^s$

We also write $\hat{u}(\xi)=|\xi|^{-s}(|\xi|^s\hat{u}(\xi))$. By using the symbol of a singular integral operator, we write Λ^s as

$$(\Lambda^s u)\hat{}(\xi)=|\xi|^s\hat{u}(\xi). \tag{7.6}$$

On the other hand, by (3.78), we have

$$\mathscr{F}[|\xi|^{-s}]=c\frac{1}{|x|^{n-s}}.$$

From this, and by lemma 2.14, we have

$$u=\frac{c}{|x|^{n-s}}*(\Lambda^s u).$$

Applying corollary 2 of lemma 7.1, we have

$$\|u\|_{L^p}=c\left\|\frac{1}{|x|^{n-s}}*(\Lambda^s u)\right\|_{L^p}\leqslant c\,(s,n)\|\Lambda^s u\|_{L^2}$$

where $1/p=\frac{1}{2}-(s/n)$. By Plancherel's theorem

$$\|\Lambda^s u\|_{L^2}=\||\xi|^s\hat{u}(\xi)\|_{L^2}\leqslant\|(1+|\xi|)^s\hat{u}(\xi)\|_{L^2}=\|u\|_s.$$

Hence, (7.3) is valid.

(2) Let $u\in\mathscr{D}_{L^2}{}^s$ $(s>\frac{1}{2}n)$. By the Schwartz inequality, we have

$$|u(x)|\leqslant\int|\hat{u}(\xi)|\,\mathrm{d}\xi\leqslant\|(1+|\xi|)^s\hat{u}(\xi)\|_{L^2}\cdot\|(1+|\xi|)^{-s}\|_{L^2},$$

i.e. $|u(x)|\leqslant c(s,n)\|u\|_s$. Let us now demonstrate that Hölder continuity is also available in this case. First, we note that

$$u(x)-u(x')=\int\exp(2\pi\mathrm{i}x\xi)\,\{1-\exp(2\pi\mathrm{i}(x'-x)\xi)\}\hat{u}(\xi)\,\mathrm{d}\xi.$$

We write

$$M_\sigma=\sup_{-\infty<\lambda<+\infty}\left|\frac{\mathrm{e}^{i\lambda}-1}{|\lambda|^\sigma}\right|$$

for $0<\sigma\leqslant 1$. Then, we put $\lambda=2\pi(x-x')\xi$,

$$\begin{aligned}\frac{|u(x)-u(x')|}{|x-x'|^\sigma}&\leqslant(2\pi)^\sigma M_\sigma\int|\xi|^\sigma|\hat{u}(\xi)|\,\mathrm{d}\xi\\&\leqslant(2\pi)^\sigma M_\sigma\|(1+|\xi|)^s\hat{u}(\xi)\|_{L^2}\cdot\|(1+|\xi|)^{\sigma-s}\|_{L^2}.\end{aligned}$$

From $\sigma-s<-\frac{1}{2}n$, we conclude (7.5).

Q.E.D.

COROLLARY 1. Let $u\in\mathscr{D}_{L^2}^{[\frac{1}{2}n]+1}$. For $1\leqslant|\nu|\leqslant[\frac{1}{2}n]+1$, we have $(\partial/\partial x)^\nu u\in L^p$ where p satisfies

$$\left.\begin{cases}\dfrac{1}{p}\in\left[\dfrac{|\nu|}{n}-\dfrac{1}{n},\dfrac{1}{2}\right]\setminus\{0\} & (n \text{ is even}),\\[2ex] \dfrac{1}{p}\in\left[\dfrac{|\nu|}{n}-\dfrac{1}{2n},\dfrac{1}{2}\right] & (n \text{ is odd}).\end{cases}\right\} \tag{7.7}$$

Then, we have

$$\left\|\left(\frac{\partial}{\partial x}\right)^\nu u\right\|_{L^p}\leqslant c(\nu,n,p)\|u\|_{[\frac{1}{2}n]+1}.$$

Proof. We note that if $u\in\mathscr{D}_{L^2}^{[\frac{1}{2}n]+1}$ then $(\partial/\partial x)^\nu u\in\mathscr{D}_{L^2}^s$ $(0\leqslant s\leqslant[\frac{1}{2}n]+1-|\nu|)$. In this case, from $|\nu|\geqslant 1$, we have $[\frac{1}{2}n]+1-|\nu|\leqslant\frac{1}{2}n$ where the equality holds if n is an even number and $|\nu|=1$. But then

$$\frac{1}{p}=\frac{1}{2}-\frac{[\frac{1}{2}n]+1-|\nu|}{n}\begin{cases}=\dfrac{|\nu|-1}{n} & (n \text{ is even}),\\[2ex] =\dfrac{|\nu|}{n}-\dfrac{1}{2n} & (n \text{ is odd}).\end{cases}$$

So that (7.7) follows.

Q.E.D.

COROLLARY 2. Let $u\in\mathscr{D}_{L^2}^{[\frac{1}{2}n]+N}$ $(N\geqslant 1)$. Then, for $N\leqslant|\nu|\leqslant[\frac{1}{2}n]+N$, we have $(\partial/\partial x)^\nu u\in L^p$, where p satisfies

$$\left.\begin{cases}\dfrac{1}{p}\in\left[\dfrac{|\nu|}{n}-\dfrac{N}{n},\dfrac{1}{2}\right]\setminus\{0\} & (n \text{ is even}),\\[2ex] \dfrac{1}{p}\in\left[\dfrac{|\nu|}{n}-\dfrac{2N-1}{2n},\dfrac{1}{2}\right] & (n \text{ is odd}).\end{cases}\right\} \tag{7.8}$$

Then we have

$$\left\|\left(\frac{\partial}{\partial x}\right)^\nu u\right\|_{L^p}\leqslant c\,(\nu,n,N,p)\|u\|_{[\frac{1}{2}n]+N}. \tag{7.9}$$

We omit the proof which is similar to that of the previous corollary.

We have the following theorem:

THEOREM 7.1 *(Estimation of a product of functions by the L^2-norm).*

(1) Let $u_1(x), \ldots, u_l(x) \in \mathscr{D}_{L^2}^{[\frac{1}{2}n]+1}$. If $\sum_{j=1}^{l} |\nu_j| \leqslant [\frac{1}{2}n]+1$, then

$$\left(\frac{\partial}{\partial x}\right)^{\nu_1} u_1 \cdot \left(\frac{\partial}{\partial x}\right)^{\nu_2} u_2 \cdot \cdots \cdot \left(\frac{\partial}{\partial x}\right)^{\nu_l} u_l \in L^2,$$

and

$$\left\| \left(\frac{\partial}{\partial x}\right)^{\nu_1} u_1(x) \cdot \cdots \cdot \left(\frac{\partial}{\partial x}\right)^{\nu_l} u_l(x) \right\|_{L^2} \leqslant c \prod_{j=1}^{l} \|u_p\|_{[\frac{1}{2}n]+1}, \tag{7.10}$$

where c depends on n, $\nu_1, \ldots, \nu_l$.

(2) Let $|\nu_1| \geqslant |\nu_j|$ $(j=2, \ldots, l)$. We assume that $u_1 \in \mathscr{D}_{L^2}^{[\frac{1}{2}n]+N+1}$ $(N \geqslant 1)$ and $u_2, \ldots, u_l \in \mathscr{D}_{L^2}^{[\frac{1}{2}n]+N}$. In this case, if $\sum |\nu_j| \leqslant [\frac{1}{2}n]+N+1$, then

$$\left(\frac{\partial}{\partial x}\right)^{\nu_1} u_1(x) \cdot \cdots \cdot \left(\frac{\partial}{\partial x}\right)^{\nu_l} u_l(x) \in L^2,$$

and also

$$\left\| \left(\frac{\partial}{\partial x}\right)^{\nu_1} u_1 \cdot \cdots \cdot \left(\frac{\partial}{\partial x}\right)^{\nu_l} u_l \right\|_{L^2} \leqslant c \|u_1\|_{[\frac{1}{2}n]+N+1} \prod_{j=2}^{l} \|u_j\|_{[\frac{1}{2}n]+N}. \tag{7.11}$$

Proof.

(1) If there is a term where $\nu_j = 0$, then by (7.4), the term is negligible. So that, without loss of generality, we can assume that $|\nu_j| \geqslant 1$ and $l \geqslant 2$.

By (7.7), $u_j \in L^{p_j}$ and

$$\left.\begin{aligned} &\frac{1}{p_p} \in \left[\frac{|\nu_p|}{n} - \frac{1}{n}, \frac{1}{2}\right] \setminus \{0\} \quad (n \text{ is even}), \\ &\frac{1}{p_j} \in \left[\frac{|\nu_j|}{n} - \frac{1}{2n}, \frac{1}{2}\right] \quad (n \text{ is odd}). \end{aligned}\right\} \tag{7.7$'$}$$

On the other hand, if we write the largest lower bound of $1/p_j$ satisfying the condition (7.7′) as $1/P_j$, then, for an even n, we have

$$\sum \frac{1}{P_j} = \sum \left(\frac{|\nu_j|}{n} - \frac{1}{n}\right) \leqslant \frac{\frac{1}{2}n+1}{n} - \frac{l}{n} \leqslant \frac{1}{2} - \frac{1}{n},$$

(we use $l \geqslant 2$). If n is odd, we have $\sum 1/P_j \leqslant \frac{1}{2} - 1/2n$. Therefore, in either case, we have $\sum 1/P_j < \frac{1}{2}$.

From this, we see that we can choose $p_1, \ldots, p_l$ satisfying $\sum_{j=1}^{l} 1/p_j = \frac{1}{2}$. Let us fix such $p_1, \ldots, p_l$. By Hölder's inequality (3.61), we have

$$\int \left|\left(\frac{\partial}{\partial x}\right)^{\nu_1} u_1(x) \cdot \cdots \cdot \left(\frac{\partial}{\partial x}\right)^{\nu_l} u_l(x)\right|^2 \mathrm{d}x \leqslant \prod_j \left(\int \left|\left(\frac{\partial}{\partial x}\right)^{\nu_j} u_j(x)\right|^{2 \cdot \frac{1}{2} p_j} \mathrm{d}x\right)^{2/p_j}$$

$$= \prod_j \left\|\left(\frac{\partial}{\partial x}\right)^{\nu_j} u_j\right\|_{L^{p_j}}^2 \leqslant c \prod_j \|u_j\|_{[\frac{1}{2}n]+1,}$$

(from (7.9)).

(2) There are three cases as follows:

$$(a) \quad |\nu_1| \leqslant N-1, \qquad (b) \quad |\nu_1| = N, \qquad (c) \quad |\nu_1| \geqslant N+1.$$

(*a*) Let

$$f(x) = \left(\frac{\partial}{\partial x}\right)^{\nu_1} u_1(x) \cdot \cdots \cdot \left(\frac{\partial}{\partial x}\right)^{\nu_l} u_l(x).$$

From $|\nu_j| \leqslant N-1$ $(j=2, \ldots, l)$, we have

$$\|f(x)\|_{L^2} \leqslant \left\|\left(\frac{\partial}{\partial x}\right)^{\nu_1} u_1(x)\right\| \prod_{j=2} \sup \left|\left(\frac{\partial}{\partial x}\right)^{\nu_j} u_j(x)\right|.$$

We use (7.4) for

$$(\partial/\partial x)^{\nu_j} u_j(x).$$

(*b*) We have $\sum_{j=2}^{l} |\nu_j| \leqslant [\frac{1}{2}n]+1$ and

$$\|f(x)\|_{L^2} \leqslant \sup \left|\left(\frac{\partial}{\partial x}\right)^{\nu_1} u_1(x)\right| \cdot \left\|\left(\frac{\partial}{\partial x}\right)^{\nu_2} u_2 \cdot \cdots \cdot \left(\frac{\partial}{\partial x}\right)^{\nu_l} u_l\right\|_{L^2}.$$

From (1) we have the desired result.

(*c*) We can neglect the terms such that $|\nu_j| \leqslant N-1$. Without loss of generality, we can assume $|\nu_1| \geqslant N+1$, $|\nu_j| \geqslant N$ $(j=2, \ldots, l)$. We have $u_j \in L^{p_j}$. By (7.8), if n is even, we have

$$\left.\begin{aligned} &\frac{1}{p_1} \in \left[\frac{|\nu_1|}{n} - \frac{N+1}{n}, \frac{1}{2}\right] \setminus \{0\}, \\ &\frac{1}{p_j} \in \left[\frac{|\nu_j|}{n} - \frac{N}{n}, \frac{1}{2}\right] \setminus \{0\} \ (j=2, \ldots, l). \end{aligned}\right\} \tag{7.8}'$$

Similarly to (1), we have

$$\sum_{j=1}^{l} \frac{1}{P_j} = \sum_{j=1}^{l} \left(\frac{|\nu_j|}{n} \right) - \frac{N+1}{n} - \sum_{j=2}^{l} \frac{N}{n} < \frac{1}{2}.$$

In the case of odd n, we have an analogous argument, so that we can establish a result which is similar to (1). Hence, (7.11) is true.

Q.E.D.

3 *Smoothness of a composed function*

Let $\Omega = R^n \times [0, T]$, and let $f(x, t; u)$ be a $([\frac{1}{2}n]+1)$-times continuously differentiable function defined on $\Omega \times C$. We write

$$\alpha(x) f(x, t; u) = \tilde{f}(x, t; u) \tag{7.12}$$

where $\alpha \in \mathscr{D}$, and $\alpha(x)$ can be regarded as a function localizing f in x-space. For simplicity, we write $(\partial/\partial x, \partial/\partial u)^\beta$ which means a derivative of order $|\beta|$ with respect to (x, u). Also, we write $F(x, t)$, $\tilde{F}(x, t)$, $G(x, t)$... for $f(x, t; u(x, t))$, $\tilde{f}(x, t; u(x, t))$, $g(x, t; u(x, t))$..., and

$$U = \{(x, t, u); (x, t) \in \Omega, \quad |u| \leqslant \sup_{\Omega} |u(x, t)|\}. \tag{7.13}$$

Hereafter $c_1(n)$, $c_2(n)$,... signify constants depending only on n.

LEMMA 7.2. Let $u(x, t) \in \mathscr{E}_t^0(\mathscr{D}_{L^2}{}^{[\frac{1}{2}n]+1})$ $(0 \leqslant t \leqslant T)$. Then,

$$\tilde{F}(x, t) \equiv \alpha(x) f(x, t; u(x, t)) \in \mathscr{E}_t^0(\mathscr{D}_{L^2}{}^{[\frac{1}{2}n]+1}) \quad (0 \leqslant t \leqslant T),$$

and

$$\|\tilde{F}(t)\|_{[\frac{1}{2}n]+1} \leqslant c_1(n) M \{1 + \|u(t)\|^k_{[\frac{1}{2}n]+1}\} \quad (k = [\tfrac{1}{2}n]+1), \tag{7.14}$$

where

$$M = \max_{|\beta| \leqslant [\frac{1}{2}n]+1} \sup_U \left| \left(\frac{\partial}{\partial x}, \frac{\partial}{\partial u} \right)^\beta \tilde{f}(x, t; u) \right|. \tag{7.15}$$

Proof. Let us use a mollifier $u_\delta(x, t) = \varphi_\delta(x) * u(x, t)$. Obviously,

$$|u_\delta(x, t)|_{\mathscr{B}_{x0}} \leqslant |u(x, t)|_{\mathscr{B}_{x0}}. \tag{7.16}$$

If we write $\tilde{F}_\delta(x, t) = \tilde{f}(x, t; u_\delta(x, t))$, then $\tilde{F}_\delta(x, t) \to \tilde{F}(x, t)$ as $\delta \to 0$. This convergence takes place in $L^2(R^n)$. In fact,

$$\|\tilde{F}_\delta(x,t)-\tilde{F}(x,t)\|_{L^2}=\left\|\left[\frac{\partial \tilde{f}}{\partial u}\right](x,t;\tilde{u}(x,t))\,(u_\delta(x,t)-u(x,t))\right\|$$
$$\leqslant M\cdot\|u_\delta-u\|_{L^2}\to 0 \quad (\delta\to 0).$$

Next, we show that if $|\nu|\leqslant[\frac{1}{2}n]+1$ then

$$\left(\frac{\partial}{\partial x}\right)^\nu \tilde{F}_\delta(x,t)\to\left(\frac{\partial}{\partial x}\right)^\nu \tilde{F}(x,t),$$

with the L^2-topology, where the derivative is to be understood in the sense of distribution. Note that in this case the convergence is also valid in the sense of distribution (this is obvious from the definition of derivative). We have

$$\left(\frac{\partial}{\partial x}\right)^\nu \tilde{F}_\delta(x,t)=\sum_{\substack{\Sigma|\rho_j|\leqslant|\nu|\\ l\leqslant|\nu|}} c_{\rho_1\cdots\rho_l}g_{\rho_1\cdots\rho_l}(x,t;u_\delta(x,t))\prod_{j=1}^{l}\left(\frac{\partial}{\partial x}\right)^{\rho_j}u_\delta(x,t) \tag{7.17}$$

after the differentiation of a composed function where $g_{\rho_1\cdots\rho_l}(x,t;u)$ indicates one of

$$\left(\frac{\partial}{\partial x},\frac{\partial}{\partial u}\right)^\beta \tilde{f}(x,t;u) \quad (|\beta|\leqslant|\nu|).$$

From $\sum|\rho_j|\leqslant|\nu|\leqslant[\frac{1}{2}n]+1$, we have

$$g_{\rho_1\cdots\rho_l}(x,t;u(x,t))\prod_{j=1}^{l}\left(\frac{\partial}{\partial x}\right)^{\rho_j}u(x,t)\equiv G_{\rho_1\cdots\rho_l}(x,t)\prod_{j=1}^{l}\left(\frac{\partial}{\partial x}\right)^{\rho_j}u(x,t)$$

which belongs to $L^2(R^n)$ (by theorem 7.1). Next, we put

$$J_\delta(x,t)=G_{\rho_1\cdots\rho_l}{}^{(\delta)}(x,t)\prod_j\left(\frac{\partial}{\partial x}\right)^{\rho_j}u_\delta(x,t)-G_{\rho_1\cdots\rho_l}(x,t)\prod_j\left(\frac{\partial}{\partial x}\right)^{\rho_j}u(x,t).$$

Then, we see

$$\|J_\delta\|_{L^2}\leqslant\varepsilon(\delta)\left\|\prod_j\left(\frac{\partial}{\partial x}\right)^{\rho_j}u_\delta(t)\right\|_{L^2}+M\sum_j\left\|\left(\frac{\partial}{\partial x}\right)^{\rho_1}u(t)\cdot\cdots\cdot\left(\frac{\partial}{\partial x}\right)^{\rho_j}\{u_\delta(t)-u(\ )\}\right.$$
$$\left.\times\left(\frac{\partial}{\partial x}\right)^{\rho_{j+1}}u_\delta(t)\cdot\cdots\cdot\left(\frac{\partial}{\partial x}\right)^{\rho_l}u_\delta(t)\right\|_{L^2} \quad (\varepsilon(\delta)\to 0).$$

Therefore, by (7.10), we have an inequality

$$\|J_\delta(t)\|_{L^2} \leqslant c_2(n)M\|u_\delta - u\|_{[\frac{1}{2}n]+1}\|u\|_{[\frac{1}{2}n]+1}^{l-1}.$$

Hence, $J_\delta(t) \to 0$ in L^2, and this convergence is uniform with respect to t. From this and by (7.17), we have

$$\left(\frac{\partial}{\partial x}\right)^{\nu} \tilde{F}(x, t) = \sum c_{\rho_1 \cdots \rho_l} g_{\rho_1 \cdots \rho_l}(x, t; u(x, t)) \prod \left(\frac{\partial}{\partial x}\right)^{\rho_j} u(x, t). \tag{7.18}$$

If we examine the proof, we see that (7.14) is also true.

Q.E.D.

Note. Under the assumption of the theorem, the representation of the derivative (in the sense of distribution) of a composed function can be calculated by the classical method.

For the purpose of future applications, we state some facts which we can derive from the above argument. Let $f(x, t; u_1, \ldots, u_N) \equiv f(x, t; u)$ be m-times continuously differentiable in $\Omega \times C^N$, and let $m \geqslant [\frac{1}{2}n]+1$. Let $U = \{(x, t, u_1, \ldots, u_N); (x, t) \in \Omega, |u| \leqslant \sup |u(x, t)|\}$ where $|u|^2 = \sum_{i=1}^{N} |u_i|^2$. Furthermore, we write

$$M_m = \max_{|\beta| \leqslant m} \sup_{U} \left|\left(\frac{\partial}{\partial x}, \frac{\partial}{\partial u}\right)^{\beta} f(x, t; u)\right|.$$

We have the following:

THEOREM 7.2. If $u(t) \equiv (u_1(x, t), \ldots, u_N(x, t)) \in \mathscr{E}_t^0(\mathscr{D}_{L^2}{}^m)$ $(0 \leqslant t \leqslant T)$, then

$$F(x, t) \equiv \alpha(x) f(x, t; u(x, t)) \in \mathscr{E}_t^0(\mathscr{D}_{L^2}{}^m).$$

(1) For $m = [\frac{1}{2}n]+1$, we have

$$\|\tilde{F}(t)\|_{[\frac{1}{2}n]+1} \leqslant C_k M_k\{1 + \|u(t)\|_{[\frac{1}{2}n]+1}^k\} \quad (k = [\tfrac{1}{2}n]+1). \tag{7.19}$$

(2) For $m \geqslant [\frac{1}{2}n]+2$, we have

$$\|\tilde{F}(t)\|_m \leqslant C_m M_m\{1 + (1 + \|u(t)\|_{m-1}^{m-1})\|u(t)\|_m\} \tag{7.20}$$

where $\|u(t)\|_m$ in the right hand term is an abbreviation of $\|u(t)\|_m{}^2 = \sum_{j=1}^{N} \|u_j(t)\|_m{}^2$.

The proof of the theorem is not difficult. (1) is obvious, and for (2) we use (7.11) and the argument of lemma 7.2.

Q.E.D.

4 *Existence theorem* 1 *(the case of the hyperbolic system)*

We consider a semi-linear equation

$$M[u]=\frac{\partial}{\partial t}u-\sum A_k(x,t)\frac{\partial}{\partial x_k}u=f(x,t;u) \tag{7.21}$$

defined in $\Omega=R^n\times[0,T]$. For simplicity we assume that:
(*a*) $t\to A_k(x,t)\in\mathscr{B}^{[\frac{1}{2}n]+2}$, $t\to(\partial/\partial t)A_k(x,t)\in\mathscr{B}^0$ are continuous, and
(*b*) $f(x,t;u)$ is $([\frac{1}{2}n]+3)$-times continuously differentiable in $\Omega\times C^N$.
Or, more generally,
(*a'*) $t\to A_k(x,t)\in\mathscr{B}^m$, $t\to(\partial/\partial t)A_k\in\mathscr{B}^0$ and
(*b'*) $f\in\mathscr{E}^{m+1}$ where $m\geqslant[\frac{1}{2}n]+2$, and $M[u]$ is to be understood as one of the following type:
(1) $A_k(k=1,2,\dots,n)$ are Hermitian in Ω, or
(2) Ω is regularly hyperbolic (see section 9, chapter 6) which we treated in the previous chapter.

Our main interest is whether there exists a global solution for the system or not. To this end we proceed as follows. First, we decompose f as

$$f(x,t;u)=f(x,t;0)+(f(x,t;u)-f(x,t;0))=f(x,t;0)+g(x,t;u)$$

where we note that $g(x,t;0)\equiv0$. Then, we define $\tilde{f}$ as

$$\tilde{f}(x,t;u)=\alpha(x)g(x,t;u)+\beta(x)f(x,t;0) \tag{7.22}$$

where $\alpha,\beta\in\mathscr{D}$. We replace f in (7.21) by $\tilde{f}$, considering that the system is hyperbolic.

Note that because $\tilde{f}(x,t;u)$ has a compact support in x-space, if the initial value u_0 also has a compact support, and if there exists a solution for the system, then the solution has a compact support. We look at

$$M[u]=\tilde{f}(x,t;u) \tag{7.23}$$

instead of (7.21). We shall solve this by the method of successive approximations. Given $u_0\in\mathscr{D}_{L^2}{}^{[\frac{1}{2}n]+2}$, there exists a unique solution satisfying

$$M[\psi]=\beta(x)f(x,t;0)\quad(\psi(0)=u_0) \tag{7.24}$$

(by theorems 6.4 and 6.9). In this case, there exists a positive constant γ_0 and

$$\|\psi(t)\|_m \leqslant \gamma_0\{\|u_0\|_m + \sup_{0\leqslant t\leqslant T} \|\beta(x)f(x, t; 0)\|_m\} \quad (m=[\tfrac{1}{2}n]+1, [\tfrac{1}{2}n]+2). \tag{7.25}$$

Therefore, the Cauchy problem of (7.23) can be stated as follows.

We seek a solution $u(t)\in\mathscr{E}_t^0(\mathscr{D}_{L^2}^{[\frac{1}{2}n]+2})$ $(t\geqslant 0)$ satisfying

$$M[u]=\tilde{g}(x, t; \psi+u)\equiv\alpha(x)\{f(x, t; u+\psi)-f(x, t; 0)\} \tag{7.26}$$

with initial condition $u(0)=0$. We shall determine the domain of the existence of solution $R^n\times\{0\leqslant t\leqslant h\}$ in relation to the initial value $u_0\in\mathscr{D}_{L^2}^{[\frac{1}{2}n]+2}$. For the time being, we fix α and β in (7.26). In order to solve (7.26), we define a sequence of functions $\{u_j\}$ as follows:

$$\begin{aligned}
M[u_1]&=\tilde{g}(x, t; \psi) && (u_1(0)=0).\\
M[u_2]&=\tilde{g}(x, t; \psi+u_1) && (u_2(0)=0),\\
&\;\vdots && \quad\vdots\\
M[u_j]&=\tilde{g}(x, t; \psi+u_{j-1}) && (u_j(0)=0).
\end{aligned}$$

We note that $\psi(t)\in\mathscr{E}_t^0(\mathscr{D}_{L^2}^{[\frac{1}{2}n]+2})$, so that, by theorem (7.2), $\tilde{g}(x, t; \psi+u_1)\in \in\mathscr{E}_t^0(\mathscr{D}_{L^2}^{[\frac{1}{2}n]+2})$. By the results of theorems 6.4 and 6.9, we conclude $u_1(t)$ belongs to the same space. Repeating the same argument we have

$$u_j(t)\in\mathscr{E}_t^0(\mathscr{D}_{L^2}^{[\frac{1}{2}n]+2}) \ (0\leqslant t\leqslant T;\ \ j=1, 2, 3, \ldots).$$

Proposition 7.2. There exists a certain non-increasing function $\varphi(\xi)$, $\xi>0$ having the following property. For an arbitrary $u_0\in\mathscr{D}_{L^2}^{[\frac{1}{2}n]+2}$, and $h=\varphi(\|u_0\|_{[\frac{1}{2}n]+1})$, $\{u_j(t)\}$ (which we have just defined) is a bounded sequence in $0\leqslant t\leqslant h$ by the topology of $\mathscr{D}_{L^2}^{[\frac{1}{2}n]+1}$, i.e. $\{\sup_{0\leqslant t\leqslant h}\|u_j(t)\|_{[\frac{1}{2}n]+1}\}$ is a bounded sequence.

Proof. Let $\gamma=\sup_{\Omega}|\psi(x, t)|$. By (7.25), we have

$$\gamma\leqslant c_0+c_1\|u_0\|_{[\frac{1}{2}n]+1}, \tag{7.27}$$

where c_0 and c_1 are constants. Let b be a fixed positive number, and let F be defined by

$$F=\{(x, t; u);\ (x, t)\in\Omega,\ |u|\leqslant b+\gamma\},$$

and

$$M = \sup_{|\alpha| \leqslant [\frac{1}{2}n]+2,\, F} \left| \left(\frac{\partial}{\partial x}, \frac{\partial}{\partial u} \right)^{\alpha} \tilde{g}(x, t; u) \right| \tag{7.28}$$

where M is an increasing function which increases as the parameter $b+\gamma$ increases.

If $u(t)$ satisfies $|u(x, t)| \leqslant b$, then, from (7.19) we have

$$\|\tilde{g}(x, t; (\psi+u)(x, t))\|_{[\frac{1}{2}n]+1} \leqslant MC\{1+\|\psi(t)+u(t)\|^k_{[\frac{1}{2}n]+1}\}. \tag{7.29}$$

On the other hand, from $u_j(0)=0$ we have

$$\|u_j(t)\|_{[\frac{1}{2}n]+1} \leqslant C(T) \int_0^t \|\tilde{g}(x, s; (\psi+u_{j-1})(x, s))\|_{[\frac{1}{2}n]+1}\, ds,$$

so that

$$\|u_j(t)\|_{[\frac{1}{2}n]+1} \leqslant MCC(T) \int_0^t (1+\|\psi(s)+u_{j-1}(s)\|^k_{[\frac{1}{2}n]+1})\, ds$$
$$(k=[\tfrac{1}{2}n]+1, j=1, 2, 3, \ldots), \tag{7.30}$$

where we have assumed $|u_{j-1}(x, t)| \leqslant b$.

For a sufficiently small h, and an arbitrary j we shall demonstrate that (7.30) is valid in $0 \leqslant t \leqslant h$. To this end, we put

$$c_2 = MC\cdot C(T); \quad \gamma_1 = 1+2^k \sup_{0\leqslant t\leqslant T} \|\psi(t)\|^k_{[\frac{1}{2}n]+1}. \tag{7.31}$$

From

$$\|\psi+u_{j-1}\|^k_{[\frac{1}{2}n]+1} \leqslant 2^k\{\|u_{j-1}\|^k_{[\frac{1}{2}n]+1} + \|\psi\|^k_{[\frac{1}{2}n]+1}\},$$

we can write (7.30) as

$$\|u_j(t)\|_{[\frac{1}{2}n]+1} \leqslant 2^k c_2 \int_0^t \{\gamma_1 + \|u_{j-1}(s)\|^k_{[\frac{1}{2}n]+1}\}\, ds \quad (j=1, 2, \ldots), \tag{7.32}$$

where we put $u_0(t) \equiv 0$.

Let $c_s(n)$ be Sobolev's constant, i.e.

$$\sup |\varphi(x)| \leqslant c_s(n) \|\varphi\|_{[\frac{1}{2}n]+1}.$$

Let

$$b' = b/c_s(n). \tag{7.33}$$

Write $2^k c_2(\gamma_1 + b'^k)$ as $\tilde{M}$, then

$$h = \frac{b'}{\tilde{M}} \left(= \frac{b'}{2^k(\gamma_1+b'^k)} \cdot \frac{1}{c_2} \right). \tag{7.34}$$

In this case, $|u_j(x, t)| \leqslant b$ in $R^n \times [0, h]$. In fact, a sequence of functions $\{y_j(t)\}$ defined as

$$y_j(t) = c_3 \int_0^t \{\gamma_1 + y_{j-1}{}^k(s)\}\, ds \qquad (t \geqslant 0,\ y_0(t) \equiv 0,\ c_3 = 2^k c_2)$$

satisfies $y_j(t) \geqslant \|u_j(t)\|_{[\frac{1}{2}n]+1}$. Therefore,

$$y_1(t) \leqslant c_3 \gamma_1 t \leqslant \tilde{M} t \leqslant \tilde{M} h = b';\ y_2(t) \leqslant \tilde{M} t \leqslant \tilde{M} h = b', \ldots,$$

hence,

$$\|u_j(t)\|_{[\frac{1}{2}n]+1} \leqslant b' \qquad (0 \leqslant t \leqslant h), \tag{7.35}$$

i.e. $\sup|u_j(x, t)| \leqslant c_s(n) b' = b \quad (0 \leqslant t \leqslant h)$ (by (7.33)).

Finally, we can write (7.34) as

$$\frac{1}{h} = C(n, T) \frac{b^k + c_1(n) + c_2(n) \|\psi(t)\|^k{}_{[\frac{1}{2}n]+1}}{b} \cdot M, \tag{7.36}$$

where $M = M(\gamma + b)$, and $M(\xi)$ is an increasing function of ξ. Therefore, if $\|u_0\|_{[\frac{1}{2}n]+1}$ runs over a bounded set, then h has a positive largest lower bound for a fixed b. So that we have proved the proposition.

Q.E.D.

Note. If we obtain the initial value by putting $t = t_0 (0 \leqslant t_0 < T)$ instead of $t = 0$, this means that we define $\psi(t; t_0)$ instead of $\psi(t)$ in the above argument. $\|\psi(t; t_0)\|_{[\frac{1}{2}n]+1}$ can be estimated by $c_0 + c_1 \|u_0\|_{[\frac{1}{2}n]+1}$ where c_0 and c_1 are taken independent of t_0. Therefore, (7.36) is also valid in this case. Hence, if the initial value u_0 runs over a bounded set of $\mathscr{D}_{L^2}{}^{[\frac{1}{2}n]+1}$, then h can be taken independent of t_0. From this, we may assume that h is independent of t_0 (in proposition 7.2).

Next, we shall demonstrate that $\{u_j(t)\}_{j=1,2,\ldots}$ is a Cauchy sequence in $\mathscr{E}_t{}^0(\mathscr{D}_{L^2}{}^{[\frac{1}{2}n]+2})$ $(0 \leqslant t \leqslant h)$. To see this, we note that

$$M[u_{j+1} - u_j] = \tilde{g}(x, t; u_j + \psi) - \tilde{g}(x, t; u_{j-1} + \psi),$$

so that

$$\|u_{j+1} - u_j\|_{[\frac{1}{2}n]+2} \leqslant C(T) \int_0^t \|\tilde{g}(x, s; u_j + \psi) - \tilde{g}(x, s; u_{j-1} + \psi)\|_{[\frac{1}{2}n]+2}\, ds.$$

The integrand on the right hand side can be estimated by

$$C \|u_j - u_{j-1}\|_{[\frac{1}{2}n]+2} \{1 + \|u_j + \psi\|^k{}_{[\frac{1}{2}n]+1} + \|u_{j-1} + \psi\|^k{}_{[\frac{1}{2}n]+1}\}$$

because of (7.20). By the boundedness which we have just shown, this can be estimated from $C'\|u_j-u_{j-1}\|_{[\frac{1}{2}n]+2}$. From

$$\|u_{j+1}(t)-u_j(t)\|_{[\frac{1}{2}n]+2}\leqslant C'\int_0^t \|u_j(s)-u_{j-1}(s)\|_{[\frac{1}{2}n]+2}\,\mathrm{d}s$$

it is obvious that $\{u_j(t)\}$ forms a Cauchy sequence in $\mathscr{E}_t^0(\mathscr{D}_{L^2}^{[\frac{1}{2}n]+2})(0\leqslant t\leqslant h)$.

Let $u(t)$ be the limit of $u_j(t)$. Then $u(t)\in\mathscr{E}_t^0(\mathscr{D}_{L^2}^{[\frac{1}{2}n]+2})(0\leqslant t\leqslant h)$, and this is a solution of (7.23). Finally, if $m\geqslant[\frac{1}{2}n]+3$, then the existence of a solution for an arbitrary $u_0\in\mathscr{D}_{L^2}^m$ can be established in $\mathscr{D}_{L^2}^m$ by imposing the conditions (a') and (b') (see the beginning of this section) on M and f. To see this, it is sufficient to show that $\{u_j(t)\}$ is bounded in $\mathscr{D}_{L^2}^{m-1}$ for $0\geqslant t\geqslant h$. But, this is obvious from (7.20) (use an induction process with respect to m).

Summing up, we have:

THEOREM 7.3 *(Local existence theorem)*. For an arbitrary initial value $u_0\in\mathscr{D}_{L^2}^m$, $m\geqslant[\frac{1}{2}n]+2$ and an arbitrary initial time $t_0(0\leqslant t_0<T)$, there exists a solution $u(t)\in\mathscr{E}_t^0(\mathscr{D}_{L^2}^m)$ $(t_0\leqslant t\leqslant t_0+h)$ for

$$M[u]=\tilde{f}(x,t;u)=\beta(x)f(x,t;0)+\alpha(x)\{f(x,t;u)-f(x,t;0)\}, \quad (7.23)$$

where h can be taken as a constant which is independent of t_0 if $\|u_0\|_{[\frac{1}{2}n]+1}$ moves within a bounded set.

Finally, we shall give an existence theorem for a global solution under a certain condition because the theorem is not valid for the general case. As the condition we assume the *a priori* estimation below. Then, from the local existence theorem, we have the desired theorem. Let $\beta\in\mathscr{D}$. We consider

$$M[u]=\beta(x)f(x,t;0)+(f(x,t;u)-f(x,t;0)). \quad (7.37)$$

DEFINITION 7.1 *(a priori estimation)*. Let $u_0\in\mathscr{D}_{L^2}^{[\frac{1}{2}n]+2}\cap\mathscr{E}'$. We assume that there exists a solution $u\in\mathscr{D}_{L^2}^{[\frac{1}{2}n]+2}$ of (7.37) for the initial value u_0. By saying that u_0 has an *a priori estimation* we mean that there exists a constant C which is dependent on u_0, and

$$\|u(t)\|_{[\frac{1}{2}n]+1}\leqslant C. \quad (7.38)$$

THEOREM 7.4 *(Existence theorem of a global solution)*. Let M satisfy the conditions (a') and (b') which we defined at the beginning of this section.† Furthermore, we assume that for an arbitrary initial value an *a priori* estimation can be carried out in the sense of definition 7.1. In this case, for an arbitrary $u_0 \in \mathscr{E}_{L^2(\mathrm{loc})}{}^m$ $(m \geqslant [\frac{1}{2}n]+2)$, there exists a solution $u(t)$ of (7.21) in $0 \leqslant t \leqslant T$, and

$$t \to u(t) \in \mathscr{E}_{L^2\,(\mathrm{loc})}{}^m, \quad t \to \frac{\partial}{\partial t} u(t) \in \mathscr{E}_{L^2\,(\mathrm{loc})}{}^{m-1}$$

are both continuous, where $u(t)$ is an actual solution.

Proof. As we mentioned in the previous chapter (cf. section 12), there exists a certain fixed backward cone K, and the value of the solution $u(x, t) \in C^1$ of $M[u]=f$ at (x_0, t_0) can be determined by the value of f in $(x_0, t_0)+K$ and the value of the initial value $u_0(x)$ on the intersecting set $K \cap \{t=0\}$.

Let D be a set swept by $(x, T)+K$ as x moves around the sphere $\{|x| \leqslant R\}$. Then, let D_0 be $D \cap \{t=0\}$. We write β as a function: $\beta \in \mathscr{D}$ which is identically 1 for $x \in D_0$. Given a $u_0 \in \mathscr{E}_{L^2(\mathrm{loc})}{}^m$, we consider the Cauchy problem of the following equation

$$\begin{aligned} M[u_1] &= \beta(x) f(x, t; 0) + (f(x, t; u_1) - f(x, t; 0)) \\ u_1(x, 0) &= \beta(x) u_0(x) \in \mathscr{D}_{L^2}{}^m. \end{aligned} \tag{7.39}$$

By our assumption, the solution of (7.39) $u_1(x, t)$ has an *a priori* estimation $\|u_1(t)\|_{[\frac{1}{2}n]+1} \leqslant C$. On the other hand, if $u_1(x, t)$ exists in $0 \leqslant t \leqslant T$, then it has a bounded support.

Let $\alpha(x)$ be a function: $\alpha \in \mathscr{D}$ which has a property $\alpha(x) \equiv 1$ for $|x| \leqslant R_1$. Then, for a sufficiently large R_1, we have

$$M[u_1] = \beta(x) f(x, t; 0) + \alpha(x)\,(f(x, t; u_1) - f(x, t; 0)) \tag{7.40}$$

which is equivalent to (7.39). u_1 has an *a priori* estimation, so by theorem 7.3 and by using the existence theorem of a local solution repeatedly we have a solution $u(t)$ in $0 \leqslant t \leqslant T$. Obviously, if $(x, t) \in D$, then $u_1(x, t)$ satisfies $M[u]=f$.

Q.E.D.

† We assume Hölder continuity of $t \to \partial A_k/\partial t \in \mathscr{B}^\sigma (\sigma > 0)$, because we intend to discuss a dependence domain. In this case, if we assume regular hyperbolicity, then, for the hyperbolic system, we can replace the global differentiability of A_k and B with a local one.

5 *Existence theorem* 2 *(case of a single equation)*

In the previous section we have given an existence theorem for hyperbolic systems. By the theorem, in order to show the existence of a global solution, it is necessary to perform an *a priori* estimation of the solution. On the other hand, in the general case when the dimension n of the space is large (even $n=2$ or 3) there are tremendous difficulties in obtaining an existence theorem. So that for the moment we shall satisfy ourselves with demonstrating that there is an existence theorem in the case of a single equation. In the following section we shall obtain an *a priori* estimation for a simple case.

Let a regularly hyperbolic operator defined in $\Omega=R^n\times[0, T]$ be

$$M=\left(\frac{\partial}{\partial t}\right)^m+\sum_{\substack{|\nu|+p\leqslant m\\ p<m}}\alpha_{\nu,j}(x,t)\left(\frac{\partial}{\partial x}\right)^{\nu}\left(\frac{\partial}{\partial t}\right)^{j}.$$

Let us consider a semi-linear hyperbolic equation

$$M[u]=f\left(x,t;\left(\frac{\partial}{\partial x}\right)^{\alpha_1}\left(\frac{\partial}{\partial t}\right)^{j_1}u,\dots,\left(\frac{\partial}{\partial x}\right)^{\alpha_s}\left(\frac{\partial}{\partial t}\right)^{j_s}u\right)$$

having M as its principal part, where $|\alpha_k|+j_k\leqslant m-1$. For the coefficients we assume that

$$t\to a_{\nu,j}(x,t)\in\mathscr{B}^N,\quad t\to\frac{\partial}{\partial t}a_{\nu,j}\in\mathscr{B}^0 \tag{7.41}$$

are continuous with respect to t, $f(x,t;v_1,\dots,v_s)$ is $(N+1)$-times continuously differentiable in $\Omega\times C^s$, and $N\geqslant[\frac{1}{2}n]+2$.

Furthermore, we write

$$U=\left\{(x,t;v_1,\dots,v_s);(x,t)\in\Omega,|v_k|\leqslant\sup_{\Omega}\left|\left(\frac{\partial}{\partial x}\right)^{\alpha_k}\left(\frac{\partial}{\partial t}\right)^{j_k}u(x,t)\right|,k=1,\dots,s\right\},$$

$\tilde{f}=\alpha(x)f$ for $\alpha\in\mathscr{D}$, and

$$M_N=\max_{|\beta|\leqslant N}\sup_{U}\left|\left(\frac{\partial}{\partial x},\frac{\partial}{\partial v}\right)^{\beta}\tilde{f}(x,t;v_1,\dots,v_s)\right|.$$

Then, we have:

LEMMA 7.3. Let us write $\tilde{F}(x,t)$ for a function which is obtained by substituting v_i for $(\partial/\partial x)^{\alpha_i}(\partial/\partial t)^{j_i}$ in $\tilde{f}=\alpha(x)f(x,t;v_1,\ldots,v_s)$ where $|\alpha_i|+j_i\leqslant m-1$. In this case, if $\tilde{u}(t)\in\mathscr{E}_t^0(\tilde{\mathscr{D}}_{L^2}{}^N)$ $(0\leqslant t\leqslant T)$, $N\geqslant[\frac{1}{2}n]+1$ (see (6.97)–(6.99)), then $\tilde{F}(t)\in\mathscr{E}_t^0(\mathscr{D}_{L^2}{}^N)$ $(0\leqslant t\leqslant T)$ and

$$\|\tilde{F}(t)\|_{[\frac{1}{2}n]+1}\leqslant c(k,n)M_k\{1+\|u(t)\|_{[\frac{1}{2}n]+1}^{k}\}\qquad(k=[\tfrac{1}{2}n]+1),$$
$$\|\tilde{F}(t)\|_N\leqslant c(N,n)M_N\{1+(1+\|u(t)\|_{N-1}{}^{N-1})\|u(t)\|_N\}\qquad(N\geqslant[\tfrac{1}{2}n]+2),$$

where $\|\tilde{u}(t)\|_N$ is defined by (6.100).

Proof. We demonstrate the first inequality. First we write

$$\left(\frac{\partial}{\partial t}\right)^{j_k}\left(\frac{\partial}{\partial x}\right)^{\alpha_k}u(x,t)=v_k(t).$$

In this case

$$\|v_k(t)\|_{[\frac{1}{2}n]+1}\leqslant C\|\tilde{u}(t)\|_{[\frac{1}{2}n]+1}\qquad(k=1,2,\ldots,s).$$

In fact,

$$\|v_k(t)\|_{[\frac{1}{2}n]+1}=\left\|\left(\frac{\partial}{\partial x}\right)^{\alpha_k}\left(\frac{\partial}{\partial t}\right)^{j_k}u(t)\right\|_{[\frac{1}{2}n]+1}\leqslant C\left\|\left(\frac{\partial}{\partial t}\right)^{j_k}u(t)\right\|_{[\frac{1}{2}n]+|\alpha_k|+1},$$

where $j_k+|\alpha_k|\leqslant m-1$. So that

$$[\tfrac{1}{2}n]+|\alpha_k|+1\leqslant[\tfrac{1}{2}n]+1+(m-1-j_k).$$

Hence,

$$\|v_k(t)\|_{[\frac{1}{2}n]+1}\leqslant C\left\|\left(\frac{\partial}{\partial t}\right)^{j_k}u(t)\right\|_{[\frac{1}{2}n]+1+(m-1-j_k)}\leqslant C\|u(t)\|_{[\frac{1}{2}n]+1}.$$

From this and by (7.10) we have the results (Use lemma 7.2). For the second inequality, we use (7.11).

Q.E.D.

As in the previous section, we take α, and $\beta\in\mathscr{D}$, and

$$\tilde{f}(x,t;v_1,\ldots,v_s)=\beta(x)f(x,t;0,\ldots,0)$$
$$+\alpha(x)\{f(x,t;v_1,\ldots,v_s)-f(x,t;0,\ldots,0)\}.$$

We consider

$$M[u]=\tilde{f}\left(x,t;\left(\frac{\partial}{\partial x}\right)^{\alpha_1}\left(\frac{\partial}{\partial t}\right)^{j_1}u,\ldots,\left(\frac{\partial}{\partial x}\right)^{\alpha_s}\left(\frac{\partial}{\partial t}\right)^{j_s}u\right)\tag{7.42}$$

instead of (7.41).

THEOREM 7.5 *(Existence theorem of a local solution).* Given an arbitrary initial value $\tilde{u}_0 \in \tilde{\mathscr{D}}_{L^2}{}^N (N \geqslant [\frac{1}{2}n]+2)$, i.e. $(u_0, u_1, \ldots, u_{m-1})$, $u_j \in \mathscr{D}_{L^2}{}^{N+m-j-1}$, if we take an initial time as $t_0 (0 \leqslant t_0 < T)$, then there exists a solution $u(x, t) = u(t)$ of (7.42) for $t_0 \leqslant t \leqslant t_0 + h$, where $u(t) \in \mathscr{E}_t{}^0 (\tilde{\mathscr{D}}_{L^2}{}^N)$ $(t_0 \leqslant t \leqslant t_0 + h)$ and $\partial u / \partial t \in \mathscr{E}_t{}^0 (\tilde{\mathscr{D}}_{L^2}{}^{N-1})$ $(t_0 \leqslant t \leqslant t_0 + h)$, and h can be taken independent of t_0 if $\|\tilde{u}_0\|_{[\frac{1}{2}n]+1}$ is within a bounded set.

More explicitly, there is a positive non-increasing function $\varphi(\xi)$ $(\xi > 0)$ such that we can represent h as $h = \varphi(\|\tilde{u}_0\|_{[\frac{1}{2}n]+1})$.

Proof. The proof is left to the reader. (Use the arguments of theorems 6.12, and 7.3). To show the existence of a global solution by using the foregoing existence theorem of a local solution, it is necessary to assume the existence of the following *a priori estimation.*

Assumption. Let $\beta \in \mathscr{D}$ be an arbitrary function of x. If the solution $u(t)$ of the equation

$$M[u] = \beta(x) f(x, t; 0, \ldots, 0) + \{f(x, t; v_1, \ldots, v_k) - f(x, t; 0, \ldots, 0)\} \quad \left(v_k = \left(\frac{\partial}{\partial x}\right)^{\alpha_k} \left(\frac{\partial}{\partial t}\right)^{j_k} u\right)$$

exists for an arbitrary initial value $\tilde{u}(0) \in \tilde{\mathscr{D}}_{L^2}{}^{[\frac{1}{2}n]+2} \cap \mathscr{E}'$, then it has an estimation such that

$$\|u(t)\|_{[\frac{1}{2}n]+m} + \left\|\frac{\partial}{\partial t} u(t)\right\|_{[\frac{1}{2}n]+m-1} + \cdots + \left\|\left(\frac{\partial}{\partial t}\right)^{m-1} u(t)\right\|_{[\frac{1}{2}n]+1} \leqslant C \quad (0 \leqslant t \leqslant T), \tag{7.43}$$

where C is a constant dependent only on the initial condition.

THEOREM 7.6 *(Existence theorem of a local solution).* Let H be regularly hyperbolic in Ω, let $t \to a_{\nu,j} \in \mathscr{B}^N$ be continuous, and let $a_{\nu,j}(x, t) \in \mathscr{E}_{x,t}{}^{1+\sigma}$ $(\alpha > 0$.† We assume the *a priori* estimation (7.43). In this case, for an arbitrary

† I.e. we assume that $a_{\nu,j}$ is continuously differentiable to first order, and its first derivative is locally Hölder continuous.

initial condition $(u_0, u_1, \ldots u_{m-1})$, $u_j \in \mathscr{E}_{L^2(\mathrm{loc})}{}^{N+m-j-1}$ $(N \geqslant [\frac{1}{2}n]+2)$, there exists a unique solution of (7.41) and

$$\left(u(t), \frac{\partial}{\partial t} u(t), \ldots, \left(\frac{\partial}{\partial t}\right)^m u(t)\right)$$
$$\in \left(\mathscr{E}_{L^2\,(\mathrm{loc})}{}^{N+m-1}, \mathscr{E}_{L^2\,(\mathrm{loc})}{}^{N+m-1}, \ldots, \mathscr{E}_{L^2\,(\mathrm{loc})}{}^{N-1}\right)$$

is continuous at $0 \leqslant t \leqslant T$. Also, $u(x, t)$ is a genuine solution.

Proof. We omit the proof because it is similar to that of theorem 7.4.

Q.E.D.

Note. The essence of the proof is whether we can obtain the *a priori* estimation (7.43) or not. If the order of the derivative which appears in $f(x, t; v_1, \ldots, v_s)$ and $v_k = (\partial/\partial x)^{\alpha_k} (\partial/\partial t)^{j_k} u$, on the right hand side of (7.41) is $j + |\alpha_k|$, then the following criterion is useful. That is (7.43) is equivalent to

$$\left\| \left(\frac{\partial}{\partial x}\right)^{\alpha_k} \left(\frac{\partial}{\partial t}\right)^{p_k} u(t) \right\|_{[\frac{1}{2}n]+1} \leqslant C \quad (k = 1, 2, \ldots, s). \tag{7.44}$$

In fact, if (7.44) is true, we put $g = f(x, t; v_1, \ldots, v_s) - f(x, t; 0, \ldots, 0)$, then

$$\left\| g\left(x, t; \left(\frac{\partial}{\partial x}\right)^{\alpha_1} \left(\frac{\partial}{\partial t}\right)^{p_k} u(x, t), \ldots\right) \right\|_{[\frac{1}{2}n]+1} \leqslant C'.$$

Therefore, by (6.101) of theorem 6.12, we have (7.43).

6 *Example (semi-linear wave equation)*†

Not many interesting results have been discovered so far about the existence problem of the global solution in the case of more than one-dimensional space. The reason is that in general it is difficult to obtain an *a priori* estimation. In this section, we consider a semi-linear wave equation

$$\left(\frac{\partial}{\partial t}\right)^2 u - \Delta u + f(u) = 0 \tag{7.45}$$

which has an *a priori* estimation. For simplicity, we assume that $f(u)$ is a

† For this section see K. Jörgens: 'Das Anfangwertproblem im Grossen für eine Klasse nichtlinearer Wellengleichungen', *Math. Zeit.* 77 (1961), 295–308.

real-valued function defined for $u \in (-\infty, +\infty)$ and $f \in C^\infty$. Furthermore, we assume $f(0)=0$ for the time being. We have the following lemma for the *a priori* estimation of the solution of the equation.

LEMMA 7.4.

(1) The case $n=1$:

$$F(u) \equiv \int_0^u f(u)\,du \geqslant -L_0, \tag{7.46}$$

(2) The case $n=2$:

$$|f'(u)| \leqslant \alpha(1+|u|)^k, \tag{7.47}$$

where α, and k are certain positive numbers.

(3) The case $n=3$: the condition (7.46) and

$$|f'(u)| \leqslant \alpha(1+u^2), \tag{7.48}$$

where α is a certain positive number.

Proof. It is enough to show

$$\|u(t)\|_{[\frac{1}{2}n]+1} \leqslant C \tag{7.49}$$

because of (7.44). Let us assume that our initial condition is

$$(u_0(x), u_1(x)) \in (\mathscr{D}_{L^2}{}^{[\frac{1}{2}n]+3}, \mathscr{D}_{L^2}{}^{[\frac{1}{2}n]+2}),$$

and its support is contained in $|x| \leqslant R_0$. Then, for a fixed $T(>0)$ and for the above initial condition, it is enough to show that for the solution of (7.45) satisfying

$$u(t) \in \mathscr{D}_{L^2}{}^{[\frac{1}{2}n]+3}, \qquad \frac{\partial}{\partial t} u(t) \in \mathscr{D}_{L^2}{}^{[\frac{1}{2}n]+2}, \qquad \frac{\partial^2}{\partial t^2} u(t) \in \mathscr{D}_{L^2}{}^{[\frac{1}{2}n]+1},$$

we have an *a priori* estimation (7.49).

We note that the support of $u(x, t)$ lies in $|x| \leqslant R_0 + t \leqslant R_0 + T = R$. Let $C > L_0$, and let

$$E_1(t) = \int_{|x| \leqslant R} \left[\left(\frac{1}{2}\right)\left\{\left(\frac{\partial u}{\partial t}\right)^2 + \sum_{j=1}^n \left(\frac{\partial u}{\partial x_j}\right)^2\right\} + F(u) + C\right] dx.$$

Then,

$$E_1'(t)=\int_{|x|\leqslant R}\left\{\frac{\partial u}{\partial t}\frac{\partial^2 u}{\partial t^2}+\sum_j \frac{\partial u}{\partial x_j}\frac{\partial^2 u}{\partial x_j\partial t}+f(u)\frac{\partial u}{\partial t}\right\}\mathrm{d}x,$$

$$\int \frac{\partial u}{\partial x_j}\frac{\partial^2}{\partial x_j\partial t}\,\mathrm{d}x=-\int\frac{\partial^2}{\partial x_j^2}u\cdot\frac{\partial}{\partial t}u\,\mathrm{d}x.$$

By integration by parts we have

$$E_1'(t)=\int_{|x|\leqslant R}\frac{\partial u}{\partial t}(\square u+f(u))\,\mathrm{d}x=0\qquad\left(\square=\frac{\partial^2}{\partial t^2}-\Delta\right).$$

Therefore,

$$\frac{1}{2}\int\left\{\left(\frac{\partial u}{\partial t}\right)^2+\sum_j\left(\frac{\partial u}{\partial x_j}\right)^2\right\}\mathrm{d}x\leqslant E_1(t)=E_1(0).$$

By Poincaré's inequality (3.38), we have

$$\|u(t)\|_1\leqslant c(R)E_1(0). \tag{7.50}$$

where $c(R)$ is a constant depending only on R, i.e. if $n=1$, then (7.49) is true.

Next we consider the estimation of $\|u\|_2$ in the cases $n=2$ and 3. If we differentiate (7.45) with respect to x_j, then we have

$$\square u_j+f'(u)u_j=0\qquad\left(u_j=\frac{\partial}{\partial x_j}u\right). \tag{7.51}$$

Let us write $\partial^2 u_{jk}/\partial x_j\partial x_k$ and consider

$$E_2(t)=\sum_{j=1}^n\frac{1}{2}\int(u_{jt}^2+\sum_{k=1}^n u_{jk}^2)\,\mathrm{d}x.$$

We have

$$E_2'(t)=\sum_j\int(u_{jt}\cdot u_{jtt}+\sum_k u_{jk}\cdot u_{jkt})\,\mathrm{d}x,$$
$$=\sum_j\int(\square u_j)u_{jt}\,\mathrm{d}x=-\sum_j\int f'(u)u_ju_{jt}\,\mathrm{d}x,$$

integrating by parts. By proposition 7.1 and (7.3), we have

$$\|u\|_{L^p}\leqslant C\|u\|_1, \tag{7.52}$$

and thus $\|u_j\|_{L^p}<C\|u\|_2$. where p satisfies

$$\left.\begin{aligned}&(a)\qquad 0<\frac{1}{p}\leqslant\tfrac{1}{2}\quad\text{if}\quad n=2,\\&(b)\qquad \frac{1}{p}=\tfrac{1}{2}-\tfrac{1}{3}=\tfrac{1}{6}\quad\text{if}\quad n=3,\end{aligned}\right\}$$

and C is independent of u but dependent on p when $n=2$. In case $n=3$, by Hölder's inequality,

$$|\int f'(u)\, u_j u_{jt}\, \mathrm{d}x| \leqslant \|u_{jt}\|_{L^2} \|u_j\|_{L^6} \|f'(u)\|_{L^3}.$$

By our hypothesis (7.48), we have

$$\int_{|x|\leqslant R} |f'(u)|^3\, \mathrm{d}x \leqslant \alpha'^3 \int_{|x|\leqslant R} (u^6+1)\, \mathrm{d}x = \alpha'^3 \|u\|_{L^6}{}^6 + C_1(\alpha', R).$$

By (7.52), this is $\leqslant C_2(\alpha', R)(\|u\|_1{}^6+1)$. On the other hand, by (7.50), $\|u(t)\|_1$ has an *a priori* estimation, so that the last quantity is less than γ'. Therefore, $\mathrm{d}E_2(t)/\mathrm{d}t \leqslant \gamma E_2(t)$, i.e. $E_2(t) \leqslant \mathrm{e}^{\gamma t} E_2(0)$. Hence

$$\|u(t)\|_2 < C E_2(0) \quad (0 \leqslant t \leqslant T). \tag{7.53}$$

That is (7.49) is true for $n=3$. It is easy to prove the case $n=2$; if we consider

$$|\int f'(u)\, u_j u_{jt}\, \mathrm{d}x| \leqslant \|u_{jt}\|_{L^2} \|u_j\|_{L_j} \|f'(u)\|_{L^q} \quad ((1/p)+(1/q)=\tfrac{1}{2})$$

(the pair of p and q are fixed for the time being), we can apply (7.52) to this inequality, so that we have a similar argument as above.

Finally, we consider the case $f(0)=c_0 \neq 0$. By the assumption of the existence of an *a priori* estimation, this is equivalent to performing an *a priori* estimation of the solution of $\Box u + c_0\beta(x) + (f(u)-c_0) = 0$ for an arbitrary $\beta \in \mathscr{D}$. Let us write

$$E_1(t) = \int_{|x|\leqslant R} \left[\frac{1}{2}\left\{\left(\frac{\partial u}{\partial t}\right)^2 + \sum_p \left(\frac{\partial u}{\partial x_p}\right)^2\right\} + F(u) - c_0 u + \gamma(u^2+1)\right] \mathrm{d}x$$

with u as before, where we choose $(\gamma>0)$ such that $F(u)-c_0u+\gamma(u^2+1) \geqslant 0$. In this case we have

$$\begin{aligned} E_1'(t) &= \int_{|x|\leqslant R} \left\{\frac{\partial u}{\partial t} \Box u + (f(u)-c_0)\frac{\partial u}{\partial t} + 2\gamma u \frac{\partial u}{\partial t}\right\} \mathrm{d}x \\ &= \int_{|x|\leqslant R} 2\gamma u \frac{\partial u}{\partial t} - c_0\beta(x)\frac{\partial u}{\partial t}\, \mathrm{d}x \leqslant C E_1(t). \end{aligned}$$

From $\Box u_j + f'(u) u_j + c_0 \partial\beta/\partial x_j = 0$, we see that the lemma is true in the case of $E_2(t)$.

Q.E.D.

From lemma 7.4 and theorem 7.6 we have the following:

THEOREM 7.7. Let us impose the same assumption on $f(u)$ as we did in lemma 7.4. In this case, for an arbitrary initial condition

$$(u_0(x), u_1(x)), \qquad u_0 \in \mathscr{E}_{L^2(\mathrm{loc})}{}^m, \qquad u_1 \in \mathscr{E}_{L^2(\mathrm{loc})}{}^{m-1} \qquad (m \geqslant [\tfrac{1}{2}n] + 3),$$

there exists a unique global solution of (7.45), $u(t) = u(x, t)$, $0 \leqslant t < +\infty$, and

$$u(t) \in \mathscr{E}_{L^2(\mathrm{loc})}{}^m, \qquad \frac{\partial}{\partial t} u \in \mathscr{E}_{L^2(\mathrm{loc})}{}^{m-1}, \qquad \frac{\partial^2}{\partial t^2} u \in \mathscr{E}_{L^2(\mathrm{loc})}{}^{m-2} \qquad (m \geqslant [\tfrac{1}{2}n] + 3),$$

are all continuous functions of t in $0 \leqslant t < +\infty$.

Chapter 8

Green's functions and spectra

1 *Introduction*

In chapter 3 we considered equations of the elliptic type. In that case the Green's operator G_t played an important rôle. In this chapter we shall show that Green's function in the traditional sense is, in fact, nothing but G_t. More precisely, G_t is a representation of a solution with a Green's kernel. The reason why we consider the relationship is that, firstly, we should like to establish a historical link between the new notion and the old, and, secondly, by doing so, we may obtain useful information about continuous spectra which appear in certain problems lying outside of the scope of chapter 3.

2 *Green's functions and compensating functions*

To simplify the problem we look at boundary value problems of the first and third kind for

$$L[u] \equiv -\Delta u + \sum_{i=1}^{n} a_i \frac{\partial}{\partial x_i} u + cu = f, \tag{8.1}$$

where a_i and c are constants, S is a compact hypersurface of C^2-class, and Ω is an interior or exterior domain in relation to S. First, we shall define a Green's function $G(x, \xi)$.

DEFINITION 8.1 *(Green's function)*. $G(x, \xi)$ is a distribution with a parameter $\xi \in \Omega$ defined in Ω such that

(1)

$$L_x[G(x, \xi)] = \delta_{x-\xi} \tag{8.2}$$

where $\delta_{x-\xi}$ is a Dirac measure at a point ξ.

(2) (Boundary condition) For $\xi \in \Omega$, we pick $\alpha \in \mathscr{D}(\Omega)$ such that $\alpha(x)=1$ in the neighbourhood of ξ. In this case, if ξ is regarded as a parameter, then

as a function of x we have $G(x, \xi)$ such that

$$(1-\alpha(x))G(x, \xi)\in N, \tag{8.3}$$

where N satisfies $N=\mathscr{D}_{L^2}{}^1(\Omega)$ for a Dirichlet problem and

$$N=\left\{\psi(x);\ \psi\in\mathscr{E}_{L^2}{}^2(\Omega), \left(\frac{\mathrm{d}}{\mathrm{d}n}+h(x)\right)\psi(x)=0\right\}$$

for a boundary value problem of the third kind (see chapter 3).†

Let us demonstrate that if a Green's operator G exists for L then G has a representation by a Green's kernel. To show this, we shall use an elementary solution for L. As is well known‡ we have

$$L\cdot E(x)=\delta \tag{8.4}$$

(the solution may not be unique). In this case, we have

$$E\in\mathscr{E}^\infty(R^n\setminus\{0\}), \quad \left(E(x), \frac{\partial}{\partial x_j}E(x)\in L^1{}_{\text{loc}}\right). \tag{8.5}$$

Let us fix E which satisfies (8.4) and (8.5). We take $\alpha_\varepsilon\in\mathscr{D}$ such that it has a support in $|x|\leqslant\varepsilon$; $\alpha_\varepsilon(x)\equiv 1$ in $|x|\leqslant\frac{1}{2}\varepsilon$, and α is a function of $|x|$ only. Then we have the following:

LEMMA 8.1. Let $L[u]=f$, $f\in L^1{}_{\text{loc}}(\Omega)$. In this case for $x\in\Omega_{2\varepsilon}$, where $\Omega_{2\varepsilon}$ is the subset of Ω whose points are a distance larger than 2ε from the boundary of Ω, we have

$$u(x)=(\alpha_\varepsilon E)*f+L\{(1-\alpha_\varepsilon(x))\,E(x)\}*u\,. \tag{8.6}$$

Proof. Let $\operatorname{supp}[\varphi(x)]\subset\Omega_{2\varepsilon}$.
We consider

$$\langle\varphi(x), L\{(1-\alpha_\varepsilon(x))\,E(x)\}*u\rangle. \tag{8.7}$$

We note that $L\{(1-\alpha_\varepsilon(x))E(x)\}=\Phi_\varepsilon(x)$ is a C^∞-function (by Leibniz's formula), and its support is contained in $|x|\leqslant\varepsilon$. Therefore, by the definition of

† Even if there exists $G(x, \xi)$ satisfying the conditions of the definition, it does not immediately mean the existence of a solution of (8.1).

‡ See section 12, chapter 2.

convolution, we can see that in (8.7) if we take $\zeta \in \mathscr{D}(\Omega)$ such that $\zeta(x) \equiv 1$ in Ω_ε, and if we replace u by ζu, then we have

$$\Phi_\varepsilon(x) * u(x) = \Phi_\varepsilon(x) * (\zeta u)(x),$$

for $x \in \Omega_{2\varepsilon}$. Therefore, we can write (8.7) as

$$\langle \varphi(x), L \cdot \{(1-\alpha_\varepsilon(x))E(x)\} * \zeta(x)u(x) \rangle .$$

But, this is equal to

$$\langle \varphi(x), L \cdot E(x) * \zeta(x)u(x) \rangle - \langle \varphi(x), L \cdot [\alpha_\varepsilon(x)E(x)] * \zeta u \rangle .$$

We see that the first term is equal to

$$\langle \varphi(x), \zeta(x)u(x) \rangle = \langle \varphi(x), u(x) \rangle$$

because $LE = \delta$. The second term is

$$L \cdot [\alpha_\varepsilon(x)E(x)] * \zeta u = [\alpha_\varepsilon(x)E(x)] * L[\zeta u] .$$

Obviously, for $x \in \Omega_{2\varepsilon}$ the last term is $\alpha_\varepsilon(x)E(x) * (Lu)(x)$. Therefore, finally, the second term is equal to

$$\langle \varphi(x), \alpha_\varepsilon(x)E(x) * f(x) \rangle .$$

Hence,

$$L\{(1-\alpha_\varepsilon(x))E(x)\} * u = u(x) - (\alpha_\varepsilon(x)E) * f \tag{8.8}$$

is true for $x \in \Omega_{2\varepsilon}$

Q.E.D.

Now, we write†

$$L[(1-\alpha_\varepsilon(|x|))E(x)] = \Phi_\varepsilon(x) . \tag{8.9}$$

Then, (8.6) becomes

$$u(x) = (\alpha_\varepsilon E) * f + \Phi_\varepsilon(x) * u . \tag{8.10}$$

From this, $u \in \mathscr{E}^\infty(\Omega)$ when $f \in \mathscr{E}^\infty(\Omega)$. Next, since $E'(x) = E(-x)$ satisfies

$$^tL \cdot E'(x) = \delta \tag{8.11}$$

In this case if we define

$$^tL \cdot [(1-\alpha_\varepsilon(|x|))\, E'(x)] = \Phi_\varepsilon'(x) \tag{8.12}$$

† Where $\Phi_\varepsilon(x)$ is a C^∞-function whose support lies in $\frac{1}{2}\varepsilon \leqslant |x| \leqslant \varepsilon$.

similarly to (8.9), we see now

$$\Phi_\varepsilon'(x)=\Phi_\varepsilon(-x). \tag{8.13}$$

COROLLARY. If ${}^tL[v]=g$, $g\in L^1{}_{\text{loc}}(\Omega)$, then for $x\in\Omega_{2\varepsilon}$ we have

$$v(x)=(\alpha_\varepsilon E')*g+\Phi_\varepsilon'(x)*v. \tag{8.14}$$

From this, if $g\in\mathscr{E}^\infty(\Omega)$ then $v\in\mathscr{E}^\infty(\Omega)$.

THEOREM 8.1. A Green's operator G has a kernel representation, i.e. there exists a Green's function $G(x, \xi)$ in the sense of definition 8.1, and for $f\in L^2(\Omega)$ we have

$$(Gf)(x)=\int_\Omega G(x, \xi)f(\xi)\,\mathrm{d}\xi, \tag{8.15}$$

where $G(x, \xi)$ satisfies the following conditions.

(1) For $\xi\in\Omega_{2\varepsilon}$,

$$G(x, \xi)=\alpha_\varepsilon(x-\xi)\,E(x-\xi)+\underset{x'\to x}{G}\cdot\Phi_\varepsilon(x'-\xi). \tag{8.16}$$

(2) For $x\in\Omega_{2\varepsilon}$,

$$G(x, \xi)=\alpha_\varepsilon(x-\xi)\,E(x-\xi)+\underset{\eta\to\xi}{{}^tG}\cdot\Phi_\varepsilon(x-\eta), \tag{8.17}$$

where $\underset{x'\to x}{G}$ means that for $\varphi\in L^2(\Omega)$ we have $\psi(x)=(G\varphi)(x)=\underset{x'\to x}{G}\varphi(x')$.

Proof. We prove (8.17), because Gf satisfies $L[Gf]=f$. By (8.10), we have $(Gf)(x)=(\alpha_\varepsilon E)*f+\Phi_\varepsilon*Gf$. The second of the right hand terms is

$$\int\Phi_\varepsilon(x-\xi)(Gf)(\xi)\,\mathrm{d}\xi=\langle\Phi_\varepsilon(x-\xi), (Gf)(\xi)\rangle_\xi=\langle\underset{\eta\to\xi}{{}^tG}\,\Phi_\varepsilon(x-\eta), f(\xi)\rangle_\xi.$$

Therefore, (8.17) is true.†

Next, we prove (8.16). To do this we use (8.14) and follow the same argument as in the case of (8.17). We have

$$({}^tGf)\,(x)=(\alpha_\varepsilon E')*f+\langle\underset{\eta\to\xi}{G}\;\Phi_\varepsilon'(x-\eta), f(\xi)\rangle\qquad(x\in\Omega_{2\varepsilon}).$$

† It is easy to see that ${}^tG\Phi_\varepsilon(x-\eta)=L_\varepsilon(x, \xi)\eta\to\xi\in C^\infty$ for $(x, \xi)\in\Omega_{2\varepsilon}\times\Omega$, i.e. for ξ, we can refer to the corollary of lemma 8.1, and for x, we note that $\Phi_\varepsilon(x-\eta)$ is regular with respect to the parameter x. Similarly, we can prove $G\Phi_\varepsilon'(x-\eta)=L_\varepsilon'(x, \xi)\eta\to\xi$.

Let $\alpha_\varepsilon(x-\xi)E'(x-\xi)+G\Phi_\varepsilon'(x-\eta)=K(x,\xi)$ where $K(x,\xi)\in C^\infty$ except for the case $x=\xi$, and $|K(x,\xi)|\leqslant\text{const}|x-\xi|^{-n+2}$ for $(x,\xi)\in\Omega_{2\varepsilon}\times\Omega_{2\varepsilon}$ in the neighbourhood of $x=\xi$ (see (2.19)). Consider $f, g\in\mathscr{D}(\Omega)$. If we assume the supports of f and g lie in $\Omega_{2\varepsilon}$, then

$$\begin{aligned}\langle\textstyle\int G(x,\xi)f(\xi)\,d\xi, g(x)\rangle&=\langle(Gf)(x), g(x)\rangle=\langle f(x), ({}^tGg)(x)\rangle\\&=\langle f(x), \textstyle\int K(x,\xi)\,g(\xi)\,d\xi\rangle.\end{aligned}$$

By Fubini's theorem, the last term is equal to $\langle\int K(x,\xi)f(x)dx, g(\xi)\rangle$. Hence $G(x,\xi)=K(\xi,x)$ is true for $(x,\xi)\in\Omega_{2\varepsilon}\times\Omega_{2\varepsilon}$, where ε is arbitrary, so that the same relation is valid in $\Omega\times\Omega$. From this $G(x,\xi)=\alpha_\varepsilon(x-\xi)E'\times(\xi-x)+\underset{x'\to x}{G}\,\Phi_\varepsilon'(\xi-x')$. By (8.13) and from the fact $E'(x)=E(-x)$, we have (8.16).

Q.E.D.

This representation of a Green's function (8.16) is not convenient, for actual calculation even if the related Green's operator is proved to exist. In fact, the following theorem is useful for this purpose.

THEOREM 8.2. Let us assume the existence of a Green's operator G. In this case, if $K(x,\xi)=E(x-\xi)+u(x,\xi)$ satisfies the following conditions:

(1) $L_x[u(x,\xi)]=0$ is true for $\xi\in\Omega$, and

(2) $(1-\alpha_\varepsilon(x-\xi))E(x-\xi)+u(x,\xi)\in N_x$ $(\xi\in\Omega_{2\varepsilon})$ (see definition 8.1),

then $K(x,\xi)$ is a Green's function.†

Proof. This is rather easy. First, we write

$$K(x,\xi)=\alpha_\varepsilon(x-\xi)\,E(x-\xi)+\{1-\alpha_\varepsilon(x-\xi)\}\,E(x-\xi)+u(x,\xi).$$

Then

$$L_x[(1-\alpha_\varepsilon(x-\xi))\,E(x-\xi)+u(x,\xi)]=\Phi_\varepsilon(x-\xi),$$

and for $\varphi\in N$ we have $G\circ L\varphi=\varphi$. From this, we obtain

$$\{1-\alpha_\varepsilon(x-\xi)\}\,E(x-\xi)+u(x,\xi)=\underset{x'\to x}{G\Phi_\varepsilon}(x'-\xi).$$

Therefore, from (8.16) $K(x,\xi)=G(x,\xi)$.

Q.E.D.

† Of course, in this case $K(x,\xi)$ must be understood to be different from $K(x,\xi)$ used in the proof of the previous theorem.

DEFINITION 8.2. $u(x, \xi)$ which appears in theorem 8.2 is called a *compensating function.*

3 *Green's function for $(\Delta-\lambda)$*

In the previous section we defined a compensating function. In this section we consider the properties of this function. For simplicity, we assume that the dimension of the underlying space is three. Let S be a compact surface of C^2-class. Further, we impose the following assumptions: Ω is an exterior domain and the equation in question is

$$(\Delta-\lambda^2)\,u=f \quad (\operatorname{Re}\lambda>0).^\dagger \tag{8.18}$$

Hereafter, we consider only the Neumann problem, but, as we see later, the case of the Dirichlet problem can be similarly argued.

By theorem 8.2, we can write a Green's function for the exterior Neumann problem of (8.18) as $G(x, \xi \mid \lambda)$. We put

$$G(x, \xi \mid \lambda)=-\frac{1}{4\pi}\frac{e^{-\lambda|x-\xi|}}{|x-\xi|}+K_c(x, \xi \mid \lambda) \quad (\operatorname{Re}\lambda>0),$$

and consider $K_c(x, \xi \mid \lambda)$ where K_c satisfies

$$\frac{\mathrm{d}}{\mathrm{d}n}K_c(x, \xi \mid \lambda)=\frac{\mathrm{d}}{\mathrm{d}n}\left[\frac{1}{4\pi}\frac{\mathrm{e}^{-\lambda|x-\xi|}}{|x-\xi|}\right]$$

for $x\in S$, also it satisfies $(\lambda^2-\Delta)K_c(x, \xi \mid \lambda)=0$ for $x\in\Omega$. We follow Kellogg's argument (see Kellogg: *Foundation of Potential Theory*, pp. 285–315) and his notation.

Let p, q, and r be points of S and let $P, Q, \ldots$ be points of Ω. We have

$$E(P-p \mid \lambda)=\frac{1}{2\pi}\frac{\mathrm{e}^{-\lambda|P-p|}}{|P-p|},$$

$$V(P)=\int_S \psi(q)\,E(P-q \mid \lambda)\,\mathrm{d}q,$$

$$W(P)=\int_S \varphi(q)\frac{\mathrm{d}}{\mathrm{d}n_q}E(P-q \mid \lambda)\,\mathrm{d}q.$$

† The aim is to construct the Green's function of $(\Delta-\lambda)$, but for convenience we change the parameters and consider $(\Delta-\lambda^2)$ $(\operatorname{Re}\lambda>0)$.

Furthermore, we put

$$\frac{d}{dn_q} E(p-q \mid \lambda) = K(p, q \mid \lambda).$$

More exactly,

$$2\pi K(p, q \mid \lambda) = \frac{d}{dn_q} \frac{e^{-\lambda|p-q|}}{|p-q|} = \frac{e^{-\lambda|p-q|}}{|p-q|^2} \cos(n_q, \overrightarrow{qp}) + \lambda \cdot \frac{e^{-\lambda|p-q|}}{|p-q|} \cos(n_q, \overrightarrow{qp}),$$

where n_q is a unit vector at a point $q \in S$ in the direction of the normal which is pointing outwardly to the exterior domain. As is well known, we have

$$W_-(p) = -\varphi(p) + \int_S K(p, q \mid \lambda)\, \varphi(q)\, dq, \tag{8.19}$$

$$\frac{dV}{dn_+}(p) = -\psi(p) + \int_S K(q, p \mid \lambda)\, \psi(q)\, dq, \tag{8.20}$$

where $\pm$ means limits taken along the normal from the exterior (interior) domain. Therefore, (8.19) is an integral equation corresponding to an interior Dirichlet problem. Similarly, we note that

$$W_+(p) = \varphi(p) + \int_S K(p, q \mid \lambda)\, \varphi(q)\, dq, \tag{8.21}$$

$$\frac{dV}{dn_-}(p) = \psi(p) + \int_S K(q, p \mid \lambda)\, \psi(q)\, dq. \tag{8.22}$$

As for the solubility of (8.20), we shall examine this by considering its dependence on the parameter λ. To do this, we follow Fredholm's argument.

4 *Fredholm's theorem*

Following the traditional approach we begin with the case of a continuous kernel. Let

$$\varphi(p) - \int_S K(p, q \mid \lambda)\, \varphi(q)\, dq = f(p), \tag{8.23}$$

$$\psi(p) - \int_S K(q, p \mid \lambda)\, \psi(q)\, dq = g(p), \tag{8.24}$$

and let $K(p, q \mid \lambda)$ be continuous in $S \times S \times C$. Also, we assume that $K(p, q \mid \lambda)$ is holomorphic with respect to λ. Furthermore, we assume that, for a certain domain D of λ, we have

$$|K^{(n)}(p, q \mid \lambda)| \leqslant C\theta^n \quad (0<\theta<1,\ n=1, 2, \ldots), \tag{8.25}$$

where $K^{(n)}(p, q \mid \lambda)$ is an $(n-1)$ repeating kernel of K.

On the other hand, for $\lambda \in D$, from (8.25),

$$\varphi(p)=f(p)+\int K(p, q \mid \lambda) f(q)\, dq+\int K^{(2)}(p, q \mid \lambda) f(q)\, dq+\cdots$$

represents a unique solution of (8.23). We write

$$R(p, q \mid \lambda)=K(p, q \mid \lambda)+K^{(2)}(p, q \mid \lambda)+\cdots+K^{(n)}(p, q \mid \lambda)+\cdots \quad (\lambda \in D). \tag{8.26}$$

Then, $R(p, q \mid \lambda)$ is continuous with respect to (p, q, λ) and holomorphic with respect to λ. In this case, the equations of resolvent

$$K(p, q \mid \lambda)=R(p, q \mid \lambda)-\int K(p, r \mid \lambda)\, R(r, q \mid \lambda)\, dq\,, \tag{8.27}$$

$$K(p, q \mid \lambda)=R(p, q \mid \lambda)-\int K(r, q \mid \lambda)\, R(p, r \mid \lambda)\, dr\,, \tag{8.28}$$

are true for $\lambda \in D$.

The key idea of Fredholm's theory, the analytic extension of $R(p, q \mid \lambda)$ can be established in this case. More exactly, the situation is as follows. Let us consider $\delta(\lambda)=1-\delta_1(\lambda)+\delta_2(\lambda)-\cdots+(-1)^p\delta_p(\lambda)+\cdots$ where

$$\delta_n(\lambda)=\frac{1}{n!}\int \cdots \int K\left(\begin{matrix} r_1 \cdots r_n \\ r_1 \cdots r_n \end{matrix}\,\middle|\, \lambda\right) dr_1 \ldots dr_n$$

and the function to be integrated is the so-called *Fredholm's determinant*, i.e.

$$K\left(\begin{matrix} p_1\, p_2 \cdots p_n \\ q_1\, q_2 \cdots q_n \end{matrix}\,\middle|\, \lambda\right)=\begin{vmatrix} K(p_1, q_1 \mid \lambda) & K(p_1, q_2 \mid \lambda) & \ldots & K(p_1, q_n \mid \lambda) \\ K(p_2, q_1 \mid \lambda) & K(p_2, q_2 \mid \lambda) & \ldots & K(p_2, q_n \mid \lambda) \\ \vdots & & & \\ K(p_n, q_1 \mid \lambda) & K(p_n, q_2 \mid \lambda) & \ldots & K(p_n, q_n \mid \lambda) \end{vmatrix}.$$

Next, we have

$$N(p, q \mid \lambda)=K(p, q \mid \lambda)-N_1(p, q \mid \lambda)+\ldots+(-1)^n N_n(p, q \mid \lambda)+\ldots,$$

$$N_n(p, q \mid \lambda)=\frac{1}{n!}\int \ldots \int K\left(\begin{matrix} p r_1 \ldots r_n \\ q r_1 \ldots r_n \end{matrix}\,\middle|\, \lambda\right) dr_1 \ldots dr_n\,.$$

LEMMA 8.2. $\delta(\lambda)$ and $N(p, q \mid \lambda)$ are both entire functions† of λ, and for $\lambda \in D$ we have

$$R(p, q \mid \lambda) = \frac{N(p, q \mid \lambda)}{\delta(\lambda)}. \tag{8.29}$$

That is, $R(p, q \mid \lambda)$ which is defined by the right hand side term of (8.29) is a meromorphic function of λ, and for an arbitrary λ (8.27) and (8.28) are true.

Proof. By Hadamard's inequality, $\delta(\lambda)$ and $N(p, q \mid \lambda)$ are both entire functions of λ. In fact, if we write the maximum value of $|K(p, q \mid \lambda)|$ in $|\lambda| \leqslant R$ as K_R, then in $|\lambda| \leqslant R$ we have

$$|\delta_n(\lambda)| \leqslant \frac{1}{n!} n^{\frac{1}{2}n} K_R{}^n |S|^n$$

where $|S|$ is the surface area of S. From this, we see that in $|\lambda| \leqslant K_R$, $\delta(\lambda)$ is a regular function which is the limit of the uniformly convergent sequence of regular functions. R is arbitrary, so that $\delta(\lambda)$ is an entire function of λ. Similarly $K(p, q \mid \lambda)$ is also an entire function.

As for (8.29), we shall only sketch the outline of the proof because the proof of this fact can be found in many documents. Let $\lambda \in D$. Let us fix λ (the dependence on λ can be ignored). Write $K(p, q)$ for $K(p, q \mid \lambda)$, then from (8.25),

$$|K^{(n)}(p, q)| \leqslant C\theta^n \quad (0<\theta<1). \tag{8.25$'$}$$

Now, we introduce a parameter μ, $|\mu| \leqslant 1$. We write

$$R_\mu(p, q) = K(p, q) + \mu K^{(2)}(p, q) + \cdots + \mu^n K^{(n+1)}(p, q) + \cdots$$

and we now have

$$K(p, q) = R_\mu(p, q) - \mu \int K(p, r) R_\mu(r, q)\, dq. \tag{8.30}$$

Similarly, we put

$$D(\mu) = 1 + \sum_{p=1}^{\infty} (-\mu)^p \delta_p \quad \left(\delta_p = \frac{1}{p!} \int \cdots \int K\begin{pmatrix} r_1 \ldots r_p \\ r_1 \ldots r_p \end{pmatrix} dr_1 \ldots dr_p\right), \tag{8.31}$$

and (8.30) gives

$$K(p, q) D(\mu) = R_\mu(p, q) D(\mu) - \mu \int K(p, r) R_\mu(r, q) D(\mu)\, dr.$$

† $N(p, q \mid \lambda)$ is a continuous function of (p, q, λ).

Put

$$R_\mu(p,q)D(\mu)=K(p,q)+\sum_{n=1}^{\infty}\frac{(-\mu)^n}{n!}C_n(p,q)$$

and comparing the coefficients of μ^n in the terms on both sides, we see that

$$(-1)^n\delta_n K(p,q)=\frac{(-1)^n}{n!}C_n(p,q)-\frac{(-1)^{n-1}}{(n-1)!}\int K(p,r)C_{n-1}(r,q)\,\mathrm{d}r,$$

i.e.

$$C_n(p,q)=n!\delta_n K(p,q)\\ +n\int K(p,r)C_{n-1}(r,q)\,\mathrm{d}r \quad (n=1,2,3,\ldots). \tag{8.32}$$

From this,

$$C_n(p,q)=\int\ldots\int K\begin{pmatrix}pr_1\ldots r_n\\ qr_1\ldots r_n\end{pmatrix}\mathrm{d}r_1\ldots\mathrm{d}r_n \quad (n=1,2,3,\ldots) \tag{8.33}$$

is seen to be true using induction with respect to n.

In fact, if we expand Fredholm's determinant along the first row, then we have

$$K\begin{pmatrix}pr_1\ldots r_n\\ qr_1\ldots r_n\end{pmatrix}=K(p,q)K\begin{pmatrix}r_1\ldots r_n\\ r_1\ldots r_n\end{pmatrix}-K(p,r_1)K\begin{pmatrix}r_1r_2\ldots r_n\\ qr_2\ldots r_n\end{pmatrix}\\ -K(p,r_2)K\begin{pmatrix}r_1r_2\ldots r_n\\ r_1q\ldots r_n\end{pmatrix}\cdots-K(p,r_n)K\begin{pmatrix}r_1\ldots r_n\\ r_1\ldots q\end{pmatrix}.$$

Then we integrate this over $S\times\cdots\times S$ with respect to $(r_1\cdots r_n)$. We note that Fredholm's determinant is invariant with respect to simultaneous interchange of its rows and columns. Therefore (8.33) follows from induction with respect to n.

We put $\mu=1$. From the definition of $N(p,q\mid\lambda)$, (8.25) follows.

Q.E.D.

We give the following lemma as a criterion of lemma 8.2.

LEMMA 8.3. If $\lambda=\lambda_0$ is a regular point of $R(p,q\mid\lambda)$ defined in (8.29), there exists a unique solution of (8.23) corresponding to $\lambda=\lambda_0$, and it can be represented by

$$\varphi(p)=f(p)+\int R(p,q\mid\lambda_0)f(q)\,\mathrm{d}q. \tag{8.34}$$

Also, a unique solution of (8.24) exists and can be represented by

$$\psi(p)=g(p)+\int R(q,p\mid\lambda_0)g(q)\,dq\,. \tag{8.35}$$

Proof. We shall give only the proof of (8.34).

(1) Let us demonstrate that (8.34) is, in fact, a solution of (8.23). We operate with K in (8.34) from the left. Then, we have

$$\int K(p,q\mid\lambda_0)\varphi(q)\,dq=\int K(p,q\mid\lambda_0)f(q)\,dq \\ +\int K(p,q\mid\lambda_0)\,dq\int R(q,r\mid\lambda_0)f(r)\,dr\,.$$

We note that the second integral of the right hand side can be written

$$\int\left[\int K(p,q\mid\lambda_0)R(q,r\mid\lambda_0)\,dq\right]f(r)\,dr\,.$$

Consider the equation of resolvent (8.27). The right hand term of the first equation is equal to $\int R(p,q\mid\lambda_0)f(q)\,dq$. Also, we consider (8.34). We see that (8.34) is equal to $\varphi(p)-f(p)$. Therefore, (8.34) is a solution of (8.23).

(2) The uniqueness of the solution follows from (8.28). In fact, if

$$\varphi_1(p)=f(p)+\int K(p,q\mid\lambda_0)\varphi_1(q)\,dq\,,$$

then, by the substitution of $f(p)$ in (8.34), we have

$$\begin{aligned}\varphi(p)&=\varphi_1(p)-\int K(p,q\mid\lambda_0)\varphi_1(q)\,dq\\&\quad+\int R(p,q\mid\lambda_0)\left[\varphi_1(q)-\int K(q,r\mid\lambda_0)\varphi_1(r)\,dr\right]dq\\&=\varphi_1(p)-\int\left[K(p,q\mid\lambda_0)-R(p,q\mid\lambda_0)\right.\\&\quad\left.+\int R(p,r\mid\lambda_0)K(r,q\mid\lambda_0)\,dr\right]\varphi_1(q)\,dq\,.\end{aligned}$$

The kernel which appears in the last integral is 0 (by (8.28)), therefore $\varphi(p)\equiv\varphi_1(p)$.

Q.E.D.

Next, we shall examine the structure of $R(p,q\mid\lambda)$ in the neighbourhood of λ_0 when $\lambda=\lambda_0$ is a pole of $R(p,q\mid\lambda)$.

(1) If $\lambda=\lambda_0$ is a simple pole, then

$$R(p,q\mid\lambda)=\frac{A(p,q)}{\lambda-\lambda_0}+B(p,q\mid\lambda)\,,$$

where $A(p,q)\not\equiv 0$, and $B(p,q\mid\lambda)$ is regular in the neighbourhood of $\lambda=\lambda_0$.

We re-write the equations of resolvent (8.27) and (8.28) using the above rule. We have

$$K(p, q \mid \lambda)=\frac{A(p, q)}{\lambda-\lambda_0}+B(p, q \mid \lambda)-\int K(p, r \mid \lambda)\left[\frac{A(r, q)}{\lambda-\lambda_0}+B(r, q \mid \lambda)\right] \mathrm{d}r$$

$$=\frac{A(p, q)}{\lambda-\lambda_0}+B(p, q \mid \lambda)-\int\left[\frac{A(p, r)}{\lambda-\lambda_0}+B(p, r \mid \lambda)\right] K(r, q \mid \lambda)\, \mathrm{d}r.$$

By the comparison of the coefficients of $(\lambda-\lambda_0)^{-1}$ of the terms of both sides, we have

$$A(p, q)=\int K(p, r \mid \lambda_0) A(r, q)\, \mathrm{d}r, \tag{8.36}$$

$$A(p, q)=\int A(p, r) K(r, q \mid \lambda_0)\, \mathrm{d}r. \tag{8.37}$$

From $A(p, q) \not\equiv 0$, we see that $\varphi(p)=\int K(p, q \mid \lambda_0)\varphi(q)\, \mathrm{d}q$ has a non-trivial solution.

If $\varphi_1(p), \ldots, \varphi_n(p)$ form a linearly independent basis of an eigenspace, then we can write

$$A(p, q)=\varphi_1(p)\psi_1(q)+\cdots+\varphi_n(p)\psi_n(q). \tag{8.38}$$

Now, we see that $A(p, q)$ is continuous, and so is $\psi_j(q)$. In this case, if $\psi_i(q) \equiv 0$, we omit the term, and, if $\psi_i(q)$ are not linearly independent, then we can pick $\psi_1(q), \ldots, \psi_n(q)$ which are independent of each other. We put (8.38) in (8.37) to obtain

$$\sum_i \varphi_i(p)\left[\psi_i(q)-\int \psi_i(r) K(r, q \mid \lambda_0)\, \mathrm{d}r\right] \equiv 0,$$

therefore,

$$\psi_i(q)=\int \psi_i(r) K(r, q \mid \lambda_0)\, \mathrm{d}r \quad (i=1, 2, \ldots, n).$$

(2) If $\lambda=\lambda_0$ has a pole of a higher order, we put

$$R(p, q \mid \lambda)=\frac{A_m(p, q)}{(\lambda-\lambda_0)^m}+\frac{A_{m-1}(p, q)}{(\lambda-\lambda_0)^{m-1}}+\cdots+B(p, q \mid \lambda) \quad (A_m(p, q) \not\equiv 0).$$

As before, we have

$$A_m(p, q)=\int K(p, r \mid \lambda_0) A_m(r, q)\, \mathrm{d}r=\int A_m(p, r) K(r, q \mid \lambda_0)\, \mathrm{d}r.$$

Similarly, we can write

$$A_m(p, q)=\varphi_1(p)\psi_1(q)+\cdots+\varphi_n(p)\psi_n(q).$$

Summing up, we have

PROPOSITION 8.1. If $\lambda=\lambda_0$ is a pole of mth order of $R(p, q \mid \lambda)$, then, in the neighbourhood of $\lambda=\lambda_0$, we can write

$$R(p, q \mid \lambda)=\frac{\varphi_1(p)\psi_1(q)+\cdots+\varphi_n(p)\psi_n(q)}{(\lambda-\lambda_0)^m}+\cdots,$$

where $\{\varphi_i(p)\}_{i,p=1,2,\ldots,n}$, $\{\psi_i(q)\}_{i=1,2,\ldots,n}$ are both independent sets and continuous and satisfy

$$\varphi_i(p)=\int K(p, q \mid \lambda_0)\varphi_i(q)\,dq, \quad \psi_i(p)=\int K(q, p \mid \lambda_0)\psi_i(q)\,dq.$$

Resolvent kernel for a discontinuous kernel

The kernel $K(p, q \mid \lambda)$ which appears in (8.19) and (8.20) is not a continuous kernel, but, by using the idea of a repeating kernel, we have the same results as we obtained for continuous kernels. As we shall see later,† we have the following properties of $K(p, q \mid \lambda)$.

(1) If $p \neq q$, then $K(p, q \mid \lambda)$, $K^{(2)}(p, q \mid \lambda)$ are both continuous with respect to (p, q, λ), regular with respect to λ, and

$$|K(p, q \mid \lambda)| \leqslant C|p-q|^{-1}, \quad |K^{(2)}(p, q \mid \lambda)| \leqslant C+C_1 \log\frac{1}{|p-q|}$$

where C and C_1 are dependent on λ, but, if λ is bounded, C and C_1 can be constants.

(2) $K^{(3)}(p, q \mid \lambda)$ is continuous in $S \times S \times C$, and regular with respect to λ.

In general, $K^{(n)}(p, q \mid \lambda)$ $(n \geqslant 3)$ has the same properties, and satisfies the condition (8.25). If we assume that the above properties (1) and (2) are true, then (8.26) is meaningful, and $R(p, q \mid \lambda)$ satisfies (8.27) and (8.28). In this case, lemma 8.2 gives some difficulty because as it stands Fredholm's determinant has no meaning. To avoid this situation, we use the idea of a 'repeating kernel'.

Before proceeding further, we change our old notation of an n-times repeating kernel of K for a new notation $K_n(p, q \mid \lambda)$, i.e. $K^{(n+1)}(p, q \mid \lambda) \equiv K_n(p, q \mid \lambda)$. First we note that $K_2(p, q \mid \lambda) \equiv K^{(3)}(p, q \mid \lambda)$ has been treated before, so that, for $\lambda \in D$, we write the resolvent kernel of K_2 as $R_2(p, q \mid \lambda)$. Then, we have

$$R_2(p, q \mid \lambda)=K_2(p, q \mid \lambda)+K_5(p, q \mid \lambda)+K_8(p, q \mid \lambda)+\cdots \quad (\lambda \in D).$$

† See the last half of section 6.

Obviously,

$$\int R_2(p, r \mid \lambda)K(r, q \mid \lambda)\,\mathrm{d}r = K_3(p, q \mid \lambda) + K_6(p, q \mid \lambda) + \cdots$$

$$\int R_2(p, r \mid \lambda)K_1(r, q \mid \lambda)\,\mathrm{d}r = K_4(p, q \mid \lambda) + K_7(p, q \mid \lambda) + \cdots$$

Therefore, for $\lambda \in D$, we can write

$$R(p, q \mid \lambda) = K(p, q \mid \lambda) + K_1(p, q \mid \lambda) + R_2(p, q \mid \lambda) + \int R_2(p, r \mid \lambda)\,[K(r, q \mid \lambda) + K_1(r, q \mid \lambda)]\,\mathrm{d}r. \tag{8.39}$$

By lemma 8.2,

$$R_2(p, q \mid \lambda) = \frac{M(p, q \mid \lambda)}{\eta(\lambda)},$$

where M is a continuous function of (p, q, λ), and a regular function with respect to λ as η is a regular function. Therefore, by (8.39), we have

$$\begin{aligned} R(p, q \mid \lambda) &= K(p, q \mid \lambda) + K_1(p, q \mid \lambda) \\ &\quad + \frac{1}{\eta(\lambda)}\{M(p, q \mid \lambda) + \int M(p, r \mid \lambda)\,[K(r, q \mid \lambda) + K_1(r, q \mid \lambda)]\,\mathrm{d}r\}) \\ &= K(p, q \mid \lambda) + K_1(p, q \mid \lambda) + \frac{M_1(p, q \mid \lambda)}{\eta(\lambda)}, \end{aligned} \tag{8.40}$$

where $M_1(p, q \mid \lambda)$ is continuous in $S \times S \times C$ and regular with respect to λ.

We see that the right hand side of (8.40) defines an analytic continuation of $R(p, q \mid \lambda)$, and the pole of $R(p, q \mid \lambda)$ is that of $M_1(p, q \mid \lambda)/\eta(\lambda)$. Therefore, we have:

PROPOSITION 8.2. If $K(p, q \mid \lambda)$ satisfies the above conditions (1) and (2), then proposition 8.1 is true.

5 *Concrete construction of a Green's function*

Let us go back to section 3 where we considered

$$W_-(p) = -\varphi(p) + \int_S K(p, q \mid \lambda)\varphi(q)\,\mathrm{d}q, \tag{8.19}$$

$$\frac{\mathrm{d}V}{\mathrm{d}n_+}(p) = -\psi(p) + \int_S K(q, p \mid \lambda)\psi(q)\,\mathrm{d}q. \tag{8.20}$$

If $\lambda=\lambda_0$ is a pole of the resolvent kernel of $K(p, q \mid \lambda)$, then both

$$\varphi(p)=\int K(p, q \mid \lambda_0)\varphi(q)\,\mathrm{d}q \tag{8.41}$$

and

$$\psi(p)=\int K(q, p \mid \lambda_0)\psi(q)\,\mathrm{d}q \tag{8.42}$$

have non-trivial continuous solutions. The inverse of this statement is also true.

Therefore, we call λ_0 an *eigenvalue* of $K(p, q \mid \lambda)$, and call $\varphi(p)$ and $\psi(p)$ the *eigenfunctions* of $K(p, q \mid \lambda)$, $K(q, p \mid \lambda)$ corresponding to $\lambda=\lambda_0$ respectively. Obviously, in this sense an eigenfunction has a Hölder continuity. To make use of Fredholm's alternative theorem, we prepare the following lemma. All the following lemmas are in the three-dimensional case.

LEMMA 8.4. If u is a function such that $(\lambda+\Delta)u(x)=0$ $(\lambda>0)$ in the exterior domain of S, and $u\in L^2$, if $|x|\to+\infty$, then $u(x)\equiv 0$ $(x\in\Omega)$.

Proof. Let $Y_n(\theta, \varphi)\equiv Y_n(\omega)$ be a normalized nth order spherical function, and let its Fourier's coefficient be

$$Z(r)=\int_{|x|=r} u(r\omega)Y_n(\omega)\,\mathrm{d}\omega\,.$$

Obviously,

$$\frac{1}{r^2}\frac{\mathrm{d}}{\mathrm{d}r}\left(r^2\frac{\mathrm{d}Z}{\mathrm{d}r}\right)+\left(\lambda-\frac{n(n+1)}{r^2}\right)Z=0.$$

From the well-known property of a Bessel function, this equation has an independent solution in the neighbourhood of $r=+\infty$ having an asymptotic expansion in the form of

$$\sim\cos\lambda^{\frac{1}{2}}r\left[\frac{1}{r}+\frac{a_2}{r^2}+\cdots\right],$$

$$\sim\sin\lambda^{\frac{1}{2}}r\left[\frac{1}{r}+\frac{b_2}{r^2}+\cdots\right].$$

So that, there exist certain constant c_1 and c_2 such that

$$Z(r)=\frac{1}{r}(c_1\cos\lambda^{\frac{1}{2}}r+c_2\sin\lambda^{\frac{1}{2}}r)+\frac{B(r)}{r^2}\quad(r\geqslant r_0),$$

where $B(r)$ is bounded. From

$$Z(r)^2 = [\int u(\omega r) Y_n(\omega)\, \mathrm{d}\omega]^2 \leqslant \int_{|x|=r} u^2(\omega r)\, \mathrm{d}\omega$$

it follows that

$$\int_{r_0}^{+\infty} r^2 Z(r)^2\, \mathrm{d}r < +\infty .$$

On the other hand,

$$\int_{r_0}^{R} r^2 Z(r)^2\, \mathrm{d}r = c_1{}^2 \int_{r_0}^{R} \cos^2 \lambda^{\frac{1}{2}} r\, \mathrm{d}r + c_2{}^2 \int_{r_0}^{R} \sin\ \lambda\ r\, \mathrm{d}r + O(\log R)$$

From this we have $c_1 = c_2 = 0$. Hence, $Z(r) = 0 (r \geqslant r_0)$. Y_n is arbitrary, so that $u(x) \equiv 0$. $u(x)$ is a solution of $(\lambda + \Delta) u(x) = 0$ such that it is an analytic function in Ω. From this we have the lemma.

Q.E.D.

LEMMA 8.5. Let Ω be the exterior domain of a compact surface S of C^2-class. Let $\varphi(s)$ be a function having Hölder continuity. In this case, if a simple layer potential function

$$u(x) = \int_S \varphi(s) \frac{\mathrm{e}^{ik|x-s|}}{|x-s|}\, \mathrm{d}S \qquad (k > 0)$$

defined for $x \in \Omega$ satisfies $u(s)_+ \equiv 0$ on S or $\mathrm{d}u(s)/\mathrm{d}n_+ \equiv 0$ on S, then $u(x) \equiv 0$.

Proof. Let us first demonstrate that $u(x)$ satisfies Sommerfeld's radiation conditions:

$$u(x) = O(|x|^{-1}), \qquad \frac{\mathrm{d}}{\mathrm{d}r} u(x) - iku(x) = o(|x|^{-1}) \qquad (|x| \to +\infty).$$

Because $\varphi(s)$ has Hölder continuity, u is continuous in Ω (including its boundary), so is the first derivative of u. By a simple calculation, we see that

$$\frac{\mathrm{d}}{\mathrm{d}r} |x-s| = 1 + O\left(\frac{1}{|x|}\right), \qquad |x-s| = |x| - \left\langle \frac{x}{|x|} \cdot s \right\rangle + O\left(\frac{1}{|x|}\right).^{\dagger}$$

From this, we have

$$\frac{\mathrm{d}}{\mathrm{d}r} u(x) - iku(x) = O(|x|^{-2}).$$

† $\mathrm{d}/\mathrm{d}r$ represents a differential in the direction of the radius, i.e. $\mathrm{d}/\mathrm{d}|x|$.

Hence, we can apply Rellich's uniqueness theorem to prove the lemma.

Alternatively, we can prove the lemma directly as follows. First, from Green's theorem, we have

$$0=\int_{|x|=R}\left(u\frac{\mathrm{d}}{\mathrm{d}r}\bar{u}-\bar{u}\frac{\mathrm{d}}{\mathrm{d}r}u\right)\mathrm{d}S=-2ik\int_{|x|=R}|u|^2\,\mathrm{d}S+O(R^{-1}).$$

Therefore,

$$\int_{|x|=R}|u|^2\,\mathrm{d}S=O(R^{-1}).$$

On the other hand, we can write

$$u(x)=\frac{\exp(ik|x|)}{|x|}\int_S\varphi(s)\exp\left(-ik\left\langle\frac{x}{|x|},s\right\rangle\right)\left\langle\frac{x}{|x|},s\right\rangle\mathrm{d}S+O(|x|^{-2}).$$

So that the first term of the right hand side is identically 0. Hence, $u\in L^2(\Omega)$, and $u(x)$ satisfies $(k^2+\Delta)u(x)=0$. Hence, we can apply the previous lemma to get $u(x)\equiv 0$.

Q.E.D.

LEMMA 8.6. Let Ω be the same as the previous lemma. Let $\Delta u(x)=0(x\in\Omega)$ where u is continuous in Ω and on its boundary, and let the first derivative have the same property. Further, we assume that $u(x)_+\equiv 0$ or $\mathrm{d}u(x)/\mathrm{d}n_+\equiv 0$ on S, and $|x|u(x)$, $|x|^2\partial u(x)/\partial x_j$ are bounded as $|x|\to\infty$. In this case, we have $u(x)\equiv 0$.

Proof. From Green's theorem, we have

$$0=\int_D u\Delta u\,\mathrm{d}x=-\int_D\sum_j\left|\frac{\partial u}{\partial x_j}\right|^2\mathrm{d}x+\int_{|x|=R}u\frac{du}{dr}\,\mathrm{d}S,$$

where D is a part of Ω such that $|x|\leqslant R$. From this, $\partial u(x)/\partial x_j\equiv 0(j=1,2,3)$. Therefore, $u(x)\equiv 0$.

Q.E.D.

THEOREM 8.3. Let $\mathrm{Re}\lambda\geqslant 0$ in the equations (8.19) and (8.20). If $\lambda=\lambda_0$ is an eigenvalue of $K(p,q\mid\lambda)$, then $\lambda_0{}^2$ is an eigenvalue of the interior Dirichlet problem with respect to Δ.

Proof. Let us write $\psi_0(p)$ for one of the proper solutions of (8.42) with respect to λ_0, and let us consider a single layer potential

$$V(P)=\int_S \varphi_0(q)E(P-q \mid \lambda_0)\, \mathrm{d}q\,.$$

First, we note that $\mathrm{d}V(P)/\mathrm{d}n_+ \equiv 0$. On the other hand, if $Re\lambda_0>0$, then $V\in L^2(\Omega)$, and $V(P)$ and its first derivatives are continuous in S including its boundary.† Therefore, $V(P)\equiv 0$. (This can be seen from the result of chapter 3, or directly from Green's theorem.) Next, if $\mathrm{Re}\lambda_0=0$, then $V(P)\equiv 0$ (use lemmas 8.5 and 8.6). Suppose $\lambda_0{}^2$ is not an eigenvalue for the interior Dirichlet problem. Because $V(P)$ is continuous in the entire space, $V(P)\equiv 0$ on S, $(\lambda_0{}^2-\Delta)V=0$ the interior of S, and $V(P)$ and its first derivative are continuous in S including its boundary. Obviously, $V\in\mathscr{D}_{L^2}{}^1(\Omega')$ where Ω is an interior domain, so that $V(P)\equiv 0$ in the interior of S. On the other hand, from (8.20) and (8.22) of section 3,

$$\left(\frac{\mathrm{d}}{\mathrm{d}n_-}-\frac{\mathrm{d}}{\mathrm{d}n_+}\right)V(p)=2\psi_0(p)\,.$$

Hence, $\psi_0(p)\equiv 0$, which is a contradiction.

Q.E.D.

We apply the same argument to (8.21) and (8.22). We then have the following:

THEOREM 8.3′. Let $\mathrm{Re}\lambda\geqslant 0$ in (8.21) and (8.22). Let $\lambda=\lambda_0{}'$ be an eigenvalue of $-K(p,q\mid\lambda)$. In this case, $\lambda_0{}'^2$ is an eigenvalue for the interior Neumann problem with respect to Δ.

Let us consider the eigenvalues for the interior Dirichlet problem with respect to Δ,

$$-\lambda_0>-\lambda_1>-\lambda_2>\cdots\to-\infty \quad (\lambda_0>0). \tag{8.43}$$

From the previous theorem and by lemma 8.3, we have the following:

PROPOSITION 8.3. Let λ_0 be such that $\mathrm{Re}\lambda_0\geqslant 0$ and $\lambda_0\notin\{\pm\mathrm{i}\sqrt{\lambda_\nu}\}_{\nu=1,2,\ldots}$.

† Since the density function $\varphi_0(p)$ has Hölder continuity, the first derivative of $V(P)$ has a uniform Hölder continuity in the neighbourhood of S.

In this case, the integral equations (8.19) and (8.20) are uniquely soluble, and in the neighbourhood of λ_0, we have representations

$$\varphi(p;\lambda) = -W_-(p) - \int R(p,q\,|\,\lambda) W_-(q)\,\mathrm{d}q \tag{8.44}$$

$$\psi(p;\lambda) = -\frac{\mathrm{d}}{\mathrm{d}n_+} V(p) - \int R(q,p\,|\,\lambda) \frac{\mathrm{d}}{\mathrm{d}n_+} V(q)\,\mathrm{d}q\,. \tag{8.45}$$

In general, we can express compensating functions by using these representations, but for the Dirichlet problem, obtaining expressions is not simple; we have to use some sharper results for double layer potentials. For this purpose, we first assume that the smoothness of the surface S is of $C^{2+\sigma}$-class ($\sigma>0$). That is S is representable as $f(x)=0$ at each point, and such that $f_x \neq 0, f \in C^{2+\sigma}$.

LEMMA 8.7. Let S be of $C^{2+\sigma}$-class ($\sigma>0$). In this case, the solution $\varphi(p)$ of (8.41) is of $C^{1+\sigma'}$-class ($\sigma'>0$).

LEMMA 8.7′. Let S be of $C^{2+\sigma}$-class. Let $\varphi \in C^{1+\sigma'}$. If the value of

$$W(P) = \int \frac{\mathrm{d}}{\mathrm{d}n_q}\left(\frac{\mathrm{e}^{\lambda|P-q|}}{|P-q|}\right)\varphi(q)\,\mathrm{d}q$$

(λ is an arbitrary complex parameter) is defined as a limit within the domain of S, then, both in the interior and exterior domains, it has a uniform Hölder continuity† in the neighbourhood of S. The same statement is true for its first derivative.

Note. As we have seen in theorem 8.3, it is true to say that, for a single layer potential, Hölder's continuity of $\varphi(p)$ is a sufficient condition for a similar result to the previous lemma. For example, see Gunther's paper (*loc. cit.*).

We consider the eigenfunction for $\lambda_0 = \pm \mathrm{i}\sqrt{\lambda_\nu}$ corresponding to (8.43). Let us fix $\lambda_0 = \mathrm{i}\sqrt{\lambda_\nu}$.

PROPOSITION 8.4. Let $\{\varphi_i(p)\}_{i=1,\ldots,n}$ $\{\psi_i(p)\}_{i=1,\ldots,n}$ be linearly indepen-

† For example, see Gunther: *La théorie du potentiel* (1934), p. 76.

dent solutions for $\lambda_0 = i\sqrt{\lambda_\nu}$ in (8.41) and (8.42) respectively. In this case, if we write

$$\Phi_i(P) = \int \varphi_i(q) \frac{d}{dn_q} E(P-q \mid \lambda_0)\, dq \tag{8.46}$$

$$\Psi_i(P) = \int \psi_i(q) E(P-q \mid \lambda_0)\, dq, \tag{8.47}$$

the following statements are true.

(1) In the exterior domain Ω of S, $\Psi_i(P) \equiv 0$ $(i = 1, \dots, n)$.

(2) In the interior domain Ω' of S, $\{\Phi_i(P)\}$, $\{\Psi_i(P)\}$ are linearly independent eigenfunctions for the eigenvalue $-\lambda_\nu (\equiv \lambda_0^2)$ corresponding to the interior Dirichlet problem of Δ.

Proof. The proof of (1) is exactly the same as that of the first part of theorem 8.3. Therefore, we prove only (2). As for $\Psi_i(P)$ we proceed in exactly the same way as in theorem 8.3. Therefore, we consider $\Phi_i(P)$. Let us write one of $\Phi_i(P)$ as $\Phi(P)$. From the previous lemma, we see that on Ω $\Phi(P)$ and its first derivative are continuous, and $\Phi(P) \equiv 0$ on S. So that it is enough to show $\Phi(P) \not\equiv 0$. If $\Phi(P) \equiv 0$, then $d\Phi(p)/dn_- \equiv 0$. As is well known† $d\Phi(p)/dn_+ = d\Phi(p)/dn_-$, hence, $d\Phi(p)/dn_+ \equiv 0$.

On the other hand, lemma 8.5 is true. (Consider the case of a double layer potential. The reader must justify this.) Hence, $\Phi(P) \equiv 0 (P \in \Omega)$. By (8.19) and (8.21), we have $\Phi_+(p) - \Phi_-(p) = 2\varphi(p) \equiv 0$, this is a contradiction.

Q.E.D.

Note.‡ This proposition is true for $\lambda_0 = -i\sqrt{\lambda_\nu}$.

From the observation we conclude that the value of a compensating function $K_c(P, Q \mid \lambda)$ can be obtained without much complication.

Now, let $\mathrm{Re}\,\lambda > 0$. We consider the exterior Neumann problem of $(\Delta - \lambda^2)$. Our boundary condition is

$$\frac{d}{dn_+} K_c(p, Q \mid \lambda) = \frac{d}{dn}\left[\frac{1}{4\pi} \frac{e^{-\lambda|p-Q|}}{|p-Q|}\right] = \frac{1}{2} \frac{d}{dn} E(p-Q \mid \lambda),$$

where d/dn is a derivative taken along an outer normal at S. From (8.45) of

† For example, see O. D. Kellogg: *Foundations of potential theory*, Springer, 1929, p. 170.

‡ We note that if we replace $e^{ik|x-s|}$ by $e^{-ik|x-s|}$ in the integral defining $u(x)$ in lemma 8.5 we have the same result.

lemma 8.3, we have the desired compensating function

$$K_c(P, Q \mid \lambda) = -\frac{1}{2}\int E(P-q \mid \lambda)\frac{\mathrm{d}}{\mathrm{d}n_q}E(q-Q \mid \lambda)\,\mathrm{d}q$$
$$-\frac{1}{2}\int E(P-q \mid \lambda)\,\mathrm{d}q \int R(r, q \mid \lambda)\frac{\mathrm{d}}{\mathrm{d}n_r}E(r-Q \mid \lambda)\,\mathrm{d}r\,.$$

In fact, if we put

$$v(q; Q) = \int R(r, q \mid \lambda)\frac{\mathrm{d}}{\mathrm{d}n_r}E(r-Q \mid \lambda)\,\mathrm{d}r\,,$$

then from the form of R (see (8.40)) $v(q; Q)$ has Hölder's continuity on the surface S as a function of q. Therefore, $K_c(P, Q \mid \lambda)$ and its first derivative are continuous as a function of P in Ω including its boundary, so that $K_c(P, Q \mid \lambda)$ satisfies a necessary condition for being a compensating function. Also, $K_c(P, Q \mid \lambda)$, its first derivative, and $-\mathrm{e}^{-\lambda|P-Q|}/4\pi|P-Q|$ tend to 0 with exponential order as $|P| \to +\infty$. So that, by theorem 8.2, $K_c(P, Q \mid \lambda)$ is in fact a compensating function, which is continuous with respect to (P, Q, λ) and a regular function of λ in $\mathrm{Re}\,\lambda > 0$.

Next, we shall obtain a Green's function for the interior Dirichlet problem of $(\Delta - \lambda^2)$. The boundary condition is

$$K_c(p, Q \mid \lambda) = \frac{1}{4\pi}\frac{\mathrm{e}^{-\lambda|p-Q|}}{|p-Q|} = \frac{1}{2}E(p-Q \mid \lambda) \quad (p \in S).$$

So that, by (8.44), the compensating function is given by

$$K_c(P, Q \mid \lambda) = -\frac{1}{2}\int \frac{\mathrm{d}}{\mathrm{d}n_q}E(P-q \mid \lambda)E(q-Q \mid \lambda)\,\mathrm{d}q$$
$$-\frac{1}{2}\int \frac{\mathrm{d}}{\mathrm{d}n_q}E(P-q \mid \lambda)\,\mathrm{d}q \int R(q, r \mid \lambda)E(r-Q \mid \lambda)\,\mathrm{d}r \qquad (8.49)$$

which is valid for $\mathrm{Re}\,\lambda \geqslant 0$, $\lambda \notin \{\pm \mathrm{i}\sqrt{\lambda_v}\}$. Let us show that K_c satisfies the conditions of theorem 8.2. If S is of $C^{2+\sigma}$-class, then, by a similar argument to that of lemma 8.7, we see that

$$v(q) = \int R(q, r \mid \lambda)E(r-Q \mid \lambda)\,\mathrm{d}r \in C^{1+\sigma'} \; (\sigma' > 0)\,.$$

By lemma 8.7′, as functions of P, $K_c(P, Q \mid \lambda)$ and its first derivative are continuous in Ω including its boundary. Therefore, by theorem 8.2, K_c is a compensating function for the interior Dirichlet problem. On the other hand, this Green's function for the interior Dirichlet problem has been already obtained in chapter 3. The result was (3.35) in theorem 3.6, i.e. if $\lambda^2 \notin \{-\lambda_\nu\}$, then

$$G_{\lambda^2} \cdot f = \sum_{\nu=1}^{\infty} \frac{(f(x), u_\nu(x))}{\lambda^2 + \lambda_\nu} u_\nu(x) \tag{8.50}$$

where $u_\nu(x)$ is an eigenfunction for the eigenvalue $-\lambda_\nu$ of the interior Dirichlet problem.

The right hand series of (8.50) is convergent in $L^2(\Omega')$ where Ω' is the interior domain of S, the series has the value $L^2(\Omega')$. Also, it has a simple pole at $\lambda = \{\pm i\sqrt{\lambda_\nu}\}$. From this, we have the following important fact about the resolvent kernel $R(p, q \mid \lambda)$.

PROPOSITION 8.5. $R(p, q \mid \lambda)$ has a simple pole at $\lambda_0 = \pm i\sqrt{\lambda_\nu}$ $(\nu = 1, 2, \ldots)$.

Proof. From (8.50) it is clear that $R(p, q \mid \lambda)$ has a pole. Now, by propositions 8.2, and 8.1, we have

$$R(p, q \mid \lambda) = \frac{\varphi_1(p)\,\psi_1(q) + \cdots + \varphi_n(p)\,\psi_n(q)}{(\lambda - \lambda_0)^m} + \cdots$$

in the neighbourhood of $\lambda = \lambda_0$. By (8.49), we have

$$K_c(P, Q \mid \lambda) = -\frac{1}{2} \frac{\Phi_1(P)\,\Psi_1(Q) + \cdots + \Phi_n(P)\,\Psi_n(Q)}{(\lambda - \lambda_0)^m} + \frac{\kappa(P, Q \mid \lambda)}{(\lambda - \lambda_0)^{m-1}} \tag{8.51}$$

in the neighbourhood of $\lambda = \lambda_0$. Therefore, for an arbitrary $f \in \mathscr{D}(\Omega')$,

$$-(G_{\lambda^2} f)(P) = -\frac{1}{4\pi} \frac{e^{-\lambda P}}{|P|} * f(P) + \int K_c(P, Q \mid \lambda) f(Q)\, dQ.$$

But $\int \kappa(P, Q \mid \lambda) f(Q)\, dQ$ is a regular function of λ in the neighbourhood $\lambda = \lambda_0$. From this $m = 1$. In fact, if $m > 1$, then, comparing this fact with (8.50), we have

$$\sum \Phi_i(P) \int \Psi_i(Q) f(Q)\, dQ \equiv 0 \quad (P \in \Omega') \quad (\text{cf. } (8.50)).$$

But $\Phi_i(P)$ are linearly independent (see proposition 8.4), so that

$$\int \Psi_i(Q) f(Q)\, dQ = 0 \quad (i = 1, 2, \ldots, n).$$

f is arbitrary, hence, $\psi_i(Q) \equiv 0$, which is a contradiction.

Q.E.D.

Recall the fact that

$$G(P, Q \mid \lambda) = -\frac{e^{-\lambda|P-Q|}}{4\pi|P-Q|} + K_c(P, Q \mid \lambda) \tag{8.52}$$

was a Green's function of the operator $(\Delta - \lambda^2)$ for the exterior Neumann problem in $\mathrm{Re}\,\lambda > 0$, where $K_c(P, Q \mid \lambda)$ was defined in (8.48) $K_c(P, Q \mid \lambda)$ is a meromorphic function of λ because $R(p, q \mid \lambda)$ is. Therefore, if $\lambda_0 \notin \{i\sqrt{\lambda_\nu}\}$, then, by propositions 8.3 and 8.5, K_c is regular in the neighbourhood of λ_0. On the other hand, we can see that K_c is also regular in the neighbourhood of $\lambda_0 \in \{i\sqrt{\lambda_\nu}\}$.

THEOREM 8.4. $K_c(P, Q \mid \lambda)$ is regular even in the neighbourhood of $\lambda_0 = \pm i\sqrt{\lambda_\nu}$ with respect to λ.

Proof. By the previous proposition, in the neighbourhood of $\lambda = \lambda_0$, we can write

$$R(p, q \mid \lambda) = \frac{\psi_1(p)\,\psi_1(q) + \cdots + \varphi_n(p)\,\psi_n(q)}{\lambda - \lambda_0} + B(p, q \mid \lambda),$$

where $B(p, q \mid \lambda)$ is continuous and regular in the neighbourhood of $\lambda = \lambda_0$ with respect to λ. By (8.52), we have

$$K_c(P, Q \mid \lambda) = -\frac{1}{2}\,\frac{\Phi_1(P)\,\Psi_1(Q) + \cdots + \Phi_n(P)\,\Psi_n(Q)}{\lambda - \lambda_0} + G_1(P, Q \mid \lambda).$$

By proposition 8.4, in $P \in \Omega$ (the exterior domain of S) $\Psi_i(Q) \equiv 0$. Therefore, $K_c(P, Q \mid \lambda) = G_1(P, Q \mid \lambda)$.

Q.E.D.

Note. We have exactly the same result for the Green's function of $(\Delta - \lambda^2)$ in relation to the exterior Dirichlet problem. More concretely, the situation is the same for a compensating function. We sketch the argument briefly before proceeding to section 6. We start from an integral equation

$$W_+(p) = \varphi(p) + \int_S K(p, q \mid \lambda)\,\varphi(q)\,\mathrm{d}q \tag{8.21}$$

We note that we have the same result about the analytic continuation of

$$R'(p, q \mid \lambda) = K(p, q \mid \lambda) - K^{(2)}(p, q \mid \lambda) + K^{(3)}(p, q \mid \lambda) - \cdots$$
$$\cdots + (-1)^{n-1} K^{(n)}(p, q \mid \lambda) + \cdots \quad (\lambda \in D)$$

as we stated in section 4. Also, in $\mathrm{Re}\,\lambda \geqslant 0$, the pole of $R'(p, q \mid \lambda)$ is nothing but the eigenvalue λ^2 for the interior Neumann problem of Δ which includes $\lambda = 0$. Notice that $\lambda = 0$ is a double pole of $R'(p, q \mid \lambda)$, but because the coefficient of λ^{-1} is 0, theorem 8.4 is valid as it stands in this case.

We derive the representation of K_c. To do this, we first note that the solution of (8.21) is

$$\varphi(p) = W_+(p) + \int_S R'(p, q \mid \lambda)\, W_+(q)\, \mathrm{d}q\,.$$

From this, the compensating function of the Green's function of $(\Delta - \lambda^2)$ for the exterior Dirichlet problem is given by

$$\begin{aligned} K_c'(P, Q \mid \lambda) = {} & \frac{1}{2} \int \frac{\mathrm{d}}{\mathrm{d}n_q} E(P-q \mid \lambda)\, E(q-Q \mid \lambda)\, \mathrm{d}q \\ & + \frac{1}{2} \int \frac{\mathrm{d}}{\mathrm{d}n_q} E(P-q \mid \lambda)\, \mathrm{d}q \int R'(q, r \mid \lambda)\, E(r-Q \mid \lambda)\, \mathrm{d}r \end{aligned} \tag{8.53}$$

where $\mathrm{Re}\,\lambda > 0$. Similarly, for the interior Neumann problem it is given by

$$\begin{aligned} K_c'(P, Q \mid \lambda) = {} & \frac{1}{2} \int E(P-q \mid \lambda) \frac{\mathrm{d}}{\mathrm{d}n_q} E(q-Q \mid \lambda)\, \mathrm{d}q \\ & + \frac{1}{2} \int E(P-q \mid \lambda) \int R'(r, q \mid \lambda) \frac{\mathrm{d}}{\mathrm{d}n_r} E(r-Q \mid \lambda)\, \mathrm{d}r\,. \end{aligned} \tag{8.54}$$

Summing up we have:

THEOREM 8.5. If $\lambda \notin (-\infty, 0]$, the Green's function $G(P, Q; \lambda)$ for the exterior Neumann problem of the operator $(\Delta - \lambda)$ is given by

$$G(P, Q; \lambda) = -\frac{\exp(-\lambda^{\frac{1}{2}}|P-Q|)}{4\pi|P-Q|} + K_c(P, Q \mid \sqrt{\lambda}), \tag{8.55}$$

where K_c is defined by (8.48) and (8.53).† In this case $G(P, Q; \lambda)$ is a regular function of λ such that the following are true.

(1) $G(P, Q; \lambda)$ is continuous with respect to (P, Q, λ) if $P \neq Q$. Also, $G(P, Q; \lambda)$ can be extended with analytical continuations from the upper

† We determine the value of $\sqrt{\lambda}$ such that if λ is positive, then $\sqrt{\lambda}$ is positive.

half-plane to the lower half-plane as a function of λ beyond the negative real axis, and vice versa.

(2) As a function of λ after an analytic continuation, G becomes a two-valued function having at most poles and only a single branch point $\lambda=0$.

From the theorem, the Green's function in the wider sense (see below) can be defined for the exterior problem of $(\Delta+k)$ $(k\geqslant 0)$.

DEFINITION 8.3 *(Green's function in the wider sense).* In (8.55) we call

$$G(P, Q; -k\pm i0) = -\frac{\exp(\pm ik^{\frac{1}{2}}|P-Q|)}{4\pi|P-Q|} + K_c(P, Q \mid \pm i\sqrt{k}) \quad (8.56)$$

($\pm$ signs taken respectively) the *Green's function in the wider sense*† for the exterior Neumann problem (or the exterior Dirichlet problem), with respect to the operator $(\Delta+k)$ $(k\geqslant 0)$.

Note. According to the definition, if $k>0$, there exist two Green's functions depending on the way the signs are taken. This seems to be irrational, but, physically, it has meaning. Mathematically speaking, in this case, the Green's operator (defined in chapter 3) does not exist. In the next section we shall obtain some properties of the Green's function in the wider sense and the Green's function for the interior problem in a classical sense.

6 *Properties of Green's functions*

The results of the section can be generally applied to Green's functions in the interior or exterior domain for Dirichlet or Neumann condition. As is well known,

$$G(P, Q; \lambda) = G(Q, P; \lambda) \quad (8.57)$$

(symmetricity), and

$$G(P, Q; \bar{\lambda}) = \overline{G(P, Q; \lambda)} \quad (8.58)$$

where $\lambda \notin (-\infty, 0)$. For the ordinary Green's function, (8.57) follows from the fact that the Green's operator for $(\lambda I-\Delta)$ satisfies ${}^tG_\lambda = G_\lambda$ (see chapter 3). For the Green's function in the wider sense, it follows from a limiting process. (8.58) is obvious. From (8.58), it follows that $G(P, Q; -k+i0) = \overline{G(P, Q; -k-i0)}$.

† K_c should be understood as in the sense of (8.48) and (8.53).

Let us estimate the Green's functions. (This is valid for Green's functions in a wider sense.)

THEOREM 8.6.

$$|G(P, Q; \lambda)| \leqslant C|P-Q|^{-1}, \quad |G_i(P, Q; \lambda)| \leqslant C|P-Q|^{-2} \tag{8.59}$$

where G_i is Gs first derivative with respect to P (or Q), and C is dependent on λ but the same C satisfies the exterior and interior domains at the same time. For a Green's function in the wider sense, if P and Q lie within a bounded set, C can be taken as a constant.

Proof.† *(First step)* First, we shall state some general properties from the standpoint of potential theory.

$$\int \frac{|\cos(\nu_q, \overrightarrow{qP})|}{|P-q|^2} \, dq \leqslant \text{const}$$

where ν_q is a unit norm at a point q on S.

(2) Let

$$|f(P, q)| \leqslant \text{const}\,|P-q|^{-2}, \quad \int |f(P, q)| \, dq \leqslant \text{const},$$
$$|g(q, Q)| \leqslant \text{const}\,|q-Q|^{-1},$$

then for

$$F(P, Q) = \int f(P, q)\, g(q, Q)\, dq,$$

we have

$$|F(P, Q)| \leqslant \frac{\text{const}}{R(P, Q)}, \tag{8.60}$$

where $R(P, Q)$ is the minimum value of $|P-q|+|Q-q|$ when q runs through the surface S. We prove these facts.

Let $R(P, Q)=\varepsilon$. We separate the defining integral of $F(P, Q)$ into

$$[\text{I}]: |Q-q| \leqslant \tfrac{1}{2}\varepsilon, \quad [\text{II}]: |Q-q| > \tfrac{1}{2}\varepsilon.$$

† We follow the proof found in H. Weyl: 'Das asymptotische Verteilungsgesetz der Eigenschwingungen eines beliebig gestalten elastischen Körpers', *Rendiconti del Circolo Matematico di Palermo* (1915).

For [I], we have $|P-q|\geqslant\frac{1}{2}\varepsilon$, so that

$$\left|\int_{[I]} f(P,q)\,g(q,Q)\,dq\right|\leqslant\frac{\text{const}}{\varepsilon^2}\int_{[I]}\frac{dq}{|q-Q|}\leqslant\frac{\text{const}}{\varepsilon}.$$

In fact, in general,

$$\int_{|q-Q|\leqslant\varepsilon}\frac{dq}{|q-Q|}\leqslant\text{const}\cdot\varepsilon$$

where the constant can be taken as independent of Q.

Next, for [II], we have $|g(q,Q)|\leqslant\text{const}\cdot\varepsilon^{-1}$, so that

$$\left|\int_{[II]}\ldots dq\right|\leqslant\text{const}\cdot\varepsilon^{-1}\int|f(P,q)|\,dq\leqslant\text{const}\cdot\varepsilon^{-1}.$$

(3) In (2) we fix f. For g, if we assume

$$|g(q,Q)|\leqslant\text{const}\,|q-Q|^{-2},\qquad\int|g(q,Q)|\,dq\leqslant\text{const},$$

then

$$|F(P,Q)|\leqslant\frac{\text{const}}{R(P,Q)^2}.\tag{8.61}$$

The proof is just as in the case before.

(4) If

$$|f(P,q)|\leqslant\text{const}\,|P-q|^{-2},\qquad\int|f(P,q)|\,dq\leqslant\text{const}$$

and

$$|g(q,r)|\leqslant\text{const}\,|q-r|^{-2+\alpha}\qquad(0<\alpha\leqslant1),$$

we have

$$|F(P,q)|\leqslant\frac{\text{const}}{|P-q|^{2-\alpha}}.\tag{8.62}$$

To prove this we proceed as follows. Let us take an arbitrary (P, q). Let us say [I] for the range of r satisfying $|q-r|\leqslant\frac{1}{2}|P-q|$ otherwise say [II]. Obviously

$$\int_{|q-r|\leqslant\varepsilon}|g(r,q)|\,dr\leqslant\text{const}\cdot\varepsilon^{\alpha},$$

so that, in [I],

$$\left|\int_{[I]} f(P,r)\,g(r,q)\,dr\right|\leqslant\frac{\text{const}}{|P-q|^2}\int_{[I]}|g(r,q)|\,dr\leqslant\frac{\text{const}}{|P-q|^{2-\alpha}}$$

(consider $|P-r|\geqslant\frac{1}{2}|P-q|$). Also, in [II],

$$\left|\int_{[II]}\ldots dr\right|\leqslant\frac{\text{const}}{|P-q|^{2-\alpha}}\int|f(P,r)|\,dr\leqslant\frac{\text{const}}{|P-q|^{2-\alpha}}$$

(consider $|q-r|>\frac{1}{2}|P-q|$).

(Second step). Now, the estimation of the compensating function $K_c(P, Q\mid\sqrt{\lambda})$ is enough to obtain our desired result. We consider only a compensating function for the interior Dirichlet problem defined by

$$\begin{aligned}K_c(P, Q\mid\lambda)=&-\frac{1}{2}\int\frac{d}{dn_q}E(P-q\mid\lambda)\,E(q-Q\mid\lambda)\,dq\\&-\frac{1}{2}\int\frac{d}{dn_q}E(P-q\mid\lambda)\,dq\int R(q,r\mid\lambda)\,E(r-Q\mid\lambda)\,dr\end{aligned}\tag{8.49}$$

because other cases are exactly the same as this. For the estimation of the first term of the right hand side we recall the property (2) of the first step; $R(P, Q)\geqslant|P-Q|$, so that this gives no problem.

For the estimation of the last half of the second term, $\int R(q,r\mid\lambda)E(r-Q\mid\lambda)\,dr$, we apply (4) of the first step together with the representation (8.40) of $R(p,q\mid\lambda)$. We see that the absolute value of this part does not exceed $\text{const}\,|q-Q|^{-1}$. Then, we apply (8.60) to this result. We now have $|K_c(P, Q\mid\lambda|\leqslant\text{const}\cdot R(P, Q)^{-1}$. Finally, for the first derivative of K_c, we consider $K_c(P, Q\mid\lambda)=K_c(Q, P\mid\lambda)$. We see that it is enough to estimate the derivative with respect to Q. To do this, we combine results (3) and (4) of the first step.

Q.E.D.

THEOREM 8.7. Let $f(Q)$ be continuous and bounded for $Q\in\Omega$. Let us assume that the support of $f(P)$ is bounded if G is a Green's function in the wider sense. In this case,

$$u(P;\lambda)=\int G(P, Q;\lambda)\,f(Q)\,dQ\tag{8.63}$$

satisfies $(\Delta-\lambda)u=f$ as a distribution defined in Ω, u and its first derivative are continuous in Ω including its boundary and satisfy the given boundary condition. If $f(P)$ has Hölder's continuity, then $u\in C^2$ and it is a genuine solution.

Proof. Let us write $f_\varepsilon(Q)$ for the function which is obtained from $f(Q)$ by replacing the value of $f(Q)$ with 0, at the points within a distance less than ε from the boundary of Ω. In this case, it is obvious that

$$u_\varepsilon(P;\lambda)=\int_\Omega G(P,Q;\lambda)\,f_\varepsilon(Q)\,dQ$$

satisfies the condition of the theorem, and as $\varepsilon\to 0$, $u_\varepsilon(P;\lambda)$ and its first derivative are uniformly convergent in a compact set of $\bar{\Omega}$ (if $\bar{\Omega}$ is bounded, $\bar{\Omega}$). This is evident by the estimation of the Green's function in the previous theorem. The last half of the theorem is a well-known fact, so we omit the proof.

Q.E.D.

Note. Let us consider (8.63) for the Green's function defined in (8.56). We assume that the support of $f(Q)$ is bounded. In this case, if the Green's function in the wider sense defined in definition 8.5 takes the + sign, then it satisfies Sommerfeld's radiation condition (see lemma 8.5).† Also, if $k=0$ in (8.56), then $u(P)$ satisfies lemma 8.6.

Conversely, the solution $u\in C^2(\Omega)\cap C^1(\bar{\Omega})$ of the equation $(\Delta+k)u(P)=f(P)$ satisfying the (inward) radiation condition as $|P|\to+\infty$ and the boundary condition on S, can be represented by

$$u(P)=\int_\Omega G(P,Q;-k\pm i0)\,f(Q)\,dQ,$$

(the signs should be taken respectively) where $f(P)$ is assumed to be continuous, bounded with the bounded support.

The estimation of the iterated kernel of $K(p,q\mid\lambda)$.

In section 4, the description of Fredholm's theory, we assumed that for a discontinuous $K(p,q\mid\lambda)$, (8.25) is satisfied for $n\geqslant 3$ (see section 4, resolvent kernel). Here we give the proof.

(1) $|K^{(2)}(p,q\mid\lambda)|\leqslant C_1+C_2\log|p-q|^{-1}$ where C_1 and C_2 are independent of λ, for example, for $|\mathrm{Im}\,\lambda|\leqslant 1$, $\mathrm{Re}\,\lambda>0$.

† If it takes the − sign, the inward radiation condition is satisfied $u(x)=O(|x|^{-1})$, $du/dn+iku=o(|x|^{-1})$.

Proof of (1). We write

$$2\pi K(p, q \mid r) = \frac{e^{-\lambda|p-q|}}{|p-q|^2} \cos(n_q, \overrightarrow{qp}) + \lambda \frac{e^{-\lambda|p-q|}}{|p-q|} \cos(n_q, \overrightarrow{qp})$$
$$\equiv K_0(p, q \mid \lambda) + \lambda K_1(p, q \mid \lambda).$$

Obviously,

$$|K_0(p, q \mid \lambda)| \leqslant C_0 \frac{e^{-\mathrm{Re}\,\lambda|p-q|}}{|p-q|}, \qquad |K_1(p, q \mid \lambda)| \leqslant C_1 e^{-\mathrm{Re}\,\lambda|p-q|} \tag{8.64}$$

and

$$K^{(2)}(p, q \mid \lambda) = K_0{}^{(2)}(p, q \mid \lambda) + \lambda^2 K_1{}^{(2)}(p, q \mid \lambda) +$$
$$+ \lambda \int K_0(p, r \mid \lambda)\, K_1(r, q \mid \lambda)\, dr + \lambda \int K_1(p, r \mid \lambda)\, K_0(r, q \mid \lambda)\, dr.$$

The case of $K_0{}^{(2)}$ is omitted because it can be found elsewhere (for example, see O. D. Kellogg: *Foundations of Potential Theory*, Springer, Berlin, 1929, p. 303).

The problem we now have is the estimation of the terms after the second which appear in the kernel when $|\lambda|$ is sufficiently large. To do this it is enough to assume λ is real and sufficiently large. We see that

$$|K_1{}^{(2)}(p, q \mid \lambda)| \leqslant \text{const} \int_{\mathbf{R}^2} e^{-\frac{1}{2}\lambda|p-r|}\, dr \leqslant \text{const} \cdot \lambda^{-2}.$$

$$|K_{01}(p, q \mid \lambda)| = \left|\int K_0(p, r \mid \lambda)\, K_1(r, q \mid \lambda)\, dr\right|$$
$$\leqslant \text{const} \int_{\mathbf{R}^2} \frac{e^{-\frac{1}{2}\lambda|p-r|}}{|p-r|}\, dr \leqslant \text{const} \cdot \lambda^{-1}$$

follows from (8.64) almost immediately. Therefore, (1) is true.

(2)
$$|K^{(3)}(p, q \mid \lambda)| \leqslant \theta(N) \qquad (|\mathrm{Im}\,\lambda| \leqslant 1,\ \mathrm{Re}\,\lambda \geqslant N), \tag{8.65}$$

where θ can be taken as small as possible if N is taken as sufficiently large.

Proof of (2). From result (1) and (8.64), we have

$$|K^{(3)}(p, q \mid \lambda)| \leqslant \text{const} \int \frac{e^{-\lambda|p-r|}}{|p-r|} \{c_1 + c_2 \log|r-q|^{-1}\}\, dr$$
$$+ \lambda\ \text{const} \int e^{-\lambda|p-r|} \{c_1 + c_2 \log|r-q|^{-1}\}\, dr.$$

Using the Schwartz inequality, the first term of the integral is estimated as

$$\leqslant \text{const}\left(\int \frac{e^{-2\lambda|p-r|}}{|p-r|}\,dr\right)^{\frac{1}{2}}\left(\int \frac{1}{|p-r|}\{c_1+c_2\log|r-q|^{-1}\}\,dr\right)^{\frac{1}{2}}.$$

The second factor of this expression can be estimated by a constant which is independent of (p, q), and the first factor tends to 0 as $\lambda \to +\infty$.

Now we must estimate the second term of the integral. To do this, we first take $\varepsilon(>0)$ sufficiently small, then we consider the integral within the range of $|p-r|\leqslant\varepsilon$.

From the Schwartz inequality, we see that the second term

$$\leqslant \lambda\Big(\int_{|p-r|\leqslant\varepsilon} e^{-\lambda|p-r|}dr\Big)^{\frac{1}{2}}\Big(\int_{|p-r|\leqslant\varepsilon}\{c_1+c_2\log|r-q|^{-1}\}^2\,dr\Big)^{\frac{1}{2}},$$

where the first factor is estimated by a constant times a multiple of λ^{-1}, and the second factor can be taken as small as possible, together with ε independent in relation to (p, q). Therefore, we can choose ε such that the estimation of the second term of the integral (the above expression on the right hand side of $\leqslant$) is smaller than θ. Next, for $|p-r|>\varepsilon$, the second integral is

$$\leqslant \lambda e^{-\varepsilon r}\int\{c_1+c_2\log|r-q|^{-1}\}\,dr$$

and tends to 0 as $\lambda \to +\infty$. Finally, we note that $K^{(3)}(p, q \mid \lambda)$ is continuous with respect to (p, q, λ), and regular in the entire space with respect to λ.

(3) Under the same conditions as (2) let $\psi(p)=\int K(p, q \mid \lambda)\varphi(q)\,dq$, and let $\varphi(p)$ be continuous. Then,

$$\max_{j\in S}|\psi(p)|\leqslant C\max_{j\in S}|\varphi(p)|.$$

But C can be taken as small as possible if we take N sufficiently large in (2). The proof of this fact is left to the reader as an exercise.

Q.E.D.

7 *The solution of a wave equation in the exterior domain*

From the property of a Green's function, we are able to observe the behaviour of the solution $u(x, t)$ of the initial boundary value problem for

$$\frac{\partial^2}{\partial t^2}u-\Delta u=0. \tag{8.66}$$

Let S be a compact surface of the $C^{2+\sigma}$-class ($\sigma>0$) of R^3. Let the exterior domain of S be Ω. Now, we consider the solution $u(x, t)$ of the wave equation in Ω satisfying the conditions:

$$\left.\begin{array}{l}\text{(1) The initial condition: } u(x, 0)=f_0(x),\ \partial u(x, 0)/\partial t=f_1(x).\\ \text{(2) The boundary condition:}\\ u(x)=0 \text{ (Dirichlet condition) on } S, \text{ or}\\ \qquad\qquad du/dn=0 \text{ (Neumann condition).}\end{array}\right\} \tag{8.67}$$

Let us consider the problem within the frame of L^2-space. Of course, this is an intrinsic 'frame' of necessity. The reason is that the treatment of a hyperbolic equation outside of L^2-space is essentially impossible. First, we take the domain of the definition of the operator $-\Delta$ as the set of the elements of $C^2(\Omega)\cap C^1(\bar{\Omega})$ such that they satisfy (8.67), and have a bounded support. It is obvious that $-\Delta$ is a symmetric operator, and can be extended to a unique self-adjoint operator H. The domain of the definition of H is

(1) $\mathscr{D}(H)=\{u;\ u\in\mathscr{D}_{L^2}{}^1(\Omega)\cap\mathscr{E}_{L^2}{}^2(\Omega)\}$ if it satisfies Dirichlet's condition,

(2) $\mathscr{D}(H)=\{u;\ u\in\mathscr{E}_{L^2}{}^2(\Omega),\ du/dn=0\}$ if it satisfies Neumann's condition,

where du/dn should be understood as a trace on S (see section 16, chapter 3). The hypothesis there is slightly different from this case. Nevertheless, the argument is valid. The reason is that the regularity of the solution in a neighbourhood of the boundary in $C^1(\bar{\Omega})$ is guaranteed by theorem 8.7.)

Von Neumann's theorem (we do not treat his theory in this book) says that, for the self-adjoint operator H, there exists a unique spectral decomposition $H=\int \lambda\, dE(\lambda)$. Let $f_0, f_1\in\mathscr{D}(H)$. If we write

$$u(x, t)=\int_0^{+\infty} \cos\lambda^{\frac{1}{2}}t\, dE(\lambda) f_0+\int_0^{+\infty} \frac{\sin\lambda^{\frac{1}{2}}t}{\lambda^{\frac{1}{2}}}\, dE(\lambda) f_1, \tag{8.68}$$

then we see that $u(x, t)$ is in fact the solution of a wave equation in Ω in the following sense.

(1) $t\to u(x, t)\in\mathscr{D}(H)$ is continuous. More precisely, we regard $\mathscr{D}(H)$ as a (closed) subspace of $\mathscr{E}_{L^2}{}^2(\Omega)$. Furthermore, $\partial u/\partial t$, $\partial^2 u/\partial t^2$ are continuous functions of $L^2(\Omega)$, i.e. $u\in\mathscr{E}_t{}^2(L^2)$.

(2) As $t\to 0$, we have

$$u(x, t)\to f_0(x) \quad\text{in}\quad \mathscr{D}(H), \qquad \frac{\partial}{\partial t}u(x, t)\to f_0(x) \quad\text{in}\quad L^2(\Omega).$$

(3) $u(x, t)$ satisfies a wave equation in Ω.

To see the uniqueness of the $u(x, t)$, we proceed as follows. Let $w(x, t)$

satisfy (1), (3) and be such that

$$w(x,0)=\frac{\partial}{\partial t}w(x,0)=0.$$

Also, we consider a mollifier $\varphi_\delta(t)$ such that $w_\delta(t)=\varphi_\delta(t)\underset{t}{*}w(t)$. Obviously,

$$\Box\, w_\delta=0,\quad w_\delta,\ \frac{\partial}{\partial t}w_\delta\in\mathscr{D}(H),$$

and $\partial w_\delta(t)/\partial x_i$ is continuously differentiable with respect to t in $L^2(\Omega)$. Write

$$E(t;w_\delta)=\frac{1}{2}\left\{\left\|\frac{\partial}{\partial t}w_\delta(t)\right\|^2_{L^2(\Omega)}+\sum_{i=1}^{3}\left\|\frac{\partial}{\partial x_i}w_\delta(t)\right\|^2_{L^2(\Omega)}\right\},$$

then, from our observation, we have $\mathrm{d}E(t;w_\delta)/\mathrm{d}t=0$. On the other hand, if $\delta\to+0$, then $E(t;w_\delta)\to E(t;w)$, so that

$$E(t;w)=\lim_{\delta\to 0}E(t;w_\delta)=\lim_{\delta\to 0}E(0;w_\delta)=E(0;w)=0.$$

Hence, if $E(t;w)=0$ then $w(t)=0$. Since $u(x,t)\in\mathscr{E}_{L^2}{}^2(\Omega)$, $u(x,t)$ is bounded and continuous[†] in $\bar{\Omega}$.

For simplicity, we assume that:

$$\left.\begin{array}{l}\text{for } f_0, f_1\in C^2(\bar{\Omega}),\ u(x,t)\text{ satisfies}\\ \qquad\text{(2) of (8.67), and it has a bounded support.}\end{array}\right\}\tag{8.69}$$

Our object is to prove $u(x,t)\to 0$ $(t\to+\infty)$. Carleman's proof is well known,[‡], but here we shall prove it in a slightly different way. Let us look at Carleman's argument. He proved that under the above conditions $E(\lambda)$ has a representation in terms of $\theta(x,y\mid\lambda)$, i.e. (8.68) can be written as

$$\begin{aligned}u(x,t)={}&\int_0^{+\infty}\cos\lambda^{\frac{1}{2}}t\,\mathrm{d}_\lambda\int_\Omega\theta(x,y\mid\lambda)f_0(y)\,\mathrm{d}y\\&+\int_0^{+\infty}\frac{\sin\lambda^{\frac{1}{2}}t}{\lambda^{\frac{1}{2}}}\,\mathrm{d}_\lambda\int_\Omega\theta(x,y\mid\lambda)f_1(y)\,\mathrm{d}y.\end{aligned}\tag{8.70}$$

We note that this is the representation of $L^2(\Omega)$, and is meaningful in the

† By Sobolev's lemma (three-dimensional case).

‡ See T. Carleman: *Sur les équations intégrales singulières à noyau réel et symétrique* (1923).

sense of point-wise convergence at every point. More exactly, we have

$$\int_0^{+\infty} \mathrm{d}_\lambda \left| \int_\Omega \theta(x, y \mid \lambda) f(y)\, \mathrm{d}y \right| < +\infty$$

and the convergence of this term is uniform in a compact set of Ω.

The relation between a spectral function and a Green's function

In the previous section we expressed the Green's function corresponding to $(\Delta - \lambda)$ as $G(P, Q; \lambda)$. We consider $K(P, Q; \lambda)$, a Green's function corresponding to $(\lambda + \Delta)$, which is a more suitable form for our purposes here. We see immediately

$$K(P, Q; \lambda) = G(P, Q; -\lambda) \qquad (\lambda \notin [0, \infty)) \tag{8.71}$$

For $\lambda \notin [0, \infty)$, we have

$$\begin{aligned}(\lambda + \Delta)^{-1} f &= (\lambda - H)^{-1} f = \int K(x, y; \lambda) f(y)\, \mathrm{d}y \\ &= \int_0^{+\infty} \frac{1}{\lambda - \mu}\, \mathrm{d}_\mu \theta(x, y \mid \mu) f(y).\end{aligned}$$

We may apply Stieltjes' well known formula, i.e. for a fixed $\mu (>0)$ we integrate this from $\mu + \mathrm{i}\varepsilon$, $\varepsilon > 0$, to $\mu - \mathrm{i}\varepsilon$ along the semi-circle C centred at the origin to get $2\pi\mathrm{i} \int_0^\mu \mathrm{d}_\lambda \theta(x \mid \lambda) = 2\pi\mathrm{i}\theta(x \mid \mu)$ as $\varepsilon \to 0$, where

$$\theta(x \mid \mu) = \int_\Omega \theta(x, y \mid \mu) f(y)\, \mathrm{d}y. \tag{8.72}$$

Next, we consider the left hand side. From the property of a Green's function, this can be expressed by

$$\int_0^\mu K(x, y; \lambda - \mathrm{i}0) f(y)\, \mathrm{d}y - \int_0^\mu K(x, y; \lambda + \mathrm{i}0) f(y)\, \mathrm{d}y$$

and some Green's function in the wider sense. In fact,

$$\int K(x, y; \lambda \pm \mathrm{i}0) f(y)\, \mathrm{d}y = \int G(x, y; -\lambda \mp \mathrm{i}0) f(y)\, \mathrm{d}y.$$

By theorems 8.5, 8.6, and 8.7, $\int G(x, y; \lambda) f(y)\, \mathrm{d}y$ and its first derivatives are continuous as functions of (x, λ), and they can also be extended analytically beyond the negative axis with respect to λ. Therefore, if we differentiate this with respect to μ ($\mu>0$), we have

$$\frac{\mathrm{d}}{\mathrm{d}\mu}\theta(x\mid\mu)=\frac{1}{2\pi\mathrm{i}}\int[K(x,y;\mu-\mathrm{i}0)-K(x,y;\mu+\mathrm{i}0)]f(y)\,\mathrm{d}y$$

So that for $\lambda\geqslant 0$, defining $\vartheta(x,y;\lambda)$ by

$$\vartheta(x,y;\lambda)=\frac{1}{2\pi\mathrm{i}}[K(x,y;\lambda-\mathrm{i}0)-K(x,y;\lambda+\mathrm{i}0)]$$
$$\equiv\frac{1}{2\pi\mathrm{i}}[G(x,y;-\lambda+\mathrm{i}0)-G(x,y;-\lambda-\mathrm{i}0)],^{\dagger} \tag{8.73}$$

we have

$$\frac{\mathrm{d}}{\mathrm{d}\lambda}\theta(x\mid\lambda)=\int\vartheta(x,y;\lambda)f(y)\,\mathrm{d}y, \tag{8.74}$$

From this observation, we have:

THEOREM 8.8. Let us assume that the initial condition with respect to $(f_0(x),f_1(x))$ satisfies (8.69). In this case, the solution $u(x,t)$ tends to 0 on an arbitrary compact set of Ω as $t\to+\infty$. If D is an arbitrary bounded set of R^3, then

$$E_D(t)=\int_{\Omega\cap D}\sum\left(\frac{\partial u}{\partial x_i}(x,t)\right)^2+\left(\frac{\partial u}{\partial t}(x,t)\right)^2\mathrm{d}x$$

tends to 0 as $t\to+\infty$.

Proof. We note that the essential part of this statement is clear from the Riemann–Lebesgue theorem. For simplicity, we consider

$$u(x,t)=\int_0^{+\infty}\exp(\mathrm{i}\lambda^{\frac{1}{2}}t)\,\mathrm{d}_\lambda\theta(x\mid\lambda)$$

where $\theta(x\mid\lambda)$ is defined in (8.72). Let us choose a certain $\varepsilon(>0)$ and a compact set K of Ω. Then, we take A sufficiently large in relation to ε and K such

† Therefore, if $\omega(\lambda)$, is bounded and continuous, and if $f(x)$ satisfies (8.72),

$$\int_0^\infty\omega(\lambda)\,\mathrm{d}\lambda\int\theta(x,y|\lambda)f(y)\,\mathrm{d}y=\int_0^\infty\omega(\lambda)\left(\int\vartheta(x,y|\lambda)f(y)\,\mathrm{d}y\right)\mathrm{d}\lambda\,.$$

that

$$|\int_{A}^{+\infty} \exp(i\lambda^{\frac{1}{2}}t)\, d_\lambda\theta(x \mid \lambda)| < \varepsilon \quad (x \in K)$$

irrespective of t.

Next, we can write

$$\int_{0}^{A} \exp(i\lambda^{\frac{1}{2}}t)\, d_\lambda\theta(x \mid \lambda) = \int_{0}^{A} \exp(i\lambda^{\frac{1}{2}}t)\psi(x; \lambda)\, d\lambda,$$

where

$$\psi(x; \lambda) = \int \vartheta(x; y; \lambda) f(y)\, dy. \tag{8.75}$$

Note that ψ is continuous for $\lambda \in [0, A]$. Therefore, for a fixed $x \in K$, by the Riemann–Lebesgue theorem, $\int_{0}^{A} \exp(i\lambda^{\frac{1}{2}}t)\, \psi(x; \lambda)\, d\lambda \to 0$ as $t \to +\infty$. On the other hand, if we regard $\psi(x; \lambda)$ as a family of continuous functions defined for $\lambda \in [0, A]$ when a parameter x moves around K, then the family forms a compact set. In fact, in this case, $\psi(x; \lambda)$, $\partial\psi(x; \lambda)/\partial x$: are bounded and continuous in $K \times [0, A]$. Therefore, the convergence is uniform with respect to $x \in K$.

To prove the rest of the theorem, we write

$$u(x, t) = \int_{0}^{A} \exp(i\lambda^{\frac{1}{2}}t)\, dE(\lambda)f + \int_{A}^{\infty} \exp(i\lambda^{\frac{1}{2}}t)\, dE(\lambda)f \equiv u_0(x, t) + u_1(x, t).$$

Considering $-\Delta u_i = \int_{A}^{\infty} \exp(i\lambda^{\frac{1}{2}}t)\, dE(\lambda)\, f$ and $\|\partial u_1/\partial x_i\| \leqslant C(\|\Delta u_1\| + \|u_1\|)$,† we see that

$$E_\Omega(t; u_1) \leqslant \text{const} \int_{A}^{\infty} (\lambda^2 + 1)\, d\|E(\lambda)f\|^2.$$

So that if we take A sufficiently large we can make the right hand term of the inequality $< \varepsilon$. For this reason, we consider only u_0 for fixed A. We have

$$u_0(x, t) = \int_{0}^{A} \exp(i\lambda^{\frac{1}{2}}t)\psi(x; \lambda)\, d\lambda,$$

and

$$\frac{\partial}{\partial x_i} u_0 = \int_{0}^{A} \exp(i\lambda^{\frac{1}{2}}t)\, \frac{\partial}{\partial x_i}\psi(x; \lambda)\, d\lambda.$$

For the right hand term of the last integral, we restrict the domain of x to

† Since $u_1 \in \mathscr{D}(H)$, for such u_1 (3.109) is valid.

$x \in \Omega \cap D$. Then, $\partial\psi(x; \lambda)/\partial x_i$ is an element of $L^2(\psi \cap D)$, and obviously it is continuous with respect to λ. So, taking $L^2(\Omega \cap D)$ as a value, if we consider $\partial\psi(x; \lambda)/\partial x_i$ as a continuous function of λ, then applying the Riemann–Lebesgue theorem, we have $E_D(t; u_0) \to 0$ $(t \to +\infty)$.

Q.E.D.

8 *Discrete spectrum for Schrödinger's operator*

In this section we give an elementary account of the discrete spectrum. As a concrete example we have Schrödinger's operator

$$L = -\Delta + c(x). \tag{8.76}$$

When the sign of $c(x)$ is negative, for the extension A of the self-conjugate L, the spectrum of A appears in the negative part. In this case, if we impose an appropriate condition on $c(x)$, the spectrum of A becomes a discrete one. We can even make the whole negative spectrum finite-dimensional according to the condition imposed on $c(x)$.

Alternatively, we shall look at this situation from a general point of view, despite the existence of a direct method which is available for this purpose. The reason is that the method itself can be essentially described in terms of operators which are applicable to various general problems. After showing this we shall return to consider (8.76). Originally, we owe the following argument to Friedrichs, but, in this text, we shall follow closely the paper by M. S. Birman.†

9 *Discrete spectra and essential spectra*

Let A be a self-conjugate operator defined in a Hilbert space $\mathscr{H}$, and let the spectral decomposition of A be E_λ. Symbolically,

$$A = \int \lambda \, \mathrm{d}E_\lambda.$$

DEFINITION 8.4. By saying that a real number μ is a point of *the essential spectrum* of A we mean that, for any small interval $\delta = [a, b]$ having μ as an

† K. Friedrichs: 'Spektraltheorie halbbeschränkter Operatoren und Anwendung auf die Spektralzerlegung von Differentialoperatoren', *Math. Ann.* 109 (1934), 465–87; 685–713. M. S. Birman: 'On the spectrum of singlar boundary problems', *Math. Sbor.* (1961).

interior point, $E(\delta)\mathscr{H}$, the subspace of $\mathscr{H}$ which corresponds to the projective operator $E(\delta)=E_b-E_{a-0}$, is infinite-dimensional. We write $s(A)$ for the set of essential spectra. On the other hand, if there exists a certain interval δ having μ as an interior point, we say that μ is a point of the *discrete spectrum* of A.

Note: If μ is a point of the discrete spectrum of A, then μ is, in fact, a finite-dimensional eigenvalue of A, and μ is an isolate point of the spectrum of A. The converse of this statement is also true.

The following lemma is well known:

LEMMA 8.8. The necessary and sufficient conditions for μ to be a point of the essential spectrum of A, i.e. $\mu \in s(A)$ are that there exists an infinite sequence $\{f_j\}$ of $\mathscr{H}$ such that $\|f_j\|=1$, $\{f_j\}$ is an orthogonal sequence, and $(\mu I-A)f_j \to 0$ in $\mathscr{H}$.

Proof. Let $\mu \in s(A)$. If μ is an eigenvalue having an infinite-dimensional eigenspace, there is no problem. If not, there exists an infinite sequence of intervals $\delta_1, \delta_2, \ldots, \delta_n, \cdots \to \mu$, and $E(\delta_j) \neq 0 (j=1,2,\ldots)$. Therefore, if we take $\{f_j\}$ such that $f_j \in E(\delta_j)\mathscr{H}$ ($\|f_j\|=1$), the sequence satisfies the conditions of the lemma.

The rest of the proof of the sufficiency is exactly the same as the last half of the proof of the following lemma.

Q.E.D.

LEMMA 8.9. Let A be a self-adjoint operator. Further, we assume A has a bounded inverse, i.e.$\lambda=0$ does not belong to the spectrum of A. In this case, if $\mu \in s(A)$, then $\mu^{-1} \in s(A^{-1})$. The converse of this statement is also true.

Proof. If $\mu \in s(A)$, then there exists $\{f_j\} \in \mathscr{D}(A)$ satisfying the condition of the previous lemma. From

$$\left(\frac{1}{\mu}I-A^{-1}\right)f_j=\frac{1}{\mu}A^{-1}(A-\mu I)f_j,$$

we see that when $j \to \infty$ the right hand term tends to 0. Therefore, $\mu^{-1} \in s(A^{-1})$.

Conversely, if we let $\mu^{-1} \in s(A^{-1})$, then there exists $\{f_j\}$ satisfying the condition of the previous lemma. Let δ be an arbitrary interval such that μ^{-1} is an interior point of δ. From

$$\left(\frac{1}{\mu}I-A^{-1}\right)f_j=\int\left(\frac{1}{\mu}-\frac{1}{\lambda}\right)\mathrm{d}E_\lambda f_j,$$

we have

$$\left\|\left(\frac{1}{\mu}-A^{-1}\right)f_j\right\|^2\geqslant\int_{R^1\setminus\delta}\left(\frac{1}{\mu}-\frac{1}{\lambda}\right)^2\mathrm{d}\|E_\lambda f_j\|^2\geqslant\varepsilon\|(I-E(\delta))f_j\|^2,$$

where ε is the largest lower bound of $(\mu^{-1}-\lambda^{-1})^2$ where λ runs through $R^1\setminus\delta$. The left hand side of the inequality tends to 0 as $j\to\infty$, therefore, $(I-E(\delta))f_j\to 0$ $(j\to\infty)$.

If we assume that $E(\delta)\mathscr{H}$ is finite-dimensional, then the set $E(\delta)f_j$ is relatively compact. Hence, the set $\{f_j\}$ is also relatively compact. This contradicts our hypothesis, hence, $E(\delta)\mathscr{H}$ is infinite-dimensional, i.e. $\mu\in s(A)$ because δ is arbitrary.

Q.E.D.

From the lemma, we can say that if A^{-1} exists, then it is sufficient to know the essential spectrum of A^{-1} to obtain the essential spectrum of A.

LEMMA 8.10 *(Weyl)*. Let A, A' be bounded Hermitian operators. If $A-A'$ is a completely continuous operator, then $s(A)=s(A')$. In other words, for a bounded Hermitian operator, its essential spectrum is unchanged after adding a completely continuous operator to it.

Proof. Let $\mu\in s(A)$. There exists $\{f_j\}$ satisfying the condition of the lemma 8.8. On the other hand, if we write

$$(\mu I-A')f_p=(\mu I-A)f_p+(A-A')f_p,$$

then the sequence $\{f_j\}$ is weakly convergent to 0. To see this, we consider $(A-A')f_j\to 0(j\to\infty)$. Then, $(\mu I-A')f_j\to 0$, hence, $\mu\in s(A')$.

Q.E.D.

10 *Friedrichs' extension*

In this section we shall examine a method which we used in chapter 3. First, we repeat the same argument in a more generalized setting. Let $\mathscr{H}$ be a

Hilbert space, and let S be a symmetric operator defined on a dense set $\mathscr{D}(S)$. In this case, if we put $(Su, v)=H[u, v]$, then $H[u, v]$ is a Hermitian form defined on $u, v\in\mathscr{D}(S)$. Furthermore, if we assume that S has a lower bound, i.e. if there exists γ such that $(Su, u)\geqslant\gamma\|u\|^2$, then $H_\beta[u, v]=H[u, v]+\beta[u, v]$ is a positive definite Hermitian form where β is taken as satisfying $\beta+\gamma>0$.

In general, we consider the H_β-norm $|u|_{H_\beta}=\sqrt{(H_\beta[u, u])}$ for a given Hermitian form $H[u, v]$ which has a lower bound and is defined on a dense set $D[H]$ (β is defined above). We note that this defines exactly the same norm as long as H_β is positive definite. By the well-known property of a Hilbert space, if $D[H]$ is complete relative to the H_β-norm, then $D[H]$ is in fact a Hilbert space having $H_\beta[u, v]$ as its inner product. Using this fact, we have:

Definition 8.5. For a Hermitian form defined on $D[H]$, if $D[H]$ is complete in the sense of a H_β-norm, then we say that $H[u, v]$ is a *closed Hermitian form.*

In the case of a Hermitian form which is not closed,† we can make it a closed Hermitian form by an extension of the domain of definition as follows:

Lemma 8.11. If $H[u, v]$ satisfies the conditions:

$$H[u, v_j]\to 0 \quad (j\to\infty) \tag{8.77}$$

for an arbitrary $u\in D[H]$ and for an arbitrary sequence $v_j\to 0$ in $\mathscr{H}$ of $D[H]$, then H can be extended to a closed Hermitian form.

Proof. We pick $H_\beta[u, v]$ instead of $H[u, v]$ and assume that $H[u, v]$ itself is positive definite. We introduce an H-norm: $|u|_H=\sqrt{(H[u, u])}$ to $D[H]$ and perform the completion by the norm, i.e. we regard a Cauchy sequence $\{u_j\}$ (by the H-norm) as a point. Then, we identify $\{u_j\}$ with $\{v_j\}$ if and only if $|u_j-v_j|_H\to 0(j\to\infty)$. We define an inner product on the space (after the identification of the points) such that $(g_1, g_2)_H=\lim_{j\to\infty} H[u_j, v_j]$.

Then, we see that the space obtained in this way is a Hilbert space (the

† Here and in what follows we assume that a Hermitian form is bounded from below and its domain of definition is dense. Unless stated to the contrary, we shall impose this assumption on a Hermitian form.

completion of a pre-Hilbert space.) The important thing is that the space thus obtained can, in fact, be regarded as a subspace of $\mathscr{H}$. We demonstrate this fact as follows. For g, there exists an H-Cauchy sequence $\{u_j\}$ defining g. The H-norm is stronger† than that of $\mathscr{H}$, therefore, $\{u_j\}$ is a Cauchy sequence of $\mathscr{H}$. Hence, $u_j \to h$ in $\mathscr{H}$.

We note that h can be uniquely determined for a given g. It is obvious that the mapping: $g \to h$ is linear, in fact, it is one-to-one. To show this, it is sufficient to say that if $g \to 0$ then $g = 0$ because of the linearity. That is, if we write $g = \{u_j\}$, then, from $|u_n - u_m|_H \to 0 (n, m \to \infty)$ and $u_n \to 0$ in $\mathscr{H}$, we have $|u_n|_H \to 0\ (n \to \infty)$.

We consider

$$|u_n|_H{}^2 = H[u_n, u_n] = H[u_n, u_m] + H[u_n, u_n - u_m].$$

The first term of the right hand side tends to 0 as $m \to \infty$ by the hypothesis (8.77). The second term

$$|H[u_n, u_n - u_m]| \leqslant |u_n|_H |u_n - u_m|_H,$$

where $|u_n|_H$ is bounded. So that, if we take a fixed n sufficiently large, and make $m \to \infty$, then the least upper bound of the term can be made as small as desired. From this, it follows that $|u_n|_H \to 0\ (n \to \infty)$.

Finally, if we write $(h_1, h_2)_H = (g_1, g_2)_H$, then the whole set of the images of g forms a Hilbert space. Also, if $h_1, h_2 \in D[H]$, then $(h_1, h_2)_H = H[h_1, h_2]$. Therefore, this extension has the desired properties.

Q.E.D.

Note 1. Hereafter we call the extension of the positive definite Hermitian form $H[u, v]$ *the closed extension of H by the H-norm* and we write $\bar{H}[u, v]$. We define the extension as $\bar{H}_\beta[u, v] - \beta(u, v)$ for a Hermitian form bounded below. In this case, obviously, we have

$$\inf_{u \in D[\bar{H}]} \frac{\bar{H}[u, u]}{\|u\|^2} = \inf_{u \in D[H]} \frac{H[u, u]}{\|u\|^2}. \tag{8.78}$$

Note 2. In the previous lemma we assumed (8.77). On the other hand, if

† In the wider sense, $\|u\| \leqslant c|u|_H\ \ (c>0)$.

$H[u, v]=(Su, v)$, i.e. $H[u, v]$ is a Hermitian form corresponding to a symmetric operator S, then (8.77) is satisfied.†

Note 3. If H is a non-negative Hermitian form, i.e. $H[u, u]\geqslant 0$ $(u\in D[H])$, then $\bar{H}[u, v]$ is also a non-negative Hermitian form.

For a closed positive definite Hermitian form we have the following:

THEOREM 8.9 *(Friedrichs)*. Let $H[u, v]$ be a positive definite Hermitian form such that its domain of definition $D[H)]$ is dense. In this case, we can determine a unique self-adjoint operator A of $\mathscr{H}$ such that $(u, Av)=H[u, v]$ $(u\in D[H], v\in\mathscr{D}(A))$ is valid, where the domain of definition $\mathscr{D}(A)$ is dense in the sense of an H-norm. (Therefore, $\mathscr{D}(A)$ is, of course, dense in $\mathscr{H}$.)

Proof. We note that $D[H]$ is a Hilbert space with the inner product $H[u, v]$ (see definition 8.5). Also, we have, for a certain γ,

$$|u|_H\geqslant\gamma\|u\| \quad (\gamma>0, u\in D[H]). \tag{8.79}$$

For a fixed $h\in\mathscr{H}$, if we regard (u, h) as a linear form defined on $u\in D[H]$, then (u, h) is continuous. In fact, we have $|(u, h)|\leqslant\|u\|\,\|h\|\leqslant\gamma^{-1}\|h\|\,|u|_H$. Therefore, by Riesz theorem, there exists a unique $Bh\in D[H]$, so that we can write

$$(u, h)=H[u, Bh], \tag{8.80}$$

where B is obviously linear and a continuous operator from $\mathscr{H}$ to $D[H]$. To see this, by Riesz theorem again, we have

$$|Bh|_H=\sup_{|u|_H=1}|(u, h)|\leqslant\gamma^{-1}\|h\|.$$

Next, we see that B is a positive Hermitian operator because of‡

$$(Bu, u)=H[Bu, Bu]=|Bu|_H^{\,2} \quad (u\in\mathscr{H}).$$

Obviously, B is bijective, so that if we write $B^{-1}=A$, then A is the inverse

† To be more precise, this means that there exists a dense set A in the sense of the H-norm of $D[H]$, and if (8.77) is true for $u\in A$, then the lemma is valid.

‡ In fact, from (8.80), if $Bu=0$, then $u=0$.

operator of a bounded Hermitian operator, and is a self-adjoint operator when the domain of definition is dense.

$\mathscr{D}(A)$ is the range of B. If this is not dense in $D[H]$, then there exists a certain $f \neq 0$, $f \in D[H]$ and $H[f, Bu]=0$ for $u \in \mathscr{H}$. Therefore, in this case, $(f, u)=0 \cdot f=0$. This is a contradiction, so $\mathscr{D}(A)$, the range of B is dense in $D[H]$. $D[H]$ is also dense in $\mathscr{H}$.

An H-norm is not weaker than an $\mathscr{H}$-norm. Therefore, $\mathscr{D}(A)$ is dense in $\mathscr{H}$. We note that $AB=I$ in $\mathscr{H}$, $BA=I$ in $\mathscr{D}(A)$. In particular, if $v \in \mathscr{D}(A)$, then $(u, Av)=H[u, BAv]=H[u, v]$.

Finally, we show the uniqueness of A. Let A be an operator obtained as described above. Let us assume that there is a self-adjoint operator A' such that

$$(u, A'v)=H[u, v] \quad (u \in D[H], v \in \mathscr{D}(A')).$$

By (8.80), we have $(u, A'v)=H[u, BA'v]$. Therefore, for $v \in \mathscr{D}(A')$, we have $v=BA'v$. From this we see that $v \in$ range B. So that $\mathscr{D}(A') \subset \mathscr{D}(A)$. Hence, $A' \subset A$. We can take the conjugate relation $A' \supset A$. Hence, $A'=A$.

Q.E.D.

Note 1. Alternatively, we can write (8.80) as

$$(u, h)=H[u, A^{-1}h] \quad (u \in D[H], h \in \mathscr{H}), \tag{8.81}$$

where A^{-1} is a continuous operator from $\mathscr{H}$ to $D[H]$.

Note 2. Let $H[u, v]$ be a closed Hermitian form bounded below. If necessary, we take an appropriate β to make $H_\beta[u, v] \equiv H[u, v]+\beta(u, v)$ positive definite. Let us put A_β for A obtained by the theorem, i.e. $(u, A_\beta v) = H_\beta[u, v]$ $(u \in D[H], v \in \mathscr{D}(A_\beta))$. In this case, if

$$A=A_\beta-\beta I, \tag{8.82}$$

then

$$(u, Av)=H[u, v] \quad (u \in D[H], v \in \mathscr{D}(A_\beta)). \tag{8.83}$$

It is necessary to demonstrate that A is independent of β as follows. In general, if there are two different self-adjoint operators A_1, A_2 satisfying (8.83), for a sufficiently large γ such that $H_\gamma[u, v]$ is positive definite, we consider

$$(u, (A_i+\gamma I)v)=H_\gamma[u, v] \quad (u \in D[H], v \in \mathscr{D}(A_i), (i=1, 2)).$$

From the uniqueness of the theorem, we have $A_1+\gamma I=A_2+\gamma I$, Hence, $A_1=A_2$.

Note 3. In (8.83), if we write

$$\gamma(H)=\inf_{u\in D[H]}\frac{H[u,u]}{\|u\|^2},\qquad m(A)=\inf_{u\in\mathscr{D}(A)}\frac{(Au,u)}{\|u\|^2},$$

then we have

$$m(A)=\gamma(H). \tag{8.84}$$

Proof. Because $\mathscr{D}(A)\subset D[H]$, $m(A)\geqslant\gamma(H)$ is obvious. It is sufficient to show $m(A)\leqslant\gamma(H)$. To do this, we pick a sufficiently large β to make $H_\beta[u, v]$ positive definite. That is, we consider the case $\gamma(H)>0$. In this case, $\mathscr{D}(A)$ is dense in the sense of the H-norm in $D[H]$. Therefore, $m(A)\leqslant\gamma(H)$.

Q.E.D.

Friedrichs' extension

We summarize the results obtained in the foregoing argument. Suppose we have a symmetric operator S bounded below. Let us assume that its domain of definition $\mathscr{D}(S)$ is dense in $\mathscr{H}$. In this case, we extend $H[u, v]=(Su, v)$ $(u, v\in\mathscr{D}(S))$ to a closed Hermitian form by a H_β-norm (Use lemma 8.11). Thus we obtain $H[u, v]$, an extended form. By the *Friedrichs' extension* of S we mean a self-adjoint operator A defined by (8.82) and (8.83).

'Extension' means the situation $S\subset A$. In fact, $\mathscr{D}(S)\subset D[H]$. Let H_β be positive definite. If we write $S_\beta=S+\beta I$, then we have

$$(u, S_\beta v)=H_\beta[u, v]=H_\beta[u, BS_\beta v]\quad (u\in D[H], v\in\mathscr{D}(S)).$$

Therefore, $v=BS_\beta v$, hence, $\mathscr{D}(S_\beta)\subset\mathscr{D}(A_\beta)$ and $S\subset A$. We note that in this case from (8.78) and (8.84) we have

$$m(A)=m(S). \tag{8.85}$$

DEFINITION 8.6 *(Completely continuous Hermitian form).* We consider 'complete continuousness' from the stand-point of the theory of Hermitian forms. By saying that a bounded Hermitian form $B[u, v]$ defined in a Hilbert space H is *completely continuous* we mean that if we write $B[u, v]=(u, Qv)_H$ (see Riesz theorem), Q is a completely continuous operator of H. In this case, we say that Q is *an operator generated by* the form $B[u, v]$.

LEMMA 8.12. Let $B[u, v]$ be a non-negative Hermitian form. In this case, the necessary and sufficient condition for B to be completely continuous is that an arbitrary bounded set of H is relatively compact with the topology created by the B-norm: $|u|_B=(Bu, u)^{\frac{1}{2}}$.

Proof (Necessity). Let Q be completely continuous. For an arbitrary bounded sequence of H, we choose an appropriate subsequence $\{u_j\}$ such that $\{Qu_j\}$ is convergent. On the other hand, we have

$$|u_j-u_k|_B{}^2=(u_j-u_k, Q(u_j-u_k))_H\leqslant|u_j-u_k|_H|Q(u_j-u_k)|_H,$$

so that $\{u_j\}$ is a Cauchy sequence in the sense of a B-norm.

(Sufficiency) We can write

$$|Qu|_H{}^2=(Qu, Qu)_H=B[Qu, u]$$
$$=\tfrac{1}{4}\{|Qu+u|_B{}^2+|Qu-u|_B{}^2+\mathrm{i}|Qu+\mathrm{i}u|_B{}^2-\mathrm{i}|Qu-u|_B{}^2\}.^\dagger$$

Let $\{u_j\}$ be an arbitrary bounded sequence of H. $\{Qu_j+u_j\}$, $\{Qu_j-u_j\}$ and $\{Qu_j+\mathrm{i}u_j\}$, $\{Qu_j-\mathrm{i}u_j\}$ are also bounded sequences of H. From this, and our hypothesis, we can choose an appropriate subsequence $\{u_{j_p}\}$ such that $\{u_{j_p}\}$ is a Cauchy sequence in the sense of a B-norm. That is, $|Qu_{j_p}-Qu_{j_q}|_H\to 0$ $(p, q\to\infty)$. So that Q is a completely continuous operator of H.

Q.E.D.

Note. In definition 8.6, we defined a completely continuous Hermitian form. We note that this 'complete continuousness' is nothing to do with the structure of the underlying Hilbert space. To be more exact, if we consider another Hilbert space structure $(u, v)_{H_1}$ which is equivalent to the given one, and put $B[u, v]=(u, Q'v)_{H_1}$, then Q' is also completely continuous. In fact, $(u, v)_H=(u, Lv)_{H_1}$, so that L is a bounded operator. Therefore, $B[u, v]=(u, Qv)_H=(u, LQv)_{H_1}$. Hence, $Q'=LQ$ and Q are both completely continuous.

11 *Discrete spectrum*

In this section, we consider two Hermitian forms at the same time, so in order to avoid possible confusion we change the notation which we used in the previous section. Let A be a symmetric operator bounded from below having a dense domain of definition in a Hilbert space $\mathscr{H}$. We write $A[u, v]$

† Basic properties of the familiar Hermitian form.

$=(Au, v)$. Also, we write $D[A]$ for the domain of definition of the closed extension of $A[u, v]$ by an A_β-norm, and $\tilde{A}$ for the self-adjoint operator corresponding to this closed Hermitian form. In the previous section we write A for $\tilde{A}$. $\tilde{A}$ represents the Friedrichs' extension of A.

Using the new notation, we re-write the result obtained in the previous section. For example, (8.85) becomes $m(A)=m(\tilde{A})$. Also, we have $\mathscr{D}(A) \subset \mathscr{D}(\tilde{A}) \subset D[A]$ $(A \subset \tilde{A})$, and $(u, \tilde{A}v)=A[u, v](u \in D[A], v \in \mathscr{D}(\tilde{A}))$, where $\mathscr{D}(A)$ is dense in $D[A]$ in the sense of an A_β-norm. If A is positive definite, i.e. $m(A)>0$, then

$$(u, h)=A[u, \tilde{A}^{-1}h] \quad (u \in D[H], h \in \mathscr{H}).$$

$\tilde{A}^{-1}$ is a continuous mapping from $\mathscr{H}$ into $D[A]$. If A is non-negative, i.e. $(Au, u) \geqslant 0$, then so is $\tilde{A}$.

Now we are ready to prove the following key proposition.

PROPOSITION 8.6. Let $A[u, v]$ be a Hermitian closed form bounded from below, and let $B[u, v]$ be a completely continuous Hermitian form.† In this case, if we put $C[u, v]=A[u, v]-B[u, v]$, then C is also a Hermitian form bounded from below. We have $D[C]=D[A]$, i.e. C is a closed Hermitian form in $D[A]$. The C_β-norm and the A_β-norm are equivalent. We also have $s(\tilde{C})=s(\tilde{A})$.

Proof. Without loss of generality, we can take a sufficiently large β such that

$$A_\beta[u, v]=A[u, v]+\beta(u, v), \; C_\beta[u, v]=C[u, v]+\beta(u, v)$$

and assume that A is a closed positive definite Hermitian form. Therefore, we simply write A, C for A_β, C_β respectively. We look at $B[u, v]$ in $D[A]$. First we note that $D[A]$ is a Hilbert space with A-norm. Let $B[u, v]$ $=A[u, Qv]$,† and consider the spectral decomposition (see the Hilbert–Schmidt theorem) of Q in $D[A]$. We know that this consists of a sequence of discrete spectra at most converging to 0. Therefore, for $u \in D[A]$, we have

$$B[u, u]=B^{(\varepsilon)}[u, u]+\sum_{|\mu_i|>\varepsilon} \mu_i |A[u, u_i]|^2$$

where $\varepsilon(>0)$ will be explained later.

† See the last note of the previous section (p. 444).

We note that $Qu_i=\varepsilon_i u_i$, $|u_i|_A=1$. Let $|u_i|\leqslant c$, obviously, $|B^\varepsilon[u,u]|\leqslant\varepsilon|u|_A{}^2$. Let us write $N(\mu)$ for the number (with some duplications) of i such that $|\varepsilon_i|>\varepsilon$. On the other hand, $u_i\in D[A]$, and $\mathscr{D}(\tilde{A})$ is dense in $D[A]$, so we can put

$$|\tilde{u}_i-u_i|_A{}^2<\varepsilon c^{-1}N^{-1}, \qquad \tilde{u}_i\in\mathscr{D}(\tilde{A})\ (i=1,2,\ldots,N).$$

Then

$$\begin{aligned}|A[u,u_i]|&\leqslant|A[u,u_i-\tilde{u}_i]|+|A[u,\tilde{u}_i]|\leqslant|u|_A|u_i-\tilde{u}_i|_A+|(u,\tilde{A}\tilde{u}_i)|\\&\leqslant|u|_A|u_i-\tilde{u}_i|_A+\|u\|\ \|\tilde{A}\cdot\tilde{u}_i\|.\end{aligned}$$

Therefore,

$$\begin{aligned}\sum_{i=1}^{N}\mu_i|A[u,u_i]|^2&\leqslant 2c|u|_A{}^2\sum|u_i-\tilde{u}_i|_A{}^2+2\|u\|^2c\sum\|\tilde{A}\tilde{u}_i\|^2\\&\leqslant 2\varepsilon|u|_A{}^2+\gamma\|u\|^2.\end{aligned}$$

From this we have

$$B[u,u]\leqslant 3\varepsilon|u|_A{}^2+\gamma\|u\|^2.$$

Putting $\varepsilon=\frac{1}{6}$, we then have

$$C[u,u]\geqslant\tfrac{1}{2}A[u,u]-\gamma\|u\|^2 \quad \left(\gamma=2c\sum_{i=1}^{N}\|\tilde{A}\tilde{u}_i\|^2\right).$$

Hence $C_\gamma[u,v]$, $A_\gamma[u,v]$ are both positive definite and define equivalent norms. Hence, $D[C]=D[A]$.

Next, for $u,v\in D[C](=D[A])$, we have

$$C_\gamma[u,v]=A_\gamma[u,v]-B[u,v]=A_\gamma[u,v]-A_\gamma[u,Q_\gamma v],$$

where Q_γ is a completely continuous operator of $D[A]$. So that, for $\varphi\in\mathscr{H}$, $u\in D[C](=D[A])$, we can write

$$(u,\varphi)=C_\gamma[u,\tilde{C}_\gamma{}^{-1}\varphi]=A_\gamma[u,\tilde{C}_\gamma{}^{-1}\varphi]-A_\gamma[u,Q_\gamma\tilde{C}_\gamma{}^{-1}\varphi].$$

From $(u,\varphi)=A_\gamma[u,\tilde{A}_\gamma{}^{-1}\varphi]$, we have $\tilde{A}_\gamma{}^{-1}=\tilde{C}_\gamma{}^{-1}-Q_\gamma\tilde{C}_\gamma{}^{-1}$. As we mentioned at the beginning of this section, $\tilde{A}_\gamma{}^{-1}$, $\tilde{C}_\gamma{}^{-1}$ are both continuous mappings from $\mathscr{H}$ to $D[A](=DC])$. Because Q_γ is a completely continuous mapping in $D[A]$, so is $Q_\gamma\tilde{C}_\gamma{}^{-1}$ which maps from $\mathscr{H}$ into $\mathscr{H}$. Hence, by lemma 8.10, we have $s(\tilde{A}_\gamma{}^{-1})=s(\tilde{C}_\gamma{}^{-1})$. Furthermore, by lemma 8.9, we have $s(\tilde{A}_\gamma)=s(\tilde{C}_\gamma)$. That is, $s(\tilde{A})=s(\tilde{C})$.

Q.E.D.

By proposition 8.6, we obtain Friedrichs' theorem immediately. We leave the proof to the reader.

THEOREM 8.10. Suppose, in a Hilbert space $\mathscr{H}$ two symmetric operators A, B sharing a common dense domain of definition are defined. Further, we assume that $A \geqslant 0$, $B \geqslant 0$, $B[u, v]=(Bu, v)$ is bounded† in $D[A]$, and $B[u, v]$ is a completely continuous form in $D[A]$. In this case $C=A-B$, $\mathscr{D}(C)=\mathscr{D}(A)$ is bounded from below. Also, the spectrum of the Friedrichs' extension $\tilde{C}$ of C has the following property: if the spectrum appears as negative then all points contained in this part are discrete spectra.

12 *The finiteness of a spectrum in the negative part*

By the theorem of the previous section, section 11, we know that if we impose certain conditions on the operator B (we call this the term of perturbation for A), the spectra of $\tilde{C}$ in its negative part are all discrete. In this case, in general, it is possible that the point-spectra may have an accumulation point at $\lambda=0$, but if the term of the perturbation B is not larger, this is not possible. We discuss the situation in what follows.

DEFINITION 8.7. Let A be a self-adjoint operator, and let E_λ be its spectral decomposition which is right-continuous. For an open interval $\delta=(a, b)$, the dimension n of $(E_{b-0}-E_a)\mathscr{H}=E(\delta)\mathscr{H}$ is called the *dimension of the spectrum* of A at δ. In this case, if $E(\delta)\mathscr{H}$ is infinite-dimensional, we regard $n=+\infty$.

We shall make a remark which is of general applicability. Let $A[u, v]$ be a closed Hermitian form bounded from below defined in a Hilbert space H, and let $\tilde{A}$ be the Friedrichs' extension whose existence was shown in theorem 8.7, i.e.

$$(u, \tilde{A}v)=A[u, v] \quad (u \in D[A], v \in \mathscr{D}(\tilde{A})). \tag{8.86}$$

Using the spectral decomposition E_λ of $\tilde{A}$, we can write the term on the left hand side as

$$(u, \tilde{A}v)=\int_\gamma^\infty \lambda \, \mathrm{d}(E_\lambda u, v). \tag{8.87}$$

From this, we see that for $u \in \mathscr{D}(\tilde{A})$, $|u|_{A_\beta}{}^2$ and $\int_\gamma^\infty (1+|\lambda|)\, \mathrm{d}\|E_\lambda u\|^2$ are equivalent. $\mathscr{D}(\tilde{A})$ is dense in $D[A]$ with A_β-norm. So if we perform a limiting

† The topology of $D[A]$ is considered as having an A_1-norm (in the case $\beta=1$).

process in (8.87), we have

$$A[u, v] = \int_{\gamma}^{\infty} \lambda \, \mathrm{d}(E_\lambda u, v) \quad (u, v \in D[A]). \tag{8.88}$$

The necessary and sufficient condition for $u \in \mathscr{H}$ to belong to $D[A]$ is

$$\int_{\gamma}^{\infty} (1+|\lambda|) \, \mathrm{d}\|E_\lambda u\|^2 < +\infty. \tag{8.89}$$

From this, we deduce the following fact: if we decompose $u \in D[A]$ orthogonally into two projections, namely the one projected on $E_{0-0}\mathscr{H}$, the space which corresponds to the negative part of the spectrum of $\tilde{A}$, and the one projected on the space which corresponds to the non-negative part, then each component belongs to $D[A]$. We give a lemma concerning dimensions. Let $A[u, v]$ be a closed Hermitian form bounded from below in a Hilbert space $\mathscr{H}$, and let the dimension of the spectrum in the negative part of the Friedrichs' extension $\tilde{A}$ generated by A be n.

LEMMA 8.13

(1) In $D[A]$, there is an n-dimensional subspace such that $A[u, u] < 0$ $(u \neq 0)$, but there is no such subspace of $(n+1)$-dimensions. If $n = +\infty$, then, for any m, there is an m-dimensional subspace satisfying the same condition.

(2) In particular, if A is bounded, the same statements are true in a dense subspace of a given (arbitrary) $\mathscr{H}$.

Proof. Let n be finite. The spectral representation of $\tilde{A}$ shows there are $u_1, \ldots, u_n \in \mathscr{D}(\tilde{A})$ such that

$$(u_i, \tilde{A}u_j) = \mu_i \delta_i^j \quad (\mu_i < 0, \ i, j = 1, 2, \ldots, n). \tag{8.90}$$

Therefore, if we write

$$u = \sum \alpha_i u_i \in \mathscr{D}(\tilde{A})$$

then

$$(u, \tilde{A}u) = \sum_{i,j} (u_i, \tilde{A}u_j)\alpha_i \bar{\alpha}_j = \sum_i \mu_i |\alpha_i|^2 < 0, \tag{8.91}$$

and $(u, \tilde{A}u) = A[u, u]$. If $A[u, v]$ is bounded, then so is $\tilde{A}$. Therefore, if we call M a dense subspace of a given $\mathscr{H}$ then we can pick $\{u_i'\}_{i=1\,2\ldots,n} \in M$ such that $(u_i', \tilde{A}u_j')$ and $(u_i, \tilde{A}u_j)$ can be made arbitrarily close. So that (2) of the lemma is true.

If $n=+\infty$, then for an arbitrary m, it is obvious that there exists $u_1,\ldots, u_m \in \mathscr{D}(\tilde{A})$ satisfying (8.90). Finally, if n is finite, then we can show that there is no such $(n+1)$-dimensional space in $D[A]$. Suppose that there exists such a subspace, and let $u_1, u_2,\ldots, u_{n+1} \in D[A]$ be taken as independent in $D[A]$. In this case, we put $E_{0-0}=E(-\infty, 0)=E$, and decompose $\mathscr{H}=E\mathscr{H}\oplus \oplus(I-E)\mathscr{H}$ orthogonally. Then, we put $u_i=u_i'+u_i''(i=1,\ldots, n+1)$. $E\mathscr{H}$ is n-dimensional, there exists c_1 such that $\sum|c_i|\neq 0$. $\sum c_i u_i'=0$. For such c_i we have $u_0=\sum c_i u_i=\sum c_i u_i'' \in (I-E)\mathscr{H}$. From the note which we gave just before the previous lemma, $u_i'' \in D[A]$, so that $u_0 \in D[A]$. Therefore, by (8.88), $A[u_0, u_0]\geqslant 0$. It follows from this that $u_0\neq 0$, which is a contradiction.

Q.E.D.

We assume the same conditions as in theorem 8.10 for the operators A and B. Furthermore, we assume $A>0$, i.e. $(Au, u)>0(u\neq 0)$. In this case, $\tilde{A}\geqslant 0$. For simplicity, we give:

DEFINITION 8.8. Let $B[u, v]$ be a bounded Hermitian form defined in a Hilbert space H. In this case, we write $B[u, v]=(u, Qv)_H$ and call the spectrum of Q, *the spectrum of the form* $B[u, v]$ in $\mathscr{H}$.

(A) The dimension n of the spectrum in the negative part is the least upper bound of the numbers of the dimensions of subspaces of $\mathscr{D}(A)$ satisfying $(Cu, u)<0$. In fact, in $D[C]=D[A]$, n is equal to the least upper bound of the dimensions of the subspaces such that $C[u, u]<0(u\neq 0)$. On the other hand, this number is the number of the dimension of the spectrum of the bounded form $C[u, v]$ in its negative part in a Hilbert space $D_\beta[C]$ where $D_\beta[C]$ is understood to be $D[C]$ with C_β-norm. Then we apply (2) of the previous lemma.

(B) Let H_A be the completion of $\mathscr{D}(A)=\mathscr{D}(C)$ with an A-norm. This makes sense because in $\mathscr{D}(A)$ we have $(Au, u)>0(u\neq 0)$, so that $\mathscr{D}(A)$ is a pre-Hilbert space with an A-norm. Of course, $\mathscr{D}(A)$ is a dense subspace in H_A. We do not impose any condition on the relation between the A-norm and the $\mathscr{H}$-norm, so in general we cannot realize H_A within $\mathscr{H}$. Now, let $\bar{A}[u, v]$, $\bar{B}[u, v]$ be the continuous extensions of $A[u, v]$, $B[u, v]$ to H_A. $\bar{C}[u, v]=\bar{A}[u, v]-\bar{B}[u, v]$ is bounded in H_A. Therefore the dimension of the spectrum of $\bar{C}[u, v]$ in H_A in its negative part is, by (2) of the previous lemma, the same as that of the subspace satisfying $(Cu, u)<0(u\neq 0, u\in\mathscr{D}(A))$. On the other hand, the dimension of the spectrum of

$\bar{C}[u, v]$ in H_A in its negative part is the same as that of the spectrum of the bounded form $B[u, v]$ in H_A in $(1, +\infty)$.

From (A), (B) we have:

PROPOSITION 8.7. The dimension of the spectrum of $\bar{C}$ in its negative part coincides with that of the spectrum of the form $\bar{B}[u, v]$ in H_A in $(1, \infty)$.

From this proposition we have:

THEOREM 8.11 *(Birman)*. With the hypothesis of theorem 8.10, we assume $A>0$, and that $\bar{B}[u, v]$ is completely continuous in H_A. Then the dimensions of the spectrum of $\bar{C}$ in its negative part is finite.

13 *Self-adjoint extensions*

Up to section 12 we have considered the spectrum of the Friedrichs' extension. Now we are interested in the following question: given a symmetric operator A having a dense domain of definition $\mathscr{D}(A)$, is the self-adjoint extension of A unique? If it is, we call A *essentially self-adjoint*. In this case, the spectrum we discussed in the previous section plays the most important rôle.

But, first, we give a sufficient condition for A to be essentially self-adjoint.

LEMMA 8.14. A symmetric operator whose domain of definition is dense is essentially self-adjoint if the following condition is satisfied. $A\cdot\mathscr{D}(A)$ is dense in $\mathscr{H}$, and there exists a positive constant a such that $\|Au\|\geqslant a\|u\|\,(u\in\mathscr{D}(A))$.

Proof. Let the closed extension of A be $\bar{A}$, i.e. $\bar{A}$ is obtained by the process of taking the closure of (u, Au) with the graph topology $\|u\|^2+\|Au\|^2$. We see $\bar{A}^*\supset\bar{A}$. So it is sufficient to show $\bar{A}^*\subset\bar{A}$. Let $f\in\mathscr{D}(\bar{A}^*)$. $A\cdot\mathscr{D}(A)$ is dense, so there exists $f_n\in\mathscr{D}(A)$ such that $Af_n\to\bar{A}^*f\ (n\to\infty)$. On the other hand, from $\|A(f_n-f_m)\|\geqslant a\|f_n-f_m\|$, we have $f_n\to h$. Obviously $h\in\mathscr{D}(\bar{A})$, in fact, $h=f$. To see this, we observe that for $u\in\mathscr{D}(A)$ we have

$$(f-h, Au)=\lim_{n\to\infty}(f-f_n, Au)=(\bar{A}^*f, u)-\lim_{n\to\infty}(Af_n, u)=0.$$

Au is dense in $\mathscr{H}$, therefore, $f=h$. Hence, $f=h\in\mathscr{D}(\bar{A})$ and $\bar{A}^*\subset\bar{A}$.

Q.E.D.

Let us consider the self-adjoint extension of Schrödinger's operator,

$$A = -\Delta + c(x). \tag{8.92}$$

For simplicity, we assume that the underlying space is R^3. In (8.92), we assumed c to be a real-valued function, and continuous throughout except at the origin. Let us separate $c(x)$ into the positive part and the negative part: $c(x) = c_+(x) - c_-(x)$ $(c_\pm(x) \geqslant 0)$. In this case, we impose a condition

$$c_-(x) \leqslant \frac{c_1}{|x|^{\frac{3}{2}-\varepsilon}} + c_2 \qquad (c_1, c_2 \geqslant 0, \varepsilon > 0). \tag{8.93}$$

Further, we assume

$$c_+(x) \leqslant \frac{\text{const}}{|x|^{\frac{3}{2}-\varepsilon}} \ (|x| \leqslant 1)$$

except possibly at the origin where the value of $c_+(x)$ may well be $+\infty$. In short, c is bounded from below except at the origin where c may have a singularity whose order is at most $|x|^{-\frac{3}{2}+\varepsilon}$. As is well known, we have

$$\int \frac{|u(x)|^2}{|x|^2} \, \mathrm{d}x \leqslant 4 \int |\operatorname{grad} u|^2 \, \mathrm{d}x \tag{8.94}$$

for a function u of C^2-class having a compact support,† where $|\operatorname{grad} u|^2 = \sum |u_{x_i}|^2$. If c satisfies the above assumptions, and if

$$\mathscr{D}(A) = C_0{}^2 = C^2 \cap \mathscr{E}', \tag{8.95}$$

then obviously A is a symmetric operator. Also, A is bounded from below. To be more precise, for an arbitrary $\varepsilon (>0)$, there exists a certain β such that

$$(c_-(x)u, u) \leqslant \varepsilon \|\operatorname{grad} u\|^2 + \beta \|u\|^2. \tag{8.96}$$

In fact,

$$\int_{|x| \leqslant \delta} \frac{|u(x)|^2}{|x|^{\frac{3}{2}-\varepsilon}} \, \mathrm{d}x \leqslant \delta^{\varepsilon + \frac{1}{2}} \int_{|x| \leqslant \delta} \frac{|u(x)|^2}{|x|^2} \, \mathrm{d}x \leqslant 4\delta^{\frac{1}{2}} \|\operatorname{grad} u\|^2 \qquad (\delta < 1),$$

† In general, in $R^n (n \geqslant 3)$, if we put $\varphi(r) = \frac{1}{2}(n-2)/r$, we have

$$\int \left\{ \sum \left(\frac{\partial u}{\partial x_i} \right)^2 - \{\varphi(r)\}^2 u^2 \right\} \mathrm{d}x = \int \frac{1}{\varphi(r) r^{n-1}} \sum \left[\frac{\partial}{\partial x_i} \left(r^{\frac{1}{2}(n-1)} \{\varphi(r)\}^{\frac{1}{2}} u \right) \right]^2 \mathrm{d}x,$$

where u is a real-valued function. In the case of a complex-valued function, we separate it into the real part and the imaginary part.

Let us write $A+\beta I$ simply as A. From $(Au, u) \geqslant \varepsilon \|u\|^2$ $(\varepsilon>0)$,

$$\|A \cdot u\| \geqslant \varepsilon \|u\| \quad (u \in \mathscr{D}(A)). \tag{8.97}$$

Considering Lemma 8.14, if we show $A \cdot \mathscr{D}(A)$ is dense in $L^2(R^3)$, then A is essentially self-adjoint. We show this by deriving a contradiction.†

If it is not dense, then there exists $h \in L^2$ $(h \neq 0)$ such that $(h, Au)=0$ $(u \in \mathscr{D}(A))$. Therefore $A \cdot h = (-\Delta + c(x))h(x) = 0$ is valid in the sense of distribution. c is continuous except at the origin, so $c(x)h(x) \in L^2{}_{\text{loc}}(\text{C}0)$ and therefore, by theorem 3.22, $h \in \mathscr{E}_{L^2(\text{loc})}{}^2(\text{C}0)$. If we consider a C^∞-function $\xi(x)$ whose support lies in $|x| \leqslant 2$, and $\xi(x)=1$ in the neighbourhood ω of the origin, for $x \in \omega$, we have

$$h(x) = -\frac{1}{4\pi} \int \frac{\xi(y)c(y)h(y)}{|x-y|}\, dy + \alpha(x)$$

(consider the elementary solution $-1/4\pi|x|$ of Δ), where α is a C^∞-function in ω. Therefore, in order to examine the singularity in the neighbourhood of the origin, it is enough to consider the first term of the right hand side of this equation. From this observation, and the Schwartz inequality,‡ we have

$$|h(x)| \leqslant \text{const} \left[\int_{|y| \leqslant 2} \frac{dy}{|x-y|^2 |y|^{3-2\varepsilon}} \right]^{\frac{1}{2}} \|h\|_{L^2} \leqslant \text{const}\, |x|^{\varepsilon - 1}.$$

Using this estimation, we have

$$|h(x)| \leqslant \text{const} \int_{|y| \leqslant 2} \frac{dy}{|x-y|^2 |y|^{2\frac{1}{2}-2\varepsilon}} \leqslant \text{const} \frac{1}{|x|^{\frac{1}{2}-2\varepsilon}}.$$

Starting from this result as a hypothesis we reconsider the argument above, and conclude the boundedness of $h(x)$. Therefore, $c(x)h(x) \in L^2{}_{\text{loc}}$, so that finally we have $h \in \mathscr{E}_{L^2(\text{loc})}{}^2$. Next, we note that (8.97) is valid for an arbitrary $u \in \mathscr{E}_{L^2(\text{loc})}{}^2 \cap \mathscr{E}'$ (we use a limiting process). For $\alpha \in \mathscr{D}$,* and $u \in \mathscr{E}_{L^2(\text{loc})}{}^2$, we have

$$\begin{aligned} A(\alpha u)\bar{\alpha}\bar{u} &= (-\Delta + c(x))\,(\alpha u)\bar{\alpha}\bar{u} \\ &= \alpha^2 Au \cdot \bar{u} + \sum \alpha_{x_i}{}^2 |u|^2 + \sum \alpha\alpha_{x_i}(u\bar{u}_{x_i} - \bar{u}u_{x_i}) - \sum (\alpha\alpha_{x_i} u\bar{u})_{x_i}. \end{aligned}$$

† Here, we follow the argument of E. Wienholtz: 'Halbbeschränkte partielle Differentialoperatoren zweiter Ordnung vom elliptischen Typus', *Math. Ann.* 135 (1958), 50–80.

‡ By Sobolev's lemma, h is continuous outside an arbitrary ε-neighbourhood of the origin.

* α is a real-valued function.

We put $u=h$ and integrate the terms to get†

$$(A(\alpha h), \alpha h)=\int (\sum \alpha_{x_i}{}^2(x))\cdot|h(x)|^2\,dx$$

The left hand side $\geqslant \varepsilon\|\alpha h\|^2$ by (8.97), so that

$$\varepsilon \int \alpha^2(x)|h(x)|^2\,dx \leqslant \int (\sum \alpha_{x_i}{}^2(x))|h(x)|^2\,dx.$$

Now, for an arbitrary R, we can take the value of $\alpha(x)$ as 1 in $|x|\leqslant R$ and $\sum \alpha_{x_i}{}^2(x)<\frac{1}{2}\varepsilon$ $(x\in R^3)$. Therefore, for such $\alpha(x)$,

$$\varepsilon \int_{|x|\leqslant R} |h(x)|^2\,dx \leqslant \tfrac{1}{2}\varepsilon \int_{R^3} |h(x)|^2\,dx.$$

R can be taken as arbitrarily large, so $h(x)\equiv 0$, which, is a contradiction.

Therefore, if we impose the hypothesis mentioned above, we have:

THEOREM 8.12. $A=-\Delta+c(x)$ is bounded from below, and has a unique self-adjoint extension. Therefore, only the Friedrichs' extension $\tilde{A}$ of A gives a self-adjoint extension.

14 *Negative spectrum of $-\Delta+c(x)$*

Let us impose the same conditions on c as in the previous section. In this case, from theorem 8.12, we see that the self-adjoint extension of A is nothing but Friedrichs' extension. For simplicity, the spectrum of the operator A is called *the spectrum of $-\Delta+c(x)$*. We write

$$C=-\Delta+c(x)=(-\Delta+c_+(x))-c_-(x)\equiv A-B.$$

Obviously, $A>0$, $B\geqslant 0$. Further, assume that

$$c_-(x)\to 0 \quad \text{uniformly as} \quad |x|\to\infty. \tag{8.98}$$

In this case we have:

THEOREM 8.13. The spectrum of $C=-\Delta+c(x)$ in the negative part (if it exists) is a discrete spectrum.

† The first term of the right hand side becomes zero, the third term is immaterial because it is purely imaginary, and the last term becomes zero.

Proof. We use theorem 8.10 and lemma 8.12. That is, if we put $|u|_B^2 = (c_-(x)u, u)$, then it is enough to show that a bounded set with an A_β-norm is relatively compact with a B-norm. The topology of an A_β-norm is stronger than that of $\mathscr{D}_{L^2}{}^1$, so, more concretely, this is equivalent to the following. Suppose $\{u_j\}$ is an arbitrary sequence of u_j whose $\mathscr{D}_{L^2}{}^1$-norm is not larger than 1. In this case, we can choose an appropriate subsequence $\{u_{j_p}\}$ such that $\{u_{j_p}\}$ form a Cauchy sequence in the sense of B-norm within ε-difference, i.e.

$$\overline{\lim_{j,q\to\infty}}\ |u_{j_p} - u_{j_q}|_B \leqslant \varepsilon .$$

To see this we proceed as follows. We can write

$$|u_n - u_m|_B{}^2 = \int_{|x|\leqslant\delta} c_-(x)|u_n(x) - u_m(x)|^2 \,dx + \int_{|x|\leqslant R} \cdots + \int_{\delta\leqslant|x|\leqslant R} \cdots, \tag{8.99}$$

where we note that, for an arbitrary ε, if we take δ sufficiently small and R sufficiently large, then the first and second term of the right hand side of (8.99) are less than $\frac{1}{2}\varepsilon^2$.

In fact, from (8.93) and (8.94), the first term can be estimated by†

$$\delta^{\frac{1}{2}} \int_{|x|\leqslant\delta} \frac{|u_n - u_m|^2}{|x|^2} \,dx \leqslant \Delta\delta^{\frac{1}{2}} \,\|\text{grad}\,(u_n - u_m)\|^2 ,$$

and the second term by

$$\sup_{x\geqslant R} c_-(x) \int |u_n(x) - u_m(x)|^2 \,dx .$$

Let us fix (δ, R) which satisfies this property.

In this case, $c_-(x)$ is bounded in $\delta\leqslant|x|\leqslant R$ which is compact. So, applying Rellich's choice theorem (theorem 3.7), we see that there exists an appropriate subsequence of $\{u_{j_p}\}$ of $\{u_j\}$ such that it forms a Cauchy sequence in the sense of L^2-norm. Therefore, $\{u_{j_p}\}$ is a Cauchy sequence in the sense of a B-norm within ε-difference.

Q.E.D.

COROLLARY. If $\rho(r) = \inf_{|x|=r} c(x)$ tends to $+\infty$ as $r\to+\infty$, then the spectra of $-\Delta + c(x)$ are all discrete.

† (8.94) is valid for all elements belonging to $\mathscr{D}_{L^2}{}^1$ if we apply the limiting process.

Proof. For an arbitrary $\beta(>0)$, if we write $C-\beta I=-\Delta+c(x)-\beta$, then $(c(x)-\beta)$ has a compact support.

Q.E.D.

In the previous theorem, we assume (8.98). Let us impose a stronger condition:

$$c_-(x)=O(|x|^{-2-\varepsilon}) \quad (|x|\to\infty,\ \varepsilon>0). \tag{8.100}$$

THEOREM 8.14. Under the condition (8.100), the negative part of the spectrum of $C=-\Delta+c(x)$ has a finite dimension.

Proof. Apply theorem 8.11. Let $A=-\Delta+c_+(x)$. Write H_A for the completed $D[A]$ with A-norm $\mathscr{D}(A)(\subset D[A])$ is dense, so we can regard H_A as being obtained by the completion of $\mathscr{D}(A)$. From (8.94), we note that

$$\int_{R^3} |\text{grad}\, u|^2\, dx + \int_{|x|\leqslant R} |u(x)|^2\, dx \leqslant C(R)\, |u|_A{}^2 \quad (C(R)>0).$$

H_A can be identified with a subspace of $\mathscr{E}_{L^2(\text{loc})}{}^1$. To see this, we proceed as in lemma 8.11. If $\{u_n\}\in\mathscr{D}(A)$ is a Cauchy sequence in the sense of an A-norm, then from the above inequality, we can define h for $g=\{u_j)$ such that $u_n\to h\in\mathscr{E}_{L^2(\text{loc})}{}^1$. The mapping is bijective. To show this, it is enough to prove that from $|u_n-u_m|_A\to 0\,(n,\ m\to\infty)$, $u_n\to 0$ in $\mathscr{E}_{L^2(\text{loc})}{}^1$ it follows that $|u_n|_A\to 0$.

We have a decomposition

$$|u_n|_A{}^2=((-\Delta+c_+(x))\, u_n,\, u_n)=((-\Delta+c(x)_+)\, u_n,\, u_m),$$
$$+((-\Delta+c(x)_+)\, u_n,\, u_n-u_m)$$

so that, similarly to the proof of lemma 8.11, we can demonstrate $|u_n|_A\to 0$ $(n\to\infty)$. In this case, it is essential that $u_n\in\mathscr{D}(A)$ has a compact support. We define $|u|_{A'}{}^2=\|\text{grad}\, u\|^2$. In general, the A'-norm is weaker than the A-norm. To prove the theorem, it is enough to show that, given a bounded sequence $\{u_n\}\in H_A$ in the sense of A'-norm, there exists a Cauchy sequence in the sense of B-norm within ε-difference. We use the decomposition (8.99). Then the second term can be estimated as

$$\int_{|x|\geqslant R} c_-(x)\, |u_n-u_m|^2\, dx \leqslant \text{const} \int_{|x|\geqslant R} \frac{|u_n-u_m|^2}{|x|^{2+\varepsilon}}\, dx$$
$$\leqslant \text{const}\, \frac{1}{R^\varepsilon} \int \frac{|u_n-u_m|^2}{|x|^2}\, dx \leqslant \frac{\text{const}}{R^\varepsilon}\, |u_n-u_m|_{A'}{}^2,$$

the third term as

$$\int_{R^3} |\text{grad}\, u|^2\, dx + \int_{|x|<R'} |u|^2\, dx \leqslant C(R')\, |u|_{A'}^2 \quad (u \in H_{A'}),$$

and the first term presents no problem.

$\{u_j\}$ is therefore a bounded sequence of $\mathscr{E}_{L^2}{}^1(\Omega)$ in $\Omega = \{x; |x| < R'\} (R' > R)$. By Rellich's theorem, for $|x| < R$, a Cauchy sequence exists in the sense of L^2-norm.

Q.E.D.

Supplementary remarks

1 *General boundary value problems of high-order elliptic equations*

1. In chapter 3, we dealt with the Dirichlet problem of a high-order elliptic equation. However, we can find more general and systematic treatments of general boundary problems in existing literature such as Agmon, Douglis & Nirenberg, Schechter† We shall give a brief sketch of their arguments and some results derived from them. For simplicity, let us assume Ω to be a bounded domain of R^n and its boundary S to be formed with a finite number of sufficiently smooth hypersurfaces.

Let $A(x, \mathrm{D})$ be an mth order (m is an even integer) elliptic operator. Also, we assume that in this section boundary problems are always given in the form of

$$\left.\begin{array}{ll} A(x, \mathrm{D})\, u(x)=f(x) & (x\in\Omega) \\ B_p(x, \mathrm{D})\, u(x)=0 & (x\in S, j=1, 2, \ldots, b\,(=\tfrac{1}{2}m)) \end{array}\right\} \qquad \text{(A.1)}$$

where the boundary differential operators $\{B_j\}$ satisfy the following conditions.

(1) At every point x of S, the direction of the normal at x does *not* coincide with a characteristic direction of B_j (see below).

(2) The order m_j of B_j is such that $m_j<m$ and $m_j\neq m_k (j\neq k)$.

More precisely, the condition (1) is equivalent to saying that, if $B_{0j}(x, \xi)$ is the homogeneous part of B_j of order m_j, then, for N_x, the inner normal of unit length, $B_{0j}(x, N_x)\neq 0$, where if $m_j=0$ then $B_j=c(x)$, $c(x)\neq 0$. We call $\{B_j\}_{j=1,2,\ldots,b}$ satisfying (1), (2) a *normal system*. Furthermore, we impose the following algebraic condition, which is called complementary condition on $\{A; B_j (j=1, 2, \ldots, b)\}$ on S:

Let $\xi(\neq 0)$ be an arbitrary vector which is parallel to the tangent plane of S at x. We consider $A_0(x, \xi+zN_x)=0$, an equation in the homogeneous part $A_0(x, \mathrm{D})$ of order m of A. Suppose that the equation has roots whose imaginary parts are positive and the number of such roots is $b(=\tfrac{1}{2}m)$. We

† See the bibliography at the end of the book.

write these roots as $z_1(x, \xi), z_2(x, \xi), \ldots, z_b(x, \xi)$. Let us put

$$A_{0+}(z; x, \xi) = \prod_{j=1}^{b} (z - z_j(x, \xi)).$$

In this case, we assume that the polynomials $Q_j(z) \equiv B_{0j}(x, \xi + zN_x)$, $j = 1, 2, \ldots, b$ with z as their indeterminate are linearly independent modulo $A_{0+}(z)$, i.e.

(C) From $\sum_{j=1}^{b} c_j Q_j(z) \equiv 0 \pmod{A_{0+}(z)}$,

it follows $c_1 = c_2 = \cdots = c_b = 0$.

Note. The last condition means that for each $x \in S$ and ξ the condition (C) is satisfied. For example, all the boundary value problems which we treated in chapter 3 satisfy (C). Now, if we consider (A.1) in L^2-space, then the inverse operator A^{-1} of A can be defined as follows: We define the domain of the definition of A as

$$\mathscr{D}(A) = \{u(x) \in \mathscr{E}_{L^2}{}^m(\Omega);\ B_j(x, D)u(x) \doteq 0,\ x \in S, j = 1, \ldots, b\}, \quad \text{(A.2)}$$

where $B_j u$ is regarded as a trace on S. In this case, if A is bijective onto $L^2(\Omega)$, then we call the inverse operator $A^{-1} = G$ a *Green's operator*.

In general, if a normal system $\{B_j\}$ satisfies (C), then always

$$\|u\|_m \leqslant C(\|Au\| + \|u\|) \quad (u \in \mathscr{D}(A)) \quad \text{(A.3)}$$

(the *a priori* estimation of the solution). If A is one-to-one, we have

$$\|u\|_m \leqslant C\|Au\| \quad (u \in \mathscr{D}(A)), \quad \text{(A.4)}$$

where we assume that the coefficients of A, B_j are sufficiently smooth. (This property is also assumed in what follows.)

In general, for a normal system $\{B_j\}$, there exists another normal system $\{B_j'(x, D)\}_{j=1,\ldots,b}$, and

$$\left.\begin{array}{ll} A^*(x, D)\, v(x) = g(x) & (x \in \Omega) \\ B_j'(x, D)\, v(x) = 0 & (x \in S, j = 1, 2, \ldots, b), \end{array}\right\} \quad \text{(A.5)}$$

gives the conjugate problem of (A.1) where A^* is the formal conjugate operator of A. We have already observed a simple case of the conjugate problem in chapter 3. In this case, we define the domain of definition of A^* as

$$\mathscr{D}(A^*)=\{v\in\mathscr{E}_{L^2}{}^m(\Omega);\ B_j'(x, \mathrm{D})\,v(x)=0,\quad x\in S, j=1,\dots, b\}.$$

On the other hand, $\{A^*, B_j'\}$ – we call this the adjoint system of $\{A; B_j\}$ – can be characterized as satisfying the following necessary and sufficient condition. For an arbitrary $u\in\mathscr{D}(A)$,

$$(Au, v)=(u, A^*v)\Leftrightarrow v\in\mathscr{E}_{L^2}{}^m(\Omega) \text{ belongs to } \mathscr{D}(A^*). \tag{A.6}$$

The converse is also true. That is, for an arbitrary $v\in\mathscr{D}(A^*)$, (A.6) is true $\Leftrightarrow u\in\mathscr{E}_{L^2}{}^m(\Omega)$ belongs to $\mathscr{D}(A)$.

Furthermore, we can demonstrate that if $\{A; B_j\}$ satisfies (C) then $\{A^*; B'_j\}$ also satisfies (C). Thus, the result of Schechter can be summarized as:

THEOREM A.1.

(1) The uniquess of solution of (A.5), (i.e. A^* is a bijection from $\mathscr{D}(A^*)$ into $L^2(\Omega)$) implies the existence of A^{-1}.

(2) If A^{-1} exists (i.e. A is a bijection from $D(A)$ onto $L^2(\Omega)$) then there exists $(A^*)^{-1}$ for the conjugate problem (A.5).

(3) From $u\in\mathscr{D}(A)$, $Au\in\mathscr{E}_{L^2}{}^s(\Omega)$, $s=0, 1, 2, \dots$ it follows that $u\in\mathscr{E}_{L^2}{}^{m+s}(\Omega)$. A similar property holds for A^*.

2. In this section, we assume the existence of $G=A^{-1}$, and derive some results.

LEMMA A.1. Let us assume G exists, then, the mapping $u(x)\in L^2(\Omega) \to G^s u\in\mathscr{E}_{L^2}{}^{ms}(\Omega)$ is continuous for $s=1, 2, 3, \dots$. If G^* is the conjugate operator of G in $L^2(\Omega)$, then G^* is nothing but a Green's operator for the conjugate problem (A.5), that is, $G^*=(A^*)^{-1}$ is true. Therefore, $u(x)\in L^2(\Omega) \to G^{*s}u\in\mathscr{E}_{L^2}{}^{ms}(\Omega)$ is also continuous. Of course, G and G^* are completely continuous from $L^2(\Omega)$ into itself as mappings.

Proof. From theorem A.1, the mapping from $u(x)\in\mathscr{E}_{L^2}{}^{m+s}(\Omega)\cap\mathscr{D}(A)$ into $Au\in\mathscr{E}_{L^2}{}^s(\Omega)$ is bijective. $\mathscr{E}_{L^2}{}^{m+s}(\Omega)\cap\mathscr{D}(A)$ is a closed subspace of $\mathscr{E}_{L^2}{}^{m+s}(\Omega)$, and it is a Hilbert space. Therefore, by theorem 2.5, the inverse mapping is also continuous. We have $\|u\|_{m+s} \leqslant C\|Au\|_s (u \in \mathscr{D}(A))$, i.e. $\|Gu\|_{m+s} \leqslant C\|u\|_s$, $u \in \mathscr{E}_{L^2}{}^s(\Omega)$. The subsequent application of this inequality yields

$$\|G^s u\|_{ms} \leqslant \text{const}\,\|u\| \quad (u \in L^2(\Omega)). \tag{A.7}$$

Next, we demonstrate $G^* = (A^*)^{-1}$. In (A.6) we take $u \in L^2(\Omega)$, $v \in \mathscr{D}(A^*)$. Then,

$$(u, v) = (AGu, v) = (Gu, A^* v) = (u, G^* A^* v),$$

i.e. $G^* A^* = I$ in $\mathscr{D}(A^*)$. On the other hand, by theorem A.1, A^* is a bijection from $\mathscr{D}(A^*)$ onto $L^2(\Omega)$. So that G^* is the inverse of A^*. The complete continuousness follows directly from theorem 3.14.

Q.E.D.

LEMMA A.2. G^s has a kernel representation of Hilbert–Schmidt type for an integer s satisfying $ms > \frac{1}{2}n$, i.e. for $u \in L^2(\Omega)$,

$$(G^s u)(x) = \int_\Omega H(x, y)\, u(y)\, dy, \quad \iint |H(x, y)|^2\, dx\, dy < +\infty,$$

where the integrals are to be understood as being taken everywhere in Ω.

Proof. From (A.7) and Sobolev's lemma (theorem 3.15), we see that $u \in L^2(\Omega) \to G^s u \in \mathscr{B}^0(\bar{\Omega})$ is continuous. More strongly, it is also continuous as a mapping into $C^\sigma(\bar{\Omega})$, $0 < \sigma \leqslant 1$. From this, it is true that for each x that $(G^s u)(x) = L_x[u]$ is a continuous linear form defined on $L^2(\Omega)$. Therefore, from Riesz theorem

$$(G^s u)(x) = \int_\Omega \hat{H}(x, y)\, u(y)\, dy, \quad \int |\hat{H}(x, y)|^2\, dy \leqslant \text{const},$$

and, more strongly

$$|(G^s u)(x) - (G^s u)(x')| \leqslant c\|u\| \cdot |x - x'|^\sigma. \tag{A.8}$$

Note that we are not yet sure whether $\hat{H}(x, y)$ is measurable with respect to (x, y).

To clarify this point, we consider a function in the form of $f(x, y) = \sum v_i(x) u_i(y)$, where u_i, v_i are both elements of $L^2(\Omega)$ and the sum should be understood as a finite sum. The set which consists of such $f(x, y)$ forms a dense subspace of $L^2(\Omega\times\Omega)$. Then, we define

$$\langle K, f\rangle=\sum_i\int_\Omega v_i(x)\,L_x[u_i]\,\mathrm{d}x\,.$$

K determines a linear form defined on a dense subspace of $L^2(\Omega\times\Omega)$ which is continuous with respect to K. In fact,

$$|\langle K, f\rangle|\leqslant|\int_\Omega L_x[f(x,y)]\,\mathrm{d}x|\leqslant\text{const}\int(\int|f(x,y)|^2\,\mathrm{d}y)^{\frac{1}{2}}\,\mathrm{d}x$$
$$\leqslant\text{const mes}(\Omega)^{\frac{1}{2}}\|f(x,y)\|\,.$$

Therefore, K has a unique extension to a continuous linear form of $L^2(\Omega\times\Omega)$. By Riesz theorem, there exists $H(x,y)\in L^2(\Omega\times\Omega)$ and

$$\langle K, f\rangle=\iint H(x,y)\,f(x,y)\,\mathrm{d}x\,\mathrm{d}y\,.$$

In particular, if $f=v(x)u(y)$, we have

$$\int_\Omega v(x)\,L_x[u]\,\mathrm{d}x=\int_\Omega v(x)\,(\int_\Omega H(x,y)\,u(y)\,\mathrm{d}y)\,\mathrm{d}x\,.$$

Q.E.D.

Moreover, we have the following:

LEMMA A.3. Let $T=G^sBG^{s'}$. if $s, s'>n/2m$ (where s, s' are integers) and if B is a bounded operator of $L^2(\Omega)$, then T has a continuous kernel representation, i.e.

$$(Tf)\,(x)=\int_\Omega T(x,y)\,f(y)\,\mathrm{d}y\,,\qquad T(x,y)\in B^0(\bar\Omega\times\bar\Omega)\,.$$

In this case, for a linear mapping $B\to T(x,y)$,

$$\sup_{\Omega\times\Omega}|T(x,y)|\leqslant c(s,s')\,\|B\| \tag{A.9}$$

where $\|B\|$ represents an operator norm of B in $L^2(\Omega)$.

Proof. As demonstrated in the proof of the previous lemma we have

$$(G^su)\,(x)=\int_\Omega \hat{H}(x,y)\,u(y)\,\mathrm{d}y\qquad(u\in L^2(\Omega))\,.$$

On the other hand, $(BG^{s'})^*=G^{*s'}B^*$ is a continuous operator from $L^2(\Omega)$

into $\mathscr{B}^0(\bar{\Omega})$ (see lemma A.1 and the proof of the previous lemma). So that $(BG^{s'})^*f=\int \hat{K}(x, y)f(y)\,dy$. Also, $(BG^{s'})^*f=\int K(x, y,)\,f(y)\,dy$ for x almost everywhere, where $K(x, y)\in L^2(\Omega\times\Omega)$. Therefore,

$$(Tf)\,(x)=\int_\Omega \hat{H}(x, z)(\int \overline{K(y, z)}\,f(y)\,dy)\,dz\,.$$

From Fubini's theorem,

$$(Tf)(x)=\int_\Omega \Phi(x, y)\,f(y)\,dy\,,$$

$$\Phi(x, y)=\int_\Omega \hat{H}(x, z)\,\overline{K(y, z)}\,dz\,.$$

On the other hand, as a function of $y(\in\Omega)$

$$T(x, y)=\int_\Omega \hat{H}(x, z)\,\overline{\hat{K}(y, z)}\,dz$$

almost everywhere in Ω. T is a function of (x, y) and continuous with respect to (x, y) in $\bar{\Omega}\times\bar{\Omega}$. In fact, if we decompose

$$T(x, y)-T(x', y')=\int \{\hat{H}(x, z)-\hat{H}(x', z)\}\,\overline{\hat{K}(y, z)}\,dz$$
$$+\int \hat{H}(x', z)\,\{\overline{\hat{K}(y, z)}-\overline{\hat{K}(y', z)}\}\,dz$$

and we note that (A.8) is valid for $\hat{H}$ and $\hat{K}$, the last inequality is obvious.

Q.E.D.

COROLLARY. If $s\geqslant 2([n/2m]+1)$ then G^s has a continuous kernel representation.

3. In this section we shall consider the properties of an eigenvalue of A. By $\omega\in\mathscr{D}(A)$ is an eigenfunction of A we mean that ω satisfies $A\omega=c\omega$, $\omega(x)\not\equiv 0$. In this case c is an eigenvalue. If we assume the existence of the inverse G of A, this is equivalent to saying that c satisfies $(I-cG)\omega=0$. However, if G is not Hermitian, then we define an eigenfunction in a more general form such that for a certain k, $(I-cG)^k u=0$ $(u\in L^2(\Omega))$.

In this case, we call u a *generalized eigenfunction* or a *principal function* for the eigenvalue c. We note that this condition is equivalent to $(cI-A)^k u=0$ $(u\in\mathscr{D}(A^k))$, where k may be different for each u. If we apply lemma 3.8 to

$T=I-cG$ we see that the dimension of the space of principal function is finite, and there exists a (common) n and all principal functions satisfy $(I-cG)^n u=0$. We call the dimension of the space of principal functions the *degree* of an eigenvalue c.

If G is Hermitian, then the principal functions and the eigenfunctions are entirely the same. So that the degree of c is equal to the multiplicity of the eigenfunctions. Following the argument of Riesz & Nagy (pp. 178–85) we observe the following. Let the resolvent $\Gamma_G(\lambda)$ be

$$(I-\lambda G)^{-1}=I+\lambda\Gamma_G(\lambda) \tag{A.10}$$

where $\Gamma_G(\lambda)$ is a meromorphic function such that a pole $\lambda=c$ of the function has a Laurent expansion

$$\Gamma_G(\lambda)=\sum_{p=1}^{m}\frac{B_p(x,y)}{(\lambda-c)^p}+H(\lambda)$$

in the neighbourhood of $\lambda=c$, where $H(\lambda)$ is regular in the neighbourhood of $\lambda=c$, and $B_1(x,y)=\sum_{i=1}^{p}\varphi_i(x)\psi_i(y)$. $\{\varphi_i\}_{i=1,2,\dots,p}$ are linearly independent principal functions corresponding to $\lambda=c$, and span the whole space of principal functions, i.e. p represents the dimension of the eigenvalue $\lambda=c$. $B_j(x,y), j\geqslant 2$, can be similarly expressed, i.e. $B_j=\sum\varphi_i^{(j)}(x)\psi_i^{(j)}(x)$ where $\{\varphi_i^{(j)}(x)\}$ are also principal functions. Finally, we note that $\int\varphi_i(x)\psi_j(x)\,dx=-\delta_i^{\,j}$.

Proposition A.1. Let the eigenvalues of G be $\{\lambda_j\}$ and let the degree of λ_j be m_j. The eigenvalues of G^k are $\{\lambda_j^{\,k}\}$ such that the principal functions corresponding to $\lambda_j^{\,k}$ are exactly the same as those corresponding to λ_j of G. In this case, if $\lambda_{j_1}^{\,k}=\lambda_{j_2}^{\,k}=\cdots=\lambda_{j_s}^{\,k}(=c^k)$ then the principal functions corresponding to the eigenvalues c^k of G^k can be expressed as the set union of the spaces of principal functions corresponding to $\lambda_{j_1},\dots,\lambda_{j_2}$. Therefore, the degree of c^k is equal to $m_{j_1}+m_{j_2}+\cdots+m_{j_s}$.

Note. The principal functions which correspond to different eigenvalues of G are linearly independent.

Proof. First we note that

$$(I-\mu G^k)^{-1}-I=\frac{1}{2\pi i}\int_\Gamma\left\{\left(1-\frac{\mu}{\lambda^k}\right)^{-1}-1\right\}\Gamma_G(\lambda)\,d\lambda$$

where Γ is a sufficiently small circle of radius ε centred at the origin, which is described anti-clockwise, and μ satisfies $|\mu/\lambda^k|<1$. In fact, we can write $\Gamma_G(\lambda)=\sum_{j=1}^{\infty}\lambda^{j-1}G^j$ in the neighbourhood of the origin. Therefore, we have

$$\frac{1}{2\pi i}\int_\Gamma \frac{1}{\lambda^{km}}\Gamma_G(\lambda)\,d\lambda = G^{km} \qquad (m=1,2,3,\ldots).$$

The right hand side of the equation is equal to $\sum_{m=1}^{\infty}\mu^m G^{km}$. We rewrite the equation as

$$\Gamma(\mu;G^k)=\frac{-1}{2\pi i}\int_\Gamma \frac{1}{\mu-\lambda^k}\Gamma_G(\lambda)\,d\lambda.$$

Now, we replace Γ with a larger circle $|\lambda|=R$ and add the sum of residues within the circle. We have

$$\Gamma(\mu;G^k)=-\frac{1}{2\pi i}\int_{|\lambda|=R}\frac{1}{\mu-\lambda^k}\Gamma_G(\lambda)\,d\lambda+\sum_i \operatorname*{Res}_{\lambda=\lambda_i}\left[\frac{1}{\mu-\lambda^k}\Gamma_G(\lambda)\right] \tag{A.11}$$

where we take the residues for λ_i which satisfy $|\lambda_i|<R$. (A.11) is regarded as the equation giving an analytic extension of $\Gamma(\mu;G^k)$. The first term of the right hand side of the equation is regular in $|\mu|<R^k$ where R can be taken as large as we wish, so that the singularities of $\Gamma(\mu;G^k)$ occur at $\{\lambda_i{}^k\}_{i=1,2,\ldots}$ at most. We calculate the first term of the residues appearing on the right hand side. To do this, we put $\lambda_i=c$ and

$$\Gamma_G(\lambda)=\sum_{j=1}^{m}\frac{B_j}{(\lambda-c)^j}+H(\lambda)$$

around c. We have

$$\operatorname*{Res}_{\lambda=c}\left[\frac{1}{\mu-\lambda^k}\Gamma_G(\lambda)\right]=\sum_j \operatorname*{Res}_{\lambda=c}\left[\frac{1}{(\mu-\lambda^k)(\lambda-c)^j}\right]B_j+\operatorname*{Res}_{\lambda=c}\left[\frac{1}{\mu-\lambda^k}H(\lambda)\right]$$

where the last term is 0. On the other hand, from

$$\varphi_j(\mu;c)=\operatorname*{Res}_{\lambda=c}\left[\frac{1}{(\mu-\lambda^k)(\lambda-c)^j}\right]=\frac{1}{2\pi i}\int_{|\lambda-c|=\varepsilon}\frac{d\lambda}{(\mu-\lambda^k)(\lambda-c)^j},$$

we have

$$\varphi_1(\mu;c)=\frac{1}{\mu-c^k},\qquad \varphi_j(\mu;c)=\frac{1}{(j-1)!}\left(\frac{\partial}{\partial c}\right)^{j-1}\varphi_1(\mu;c),$$

so that,

$$\varphi_2(\mu; c) = \frac{kc^{k-1}}{(\mu - c^k)^2}, \ldots.$$

Therefore, for $j \geqslant 2$, $\varphi_j(\mu; c)$ has a pole of order more than 2 at $\mu = c^k$. Taking account of the remark given at the beginning of this section, the proposition is proved.

Q.E.D.

PROPOSITION A.2. Let the eigenvalues of a Green's operator G be $\{\lambda_j\}$, where we allow the duplication of λ_j the same number of times as its degree indicates. In this case, for s(integer)$>n/2m$, we have

$$\sum_j \frac{1}{|\lambda|^{2s}} < +\infty.$$

Proof. We use Schur's theorem (for example, see Goursat, pp. 481–3) which says that, for a kernel of Hilbert–Schmidt type $K(x, y)$ (whose symmetricity is *not* assumed),

$$\sum_p \frac{1}{|\lambda_p|^2} \leqslant \iint |K(x, y)|^2 \, dx \, dy \tag{A.12}$$

where $\{\lambda_j\}$ are the eigenvalues of K allowing duplications in the same way as before. From lemma A.2, for $s > n/2m$, $G^s = K$ is a kernel of Hilbert–Schmidt type and the eigenvalues of G^s can be given as $\{\lambda_j{}^s\}_{j=1,2,\ldots}$ with duplications.

Q.E.D.

2 *Completeness of a system of eigenfunctions*

1. All the boundary value problems of chapter 3 can be regarded as special cases of a general boundary problem treated in 1. Nevertheless, these problems have special properties which differentiate them from other problems. The reason for this is that for these problems theorem 3.19 is always valid and A becomes an infinitesimal generator of a parabolic semi-group, i.e. e^{tA}, $t>0$ is a parabolic semi-group. In this section 2 we restrict ourselves to the case in which A is an infinitesimal generator of a parabolic semi-group. In chapter 3 we did not touch upon the problem of the existence of eigenvalues and the more difficult problem of whether there exists an infinite number of eigenvalues. We shall demonstrate that, under certain conditions, there does indeed exist an infinite number of eigenvalues and that the entire set of generalized eigenfunctions spans the entire space $L^2(\Omega)$.

To prove this, we first pick a sufficiently large $\beta>0$ and replace A by $A-\beta I$. So we can assume that the spectrum of A lies in a fan-shaped domain Σ: $|\arg\lambda-\pi|\leqslant\varphi_0<\frac{1}{2}\pi$ with respect to λ, and

$$\|(\lambda I-A)^{-1}\|\leqslant\frac{C}{|\lambda|+1}\quad(\lambda\in C^1\setminus\Sigma).$$

In this case, the spectrum of A satisfies $\mathrm{Re}\,\lambda\leqslant-\delta$ for a certain $\delta(>0)$. From the argument of section 4 of chapter 5, we can define

$$\mathrm{e}^{tA}=\frac{1}{2\pi\mathrm{i}}\int_L(\lambda I-A)^{-1}\,\mathrm{e}^{\lambda t}\,\mathrm{d}\lambda\quad(t>0).\tag{A.13}$$

We make $\mathrm{Re}\,\lambda<0$ by taking the contour of Σ as Γ with a little alteration of the path of integration in the neighbourhood of the origin. In this case,

$$G=-\int_0^\infty\mathrm{e}^{tA}\,\mathrm{d}t.\tag{A.14}$$

In fact, from (A.13), the right hand side is equal to

$$\frac{1}{2\pi\mathrm{i}}\int_\Gamma(\lambda I-A)^{-1}\frac{\mathrm{d}\lambda}{\lambda},$$

so that it represents A^{-1}.

Now, we impose a stronger assumption. If an infinite number of eigenvalues $\{\lambda_j\}$ does exist, then they have only one asymptotic direction which is the negative real axis. More precisely, if we move λ along an arbitrary radius vector which is not the negative real axis, then, at a sufficiently distant point from the origin, there exists resolvent $(\lambda I-A)^{-1}$ and

$$\|(\lambda I-A)^{-1}\|\leqslant C/|\lambda|\quad(\lambda=r\,\mathrm{e}^{i\theta},\ r\geqslant r_0),\tag{A.15}$$

where r_0, and C may depend upon the direction of the radius vector $\theta(\neq\pi)$. It is easy to prove this for the problem we dealt with in chapter 3. (See theorem 3.19 and note.)

Henceforth, we shall consider G^{2k} where we put $k=2([n/2m]+1)$ taking account of the corollary of lemma A.3, and use a 'classical' determinant representation due to Fredholm. By the corollary just mentioned we have a continuous kernel representation of G^k. Let φ be an angle >0 satisfying $2k\varphi<\frac{1}{2}\pi$. Fix such φ. We write Σ_β for a fan-shaped domain $|\arg\lambda-\pi|\leqslant\beta$ with respect to β.

LEMMA A.4.

(1) Let us write the eigenvalues of G as $\{\lambda_j\}$. Then, the eigenvalues of G^{2k} are $\{\lambda_j{}^{2k}\}$. In this case, the principal function of G with respect to λ_j and that of G^{2k} with respect to $\lambda_j{}^{2k}$ coincide.

(2) Let $\Gamma(\mu; -G^{2k})$ be the resolvent of $-G^{2k}$. We pick α such that $2k\varphi<\alpha<\frac{1}{2}\pi$. Then $\Gamma(\mu; -G^{2k})$ is regular except for at most a finite number of poles $\mu\in C^1\setminus\Sigma_\alpha$, and $\|\Gamma(\mu; -G^{2k})\|$ tends to 0 uniformly as $|\mu|\to\infty$.

Proof. As (1) has been proved already in lemma A.1, we only prove (2). In general,

$$G^m=\frac{1}{2\pi i}\int_L (\lambda I-A)^{-1}\frac{d\lambda}{\lambda^m} \qquad (m=1, 2, \ldots).$$

From this, for a neighbourhood μ of the origin, we obtain

$$(I+\mu G^{2k})^{-1}-I=\frac{1}{2\pi i}\int_\Gamma\left(\frac{1}{1+\mu\lambda^{-2k}}-1\right)(\lambda I-A)^{-1}\, d\lambda$$

in the same way as we proved proposition A.1, where we note that

$$(\lambda I-A)^{-1}=-\Gamma_G(\lambda). \tag{A.16}$$

In fact, from $(\lambda I-A)u=f$, it follows that $u=-(I-\lambda G)^{-1}Gf$.

We consider an expansion in a neighbourhood of $\lambda=0$ to conclude $u=-\Gamma_G(\lambda)f$. From this

$$\Gamma(\mu; -G^{2k})=\frac{1}{2\pi i}\int_\Gamma\frac{1}{\mu+\lambda^{2k}}\Gamma_G(\lambda)\, d\lambda,$$

which is valid in the neighbourhood of $\mu=0$. Let us replace Γ by Γ', the contour of Σ_φ. At the same time we alter the path in the neighbourhood of the origin to make Γ' lie within the half-plane $\operatorname{Re}\lambda<0$.

We note that $\{\lambda_i\}$ lying between Γ and Γ' are at most finite in number (according to our hypothesis (A.15)). Therefore, we write

$$\Gamma(\mu; -G^{2k})=\frac{1}{2\pi i}\int_{\Gamma'}\frac{1}{\mu+\lambda^{2k}}\Gamma_G(\lambda)\, d\lambda+\sum_i \operatorname*{Res}_{\lambda=\lambda_i}\left[\frac{1}{\mu+\lambda^{2k}}\Gamma_G(\lambda)\right]. \tag{A.17}$$

On Γ' the condition (A.15) is valid. So that the first term of the right hand side is regular for $\mu \in C^1 \setminus \Sigma_\alpha$. Also in this domain it tends to 0 uniformly as $|\mu| \to \infty$.

Q.E.D.

2. Note that G^{2k} is an iterated kernel of a continuous kernel $G^k(x, y)$. We write

$$\Gamma(\lambda; -G^{2k}) = \frac{\mathrm{D}(x, y; \lambda)}{\mathrm{D}(\lambda)} \qquad ((x, y) \in \Omega \times \Omega)$$

for a representation of a Fredholm determinant. We have:

Lemma A.5. For an entire function (integral function) of λ as an indeterminant we have the following estimation

$$|\mathrm{D}(\lambda)| = O(\mathrm{e}^{\alpha|\lambda|}), \ \sup_{(x, y)} |\mathrm{D}(x, y; \lambda)| = O(\mathrm{e}^{\alpha|\lambda|}) \qquad (a > 0),$$

as $|\lambda| \to \infty$.

Proof. Let K be a continuous kernel. We write $G = K^2$, and

$$\Gamma_G(x, y; \lambda) = \mathrm{D}_G(x, y; \lambda)/\mathrm{D}_G(\lambda)$$

(Fredholm's determinant representation). Then, it is enough to show that $\mathrm{D}_G(\lambda)$, $\mathrm{D}_G(x, y; \lambda)$ have the properties described in the lemma.
From

$$(I - \lambda G)^{-1} = \{I + \lambda^{\frac{1}{2}} \Gamma_K(\lambda^{\frac{1}{2}})\}\, \{I - \lambda^{\frac{1}{2}} \Gamma_K(-\lambda^{\frac{1}{2}})\}$$

we have

$$\Gamma_G(\lambda) = \{\Gamma_K(\lambda^{\frac{1}{2}}) - \Gamma_K(-\lambda^{\frac{1}{2}})\}/\lambda^{\frac{1}{2}} - \Gamma_K(\lambda^{\frac{1}{2}})\Gamma_K(-\lambda^{\frac{1}{2}}).$$

Also, from

$$\mathrm{D}_K(\lambda^{\frac{1}{2}})\mathrm{D}_K(-\lambda^{\frac{1}{2}}) = \mathrm{D}_G(\lambda)$$

we have

$$\begin{aligned} \mathrm{D}_G(x, y; \lambda) = \{&\mathrm{D}_K(x, y; \lambda^{\frac{1}{2}})\mathrm{D}_K(-\lambda^{\frac{1}{2}}) - \mathrm{D}_K(x, y; -\lambda^{\frac{1}{2}})\mathrm{D}_K(\lambda^{\frac{1}{2}})\}/\lambda^{\frac{1}{2}} \\ &- \mathrm{D}_K(x, y; \lambda^{\frac{1}{2}})\mathrm{D}_K(x, y; -\lambda^{\frac{1}{2}}) \end{aligned}$$

For $\mathrm{D}_K(x, y; \lambda)$, $\mathrm{D}_K(\lambda)$ we have

$$\sup_{(x, y)} |\mathrm{D}_K(x, y; \lambda)| = O(\exp(a|\lambda|^2)), \ |\mathrm{D}_K(\lambda)| = O(\exp(a|\lambda|^2)) \qquad \text{(A.18)}$$

where a is a positive constant. (Use the elementary properties of Fredholm's determinant, see below.) Therefore, $\mathrm{D}_G(x, y; \lambda)$ has the desired property.

Q.E.D.

Note (*The proof of* (A.18)). First we note that

$$\mathrm{D}_K(x, y; \lambda) = \sum_{n=0}^{\infty} \frac{(-1)^n}{n!} \int_\Omega \ldots \int_\Omega K\begin{pmatrix} x & s_1 & s_2 & \cdots & s_n \\ y & s_1 & s_2 & \cdots & s_n \end{pmatrix} \mathrm{d}s_1 \ldots \mathrm{d}s_n .$$

From Hadamard's inequality, we see that the integral of the nth term does not exceed $c^n(n+1)^{\frac{1}{2}(n+1)}$ in its absolute value (c is an appropriate constant). Therefore, if we take c' larger than c, then the absolute value does not exceed $c'^n n^{\frac{1}{2}n}$ if n is sufficiently large. On the other hand, from Stirling's formula, we have $1/n! < \mathrm{e}^n n^{-n}$ if n is sufficiently large, so that

$$|\mathrm{D}_K(x, y; \lambda)| \lesssim \sum_{n=0} \frac{|c'\mathrm{e}\lambda|^n}{n^{\frac{1}{2}n}} .$$

But, we know the following asymptotic equation

$$\sum_{n=0}^{\infty} \left(\frac{r}{n^{1/\rho}}\right)^n = (1+\eta(r)) \left(\frac{2\pi\rho}{\mathrm{e}}\right)^{\frac{1}{2}} r^{\frac{1}{2}\rho} \exp \frac{r^\rho}{\rho \mathrm{e}}$$

as $r \to +\infty$ where $p > 0$ and $\eta(r) \to 0$ $(r \to +\infty)$ (Carleman [3]).

Q.E.D.

3. We shall prove the following theorem from the facts obtained so far.

THEOREM A.2. For the boundary value problem (A.1), if the normal system $\{A; B_j\}$ satisfies the condition (C) of section 1 and the condition (A.15), then the system of the generalized eigenfunctions of A spans the whole space $L^2(\Omega)$. Therefore, the number of the eigenvalues of A is infinite.†

Proof. We prove this by contradiction. If it is not true, then there exists a function $f \in L^2(\Omega)$, $f \neq 0$ and f is orthogonal with every generalized eigenfunction of A. Therefore, from lemma A.4, f is orthogonal with every principal function of A. Hence, by lemma A.4, f is orthogonal with every princi-

† Agmon [1] gives more precise results.

pal function of $-G^{2k}$. Now we take an arbitrary $g \in L^2(\Omega)$, and consider

$$F(\lambda)=(f(x), \Gamma(\lambda; -G^{2k})g(x)).$$

Obviously, F is regular in the entire space. On the other hand, from lemma A.4, in the set-complement of the fan-shaped domain Σ_α: $|\arg\lambda-\pi| \leqslant \alpha_0(<\frac{1}{2}\pi)$, $C^1 \setminus \Sigma_\alpha$, we see that F is bounded and tends to 0 as $|\lambda| \to +\infty$.

We look at F in Σ_α. It is bounded at the contour of Σ_α and

$$F(\lambda)=(f(x), \int \mathrm{D}(x, y; \lambda)g(y)\,dy)/\mathrm{D}(\lambda).$$

On the other hand, for D, there exists a sequence $r_1, r_2, \cdots \to +\infty$ (by lemma A.5), and for $|\lambda|=r_j$, $|\mathrm{D}(\lambda)| \geqslant \exp(-a|\lambda|)$. If we apply the previous lemma to $\mathrm{D}(x, y; \lambda)$, then for $|\lambda|=r_j (j=1, 2, \ldots)$, $|F(\lambda)| \leqslant$ const $\exp(2a|\lambda|)$. So that, by the Phragmén–Lindelöf theorem, F is bounded in Σ_α. Therefore, F is bounded in the entire space, hence, it is constant. $F(\lambda)=0$ as $\lambda \to +\infty$, so that, if we put $F(\lambda) \equiv 0$, $\lambda=0$, then $(f(x), G^{2k}g(x))=0$. The image of $\{G^{2k}g\}$ in $L^2(\Omega)$ forms a dense set as g runs through $L^2(\Omega)$. From this it follows $f=0$, which is a contradiction.

Q.E.D.

Guide to the literature

In this book we have been concerned with fundamental results and have explored the theory of partial differential equations. But our approach to this vast subject has relied on functional analysis. In this respect we are afraid of being partial; however, we shall give some guiding remarks here. However, we shall not give an encyclopaedic list of the existing literature. Therefore, many important works are not even mentioned. In what follows, as a principle, we shall not repeat references already given in the text.

Chapter 1 (Fourier series, Fourier transforms)

In the main, we have followed the line of argument found in Lebesgue and Bochner. We recommend that interested readers should consult Wiener, Paley & Wiener, and Carleman [5] to acquire a more complete knowledge.

Chapter 2 (Distributions)

In the main, we followed Schwartz [1, 2]. We referred to Dieudonné & Schwartz in connection with Fréchet spaces. The work of Sobolev [1] is recommendable for a thorough understanding of the theory of distributions. In this book we kept our presentation of the theory to the minimum necessary for our purpose.

For example, a more advanced topic such as *Théorie des noyaux* can only be understood on the basis of a thorough knowledge of nuclear space (*espace nucléaire*), and the tensor product of topological vector spaces, which we did not treat in this book. For this purpose, Schwartz [5, 8] and Grothendieck give the essentials.

Also, we refer to the book written by Yosida, Kawata & Iwamura (in Japanese) which treats closely related topics to ours. Also, Gelfand & Silov may be useful for the same purpose.

Finally, in connection with Laplace transforms in the several-complex case we should mention Leray [1] Schwartz [4].

Chapter 3 (Elliptic equation)

The beginning half of this chapter consists of examples in respect of which the fundamental ideas of functional analysis are very effective. Or, rather, we should say that functional analysis itself sprang from the study of these examples. This part of the chapter is based on Schwartz [6, 7]. Also, we mention Lions in which the reader may find detailed arguments. For the classical results we have Schauder [1]. Miranda also gives the classical results in detail. We note that Schauder's estimation was obtained in a general form in Agmon, Douglis & Nirenberg. The well-known text, Courant & Hilbert [1, 2], places considerable importance on this subject.

There are many Japanese books written on this subject, too. For example we have Nagumo [2], Yosida [1], Inui, etc. Nagumo [1] touches upon the boundary value problem of second-order semi-linear elliptic equations which we did not treat in this text. For the solution $u(x)$ of a linear elliptic equation with real analytic coefficients:

$$A\left(x, \frac{\partial}{\partial x}\right) \equiv \sum a_\alpha(x) \left(\frac{\partial}{\partial x}\right)^\alpha,$$

$$A\left(x, \frac{\partial}{\partial x}\right) u(x) = f(x),$$

it is known that if f is a real analytic function, then u is so too. This type of operator A is called analytic–hypoelliptic. The property is established for systems of non-linear elliptic equations of a general type, Petrowsky [1].

For the linear case, John [1] gives a proof based on fundamental solutions. For fundamental solutions (and parametrics) we refer Schwartz [1]. For the asymptotic distribution of the eigenvalues concerned with the interior problem of self-adjoint elliptic equations, a great deal of research has already been carried out. For example, we have Courant & Hilbert [2], Carleman [4], Gårding [1], Minakshisundaram, and Mizohata & Arima.

We briefly discuss some aspects of elliptic operator. Let

$$A\left(x, \frac{\partial}{\partial x}\right) = \sum a_\alpha(x) \left(\frac{\partial}{\partial x}\right)^\alpha.$$

For simplicity, we assume all $a_\alpha \in C^\infty$. Let $u(x)$ be an arbitrary distribution-solution of $A(x, \partial/\partial x)\, u(x) = g(x)$, then $u \in C^\infty$ where $g \in C^\infty$. We call A satisfying this property *hypoelliptic*. According to Hörmander [4], the statement that if A has constant coefficients, (i.e. $A(\partial/\partial x)$), then for arbitrary $|\beta|>0$, $A^{\beta}(i\xi)=o(A(i\xi))$, $|\xi|\to+\infty$, where we put $(\partial/\partial\xi)^{\beta}A(i\xi)=A^{\beta}(i\xi)$ for a vector $i\xi=(i\xi_1,\ldots,i\xi_n)$ (ξ is a real vector) is a necessary and sufficient condition for A to be hypoelliptic. The same statement is equivalent to that for certain $d(>0)$,

$$A^{\beta}(i\xi)=O(|\xi|^{-d}A(i\xi)) \quad (|\xi|\to+\infty).$$

To prove this equivalence we use an elementary lemma of algebraic geometry. In this case, if we take β such that $|\beta|$ is equal to the order of A, then $|A(i\xi)|\geqslant \geqslant c|\xi|^{d}(c>0)$, $|\xi|\to+\infty$. If A has variable coefficients, we can obtain some partially extended results of these properties (Malgrange, Hörmander [2], Treves). We note that parabolic equations are hypoelliptic.

Chapter 4 (Initial value problem)

In section 1, chapter 4 we defined a characteristic surface. Historically, the notion is derived from a physical notion, that of a 'wave front' for a hyperbolic equation. The situation is as follows. We consider an equation

$$\sum_{i,j=1}^{n} a_{ij}(x)\frac{\partial^2 u}{\partial x_i\partial x_j}+\sum_{i=1}^{n} a_i(x)\frac{\partial u}{\partial x_i}+c(x)u=0.$$

Also, we consider a surface S such that in the neighbourhood of $x=x_0\in S$, $\varphi(x)=0$ where $\Sigma|\varphi x_i|\neq 0$. In this case, let us assume that, for $m\geqslant 2$, in the neighbourhood of $x=x_0$ we have a solution satisfying

$$u(x)=\begin{cases}\varphi(x)^m v(x) & (\varphi(x)\geqslant 0),\\ 0 & (\varphi(x)\leqslant 0),\end{cases}$$

where $v(x_0)\neq 0$. By substitution, and comparison of the coefficients of $\varphi(x)^{m-2}$, we see that

$$\sum_{i,j=1}^{n} a_{ij}(x_0)\frac{\partial\varphi}{\partial x_i}(x_0)\frac{\partial\varphi}{\partial x_j}(x_0)=0$$

is a necessary condition. In particular, if the equation is of the form

$$\frac{\partial^2 u}{\partial t^2}-c^2\Delta u=0$$

(wave equation), then the equation of the wave front $\varphi(x, t)=0$ satisfies

$$\left(\frac{\partial\varphi}{\partial t}\right)^2 - c^2 \sum_{i=1}^{n} \left(\frac{\partial\varphi}{\partial x_i}\right)^2 = 0.$$

The converse of this statement is important. In general, let

$$L[u] \equiv \sum_{|\alpha| \leqslant m} a_\alpha(x) \left(\frac{\partial}{\partial x}\right)^\alpha u(x) = 0$$

where all $a_\alpha(x)$ are real-valued real analytic functions. Let us assume

$$h(x, \xi) = \sum_{|\alpha| = m} a_\alpha(x) \xi^\alpha,$$

we have a characteristic surface $\varphi(x)=0$ through $x=x_0$, and φ is real and analytic where $\Sigma|\varphi_{x_i}(x)| \neq 0$.

In this case, if we have

$$\sum_{i=1}^{n} |h_{\xi i}(x, \varphi_x)| \neq 0, \tag{S}$$

then there exists the same type of solution as in the previous general case.

In fact, if we consider a formal solution

$$u(x) = f_0(\varphi)u_0(x) + f_1(\varphi)u_1(x) + \cdots + f_p(\varphi)u_p(x) + \cdots$$

where

$$f_j(\tau) = \begin{cases} \tau^{m+j+p}/(m+s+p)! & (\tau \geqslant 0), \\ 0 & (\tau \leqslant 0), \end{cases}$$

and s is non-negative and real, then we can define appropriate $u_0(x) \neq 0$, $u_1(x)$, $u_2(x)$. ... such that in the neighbourhood of $x=x_0$ this formal solution satisfies $L[u]=0$ as a genuine solution. This idea was first conceived by Hadamard, and plays an important rôle in the study of the formation of fundamental solutions (Hadamard [2, 1], Leray [2], Ludwig, Mizohata [3]).

We should mention the theory of partial differential equations of the 'principal type' due to Hörmander, where 'principal type' means L such that for a real vector $\xi(\neq 0)$ satisfying $h(x, \xi)=0$,

$$\sum_{i=1}^{n} \left| \frac{\partial}{\partial \xi_i} h(x, \xi) \right| \neq 0.$$

Hörmander [4] gives a very detailed study of this type. His work is partially motivated by the work done by Lewy who mentioned an example of a partial differential operator without solution.

Lewy's example has caused a lot of interest and development of the study of equations of the principal type. His example was

$$-\frac{\partial u}{\partial x_1}-i\frac{\partial u}{\partial x_2}+2i(x_1+ix_2)\frac{\partial u}{\partial x_3}=\varphi'(x_3)$$

defined on R^3. If u is continuously once differentiable in the neighbourhood of the origin of R^3, then φ must be a real analytic function of x_3. So that if φ is not real analytic at the origin, then, after giving φ on the right hand side of the equation, there is no solution u for the equation in *any* neighbourhood of the origin. Hörmander gives the reason for the non-existence of the solution from an entirely different view-point (Hörmander [4]).

We discussed the system of the first-order hyperbolic equations, but as far as the 'strongly hyperbolic system' was concerned, we stopped pursuing the idea in theorem 4.14. For the characterization of a strongly elliptic system we refer to Kasahara & Yamaguti.

Chapter 5 (Evolution equation)

Mainly, we followed Hille & Phillips, together with Schwartz [9]. For details, see Yosida [2]. Parabolic semi-groups are also called analytic semi-groups. The theory of a semi-group of this type is recent. For reference, we mention Tanabe, Kato & Tanabe, Sobolevskii where fractional powers play a vital rôle. For fractional powers see Kotake & Narasimhan. Recently, the operational-theoretic approach to Navier–Stokes became possible. See, for example, Fujita & Kato.

We did not study the initial and boundary value problems of parabolic equations in detail in this book. Interested readers should refer to Levi, Itô, Eidelman [1, 2], and Arima. A common theme of this research to seek Hölder's estimation of a solution by obtaining Green's functions directly. Conversely, this way of viewing the problems may clarify the behaviour of the solution of an elliptic equation.

Chapter 6. (Hyperbolic equations)

The theory begins with Hadamard [2]. In the modern form of L^2 theory it originated with Friedrichs & Lewy, then Schauder [2] treated a general second-order hyperbolic equation, and finally, Petrowsky [2, 3] gave the current theory. But Petrowsky's original proofs were very complicated. Later, Leray [1] organized and developed the theory.

Gårding [3], starting from the view-point of Riesz, created a beautiful theory in the case of constant coefficients which was essentially the same as the original theory established by Petrowsky, but Gårding had a new methodology. Leray soon extended his method and reorganized Petrowsky's theory in the light of Laplace transforms (Leray [1]) which Gårding [4], again, reformulated in a clearer way using Fourier transforms.

On the other hand, Friedrichs [1] treated the first-order symmetric system and pointed out that this type of system is the most intrinsic one of the various types of hyperbolic systems and said that there are essential difficulties with the method by which a given system is transformed into a symmetric one.

It is difficult to say which approach (i.e. Leray & Gårding or Friedrichs) is superior to the other because both approaches have their advantages and disadvantages as well. In this book, we employed Friedrichs' method. The present author himself gave Mizohata [1, 2] based on Friedrichs' method Among these, [1] consists of the contents of the present chapter 6, but [2] is not treated here.

We must also mention the method developed by Shirota. In this book we used singular integral operators. The reader must be aware of the fact that there are some discrepancies of opinion among researchers as to whether the use of an operator of this type is appropriate to the theory of partial equations or not. Probably, its merit should be judged by future developments. Matsumura, Kumano-go are examples in which the authors used a further extension of the singular integral operator obtained by Calderón & Zygmund.

In this text, we have not discussed weakly hyperbolic equations with variable coefficients. For this type, the reader can consult A. Lax, Yamaguti, Ohya. For the initial value – boundary value problems of hyperbolic equations in general – there are Krzyzanski & Schauder, Duff, Hersh, Agmon [2]. Finally, for the unicity theorem, we have Carleman [1], Aronszajn.

Chapter 7 (Semi-linear hyperbolic equations)

We have mainly followed Sobolev [1, 2]. Recently, the existence theorem of a global solution (section 6, chapter 7) was generalized to the equation

$$\frac{\partial^2}{\partial t^2} u - \Delta u + m^2 u + g^2 u^p = 0 \quad (p \text{ is odd})$$

in an n-dimensional space (Segal, etc.). The uniqueness of the solution of the equation in general is not yet determined, although Bruhat says it is so if $p = 3$ and for an arbitrary dimension n.

Chapter 8 (Green's function and spectrum)

To define a Green's function from a Green's operator we followed Garnir. For the construction of Green's functions for the exterior problem of $(-\Delta)$, we have Kupradse. Our treatment of integral equations and the observation of the Green's function of $(\lambda-\Delta)$ are based on Mizohata [4]. As described in section 7, chapter 8, we considered an asymptotic behaviour of t as $t\to\infty$, the solution of a wave equation. There are detailed studies done by Ladyzhenskaya, Morawetz, Lax & Phillips.

We note that the constructions of Green's functions and their estimations are essential keys to acquiring precise information about solutions. These were neglected for some time, but recent studies reveal their importance. For example, the Green's function associated with the boundary value problem of Stokes' equation in hydrodynamic theory, plays an important rôle for the Navier–Stokes equation (Odqvist). In the last half of the chapter we gave some fundamental facts about Schrödinger operators. In this direction we suggest Wienholtz as a good guide for readers who need an introduction to the subject. In this book, we did not treat the perturbation of a continuous spectrum which is one of the active fields in analysis at this moment.

The object of the research is, in short, the clarification of the relation between the continuous spectrum of $H_1=H+V$ and that of H, where H is a self-adjoint operator and V is a perturbation operator in a Hilbert space. There are different approaches to this. Interested readers can see one approach in Friedrichs [2, 3], and Ladyzhenskaya & Faddeev, and a different one in Rosenblum, Kato, and Kuroda. However, there are other approaches; for example, Povzner, and Ikebe, which is an interesting development of Povzner's work. Birman & Entina is a work which is intended to unify all these approaches.

Bibliography

S. Agmon

[1] 'On the eigenfunctions and eigenvalues of general elliptic boundary value problems', *Comm. Pure Appl. Math.* 15 (1962), 119–47.

[2] 'Problèmes mixtes pour les équations hyperboliques d'ordre supérieur', *Colloques sur les équations aux dérivées partielles, C.N.R.S.* (1962), 13–18.

S. Agmon, A. Douglis, & L. Nirenberg

'Estimates near the boundary for solutions of elliptic partial differential equations satisfying general boundary value conditions, I', *Comm. Pure Appl. Math.* 12 (1959), 623–727.

R. Arima

'On general boundary value problem for parabolic equations', *J. Math. Kyoto Univ.* 4 (1964), 207–44.

N. Aronszajn

'A unique continuation theorem for solutions of elliptic partial differential equations or inequalities of second order', *J. de Math.* 36 (1957), 235–47.

S. Banach

Théorie des opérations linéaires, Warsaw, 1932.

M. S. Birman & C. B. Entina

'On the stationary approach to the abstract scattering theory', *Dokl. Akad. Nauk USSR,* 155 (1964), 506–8.

S. Bochner

Vorlesungen über Fouriersche Integrale, Leipzig, 1932.

F. Browder

'On the spectral theory of elliptic partial differential operators, I', *Math. Ann.* 142 (1961), 22–130.

Y. Bruhat

'Un théorème d'unicité de solutions faibles d'équations hyperboliques', *C. R.* 258 (1964), 3949–51.

T. Carleman

[1] 'Sur les systèmes linéaires aux dérivées partielles du premier ordre à deux variables', *C. R.* 197 (1933), 471–4.

[2] 'Sur un problème d'unicité pour les systèmes d'équations aux dérivées partielles à deux variables indépendantes', *Arkiv för Mat. Fys.* 17 (1939), 1–9.

[3] 'Sur le genre de dénominateur de Fredholm', *Arkiv för Mat. Fys.* 12 (1917), 1–7.

[4] 'Über die asymptotische Verteilung der Eigenwerte partieller Differentialgleichungen', *Ber. der Sächs. Akad. Wiss., Leipzig,* 88 (1936), 119–32.

[5] *L'intégrale de Fourier et questions qui s'y rattachent,* Uppsala, 1944.

R. Courant & D. Hilbert

[1] *Methoden der mathematischen Physik, II,* Springer, 1937.

[2] *Methods of mathematical physics, I* (1935), *II* (1962), Interscience, New York.

J. Dieudonne & L. Schwartz
'La dualité dans les espaces ($\mathscr{F}$) et ($\mathscr{LF}$)', *Ann. Inst. Fourier*, 1 (1949), 61–101.
G. F. D. Duff
'Mixed problems for linear systems of first order equations', *Can. J. Math.* 10 (1958), 127–60.
S. D. Eidelman
[1] On fundamental solutions of parabolic systems', *Mat. Sbornik* 38 (1956), 51–92, ibid. 53 (1961), 73–135.
[2] 'The theory of general boundary value problems for parabolic systems', *Dokl. Akad. Nauk USSR*, 149 (1963), 792–5.
K. O. Friedrichs
[1] 'Symmetric hyperbolic system of linear differential equations', *Comm. Pure Appl. Math.* 7 (1954), 345–92.
[2] 'Über die Spektralzerlegung eines Integraloperators', *Math. Ann*, 115 (1938), 249–72.
[3] 'On the perturbation of continuous spectra', *Comm. Pure Appl. Math.* 4 (1948), 361–406.
K. O. Friedrichs & H. Lewy
'Über die Eindeutigkeit und das Abhängigkeitsgebiet der Lösungen beim Anfangswertproblem linearer hyperbolischer Differentialgleichungen', *Math. Ann.* 98 (1928), 192–204.
H. Fujita & T. Kato
'On the Navier–Stokes initial value problem', *Arch. Rat. Mech. Anal.* 16 (1964), 269–315.
L. Gårding
[1] 'On the asymptotic distribution of eigenvalues and eigenfunctions of elliptic differential operators', *Math. Scand.* 1 (1953), 232–55.
[2] 'Dirichlet's problem for linear partial differential equations', *Math. Scand.* 1 (1953), 55–72.
[3] 'Linear hyperbolic partial differential equations with constant coefficients', *Acta Math.* 85 (1951), 1–62.
[4] 'Solution directe du problème de Cauchy pour les équations hyperboliques', *Colloques sur les équations aux derivées partielles du C.N.R.S.* (1956), 71–90.
H. G. Garnir
Les problèmes aux limites de la physique mathématique, Birkhäuser Verlag, Basel, 1958.
I. M. Gelfand & E. Silov
Generalized functions, I, II, III, Moscow, 1958.
E. Goursat
Cours d'analyse, III, Paris, 1927.
A. Grothendieck
'Produits tensoriels topologiques et espaces nucléaires', *Mem. Amer. Math. Soc.* 16 (1955).
J. Hadamard
[1] *La propagation des ondes*, Paris, 1903.
[2] *Le problème de Cauchy*, Paris, 1932.
R. Hersh
'Mixed problems in several variables', *J. Math. and Mech.* 12 (1963), 317–34.
E. Hille & R. S. Phillips
'Functional analysis and semi-groups', *Amer. Math. Soc. Coll. Publ.* 31 (1957).
L. Hörmander
[1] 'Differential operators of principal type', *Math. Ann.* 140 (1960), 124–46.
[2] 'On the interior regularity of the solutions of partial differential equations', *Comm. Pure Appl. Math.* 9 (1958), 197–218.
[3] 'On the regularity of the general boundary problems', *Acta Math.* 99 (1958), 225–64.
[4] *Linear partial differential operators*, Springer, 1963.

T. Ikebe
'Eigenfunction expansions associated with the Schrödinger operators and their applications to scattering theory', *Arch. Rat. Mech. Anal.* 5 (1960), 1–34.
T. Inui
Oyo henbibun hoteishiki (Applied partial differential equations) Iwanami, 1951.
S. Itô
'Fundamental solutions of parabolic differential equations and boundary value problems', *Japan. J. Math.* 27 (1957), 55–102.
F. John
[1] 'The fundamental solution of linear elliptic differential equations with analytic coefficients', *Comm. Pure Appl. Math.* 3 (1950), 273–304.
[2] *Plane waves and spherical means,* Interscience, New York, 1955.
K. Kasahara & M. Yamaguti
'Strongly hyperbolic systems of linear partial differential equations with constant coefficients', *Mem. Coll. Sci. Kyoto Univ., Ser. A* 32 (1959), 121–51.
T. Kato
'Oh finite-dimensional perturbations of self-adjoint operators', *J. Math. Soc. Japan,* 9 (1957), 239–49.
T. Kato & H. Tanabe
'On the abstract evolution equation', *Osaka Math. J.* 14 (1962), 107–33.
T. Kotake & M. S. Narasimhan
'Regularity theorems for fractional powers of a linear elliptic operator', *Bull. Soc. Math. France,* 90 (1962), 449–71.
M. Krzyzanski & J. Schauder
'Quasilineare Differentialgleichungen zweiter Ordnung vom hyperbolischen Typus. Gemischte Randwertaufgaben', *Studia Math.* 6 (1936), 162–89.
H. Kumano-go
'On the uniqueness for the solution of the Cauchy problem', *Osaka Math. J.* 15 (1963), 151–72.
W. D. Kupradse
Randwertaufgaben der Schwingungstheorie und Integralgleichungen, Berlin, 1956.
N. Kuroda
'Perturbation of continuous spectra by unbounded operators, I', *J. Math. Soc. Japan,* 11 (1959), 247–62.
O. A. Ladyzhenskaya
'On the asymptotic amplitude principle', *Uspehi Mat. Nauk (N.S.)* 12 (1957), 161–4.
O. A. Ladyzhenskaya & L. D. Faddeev
'On the perturbation of continuous spectra', *Dokl. Akad. Nauk USSR,* 120 (1958), 1187–90.
A. Lax
'On Cauchy's problem for partial differential equations with multiple characteristics', *Comm. Pure Appl. Math.* 9 (1956), 135–69.
P. D. Lax
'Asymptotic solutions of oscillatory initial value problems', *Duke Math. J.* 24 (1957), 627–46.
P. D. Lax & R. S. Phillips
'Scattering theory', *Bull. Amer. Math. Soc.* 69 (1963), 130–42.
H. Lebesgue
Leçons sur les séries trigonométriques, Paris, 1906.
J. Leray
[1] *Hyperbolic equations* (Princeton Lecture Note), 1954.

[2] 'Problème de Cauchy', I, *Bull. Soc. Math. France,* 85 (1957), 389–430; II, *ibid.* 86 (1958) 75–96; III, *ibid.* 87 (1959), 81–179.

E. E. Levi
'Sull'equazione del calore', *Ann. di Matematica,* 14 (1908), 187–264.

H. Lewy
'An example of a smooth linear differential equation without solution', *Ann. Math.* 66 (1957), 155–8.

J. L. Lions
Equations différentielles opérationnelles, Springer, 1961.

D. Ludwig
'Exact and asymptotic solutions of the Cauchy problem', *Comm. Pure Appl. Math.* 13 (1960), 473–508.

B. Malgrange
'Sur une classe d'opérateurs hypoelliptiques', *Bull. Soc. Math. France,* 85 (1957), 283–306.

M. Matsumura
'Existence locale de solutions pour quelques systèmes d'équations aux dérivées partielles', *Japan. J. Math.* 32 (1962), 13–49.

S. Minakshisundaram
'A generalization of Epstein Zeta functions', *Can. J. Math.* 1 (1947), 320–6.

C. Miranda
'Equazioni alle derivate parziali di tipo elliptico', *Ergebn. Math.* Berlin, 1955.

S. Mizohata
[1] 'Systèmes hyperboliques', *J. Math. Soc. Japan,* 11 (1959), 205–33.
[2] 'Le problème de Cauchy pour les systèmes hyperboliques et paraboliques', *Mem. Coll. Sci. Univ. Kyoto,* 32 (1959), 181–212.
[3] 'Solutions nulles et solutions non analytiques', *J. Math. Kyoto Univ.* 1 (1962), 271–302.
[4] 'Sur l'analyticité de la fonction spectrale de l'opérateur Δ relatif au problème extérieur', *Proc. Japan Acad.* 39 (1963), 352–7.

S. Mizohata & R. Arima
'Propriétés asymptotiques des valeurs propres des opérateurs elliptiques auto-adjoints', *J. Math. Kyoto Univ.* 4 (1964), 245–54.

C. Morawetz
'The decay of solutions of the exterior initial-boundary value problem for the wave equation', *Comm. Pure Appl. Math.* 14 (1961), 561–9.

M. Nagumo
[1] 'On principally linear elliptic differential equations of the second order', *Osaka Math. J.* 6 (1954), 207–29.
[2] *Kindaiteki henbibun hoteishiki-ron (gendai sugaku koza),* Kyoritsu Shuppan, 1957.

F. K. Odqvist
'Über die Randwertaufgaben der Hydrodynamik zäher Flüssigkeiten', *Math. Zeit.* 32 (1930), 329–75.

Y. Ohyo
'Le problème de Cauchy pour les équations hyperboliques à caractéristique multiple', *J. Math. Soc. Japan,* 16 (1964), 268–86.

R. Paley & N. Wiener
'Fourier transforms in the complex domain', *Amer. Math. Soc. Coll.,* New York, 1934.

I. G. Petrowsky
[1] 'Sur l'analyticité des solutions des systèmes d'équations différentielles', *Mat. Sbornik,* 2 (1939), 3–70.
[2] Über das Cauchysche Problem für Systeme von partieller Differentialgleichungen', *Mat. Sbornik,* 44 (1937), 815–68.
[3] Über das Cauchysche Problem für ein System linearer partieller Differentialgleich-

ungen im Gebiete der nichtanalytischen Funktionen', *Bull. Univ. État, Moscow* (1938), 1–74.

[4] On the diffusion of waves and the lacunas for hyperbolic equations', *Mat. Sbornik*, 59 (1945), 289–370.

A. Y. Povzner

'On the expansion of an arbitrary function in terms of the eigenfunctions of the operator $-\Delta u + cu$', *Mat. Sbornik*, 32 (1953), 109–56.

F. Riesz & B. Nagy

Leçons d'analyse fonctionnelle, Akademiai Kiado, Budapest, 1952.

M. Riesz

L'intégrale de Riemann-Liouville et le problème de Cauchy', *Acta Math.* 59 (1945), 289–370.

M. Rosenblum

'Perturbation of the continuous spectrum and unitary equivalence', *Pacific J. Math.* 7(1957), 999–1010.

J. Schauder

[1] Über lineare elliptische Differentialgleichiungen zweiter Ordnung', *Math. Zeit.* 38 (1934), 257–82.

[2] 'Das Anfangswertproblem einer quasilinearen hyperbolischen Differentialgleiching zweiter Ordnung', *Fund. Math.* 24 (1935), 213–46.

M. Schechter

'General boundary value problems for elliptic differential equations', *Comm. Pure Appl. Math.* 19 (1959), 457–86.

L. Schwartz

[1] *Théorie des distributions*, I, II, 1950–1, Paris.

[2] *Méthodes mathématiques pour les sciences physiques*, 1961, Paris.

[3] Les équations d'évolution liées au produit de composition', *Ann Inst. Fourier*, 2 (1950–1), 19–49.

[4] Transformation de Laplace des distributions', *Comm. Sém. Math. de l'Univ. de Lund*, (1952), 196–206.

[5] Séminaire (1953–4): *Produits tensoriels topologiques*, Inst. Henri Poincaré.

[6] Séminaire (1954–5): *Equations aux dérivées partielles*, Inst. Henri Poincaré.

[7] Séminaire (1955–6): *Problèmes mixtes pour l'équation des ondes*, Inst. Henri Poincaré.

[8] Espaces de fonctions différentiables à valeurs vectorielles', *Journal d'Analyse Math.* 4 (1954–5), 88–148.

[9] *Mixed problems in partial differential equations and representations of semi-groups*, Tata Inst., 1958.

I. E. Segal

'The global Cauchy problem for a relativistic scalar field with power interaction', *Bull. Soc. Math. France*, 91 (1963), 129–35.

T. Shirota

'On Cauchy problem for linear partial differential equations with variable coefficients', *Osaka Math. J.* 91 (1957), 43–60.

S. L. Sobolev

[1] *Some applications of functional analysis to mathematical physics*, Leningrad, 1950.

[2] *Sur les équations aux dérivées partielles hyperboliques non-linéaires*, Edizione Cremonese, Roma, 1961.

P. E. Sobolevskii

'On equations of parabolic type in Banach spaces', *Trudy Moscow Math. Soc.* 10 (1961), 297–350.

H. Tanabe

'A class of equations of evolution in a Banach space', *Osaka Math. J.* 11 (1959), 121–45.

F. Trèves
'Opérateurs différentiels hypoelliptiques', *Ann. Inst. Fourier*, 9 (1959), 1–73.
N. Wiener
The Fourier integral and certain of its applications, Dover, 1933.
E. Wienholtz
'Halbbeschränkte partielle Differentialoperatoren zweiter Ordnung vom elliptischen Typus', *Math. Ann.* 135 (1958), 50–80.
M. Yamaguti
'Le problème de Cauchy et les opérateurs d'intégrale singulière', *Mem. Coll. Sci. Kyoto*, Ser. A 32 (1959), 121–51.
K. Yosida
[1] *Kindai-Kaiseki* (*Modern Analysis*) Kyoritsu Shuppan, 1958.
[2] *Functional analysis*, Berlin, third edition, 1971.
K. Yosida, Y. Kawata & T. Iwamura *Isokaseki no kiso* (*Foundations of modern analysis*), Iwanami, 1959.

Symbols

		Page
$\left\{\begin{array}{l} D^{\alpha}f(x) \\ \left(\frac{\partial}{\partial x}\right)^{\alpha} f(x) \end{array}\right.$	$\frac{\partial^{\alpha_1+\cdots+\alpha_n}}{\partial x_1^{\alpha_1}\ldots\partial x_n^{\theta}} f(x_1, \ldots, x_n)$, where $\alpha=(\alpha_1, \alpha_2, \ldots, \alpha_n)$	28
$\left\{\begin{array}{l} \hat{f}(\xi) \\ \mathscr{F}[f(x)] \end{array}\right.$	Fourier transform of $f(x)$, i.e. $\hat{f}(\xi)=\int \exp(-2\pi i x\xi) f(x)\, dx$	24, 38, 102
$f*g$	convolution of f and g: $\int f(x-t)g(t)\, dt$	43, 109
supp $[\varphi]$	support of a function or a distribution φ	29, 47, 89
$f(t) \sqsupset F(p)$	Laplace transform: $F(p)=\int_a^{\infty} e^{-pt}f(t)\, dt$	123
Λ	$\Lambda f=\mathscr{F}[\lvert\xi\rvert \hat{f}(\xi)]$ where $\mathscr{F}$ is the inverse Fourier transform	323
Function space		
$\mathscr{B}^m$	$f(x)\in\mathscr{B}^m$ means $D^{\alpha}f(x)$ $(\lvert\alpha\rvert\leqslant m)$ are all bounded and continuous $\lvert f(x)\rvert_m=\sum_{\lvert\alpha\rvert\leqslant m} \sup_{x\in R^n} \lvert D^{\alpha}f(x)\rvert$	28
$\mathscr{B}$	we also write $\mathscr{B}^{\infty}$ which is the whole space of functions all of whose derivatives are bounded and continuous	28
$\mathscr{D}^m$	the whole space of m-times continuously differentiable functions whose supports are bounded	29
$\mathscr{D}$	we write also $\mathscr{D}^{\infty}$, which is the whole space of infinitely differentiable functions whose supports are bounded	29

$\mathscr{D}(\Omega)$	$\varphi \in \mathscr{D}(\Omega)$ means that φ is a C^∞-function defined on an open set Ω, and the support of φ is a compact set of Ω	71
$\mathscr{D}'(\Omega)$	distribution defined on an open set Ω	71
$\mathscr{E}^m$	the space of m-times continuously differentiable functions	28
$\mathscr{E}'$	the space of distributions with compact supports	89
$\mathscr{E}'(\Omega)$	the space of distributions defined on an open set Ω and their supports are a compact set of Ω	89
$\mathscr{E}_{L^p}{}^m(\Omega)$	the space of functions defined on an open set Ω, and their derivatives in the sense of distribution belong to L^p including mth order derivatives; the norm of the space is $\|f(x)\|_{m, L^p(\Omega)} = \sum_{\lvert\alpha\rvert \leqslant m} \|D^\alpha f(x)\|_{L^p(\Omega)}$, $\|f(x)\|_{m, L^2(\Omega)}{}^2 = \sum_{\lvert\alpha\rvert \leqslant m} \|D^\alpha f(x)\|_{L^2(\Omega)}{}^2$ If $p=2$ (some authors use $H_p{}^m(\Omega)$, $H_{m, L^p(\Omega)}$, $W_p{}^m(\Omega)$ instead of $\mathscr{E}_{L^p}{}^m(\Omega)$)	71
$\mathscr{D}_{L^p}{}^m(\Omega)$	a subset of $\mathscr{E}_{L^p}{}^m(\Omega)$, such that it consists of functions obtained as the limits of sequences $\{f_j(x)\}$ ($f_j \in \mathscr{D}(\Omega)$) by the topology of $\mathscr{E}_{L^p}{}^m(\Omega)$	77
$\mathscr{D}'_{L^2}{}^m(\Omega)$	the dual space of $\mathscr{D}_{L^2}{}^m(\Omega)$, it is a space of distributions whose topology is written as $\| T \|'_{m, L^2(\Omega)}$	82
$\mathscr{D}_{L^2}$	the space of $f \in (\mathscr{S}')$ such that whose Fourier transform $\hat{f}(\xi)$ satisfies $(1+\lvert\xi\rvert)^s \hat{f}(\xi) \in L^2$	191
$\mathscr{E}_t{}^m(E)$ $(a \leqslant t \leqslant b)$	E is a topological vector space. f belongs to the space if and only if f is m-times differentiable in $[a, b]$ by the topology of E	287, 291
$(\mathscr{S})$	for arbitrary k, α the whole space consists of $\varphi \in C^\infty$ such that $(1+\lvert x\rvert^2)^k D^\alpha \varphi(x)$ is bounded in R^n	91
$(\mathscr{S}')$	the space of linear continuous functionals defined on $(\mathscr{S})$	98

	[illegible]	[illegible]
	[illegible] Ω	[illegible]
	[illegible]	28
	[illegible]	69
	[illegible]	[illegible]
	[illegible]	[illegible]
	[illegible] subset [illegible]	[illegible]
	the dual space of [illegible] is a space of distributions [illegible]	[illegible]
	the space of [illegible] such that whose Fourier transform [illegible]	[illegible]
	[illegible] vector space [illegible]	287, 291
	for arbitrary [illegible] the whole space [illegible] bounded in $\mathbb{R}^n$	[illegible]
	the space of linear continuous functionals defined on [illegible]	98

Index